About the Editor

Kerry Vahala is the Ted and Ginger Jenkins Professor of Information Science and Technology and Professor of Applied Physics at the California Institute of Technology. Vahala received his bachelor's (1980) and doctoral degrees (1985) in Applied Physics from Caltech where he helped to develop the modern theory of phase noise in semiconductor lasers, including the application of lower-dimensional quantum structures to these devices. He later joined the faculty at Caltech in 1986. His research group studies the physics of optical micro-resonators, and their work has led to wafer-based devices operating in the Q regime above 100 million. Vahala is a Fellow of the OSA and the first recipient of the Hughes Richard P. Feynman Fellowship; he has also received the Presidential Young Investigator and Office of Naval Research Young Investigator Awards. He has served as topical editor for *Photonics Technology Letters*, the *Journal of the Optical Society of America*, the *Journal of Semiconductor Science and Technology* and was program co-chair for CLEO 99 and General Chair for CLEO 2001. In addition, he has been recognized three times by the Student Association at Caltech for excellence in teaching and mentoring. Vahala is also a Founder and Chairman of the Board of Xponent Photonics in Monrovia, California.

ADVANCED SERIES IN APPLIED PHYSICS

Editors-in-Charge: R. K. Chang (*Yale Univ.*) &
A. J. Campillo (*Naval Research Lab.*)

Published

Advanced
Series in
Applied
Physics
Volume 5

OPTICAL MICROCAVITIES

EDITED BY

KERRY VAHALA

CALIFORNIA INSTITUTE OF TECHNOLOGY, USA

World Scientific

NEW JERSEY • LONDON • SINGAPORE • BEIJING • SHANGHAI • HONG KONG • TAIPEI • CHENNAI

Published by

World Scientific Publishing Co. Pte. Ltd.

5 Toh Tuck Link, Singapore 596224

USA office: 27 Warren Street, Suite 401-402, Hackensack, NJ 07601

UK office: 57 Shelton Street, Covent Garden, London WC2H 9HE

British Library Cataloguing-in-Publication Data
A catalogue record for this book is available from the British Library.

OPTICAL MICROCAVITIES

ISBN 981-238-775-7

Typeset by Stallion Press
Email: sales@stallionpress.com

Printed by FuIsland Offset Printing (S) Pte Ltd, Singapore

PREFACE

The study of Optical Microcavities has over the past decade become a distinct subject of research and both the physical implementations of these devices and their applications are highly differentiated. Fabry–Perot, whispering gallery and photonic crystal resonator geometries have been demonstrated and applied in areas as diverse as optical telecommunication and cavity QED. Beyond their already important role in commercial technologies, such as vertical cavity surface-emitting lasers (VCSELs), they are destined to become an essential ingredient in emerging technologies of the twenty-first century. This text reviews research and trends in microcavity research as seen by a distinguished list of researchers in the field. In a subject that is impacting research across a wide expanse of topics, the potential for cross-fertilization of concepts and techniques from one topic to another is as important as the specific results themselves. For this reason, this text tries to provide as much breadth of topical coverage as possible. This approach has naturally limited the depth of review of specific subjects. It has also meant that some emerging topics are not treated.

The book is organized into nine chapters, which contain as broad themes: Photonics and Communications, Cavity QED and Quantum Optics, Photonic Crystal Resonators, Nonlinear Optics and Chaotic Dynamics. I thank the contributing authors for their time and energy in creating this review, including Hermann Haus who will be enormously missed by the photonics community.

I also thank Richard Chang, whose suggestion it was to revisit this subject. His book, *Optical Processes in Microcavities*, with Anthony Campillo was the inspiration for the present one. Finally, I thank Tobias Kippenberg, who, in addition to being a contributing author, has helped me to format certain chapters.

K. Vahala
California Institute of Technology
Pasadena, California

CONTENTS

Chapter 2.   Microfabricated Optical Cavities and
Photonic Crystals

M. Lončar and A. Scherer

Chapter 3.   Semiconductor Lasers for
Telecommunications

T. L. Koch

**Chapter 4. Cavity-Enhanced Single Photons from a
          Quantum Dot**   133
*J. Vučković, C. Santori, D. Fattal, M. Pelton, G. S. Solomon
and Y. Yamamoto*

**Chapter 5. Fabrication, Coupling and Nonlinear Optics
          of Ultra-High-Q Micro-Sphere and Chip-Based
          Toroid Microcavities**   177
*T. J. Kippenberg, S. M. Spillane, D. K. Armani,
B. Min, L. Yang and K. J. Vahala*

**Chapter 6.   Nonlinear Optical Properties of
Semiconductor Quantum Wells inside Microcavities**   239

*T. Meier, C. Sieh, S. W. Koch, Y.-S. Lee, T. B. Norris,
F. Jahnke, G. Khitrova and H. M. Gibbs*

Chapter 7. Polymer Microring Resonators   319

P. Rabiei and W. H. Steier

CHAPTER 1

OPTICAL RESONATORS AND FILTERS

Hermann A. Haus, Miloš A. Popović*, Michael R. Watts[†]
and Christina Manolatou

Research Laboratory of Electronics, Massachusetts Institute of Technology
77 Massachusetts Avenue, Cambridge, Massachusetts, USA
*mpopovic@alum.mit.edu
[†]mwatts@alum.mit.edu

Brent E. Little[‡] and Sai T. Chu

Little Optics, Inc., 9020 Junction Drive
Annapolis Junction, Maryland, USA
[‡]brent_little@littleoptics.com

Dielectric optical resonators of small size are considered for densely-integrated optical components. High-index-contrast microresonators of low Q are shown, using microwave design principles, to permit wavelength-sized, low-loss, reflectionless waveguide bends and low-crosstalk waveguide crossings. The analysis and synthesis of high Q, high-order microring- and racetrack-resonator channel add/drop filters are reviewed, supplemented by simulation examples. Standing-wave, distributed Bragg resonator filters are also described. The study is unified by a coupled-mode theory approach. Rigorous numerical simulations are justified for the design of high-index-contrast optical "circuits". Integrated-optical components are described within a polarization-diversity scheme that circumvents the inherent polarization dependence of high-index-contrast devices. Filters fabricated in academic and commercial research, and a review of microring resonator technology, advances and applications are presented.

1. Introduction

Integrated Optics has a long history,[1,2] yet practical applications of integrated optics are still only a few. Optical components in current use are large compared with the wavelength, and this puts a limit on their density of integration. By using structures with a large refractive-index contrast one may reduce the structure size to the order of the optical wavelength. In this limit, the structures resemble microwave components that are on the order of a single wavelength in size.

Microwave components are shielded by metallic walls and do not radiate. At optical frequencies we do not possess materials with the properties of good conductors and hence radiation has to be kept in bounds by proper layout of the structures. Structures of high index contrast enable the designer to achieve radiation quality factors[a] (Q's) that are high compared with the overall (loaded) Q of the structure. The broader the bandwidth of the signals processed by the optical "circuits", the lower is the required Q. Hence integrated optics will come into its own in the processing of signals at high bit-rates (25 Gb/s and higher).

There is a downside to optical structures of low Q and large bandwidth: in order to obtain low Q, strong coupling between the resonator and the external "access waveguides" is required. This can be achieved by evanescent coupling across very narrow gaps, but narrow lithographically-defined gaps can be a fabrication challenge.

Coupled optical resonators can be the basis of wavelength filters with flat-top drop response characteristics that are desirable in telecommunications channel add/drop filter applications. The current state of the art allows for the fabrication, in a single lithographic step, of up to sixth-order (or six-coupled-cavity) resonator structures for 25 GHz-bandwidth applications. In contrast, commercial thin-film filters for 25 GHz applications are only available with performance of fourth-order resonators, and these require 200–250 dielectric layers to be sequentially deposited.

In this chapter, we discuss the issues arising in the design of integrated optical "circuits" using optical resonators. We also review experimental and commercial research into these structures. Microwave design principles are called upon for the tentative layout of structures of desired functionality.[3,4] High-index-contrast waveguides that support very small bending radii with acceptable radiation loss are key components serving as both the resonator cavities and the optical interconnects between various structures.

---

[a]The quality factor (Q) of a resonator is defined as the ratio of the stored energy to the power flow out of the resonator due to various depletion mechanisms, times the angular resonance frequency. The Q measures the ability of a resonator to hold onto resonant energy.

Two types of resonators will be discussed:

(a) Traveling-wave waveguide ring/racetrack resonators[2,5,6]
(b) Standing-wave Bragg-reflection resonators[7-10]

Ring resonators support degenerate modes of traveling waves in opposite directions.[b] If the index contrast is high, the radii of the rings can be made small while maintaining low radiation loss, thereby providing a large free spectral range (FSR). Bragg-reflection resonators possess standing wave modes. Two standing wave resonators in cascade can simulate the performance of a traveling-wave resonator.[11] Radiation loss is an issue in both cases. In principle, one may greatly reduce the radiation loss of a Bragg resonator by proper choice of core and cladding indices,[28] but in practice the ideal situation can only be approximated.

Band-selective channel add/drop filters may be constructed using coupled resonators. The filter response is shaped by the disposition of the resonance frequencies of the coupled resonator system, and its coupling to the external waveguides. Mathematically, this leads to manipulation of the poles of the response function in the complex-frequency plane when engineering the drop-port response. In the sections that follow, various add/drop filter designs using ring and Bragg-reflection resonators are presented. The filter response is modeled using coupled-mode theory.[5,10-12,26] Numerical finite-difference time-domain (FDTD) simulations performed on the structures are shown to be consistent with this model. The FDTD simulations take radiation losses into account and thus serve as a check as to the adequacy of the resonator design.

High-index-contrast waveguides possess considerable structural birefringence.[c] Hence the response of the optical circuits is typically polarization-dependent. Since fiber communication applications require polarization independence, the device response must be made polarization-insensitive. A polarization-insensitive response can be achieved by separating the two polarizations of the incoming signal and rotating one so that both polarizations are processed in identical structures. At the output, one of the polarizations is rotated again and the two are recombined. This calls for

---

[b]When free-running, in the absence of coupling structures. Coupled external waveguides or resonators may split the degeneracy of the two standing-wave modes of a ring causing coupling between the two traveling waves. In practice, for weak coupling this is a small effect.
[c]Due to the cross-sectional waveguide shape, as contrasted against birefringence of material origin.

a broadband polarization splitter-rotator. Several integrated polarization converters have been proposed.[13-16] In Sec. 6, we describe a combined polarization splitter-rotator.[17]

The cross-section of high-index-contrast waveguides is small and efficient coupling to fibers presents challenges. Approaches have been published in the literature.[30,31] We present one design based on these proposals.

2. Microwave Circuits and Optical "Circuits"

In the heyday of radar development, it was common to simulate the transmission responses of well-known electrical circuits with sequences of microwave waveguide components. This approach enabled the designer to use circuit theory and standard tables to construct apparatus with the desired response. "Microwave Circuit Design" using components of size comparable to a wavelength gave way to MMICs (monolithic microwave integrated circuits), where all microwave components are made small compared to a wavelength and only the transmission lines interconnecting them retain dimensions comparable to a wavelength.

Optical components cannot be shrunk to sizes small compared to a wavelength because metallic conductors do not perform as good conductors in the optical regime. This fact promises a longer time-span for the concepts of "microwave circuit design" applied to integrated optics. The absence of an optical conductor also implies that optical components have to rely on dielectric discontinuities to provide wave guidance. If the effect of radiation loss is to be kept small, the radiation Q must be kept high compared with the loaded Q of the device. For a resonant frequency ω_o, the radiation Q (Q_o), is defined as ω_o times the energy W divided by the power lost to radiation. The external Q (Q_e), is $\omega_o W$ divided by the power coupled into output waveguides. The loaded Q (Q_L) is related to the other Q's by

$$\frac{1}{Q_L} = \frac{1}{Q_o} + \frac{1}{Q_e}. \tag{1}$$

The loaded Q in turn determines the processing bandwidth of the structure. For large bandwidths, the loaded Q is reduced and radiation losses pose less severe limitations. With the advent of 20 and 40 Gb/s bit-rates, the bandwidths have become sufficiently large that structures can be realized that do not suffer excessively from radiation losses.

Integrated optical circuits call for dielectric waveguide interconnects. These imply bends and crossings of waveguides. Bends have inherent radiation loss. Using structures of high index contrast, it is possible to design bends within an area of a few square wavelengths. The structures resemble microwave components and hence one can expect that the expertise gained in microwave circuit design can also be used to address problems in integrated optics design. We first apply resonator theory to the design of simple, compact waveguide bends and crossings, and then follow this with an in-depth study of resonator-based filters.

3. High-Transmission-Cavity-Waveguide (HTC-WG) Design

A lossless microwave resonator coupled to the outside by two waveguides, both with equal coupling (or external Q's), transmits from one waveguide to the other without reflection at the resonance frequency of the resonator. Figure 1(a) shows the schematic representation of such a two-port system. Increased coupling between the resonator and the outside enlarges the bandwidth of reflection-free transmission. The reflection R and transmission T at resonance are:

$$R = \left(\frac{\frac{Q_e}{2Q_o}}{1 + \frac{Q_e}{2Q_o}} \right)^2 \qquad (2)$$

$$T = \frac{1}{\left(1 + \frac{Q_e}{2Q_o}\right)^2}. \qquad (3)$$

This principle can be used to make reflection-free bends in waveguides. It can be extended to optics with the caveat that an optical resonator radiates unless special precautions are taken. Figure 1(c) shows the field in the two-dimensional, TE model of a right-angle bend constructed by "brute force", that is, with no special precautions. An FDTD simulation, where the color code is for field amplitude, shows that the performance of the bend is very poor, as expected. The transmission is only of the order of 35%, the reflection is 20% and the rest is radiation. Figure 1(d) shows a dielectric "resonator" coupled to two dielectric waveguides at an angle of 90°. The resonator has a 45° reflector which aids the mode-matching from input to output. Hence, the design is based on microwave-circuit and ray-optics considerations. The transmission is 98%, while the reflection is −40 dB.[4,18] The performance is indeed gratifying. It is one example of microwave circuit ideas applied to integrated optics.

High-index-contrast waveguides have widths of the order of one half wavelength in the medium of the guide, whereas waveguides of low

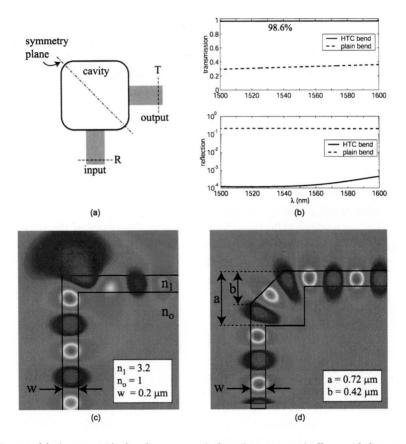

Fig. 1. (a) A waveguide bend as an optical cavity symmetrically coupled to two waveguides; (b) transmission and reflection spectral response, from 2D FDTD simulations, for a plain and high-transmission cavity (HTC) waveguide bend design; electric field (TE) plots from FDTD show (c) radiation in the plain bend; (d) low loss in the high-transmission cavity (HTC) bend.[4,18]

index contrast have widths much larger than a half wavelength. Simple perpendicular crossings of low index contrast waveguides can be easily realized without excessive cross-talk, because the field traveling in one waveguide goes through many reversals within the opening into the crossing waveguides. Thus, the overlap integral to evaluate the excitation of the perpendicular mode is small, and that mode is not excited. This is not true in the case of waveguides of high index contrast. Here, again, the resonator design principle can be invoked as shown in Figs. 2(a)–2(c). The resonant modes must be anti-symmetric with respect to the symmetry plane cutting

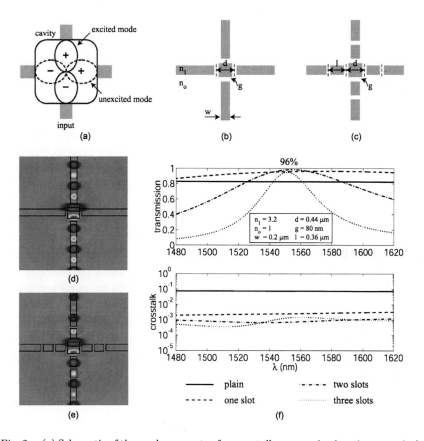

Fig. 2. (a) Schematic of the mode symmetry for crosstalk suppression in a 4-port optical cavity serving as a waveguide crossing. Reflection is suppressed by "stub-matching" using (b) one, and (c) two or more slots. 2D FDTD electric field plots are shown for (d) one-, and (e) three-slot crossings; (f) transmission and crosstalk responses of plain and slotted crossings.[4,18]

the crossing waveguides.[19] To minimize reflection, mismatch of the incident wave is removed by grooves of proper width and spacing, which serve like the matching "irises" in metallic resonators. Figures 2(d) and 2(e) show the field patterns from simulations of two-dimensional, proof-of-concept models of different crossings, and Fig. 2(f) illustrates their performance.[4,18]

In addition to transmission resonators, microwave design principles have been applied to show that wavelength-sized, square dielectric microcavities coupled to waveguides can be used to achieve wavelength-selective channel add/drop filter functionality.[4,11]

4. Add/Drop Ring- and Racetrack-Resonator Filters

Microring and racetrack resonators permit high Q, along with the benefits of a traveling-wave resonator. The latter include simple control of the
resonant mode spectrum and the inherent separation of all ports of interest in add/drop filter configurations. As a result, they occupy a prominent
position in current research.

Figure 3 shows a simulation of the 2D model of a simple channel-
dropping filter using a racetrack resonator coupled to two waveguides.
A signal traveling upward in the guide on the left is coupled to a traveling-
wave resonance of the racetrack and transferred to the other guide within
the resonance bandwidth of the resonator. A racetrack may be preferable
to a circular ring since the coupling region between the resonator and waveguides is longer, the spacing between waveguides is larger, and hence less
strict fabrication tolerances are present. On the other hand, the modal
mismatch at the straight-to-bent section interfaces must be addressed —
otherwise the loss may be excessive. This can be done by using waveguide
offsets or, in principle, the resonator concept introduced in Sec. 3. In our
example of Fig. 3, the bend radius in the racetrack resonator is gradually
varied from infinite (straight) to a minimum value and back to suppress
junction loss and reflection due to modal mismatch, while maintaining a
short cavity.

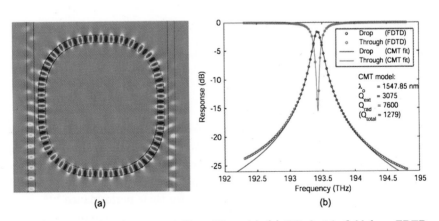

(a)   (b)

Fig. 3.   Single racetrack-resonator filter, 2D model: (a) TE electric field from FDTD
simulation, and (b) the resulting drop- and through-port spectral responses. A fit of the
FDTD results is also shown using the coupled-mode theory model of the filter, giving
the radiation Q and external Q.

An alternative approach for increasing the coupling gap involves retaining the low-loss circular ring resonator and wrapping the bus waveguide around the resonator in order to increase the coupling length and thus permit a larger gap. In this case, care must be taken to phase-match the coupler and keep the bend loss in the bus waveguide within acceptable bounds.

4.1. *Analysis by Coupled Mode Theory*

FDTD simulations in Fig. 3 fully account for radiation loss. If this loss is known or estimated by some means, the performance of the filter can also be modeled by simple coupled mode theory, the results of which are shown as solid lines in Fig. 3(b). As the physics of the filter are well captured by coupled-mode theory, we now review the coupled-mode equations.

Denote by U the amplitude of the resonator mode excited in the racetrack [see Fig. 4(a)]. The amplitude is normalized so that $|U|^2$ is equal to the mode energy. The racetrack mode couples to two waveguides, (α) and (β), and obeys the equation[5,10]

$$\frac{dU}{dt} = \left(j\omega_o - \frac{1}{\tau_{e\alpha}} - \frac{1}{\tau_{e\beta}} - \frac{1}{\tau_o} \right) U + \sqrt{\frac{2}{\tau_{e\alpha}}} a_1 + \sqrt{\frac{2}{\tau_{e\beta}}} a_4 \qquad (4)$$

where a_1 and a_4 are the incident waves in the two waveguides, normalized so that $|a_1|^2$ and $|a_4|^2$ are equal to the incident power in the two waveguides; $1/\tau_{e\alpha}$ and $1/\tau_{e\beta}$ are the coupling rates; and $1/\tau_o$ is the decay rate due to the loss (radiation and other losses combined). The resonant mode U couples

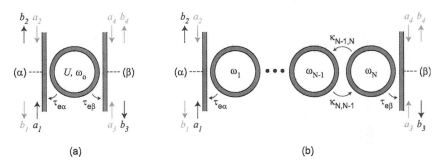

(a)   (b)

Fig. 4.   Schematics of (a) single and (b) higher order resonator channel add/drop filters, showing coupled-mode model variables. Inbound and outbound waves at each port are shown in gray and black (black ones are relevant when port 1 is excited for N odd, and are considered in the analysis).

back into the outgoing waves in the waveguides in the clockwise direction:

$$b_2 = a_1 - \sqrt{\frac{2}{\tau_{e\alpha}}} U \tag{5}$$

$$b_3 = a_4 - \sqrt{\frac{2}{\tau_{e\beta}}} U. \tag{6}$$

The coupling equations are derived under the slowly-varying envelope approximation and the coefficients are adjusted to obey energy conservation in the absence of resonator loss. The phase of U is chosen conveniently so that all coefficients are real. Note that the "backward" waves b_1 and b_4 remain unexcited if the resonant mode is a pure traveling wave. Perturbations to an ideal traveling-wave resonator may cause the actual modes to become hybrid.[20] The presence of the bus waveguides alone can, in the case of strong coupling, split the degeneracy of the standing-wave modes of a ring resonator significantly, and thus result in coupling between the traveling-wave modes.[72,73] In some cases the degeneracy may be restored by proper positioning of the two bus waveguides.

Equations (4)–(6) are used, with $a_4 = 0$, to evaluate the performance of the add/drop filter of Fig. 3. As can be seen, the form of solution of the coupled-mode equations shown in solid curves is capable of reproducing very effectively the FDTD simulation results indicated by points. The drop-port transfer characteristic of a single racetrack resonator filter is Lorentzian. For responses with flatter passbands and steeper roll-off, a cascade of coupled resonators is needed. Such a cascade is illustrated in Fig. 4(b). If the number of coupled resonators is odd, the transfer is from a_1 to b_3, whereas for an even number of resonators the transfer is from a_1 to b_4. Here, we summarize the coupled-mode equations for the case of an odd number, N, of resonators, and excited solely by a_1. For the first resonator, $n = 1$, we have

$$\frac{dU_1}{dt} = \left(j\omega_1 - \frac{1}{\tau_{e\alpha}} - \frac{1}{\tau_{o1}} \right) U_1 + j\kappa_{12}U_2 + \sqrt{\frac{2}{\tau_{e\alpha}}} a_1. \tag{7}$$

The resonators $1 < n < N$ obey the equation

$$\frac{dU_n}{dt} = \left(j\omega_n - \frac{1}{\tau_{on}} \right) U_n + j\kappa_{n,n-1}U_{n-1} + j\kappa_{n,n+1}U_{n+1} \tag{8}$$

and the last resonator satisfies

$$\frac{dU_N}{dt} = \left(j\omega_N - \frac{1}{\tau_{e\beta}} - \frac{1}{\tau_{oN}} \right) U_N + j\kappa_{N,N-1}U_{N-1}. \tag{9}$$

The coefficients of coupling between resonators obey the relation

$$\kappa_{n,n-1} = \kappa^*_{n-1,n} \tag{10}$$

imposed by energy conservation, under the assumption of bound uncoupled modes. The equations for the coupling to the waveguides are now:

$$b_2 = a_1 - \sqrt{\frac{2}{\tau_{e\alpha}}} U_1 \tag{11}$$

$$b_3 = -\sqrt{\frac{2}{\tau_{e\beta}}} U_N. \tag{12}$$

These equations may be solved to yield the transfer from guide (α) to guide (β), and correspondingly the remaining signal in (α). The solution for the latter, the "reflection" coefficient (or through-port response), is of simpler form and is expressed in terms of a continued fraction[5]:

$$\frac{b_2}{a_1} = 1 - \cfrac{\frac{2}{\tau_{e\alpha}}}{j(\omega - \omega_1) + \frac{1}{\tau_{e\alpha}} + \cfrac{|\kappa_{12}|^2}{j(\omega-\omega_2) + \cfrac{|\kappa_{23}|^2}{j(\omega-\omega_3)\cdots + \cfrac{|\kappa_{N-1,N}|^2}{j(\omega-\omega_N) + \frac{1}{\tau_{e\beta}}}}}}. \tag{13}$$

Here, the resonators are assumed lossless ($1/\tau_{on} = 0$), or else their loss decay rates may be thought of as having been absorbed into the imaginary part of complex resonator-frequency variables ω_n. The transfer (drop-port) response has a more complex form (see, e.g. Ref. 12) but, for lossless resonators, obeys power conservation with (13). The drop-port response is all-pole, and can be shown to share its pole locations with the reflection response (13).

These equations are used to analyze the filter structure based on three coupled resonators in Fig. 5. Figure 5(a) shows a field snapshot from an FDTD simulation. A pulse of finite length is launched into the simulation window which leads to the uneven appearance of the excitation in the three racetrack resonators in the snapshot. This results from the simultaneous excitation of multiple longitudinal modes of the resonators. That is, the broadband spectrum of the pulse spans several FSRs of the resonators. Since the bandwidth of the excitation is wide, the computation yields information on the transfer function over a wide spectrum, part of which is shown in Fig. 5(b) (circles).

A fit of the simulation results using the presented coupled-mode theory model, with the constraint (10), is overlaid in solid line. A better fit is possible if constraint (10) is relaxed to $\kappa_{n,n-1} = \kappa_{n-1,n}$. The latter condition

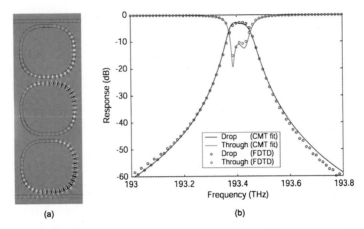

(a)   (b)

Fig. 5. Third-order racetrack resonator filter, 2D-model: (a) TE electric field from FDTD simulation excited by a short pulse, and (b) the resulting drop- and through-port spectral responses. A fit of the FDTD results using the coupled-mode theory model of the filter is also shown.

results when the theory is derived for leaky (rather than bound) uncoupled resonator modes. This variant of coupled-mode theory is used in Sec. 4.3 for the example of Fig. 8.

To account for the asymmetry in the response, the model fit requires unequal resonance frequencies even though the resonators in Fig. 5(a) are identical. This effect is due to self-coupling, expanded on in Sec. 4.3.[78]

4.2. *Synthesis of Filter Responses*

Equation (13) is also the standard form for the reflection coefficient of a ladder cascade of series and parallel resonant circuits, in the narrowband approximation,[d] as shown in Fig. 6(a). The conversion from series to parallel circuit can be accomplished by insertion of quarter-wave lines (e.g. Ref. 74). Hence, the circuit of Fig. 6(a) is also represented by parallel resonant circuits connected across the two wires, with quarter-wave sections of transmission line between them [Fig. 6(b)]. Here, the resonators are all identical and the structure starts to resemble the coupled rings of Fig. 5. The role of the quarter-wave lines is played by the phase shift of 90° contained in the coupling coefficients that are all multiplied by (the imaginary unit) j.

The circuits of Fig. 6 were studied in the 1940s and 1950s, and tables were compiled for the choice of parameters so as to produce optimal band-selective filter response characteristics.[21] For the widely-used

---

[d]It is also the standard form, exactly, of the reflection response of an LC low-pass ladder network.

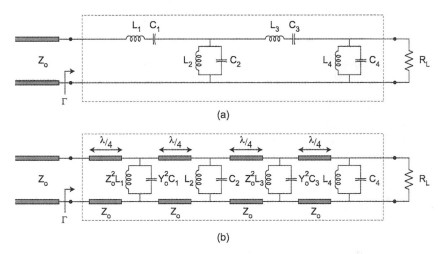

(a)

(b)

Fig. 6. Bandpass filter circuit constructed from a cascade of series and parallel resonant circuits: (a) ladder network form, and (b) equivalent form with identical resonators
spaced by quarter-wave impedance inverters. The case with four resonators is shown.
Mapping the coupled-mode model of high-order resonator optical filters to these circuit
models allows known circuit theory to be used in synthesis of responses.[5,7,12]

Butterworth (maximally flat passband) and Chebyshev (equi-ripple passband) designs, simple explicit formulas also exist.[5,74,75] Thus the coupling of
modes-in-time model comes useful in both analysis and synthesis of filters.

4.3. Design and Numerical Simulation

While two-dimensional device models (e.g. Figs. 3 and 5) suffice to explain
the physics of microring and racetrack filters, particularly in high index
contrast structures three-dimensional simulations are necessary for reliable
design.

    Microring radiation losses may be evaluated by numerically solving for
the leaky propagating mode of the ring's constituent bent waveguide, or
by directly solving for the complex-frequency leaky resonant mode of the
resonator.[72,76] In the latter case, the radiation Q is given by the ratio of the
real and imaginary parts of the complex frequency, $Q \equiv -\omega_R/2\omega_I$. High-
index-contrast resonators necessitate consideration of the full-vector field.
For an example ring resonator of $4\,\mu$m radius, Fig. 7(a) shows the radial
electric field component of the fundamental leaky resonance near 1545 nm,
over the ring cross-section. Figure 7(b) shows the dependence of bending
loss and radiation Q on radius at 1545 nm. The radiation Q is relevant for
radii which establish a resonance near 1545 nm.

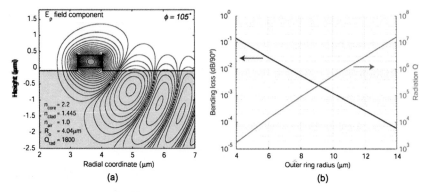

Fig. 7.   Vector-field 3D numerical modeling of the bending loss and associated radiation Q of high-index-contrast microring resonators (800×400 nm core, 100 nm overetch, other dimensions in plot): (a) radial (dominant) electric-field component of the fundamental TE-like resonant mode near 1545 nm, for 4.04 μm outer ring radius. Square-root field is shown to highlight radiation. The finite-difference numerical method employs perfectly-matched-layer (PML) boundary conditions to the right and bottom of the computation window to absorb outgoing radiation and to properly model radiation loss of the resonance.[72] (b) Bend propagation loss and radiation Q versus ring radius.

The coupled-mode theory model presented in the previous section to analyze microring filters has been advanced on the merits of the physical intuition it provides. The parameters of the model were found in Figs. 3 and 5 by fitting the response obtained from FDTD simulation. However, when derived rigorously from Maxwell's equations,[77] coupling of modes-in-time directly provides the coupling coefficients through overlap integrals defined within the theory. The latter are not unique and depend on the chosen basis. A formulation over a magnetic-field basis of two uncoupled leaky resonator modes is used to study the coupling of two microrings of the kind shown in Fig. 7, here with an 8 μm radius. Figure 8(a) shows the coupling for various spacings between the two ring resonators as calculated *ab initio* by coupled-mode theory, and as obtained from 3D FDTD simulations. The computed coupling, in the form of frequency splitting $2\sqrt{\kappa_{12}\kappa_{21}}$, that is natural to coupling of modes-in-time is converted (as in Ref. 5) to a fractional power coupled across the shared directional coupler that is more familiar in the context of traveling-wave resonators. Comparison to the FDTD result shows agreement.

However, coupled-mode theory yields less accurate predictions as the index contrast or coupling is increased (seen in Fig. 8(a)). Most importantly, it is not practical for the prediction of changes in radiation loss due to resonator coupling (i.e. coupling to radiation modes) that can be important

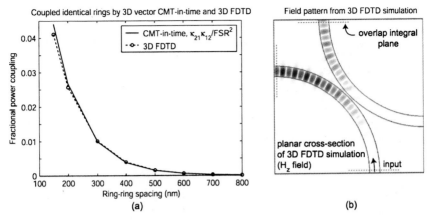

Fig. 8. Vector-field, 3D numerical evaluation of ring–ring coupling by FDTD and coupled-mode theory for two rings of Fig. 7 with outer radius of $8\,\mu m$: (a) coupling versus lateral ring–ring gap spacing using FDTD and coupled-mode theory. CMT uses a magnetic field basis of leaky resonator modes as in Fig. 7(a). CMT results in frequency splitting, which is converted to a more familiar equivalent power coupling in one pass of the directional coupler. (b) FDTD simulation field plot (H_z field component) to determine ring–ring coupling. The excitation pulse has passed the coupler.

in high-index-contrast resonator design. Thus, 3D numerical simulations are essential for reliable high-index-contrast design. Figure 8(b) shows the out-of-plane magnetic field from a 3D FDTD simulation of ring-to-ring coupling for Fig. 8(a). Relevance of the coupled-mode model advanced in Sec. 4.1 is extended by using coefficients computed from rigorous numerical simulations. An alternative model relevant only to traveling-wave resonators uses transfer matrices and waveguide modes.[58]

Finally, optimal cascaded-resonator filter responses (Sec. 4.2) are synthesized using identical resonant frequencies. An important effect to be accounted in high-index-contrast filter design is the coupling-induced frequency shift (CIFS) in resonators, which causes identical resonators to acquire slightly different resonance frequencies when coupled.[78] This frequency shift may be evaluated numerically by considering the phase shift in propagation along one microring due to the index perturbation of its neighboring structures. The coupling of modes-in-time formalism can, in principle, account for such frequency shifts with the insertion of diagonal self-coupling terms, $\kappa_{n,n}$, into Eqs. (7)–(9). These terms can be absorbed into a net resonant frequency and reveal that the solution to restoring "precoupling degeneracy" is to predistort the uncoupled resonator frequencies such that they become identical in the coupled configuration.

4.4. *Fabricated High-Order Microring Filters*

Thin-film filters and microring resonators both benefit from coupled cavity arrangements. Whereas the realization of high-order thin-film filters relies on the sequential deposition of up to three hundred dielectric layers, high-order ring cavities are fabricated all at once in a single deposition and etch step.

Examples of fabricated coupled microring resonators are shown in Fig. 9. This figure shows scanning electron micrographs of first-, second-, and

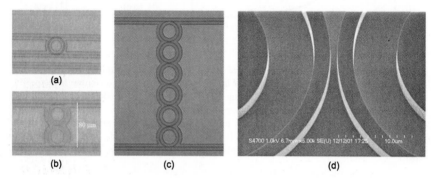

Fig. 9. Scanning electron micrographs of (a) first-, (b) second-, and (c) sixth-order microring resonators fabricated by Little Optics, Inc.; (d) close-up view of the coupling region of two rings (*courtesy of* Little Optics, Inc.).

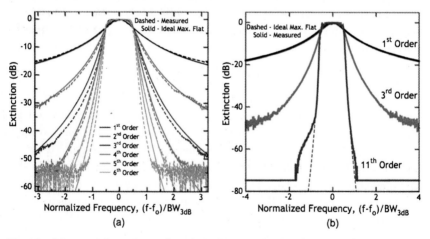

Fig. 10. Measured (experimental) and ideal maximally-flat (theoretical) responses for fabricated filters of (a) order one to six, and (b) order one, three and eleven. The 3 dB bandwidth of each filter has been normalized to allow comparison of filter shapes (*courtesy of* Little Optics, Inc.).

sixth-order cavities using rings as small as $20 \, \mu$m in radius, and were fabricated by Little Optics, Inc. The refractive index contrast in these examples was 17%.[22] Figure 10 compares the responses from fabricated filters of order one to six, and eleven. In this figure the filter bandwidths have been normalized to their respective 3 dB bandwidths in order to compare and contrast filter shapes and out-of-band signal rejection. Clearly, higher order filters give superior response, as they are intended to. Filters with linewidths ranging from 1 GHz to 100 GHz have been realized, and are suitable for many applications as highlighted in Sec. 7.

5. Distributed Feedback Resonators with Quarter-Wave Shift

A uniform Distributed Feedback (DFB) structure coupled to an input and output waveguide acts as a filter that passes electromagnetic radiation at its resonance frequencies lying outside the stop-band, and reflects radiation within its stop-band. A quarter-wave shift introduced at the center of the structure produces a high Q resonance at the center-frequency of the stop-band and the structure permits full transmission at this frequency.[23] The response is Lorentzian. Coupled quarter-wave shifted resonators can be used to construct filters with desired responses in a way similar to the filter design with ring resonators.

A quarter-wave grating resonator filter is shown in Fig. 12(a).[79,80] We denote the amplitude of the electric field in the forward wave by a, and in the backward direction by b. They are normalized so that the net power in the waveguide is

$$\langle P_z \rangle = |a|^2 - |b|^2. \tag{14}$$

We write the forward and backward waves

$$a(z) = A(z) \exp(-\pi z/\Lambda) \tag{15}$$
$$b(z) = B(z) \exp(\pi z/\Lambda) \tag{16}$$

where Λ is the grating period. The coupled mode equations for the grating are

$$\frac{dA}{dz} = -j\delta A + j\kappa B \tag{17}$$

$$\frac{dB}{dz} = j\delta B + j\kappa A \tag{18}$$

where $\delta \equiv \beta - 2\pi/\Lambda$ is the detuning from the Bragg wavelength Λ. With κ real and positive, the phase of the waves is referenced to the position of the peak of the dielectric grating, where the periodic refractive-index perturbation, $\Delta\varepsilon(x, y, z)$, is a maximum. At resonance, $\delta = 0$, and the solution is

$$A(z) = A_+ \exp(-\kappa z) + A_- \exp(\kappa z) \tag{19}$$

$$B(z) = -j[A_+ \exp(-\kappa z) - A_- \exp(\kappa z)]. \tag{20}$$

The phase of the E-field does not appear explicitly in (17) and (18) since the spatial dependence $\exp(\pm j\beta z)$ of the carrier has been factored out. However, it can be shown that the standing wave of the E-field under the exponential envelope has a maximum that is displaced by 1/8th of a wavelength from the peak of the grating corrugation, as shown in Fig. 11(a). The removal of a quarter-wave section at $z = 0$ in the pattern of Fig. 11(a) can match the field solutions decaying in opposite directions. The addition of a quarter-wave section can match the patterns if one of the patterns is reversed in sign (phase shifted by π). In either case a resonant mode is created. This trapped resonance is coupled to the incoming and outgoing waveguides if the two gratings are of finite length. When excited from one end, full transmission occurs at resonance. The transmission response is Lorentzian.

Higher order filters can be constructed by cascading such resonators. A fourth-order filter can be constructed as shown in Fig. 11(b). It is formed of five gratings with four quarter-wave gaps. The equivalent circuit is that of Fig. 6(b). The quarter-wave shift required between resonators as shown in

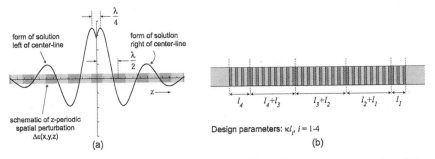

Fig. 11.   (a) Schematic of a DFB structure and forms of solution at the center of the band-gap; to match the two solutions and obtain a bound resonant mode, a quarter-wave section must be inserted or deleted in the center. (b) Schematic of four cascaded quarter-wave-shifted DFB resonators.

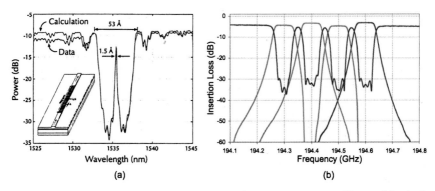

Fig. 12. Responses of fabricated DFB standing-wave-resonator filters: (a) single quarter-wave-shifted DFB-resonator filter structure and transmission response[79,80]; (b) through-port and four drop-port responses of fourth-order DFB-resonator filters fabricated by Clarendon Photonics (*courtesy of* Clarendon Photonics).

Fig. 6(b) is accomplished automatically by continuing without a break the output grating reflector of one resonator into the input grating reflector of its neighbor. The reason for this is the $1/8$th-λ displacement of the standing wave pattern from the peak of the grating corrugation. The grating length chooses the equivalent value of $Z_0^2 C_i / L_i$ whereas the product $L_i C_i$ is set by the resonance frequency.[24–27]

The performance of a four-channel Reconfigurable Optical Add/Drop Multiplexer (ROADM) developed commercially by Clarendon Photonics is shown in Fig. 12(b). The channel spacing is $100\,\text{GHz}$, bandpass widths are $50\,\text{GHz}$, and the out-of-band rejection (crosstalk) is $-30\,\text{dB}$. The filter's polarization dependent wavelength (PDW) variation is low with the central frequency of the filters differing by less than $10\,\text{GHz}$ for the two polarizations. The fiber-to-fiber insertion loss is $4\,\text{dB}$. It is surmised that each filter incorporates four resonators (this hypothesis is suggested by the four "wiggles" in the drop spectra at $-30\,\text{dB}$). The grating lengths are chosen to produce a Chebyshev bandpass transmission response.

Periodic slow-wave structures do not radiate if they are infinitely long. A finite structure with a quarter-wave shift radiates at the transitions from the uniform waveguides to the periodic structure and at the quarter-wave shift. In principle, this radiation can be suppressed in a two-dimensional slab or cylindrical DFB structure through the use of four different indices as shown in Fig. 13.[28] The indices are so chosen that transverse E-field patterns are identical within every segment of the TE mode in the slab

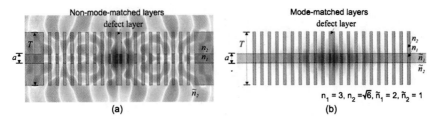

Fig. 13. (a) Conventional two-dimensional Bragg resonator, and (b) radiation-free two-dimensional Bragg resonator structure.[28] Shown in each figure is the TE electric field distribution from an FDTD simulation overlaid with the dielectric structure configuration. Square-root field amplitude is shown to highlight radiation.

case, and of the TE_{on} mode in the cylindrical case. In both cases this requires

$$\tilde{n}_1^2 - \tilde{n}_2^2 = n_1^2 - n_2^2 \tag{21}$$

where n_1 and n_2 are the core and cladding indices of one layer, respectively, and their tilde-denoted counterparts are those of the next.

6. Polarization Splitter-Rotator and Fiber-Chip Coupler

High-index-contrast structures permit smaller bends and smaller structures. However, it is difficult to make high-index-contrast structures polarization insensitive. Fiber-optic systems cannot permit polarization sensitivity since propagation along fibers causes a continuously changing polarization. We propose to achieve polarization insensitivity by splitting the incoming polarizations, rotating one of them and then processing both polarizations in parallel in two identical structures. After passage through the structures, the two polarizations are recombined. Phase delays need not be identical if a second polarization transformation is permitted. Figure 14 shows a schematic of such a scheme.[17] The symmetry of the approach inhibits polarization dependent loss.

Several proposals for integrated optic polarization splitters and rotators have been published.[13-16] However, with the exception of the polarization splitter offered by Ref. 16, all such proposals rely on mode coupling to achieve the desired result. As a result of modal (waveguide) dispersion, approaches based on mode coupling tend to be wavelength-sensitive. Moreover, in order for modes to couple efficiently, the modes must be phase-matched and waveguides precisely spaced leading to strict fabrication tolerances.

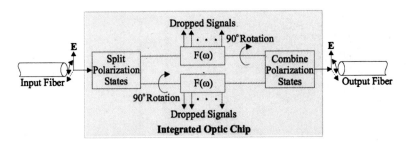

Fig. 14. Polarization-independent operation in high-index-contrast integrated optics, which are naturally polarization-sensitive, may be achieved by a polarization-diversity scheme. Illustrated here, the functional components $(F(\omega))$ in such a scheme operate in parallel on one and the same polarization state.

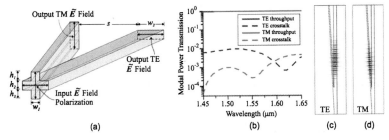

Fig. 15. Integrated polarization splitter (a) schematic of the core waveguide, (b) spectral response (transmission efficiency) computed from FDTD simulation, and (c,d) TE and TM dominant electric-field component distributions (in-plane and out-of-plane, respectively) from a pulse-excited FDTD simulation showing the polarization splitting.[29] Note: $w_1 = 0.25\,\mu\text{m}, w_2 = 0.75\,\mu\text{m}, h_{1,2,3} = 0.25\,\mu\text{m}, s = 2\,\mu\text{m}, L = 50\,\mu\text{m}, n_{\text{core}} = 2.2, n_{\text{clad}} = 1.445$.

Recently, integrated polarization splitters and rotators have been proposed that rely only on adiabatic following and therefore do not suffer from these limitations.[29] The approach works on the premise that a mode will follow and evolve along a perturbed structure so long as the coupling to other modes introduced by the perturbation is sufficiently small to allow the modes to dephase before substantial power exchange occurs. The polarization splitter and rotator are presented in Figs. 15 and 16, respectively, along with results of three-dimensional FDTD simulations.

The polarization splitter begins from a cross-shaped waveguide with degenerate polarization states and gradually shifts the two arms of the cross apart. Although the TE-like and TM-like modes of this structure are phase-matched, coupling between them is prevented by mode symmetry and the

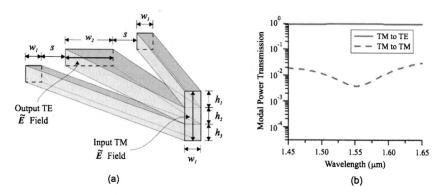

Fig. 16. Integrated polarization rotator approximating a twisted waveguide: (a) waveguide core schematic, and (b) polarization conversion and crosstalk spectral response.[29] Note: $w_1 = 0.25\,\mu m$, $w_2 = 0.75\,\mu m$, $h_{1,2,3} = 0.25\,\mu m$, $s = 0.125\,\mu m$, $L = 50\,\mu m$, $n_{core} = 2.2$, $n_{clad} = 1.445$.

TE-like and TM-like modes follow the horizontal and vertical waveguides, respectively. The polarization rotator mates up to either of the output arms of the polarization splitter depending on whether a TE or TM on-chip polarization is desired, and is essentially an approximation of a twisted waveguide. The upper and lower layers are progressively moved outward into the evanescent field of the mode while core material is added to the middle layer. The results of the FDTD simulations indicate that near perfect splitting and rotating of polarizations is achieved across the entire 1.45 μm to 1.65 μm band in structures only 50 μm long.

Finally, the task of coupling from a fiber to a single-mode high-index contrast waveguide with the cross-shaped cross-section discussed above needs to be addressed. The mode diameter of the fiber mode is on the order of 8–10 microns, whereas that of a typical high-index-contrast waveguide is between a few tenths of a micron in size in Silicon and 1–2 microns in more moderate "high-index contrast". Here again, adiabatic tapers are of great utility, as attested to by the many such proposed solutions.[30,31] Figure 17(a) shows a tapered transition to the cross waveguide proposed by Dr. G. Gorni. The wide middle layer tapers down to a small 200 × 250 nm cross-section with a mode distribution that approximates that of the fiber. The remaining 50-micron-long section of the taper introduces a top and bottom waveguide and transitions to the cross-shaped waveguide used as the input to the polarization splitter/rotator component. The performance characteristics of this taper are shown in Fig. 17(b).

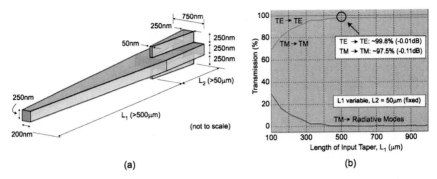

Fig. 17. Adiabatic fiber-to-chip coupler design to mate with the polarization splitter in Fig. 15: (a) three-layer core schematic, and (b) 3D mode-matching simulation results for fiber-to-chip coupling efficiency for both polarizations versus taper length. Core and cladding indices are 2.2 and 1.445 (silicon nitride in silica), and L_2 is fixed at 50 μm (*courtesy of* G. Gorni, Pirelli Labs).

7. Review of Microring Resonator Technologies

The previous six sections have addressed the physics and design of integrated optical structures using traveling- and standing-wave microresonators. Of them, in recent years the microring resonator has attracted a considerable amount of attention among researchers, fueled by telecommunications applications. It has been responsible for a large body of research advances. In this and the following section, we review advances in microring resonator technology and the emerging applications for microrings.

Although microring resonators for wavelength filters have been proposed as far back as 1969,[2] fabrication technology did not evolve sufficiently for the realistic realization of sub-100 micron structures until about the early 1990s. There are two fundamental challenges in realizing high Q ring resonators. The first is finding a high-index contrast material system in which low-loss waveguides can be fabricated. The second is the ability to pattern very narrow gaps that are essential to the coupling of rings to each other, and to the input and output bus waveguides. In this section we highlight several aspects of the ongoing research in microring resonator technology. We start with a review of the material systems used in the fabrication of microrings.

7.1. Material Systems

The fundamental requirement of a material system is that it have an index contrast sufficiently large to allow the fabrication of rings smaller than about $100\,\mu$m. This minimum index contrast is approximately 8%. Silicon waveguides with index contrasts in excess of 200% have been studied by Little *et al.*[32] Chu *et al.*[33,34] have studied the compound glass Ta_2O_5: SiO_2 having an index contrast of about 20%. Silicon Oxynitride (SiON) has been studied for conventional as well as microring resonator technology.[35] The index contrast of SiON can be adjusted over the range of 0% to about 30%. Silicon Nitride with an index contrast of 30% has been studied by Bloom *et al.*[36,37] and Klunder *et al.*[38] Compound glass is sputter-deposited, while the silicon-based materials can be deposited by chemical vapor deposition (CVD) or low-pressure chemical vapor deposition (LPCVD). Reactive Ion Etching (RIE) is the preferred method for patterning these types of materials. The foregoing materials are attractive for their long-term reliability and for their compatibility with conventional semiconductor processing.

Polymer materials have several attributes that make them attractive, including ease of fabrication and planarization, an enhanced thermo-optic coefficient, UV sensitivity, and the ability to be doped with a wide range of other materials. Chen *et al.* have studied benzocyclobutene (BCB) based rings.[39] Rabiei has studied electro-optic polymers.[40] Polymeric materials can be processed by dry etching such as RIE or by photo patterning. Photo-patternable polymers have the advantage of ultra-smooth sidewalls and thus low scattering loss. Nano-imprinting technology has also been studied as a means for mass producing rings.[41]

Semiconductor-based rings are attractive for their active and electro-optic properties. Rafizadeh *et al.*,[42,43] Hryniewicz *et al.*,[6] and Tishinin *et al.*[44] have fabricated GaAs-based rings. Grover *et al.*[45] and Rabus *et al.*[46] have studied InP-based rings. III–V semiconductor lasers integrated with rings have also been fabricated.[47,48] Semiconductor rings have been patterned by RIE and chemically-assisted ion beam etching (CAIBE). Figure 18 shows examples of first-, second- and third-order GaAs racetrack resonators fabricated at the University of Maryland.[6]

Although high-index-contrast is necessary for reducing the fundamental bending-induced losses in rings, typically a higher index contrast implies higher scattering losses. Best results are obtained for devices that use moderate index contrasts (10%–25%) and those that have the smoothest sidewalls. A model for roughness-induced scattering and junction losses in semiconductor ring resonators has been developed and compared with fabricated devices.[49]

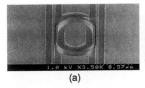 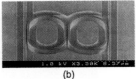 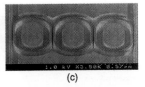

(a)   (b)   (c)

Fig. 18. Scanning electron micrographs (SEMs) of (a) first-, (b) second- and (c) third-order racetrack-resonator filters in GaAs, fabricated at the University of Maryland.[6] The racetrack and bus waveguides are in the same lithographic layer and are side-coupled (reprint from Ref. 6).

7.2. *Coupling Geometries*

Two physical geometries have been studied for the coupling of energy from bus waveguides into and out of rings. These are lateral coupling and vertical coupling. In lateral coupling the ring and bus waveguides are in the same planar layer and are evanescently side-coupled. A thin gap remains between the ring and bus as depicted in Fig. 18 and this gap determines the coupling strength. Typical gaps are on the order of 100 nm to 500 nm. These gaps are close to the resolution limit of conventional lithography and, in fact, often fall below the lithographic resolution limit. E-beam lithography has often been necessary to pattern such fine gaps. Designs that use materials with larger index contrasts necessitate narrower gaps.

Variability in the ring-to-bus coupling strength leads to variability in the line shape, which is unacceptable for telecom devices. Vertical coupling has been developed as a method for controlling the coupling strength to a higher degree than etching narrow gaps. In vertical coupling the rings and the bus waveguides are in two different layers. The separation between the bus and rings is realized by a deposition step, rather than an etch step. A further advantage of vertical coupling is that different devices with different bandwidths can be realized using the same mask set, by simply adjusting the vertical separation during fabrication. Figure 19 shows the concept of vertical coupling. In vertical coupling it is essential that all layers are planar. Suzuki *et al.*[50] achieved planarization by using an electron cyclotron CVD approach (biased-CVD) to overclad the core layers. Chu *et al.*[51,52] have developed a lift-off technique for planarization. Absil *et al.*[53] have developed a wafer bonding technique based on polymer bonding, while Tishinin *et al.*[44] have developed a wafer bonding technique based on wafer fusion. Wafer bonding allows for pristinely planar layers.

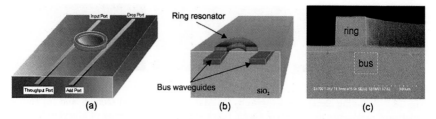

Fig. 19. Vertical ring–bus coupling scheme: (a) schematic of a vertically-coupled filter, (b) cross-sectional schematic showing vertical ring–bus coupling, and (c) scanning electron micrograph (SEM) of a fabricated arrangement (*courtesy of* Little Optics, Inc.).

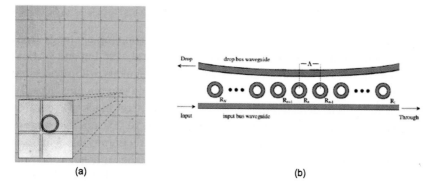

Fig. 20. (a) Cross-grid array of microring resonators for large-scale array interconnections[33]; (b) alternative optical filter design using a hybrid resonator-DFB structure.[55]

A further benefit of vertical coupling is for large-scale array interconnections as depicted in the cross-grid array of Fig. 20(a).[33] In this array, the waveguides are in one layer and the rings are in a second layer. The waveguides are arranged in a Manhattan grid pattern with rings situated at the intersections. Each layer can be optimized separately. The ring layer is optimized to have high-index-contrast in the plane, while the bus layer is optimized to have low propagation loss, lower index-contrast, and better fiber-matching dimensions.

7.3. *Devices*

Resonators are attractive building blocks because they can be used for a wide variety of photonic functions. As outlined in this chapter, they are particularly well suited to wavelength filters. Little *et al.* have studied rings for use in wavelength add/drop filters.[5,54] In Secs. 4.1 and 4.2,

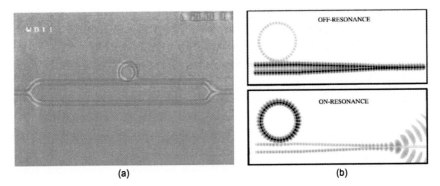

Fig. 21. Notch filter using a microring resonator in one arm of a Mach–Zehnder inter-ferometer: (a) scanning electron micrograph of a fabricated filter, and (b) the field from simulations in the off- and on-resonance states showing the notch filtering function.[59]

higher order filters were studied for improving the response shape. Rings can also be periodically coupled as shown in Fig. 20(b), yielding a hybrid resonator-distributed feedback structure,[54–56] Madsen has studied more general arrangements of rings using ring-coupled Mach–Zehnder interfer-ometers in order to achieve elliptic function responses.[57,58] Notch filters using Mach–Zehnder coupled rings have also been fabricated and tested.[59] Figure 21(a) shows a GaAs ring coupled to an ultra-compact Mach–Zehnder for use as a notch filter. Figure 21(b) shows a simulation of the opti-cal field distribution at wavelengths that are ON and OFF resonance. At wavelengths that are ON resonance the ring introduces a π-phase shift in one branch of the Mach–Zehnder, thereby imbalancing it. Filters with a second-order response, yet using a single ring can be realized by putting a controlled perturbation in the ring.[60] This type of ring supports hybrid counter-propagating modes and can be used as a narrowband reflection filter.

Rings are also very attractive for all-pass filters which can be used for fixed or tunable dispersion compensation.[61] Light that couples into the ring recirculates for an amount of time that depends on the wavelength and on the ring–bus coupling strength. The closer the wavelength is to a resonance, the longer the light dwells in the ring. There is a fundamental relationship between the amount of dispersion achievable in a specific ring, and the opti-cal bandwidth.[62] Rings are attractive for dispersion compensation because they can be cascaded in series on the same chip.

Nonlinear interactions are enhanced in resonators because the amplitude inside the cavity is larger than the amplitude that is launched from the outside of the cavity. Rings have been used to enhance the efficiency of four-wave mixing for example. It has been shown by Absil *et al.* that four-wave mixing is enhanced by the sixth power of the resonator Q.[63,64] Nonlinear switching has also recently been demonstrated.[65]

High Q rings can be switched ON or OFF by applying absorption to them.[66,67] The absorption quenches the resonance. Although a high amount of absorption in the ring switches the ring OFF, the optical signal does not itself experience high attenuation because for high Q resonators the ring and the bus are only weakly coupled.

Other novel devices include polarization rotators.[68] Polarization rotation in rings occurs when the ring waveguides have sloped sidewalls. The modes supported by rings with sloped sidewalls have hybrid polarization (a significant mixing of TE and TM field components)[e], in contrast to the bus waveguides, which support fundamental modes whose field distribution is predominantly TE-like or TM-like. An incident TE or TM mode from the bus waveguide will excite the hybrid mode in the ring. This hybrid mode in turn excites both the TE and TM modes in the bus waveguide when it couples back to the bus.

Wavelength filters need to be positioned very accurately to the particular channels they are to process. Thermal tuning can be used to dynamically position the filter resonance. Permanent resonance changes can also be induced, for example, by UV trimming.[69,70]

8. Commercial Applications of Ring Resonators

As manufacturing technology matures, microrings and other types of resonators are finding increased use in commercial applications where their performance matches or exceeds that of conventional approaches. Their compact size is highly desirable for large-scale photonic integration, and it is expected that this technology will continue to penetrate the optical layer in systems. In this section, we give examples of commercially available products based on microrings.

8.1. *Universal Demultiplexers*

Microring resonator filters are particularly effective in network systems that are based on a banded architecture. In a banded architecture the frequency

---

[e]Where terms TE and TM are used to denote the directions parallel and perpendicular to the chip surface, respectively.

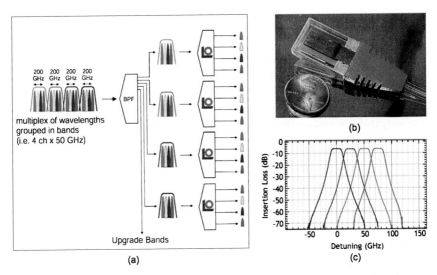

Fig. 22. Universal demultiplexers: (a) the terminal end of a system based on a frequency-banded architecture; (b) photograph; and (c) drop-port insertion-loss spectra of a fabricated universal demultiplexer based on fifth-order microring resonators (*courtesy of* Little Optics, Inc.).

spectrum is subdivided into coarse sub-bands. Each sub-band further comprises a number of more narrowly separated channels. Sub-band architectures provide for modular growth of system traffic. Figure 22(a) shows schematically the terminal end of network based on a banded architecture. In this particular architecture each sub-band is 200 GHz wide. The sub-bands carry four channels that are spaced by 50 GHz. At the terminal end a coarse (and inexpensive) band-splitting filter splits the sub-bands into independent ports supporting 200 GHz of bandwidth. A narrowband demultiplexer further separates the closely spaced channels on each sub-band. As network traffic grows, more sub-bands can be populated, or the number of channels within each sub-band can be increased, for example, by going to eight 25 GHz-spaced channels within the 200 GHz sub-bands.

Each narrowband demultiplexer within each sub-band serves the same purpose; it spatially demultiplexes the narrowly spaced channels. The difference between the demultiplexers in each sub-band is one of absolute frequency of the demultiplexed signals. An ideal demultiplexer is one which can demultiplex the four consecutive channels in any sub-band. Ring resonators are ideal for such an application. Rings have a periodic response with a periodicity that can be designed by choice of the appropriate ring radius. For example, a ring can be designed to have a periodicity of exactly 200 GHz. A demultiplexer comprised of four ring-based filters each with an FSR of 200 GHz could be used in any sub-band without having to be specialized for any one sub-band. Such a device is called a "colorless demux" or "universal demux". Further, each ring filter can be tuned independently of any other filter, and this provides wavelength trimming or tracking on a wavelength-by-wavelength basis. This is particularly important for closely spaced channels. One may compare this with an AWG for example, where all channels must be tuned simultaneously.

An example of a commercial Universal Demux is shown in Figs. 22(b) and 22(c).[22] This device is manufactured by Little Optics, Inc., and can accommodate up to 32 channels at 25 GHz (800 GHz FSR). Each filter consists of a fifth-order microring cavity with the spectral response shown. The size of the optical chip is less than 1 cm by 5 mm.

8.2. *Widely Tunable Wavelength Filters*

Two-port tunable filters are ubiquitous in optical communication systems. They are widely used in broadcast-and-select architectures where they serve essentially as tunable receivers. They also find applicability in optical power monitors and performance monitors. Typical criteria call for wide tunability (at least 32 nm of tuning range), narrowband performance, tight frequency control, repeatability and long time reliability, low power consumption, and large out-of-band signal rejection.

Microring resonator filters based on a Vernier architecture provide desirable characteristics for tunable filter applications. Thermally tuned ring filters are solid state, yielding exceptional environmental stability. They are small and therefore consume significantly less thermal power compared with conventional planar optics. The Vernier configuration uses double filtering which improves the out-of-band rejection ratio significantly, and also allows for tuning that spans a range much larger than the natural FSR of each ring filter in isolation.

The Vernier concept is illustrated in Fig. 23 for single ring resonators. An input signal is filtered by a first ring having an FSR of FSR1. The output of the first ring, which consists of a comb of transmission peaks, is subsequently filtered by a second ring having an FSR of FSR2. There will only be appreciable signal at the final output when a particular wavelength is simultaneously resonant in both rings. All other noncoincident resonant wavelength peaks are suppressed, giving rise to a net drop-port-response spectrum that has only one peak over a very large wavelength

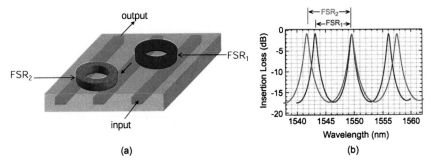

(a) (b)

Fig. 23. A simple, "Vernier-cascade" arrangement extends the effective free spectral range (FSR) of the drop-response of a filter: (a) schematic, (b) FSRs of the two individual rings show one matching resonance and two resonances which are suppressed in the total drop-port response. Small tuning of the resonance frequency of one ring permits a large tuning (in discrete steps) of the total filter as the next resonance in the comb is selected.

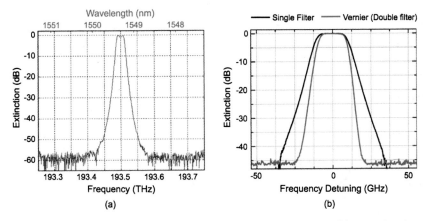

(a) (b)

Fig. 24. Drop-port spectral responses of Vernier-tunable filters: (a) cascade of two second-order filters fabricated by Lambda Crossing; (b) single and Vernier cascade of fifth-order filters fabricated by Little Optics, Inc. (*courtesy of* Lambda Crossing and Little Optics, Inc., respectively).

range. Figure 23(b) shows theoretical, stand-alone spectra from two rings having slightly different FSRs, with one particular wavelength being resonant in both rings simultaneously.

A tunable filter manufactured by Lambda Crossing is shown in Fig. 24(a). This filter is based on a Vernier architecture that is composed of cascaded second-order filters. A tunable filter manufactured by Little Optics, Inc. is shown in Fig. 24(b). This filter is based on a Vernier architecture that is comprised of cascaded fifth-order filters. Such Vernier-type filters give out-of-band rejection in excess of 60 dB. Further, the cascading of higher-order filters gives responses that have very rapid frequency roll-off.

9. Conclusion

Integrated Optics has a long history. The name originated at Bell Laboratories in the early 1970s when many of the concepts presented here were first envisioned. In the intervening three decades numerous new ideas have emerged, notably the Arrayed-Waveguide Grating (AWG) multiplexer and demultiplexer.[71] However, wide acceptance of integrated optics in commercially deployed systems is still lacking. This may change with the advent of ultra-high bit-rate fiber optic transmission. The radiation Q's of integrated optical components become adequate, and the tolerances of the components become less strict as the bandwidth of the channels increases. Linear signal processing devices for use in high bit-rate systems, like the ones described in this chapter, seem at the verge of wide-scale deployment. Already, microcavities are finding their way into commercial applications due not only to their size, but to their superior performance characteristics. This trend is expected to continue for the foreseeable future as fabrication technology improves.

Acknowledgments

The authors acknowledge gratefully the many contributions to this presentation by Dr. M. Jalal Khan and Dr. Giacomo Gorni, and by their colleagues at Little Optics, Inc., Lambda Crossing, and Clarendon Photonics.

References

1. S. E. Miller, "Integrated optics: an introduction," *The Bell Syst. Tech. J.* **48**, 2059–2069, September 1969.
2. E. A. J. Marcatili, "Bends in optical dielectric guides," *The Bell Syst. Tech. J.* **48**, 2103–2132, September 1969.
3. H. A. Haus, "Microwaves and photonics," in *OSA TOPS 23 Symp. Electro-Optics: Present and Future,* ed. H. A. Haus, pp. 2–8, Optical Society of America, Washington, DC, 1998.
4. C. Manolatou and H. A. Haus, *Passive Components for Dense Optical Integration,* Kluwer Academic Publishers, Norwell, MA, 2002.
5. B. E. Little, S. T. Chu, H. A. Haus, J. Foresi and J.-P. Laine, "Microring resonator channel dropping filters," *J. Lightwave Technol.* **15**, 998–1005, June 1997.

6. J. V. Hryniewicz, P. P. Absil, B. E. Little, R. A. Wilson and P.-T. Ho, "Higher order filter response in coupled microring resonators," *IEEE Photonics Technol. Lett.* **12**, 320–322, March 2000.

7. H. A. Haus and Y. Lai, "Narrow-band distributed feedback reflector design," *J. Lightwave Technol.* **9**, 754–760, June 1991.

8. H. A. Haus and Y. Lai, "Narrow-band optical channel-dropping filter," *J. Lightwave Technol.* **10**, 57–62, January 1992.

9. M. J. Khan, "Integrated optical filters using Bragg gratings and resonators," PhD Thesis, Department of Electrical Engineering and Computer Science, Massachusetts Institute of Technology, Cambridge, MA, February 2002.

10. H. A. Haus, *Waves and Fields in Optoelectronics,* Prentice-Hall, Englewood Cliffs, NJ, 1984.

11. C. Manolatou, M. J. Khan, S. Fan, P. R. Villeneuve, H. A. Haus and J. D. Joannopoulos, "Coupling of modes analysis of resonant channel add-drop filters," *IEEE Photonics Technol. Lett.* **35**, 1322–1331, September 1999.

12. M. J. Khan, C. Manolatou, S. Fan, P. R. Villeneuve, H. A. Haus and J. D. Joannopoulos, "Mode coupling analysis of multipole symmetric resonant add/drop filters," *IEEE J. Quant. Electron.* **35**, 1451–1460, October 1999.

13. J. Z. Huang, R. Scarmozzino, G. Nagy, M. J. Steel and R. M. Osgood, "Realization of a compact and single-mode optical passive polarization converter," *IEEE Photonics Technol. Lett.* **12**, 317–319, March 2000.

14. V. P. Tzolov and M. Fontaine, "A passive polarization converter free of longitudinally periodic structure," *Opt. Commun.* **127**, 7–13, June 1996.

15. W. W. Lui, T. Hirono, K. Yokoyama and W. P. Huang, "Polarization rotation in semiconductor bending waveguides," *J. Lightwave Technol.* **16**, 929–936, May 1998.

16. Y. Shani, C. H. Henry, R. C. Kistler, R. F. Kazarinov and K. J. Orlowsky, "Integrated optic adiabatic polarization splitter on silicon," *Appl. Phys. Lett.* **56**(2), January 8, 1990.

17. M. R. Watts, "Wavelength switching and routing through evanescently induced absorption," MS Thesis, Department of Electrical Engineering and Computer Science, Massachusetts Institute of Technology, Cambridge, MA, 2001.

18. C. Manolatou, S. G. Johnson, S. Fan, P. R. Villeneuve, H. A. Haus and J. D. Joannopoulos, "High-Density Integrated Optics," *J. Lightwave Technol.* **17**(9), September 1999.

19. S. G. Johnson, C. Manolatou, S. Fan, P. R. Villeneuve, J. D. Joannopoulos and H. A. Haus, "Elimination of crosstalk in waveguide intersections," *Opt. Lett.* **23**, 1855–1857, December 1998.

20. B. E. Little, J.-P. Laine and S. T. Chu, "Surface-roughness-induced contradirectional coupling in ring and disk resonators," *Opt. Lett.* **22**, 4–6, 1997.

21. A. Zverev, *Handbook of Filter Synthesis,* John Wiley and Sons, 1967.

22. B. E. Little, "Advances in microring resonators," *Proc. Integrated Photonics Research Conf.* 2003 (IPR 2003), pp. 165–167, 2003.

23. H. A. Haus and C. V. Shank, "Antisymmetric taper of distributed feedback lasers," *IEEE J. Quant. Electron.* **12**, 532–539, 1976.

24. H. A. Haus and R. V. Schmidt, "Transmission response of cascaded gratings," *IEEE J. Sonics Ultrasonics* **SU-24**, 94–101, March 1977.

25. J. N. Damask and H. A. Haus, "Wavelength division multiplexing using channel-dropping filters," *IEEE J. Lightwave Technol.* **11**, 424–428, 1993.

26. H. A. Haus and Y. Lai, "Theory of cascaded quarter-wave shifted distributed feedback resonators," *IEEE J. Quant. Electron.* **28**, 205–213, 1992.

27. J. N. Damask, "Practical design of side-coupled quarter-wave shifted distributed-Bragg resonant filters," *IEEE J. Lightwave Technol.* **14**, 812–821, 1996.

28. M. R. Watts, S. G. Johnson, H. A. Haus and J. D. Joannopoulos, "Electromagnetic cavity with arbitrary Q and small modal volume without a complete photonic bandgap," *Opt. Lett.* **27**(20), 1785–1787, October 2002.

29. M. R. Watts, H. A. Haus, G. Gorni and M. Cherchi, "Polarization splitting and rotating through adiabatic transitions," *Proc. Integrated Photonics Research Conf.* 2003 (IPR 2003), pp. 26–28, 2003.

30. I. Moerman, P. P. Van Daele and P. M. Demeester, "A review on fabrication technologies for the monolithic integration of tapers with III-V semiconductor devices," *IEEE J. Sel. Topics Quant. Electron.* **3**(6), 1308–1320, December 1998.

31. B. Mersali, A. Ramdane and A. Carenco, "Optical-mode transformer: a III-V circuit integration enabler," *IEEE J. Sel. Topics Quant. Electron.* **3**(6), 1321–1331, December 1998.

32. B. E. Little, J. S. Foresi, G. Steinmeyer, E. R. Thoen, S. T. Chu, H. A. Haus, E. P. Ippen, L. C. Kimerling and W. Greene, "Ultra-compact Si-SiO2 microring resonator optical channel dropping filters," *IEEE Photonics Technol. Lett.* **10**, 549–551, 1998.

33. S. T. Chu, B. E. Little, W. Pan and Y. Kokubun, "A cross-grid array of microresonators for very large scale integrated photonic circuits," *Conf. Lasers and Electro-Optics,* CLEO 1999, CPD20/1-20/2, 1999.

34. S. T. Chu, B. E. Little, W. Pan, S. Sato, T. Kaneko and Y. Kokubun, "An 8 channel add/drop filter using vertically coupled microring resonators over a cross-grid," *IEEE Photonics Technol. Lett.* **11**, 691–693, 1999.

35. R. M. de Ridder, K. Wörhoff, A. Driessen, P. V. Lambeck and H. Albers, "Silicon oxynitride planar waveguiding structures for application in optical telecommunication," Special Issue on *Silicon-Based Optoelectronics, IEEE J. Sel. Topics Quant. Electron.* **4**, 930–937, 1998.

36. F. C. Bloom, "Experimental study of integrated optics microcavity resonators: towards an all-optical switching device," *Appl. Phys. Lett.* **71**, 747–749, 1997.

37. F. C. Bloom, "A single channel dropping filter based on a cylindrical microresonators," *Opt. Commun.* **167**, 77–82, 1999.

38. D. J. W. Klunder, F. S. Tan, T. Van der Veen, H. F. Bulthuis, H. J. W. M. Hoekstra and A. Driessen, "Design and characterization of

waveguide-coupled cylindrical microring resonators in Si3Ni4," *Lasers and Electro-Optics Society 2000 Annual Meeting*, Vol. 2, pp. 13–16, 2000.

39. W.-Y. Chen, R. Grover, T. A. Ibrahim, V. Van and P.-T. Ho, "Compact single-mode benzocyclobutene microracetrack resonators," *Integrated Photonics Research* (IPR 2003) *Conf.* pp. 191–193, 2003.

40. P. Rabiei, W. H. Steier, C. Zheng and L. R. Dalton, "Polymer micro-ring filters and modulators," *IEEE J. Lightwave Technol.* **20**, 1968–1975, 2002.

41. C. Chen and L. J. Guo, "Polymer microring resonators fabricated by nonimprint technique," *J. Vac. Sci. Techol. B* **20**, 2862–2866, 2002.

42. D. Rafizadeh, J. P. Zhang, S. C. Hagness, A. Taflov, K. A. Stair, S. T. Ho and R. C. Tiberio, "Waveguide-coupled AlGaAs/GaAs microcavity ring and disk resonators with high finesse and 21.6 nm free spectral range," *Opt. Lett.* **22**, 1244–1226, 1997.

43. D. Rafizadeh, J. P. Zhang, R. C. Tiberio and S. T. Ho, "Propagation loss measurements in semiconductor ring and disk resonators," *IEEE J. Lightwave Technol.* **16**, 1308–1313, 1998.

44. D. V. Tishinin, P. D. Dapkus, A. E. Bond, I. Kim, C. K. Lin and J. O'Brien, "Vertical resonant couplers with precise coupling efficiency control fabricated by wafer bonding," *IEEE Photonic Technol. Lett.* **11**, 1003–1005, 1999.

45. R. Grover, P. P. Absil, V. Van, J. V. Hryniewicz, B. E. Little, O. King, L. C. Calhoun, F. G. Johnson and P.-T. Ho, "Vertically coupled GaInAsP-InP microring resonators," *Opt. Lett.* **26**, 506–508, 2001.

46. D. G. Rabus and M. Hamacher, "MMI-coupled ring resonators in GaInAsP-InP," *IEEE Photonic Technol. Lett.* **13**, 812–814, 2001.

47. D. G. Rabus, H. Heidrich, M. Hamacher and U. Troppenz, "Channel dropping filters based on ring resonators and integrated SOAs," *Integrated Photonics Research Conf.* IPR 2003, pp. 148–150, 2003.

48. Z. Bian, B. Liu and A. Shakouri, "InP-based passive ring-resonator-coupled lasers," *IEEE J. Quant. Electron.* **39**, 859–865, 2003.

49. V. Van, P. P. Absil, J. V. Hryniewicz and P.-T. Ho, "Propagation loss in single-mode GaAs-AlGaAs microring resonators: measurement and model," *IEEE J. Lightwave Technol.* **19**, 1734–1739, 2001.

50. S. Suzuki, K. Shuto and Y. Hibino, "Integrated-optic ring resonators with two stacked layers of silica waveguides on Si," *IEEE Photonic Technol. Lett.* **4**, 1256–1258, 1992.

51. S. T. Chu, W. Pan, T. Kaneko, Y. Kokubun, B. E. Little, D. Ripin and E. Ippen, "Fabrication of vertically coupled glass microring resonator channel dropping filters," *Optical Fiber Communication Conf.* (OFC 1999). OFC/IOOC '99. Technical Digest, Vol. 3, pp. 107–108, 1999.

52. B. E. Little, S. T. Chu, W. Pan, D. Ripin, T. Kaneko, Y. Kokubun and E. Ippen, "Vertically coupled glass microring resonator channel dropping filters," *IEEE Photonic Technol. Lett.* **11**, 215–217, 1999.

53. P. P. Absil, J. V. Hryniewicz, B. E. Little, F. G. Johnson, K. J. Ritter and P.-T. Ho, "Vertically coupled microring resonators using polymer wafer bonding," *IEEE Photonics Technol. Lett.* **13**, 49–51, 2001.

54. B. E. Little, S. T. Chu and H. A. Haus, "Micro-ring resonator channel dropping filters," *IEEE Lasers and Electro-optics Society Annual Meeting,* Vol. 2, pp. 233–234, 1995.

55. B. E. Little, S. T. Chu, J. V. Hryniewicz and P. P. Absil, "Filter synthesis for periodically coupled microring resonators," *Opt. Lett.* **25**, 344–346, 2000.

56. S. T. Chu, B. E. Little, W. Pan, T. Kaneko and Y. Kokubun, "Second-order filter response from parallel coupled glass microring resonators," *IEEE Photonics Technol. Lett.* **11**, 1426–1428, 1999.

57. C. K. Madsen, "Efficient architectures for exactly realizing optical filters with optimum bandpass designs," *IEEE Photonic Technol. Lett.* **10**, 1136–1138, 1998.

58. C. K. Madsen and J. H. Zhao, *Optical Filter Design and Analysis, a Signal Processing Approach,* Wiley, New York, 1999.

59. P. P. Absil, J. V. Hryniewicz, B. E. Little, R. A. Wilson, L. G. Joneckis and P.-T. Ho, "Compact microring notch filter," *IEEE Photonic Technol. Lett.* **12**, 398–400, 2000.

60. B. E. Little, Sai. T. Chu and H. A. Haus, "Second-order filtering and sensing with partially coupled traveling waves in a single resonator," *Opt. Lett.* **23**, 1570–1572, 1998.

61. C. K. Madsen and G. Lenz, "Optical all-pass filters for phase response design with applications for dispersion compensation," *IEEE Photonics Technol. Lett.* **10**, 994–996, 1998.

62. G. Lenz, B. J. Eggleton, C. R. Giles, C. K. Madsen and R. E. Slusher, "Dispersive properties of optical filters for WDM systems," *IEEE J. Quant. Electron.* **34**, 1390–1402, 1998.

63. P. P. Absil, J. V. Hryniewicz, B. E. Little, P. S. Cho, R. A. Wilson, L. G. Joneckis and P.-T. Ho, "Wavelength conversion in GaAs microring resonators," *Opt. Lett.* **25**, 554–556, 2000.

64. V. Van, T. A. Ibrahim, P. P. Absil, F. G. Johnson, R. Grover and P.-T. Ho, "Optical signal processing using nonlinear semiconductor microring resonators," *IEEE Sel. Topics Quant. Elec.* **8**, 705–713, 2002.

65. T. A. Ibrahim, W. Cao, Y. Kim, J. Li, J. Goldhar, P.-T. Ho and C. H. Lee, "All-optical switching in a laterally coupled microring resonator by carrier injection," *IEEE Photonics Technol. Lett.* **15**, 36–38, 2003.

66. B. E. Little, H. A. Haus, J. S. Foresi, L. C. Kimerling, E. P. Ippen and D. J. Ripin, "Wavelength switching and routing using absorption and resonance," *IEEE Photonics Technol. Lett.* **10**, 816–818, 1998.

67. R. A. Soref and B. E. Little, "Proposed N-wavelength M-fiber WDM cross-connect switch using active microring resonators," *IEEE Photonics Technol. Lett.* **10**, 1121–1123, 1998.

68. B. E. Little and S. T. Chu, "Theory of polarization rotation and conversion in vertically coupled microring resonators," *IEEE Photonics Technol. Lett.* **12**, 401–403, 2000.

69. S. T. Chu, W. Pan, S. Sato, T. Kaneko, B. E. Little and Y. Kokubun, "Wavelength trimming of a microring resonator filter by means of a UV sensitive polymer overlay," *IEEE Photonics Technol. Lett.* **11**, 688–690, 1999.

70. Y. Kokubun, H. Haeiwa and H. Tanaka, "Precise center wavelength trimming of vertically coupled microring resonator filter by direct UV irradiation to ring core," *Lasers and Electro-Optics Society Meeting* LEOS 2002, Vol. 2, pp. 746–747, 2002.

71. M. K. Smit, "New focusing and dispersive planar component based on an optical phased array," *Electron. Lett.* **24**(7), 385–386, March 1988.

72. M. Popović, "Complex-frequency leaky mode computations using PML boundary layers for dielectric resonant structures," in *Integrated Photonics Research, OSA Technical Digest,* Optical Society of America, Washington, DC, pp. 143–145, 2003.

73. S. V. Boriskina, T. M. Benson, P. Sewell and A. I. Nosich, "Effect of a layered environment of the complex natural frequencies of two-dimensional WGM dielectric ring resonators," *J. Lightwave Technol.* **20**(8), 1563–1572, August 2002.

74. R. E. Collin, *Foundations for Microwave Engineering,* 2nd edition, IEEE Press, NY, 2001.

75. A. Melloni and M. Martinelli, "Synthesis of direct-coupled-resonators bandpass filters for WDM systems," *J. Lightwave Technol.* **20**(2), 296–303, February 2002.

76. N. N. Feng, G. R. Zhou, C. Xu and W. P. Huang, "Computation of full-vector modes for bending waveguide using cylindrical perfectly matched layers," *J. Lightwave Technol.* **20**(11), 1976–1980, November 2002.

77. H. A. Haus and W.-P. Huang, "Coupled-Mode Theory (Invited)," in *Proc. IEEE,* **79**(10), 1505–1518, October 1991.

78. C. Manolatou, M. A. Popović, P. T. Rakich, T. Barwicz, H. A. Haus and E. P. Ippen, "Spectral anomalies due to coupling-induced frequency shifts in dielectric coupled-resonator filters," in *Proc. Optical Fiber Communication Conf.,* Los Angeles, CA, Session TuD5, February 2004.

79. V. V. Wong, J. Ferrera, J. N. Damask, T. E. Murphy, H. I. Smith and H. A. Haus, "Distributed Bragg grating integrated-optical filters: Synthesis and fabrication," *J. Vac. Sci. Technol. B* 2859–2864, November/December 1995.

80. J. N. Damask, "Integrated-optic grating-based filters for optical communication systems," PhD Thesis, Department of Electrical Engineering and Computer Science, Massachusetts Institute of Technology, Cambridge, MA, May 1996.

CHAPTER 2

MICROFABRICATED OPTICAL CAVITIES
AND PHOTONIC CRYSTALS

Marko Lončar* and Axel Scherer

California Institute of Technology, Pasadena, CA 91125, USA
* loncar@deas.harvard.edu

Microfabricated periodic structures with a high refractive index contrast have recently become very interesting geometries for the manipulation of light. The existence of a photonic bandgap, a frequency range within which propagation of light is prevented in all directions, is very useful where spatial localization of light is required. Ideally, by constructing three-dimensional confinement geometries, light propagation can be controlled in all three dimensions. However, since the fabrication of 3D photonic crystals is difficult, a more manufacturable approach is based on the use of one- or two-dimensional geometries. Here we describe the evolution of microcavities from 1D Bragg reflectors to 2D photonic crystals. The 1D microcavity laser (VCSEL) has already found widespread commercial use in data communications, and the equivalent 2D geometry has recently attracted a lot of research attention. 2D photonic crystal lasers, fabricated within a thin dielectric membrane and perforated with a two-dimensional lattice of holes, are very appealing for dense integration of photonic devices in telecommunications and optical sensing systems. In this chapter, we describe theory and experiments of planar photonic crystals as well as their applications towards lasers and super-dispersive elements. Low-threshold 2D photonic crystal lasers were recently demonstrated both in air and in different chemical solutions and can now be used to perform spectroscopic tests on ultra-small volumes of analyte.

1. Introduction

Photonics has recently become an attractive alternative to electronics for communications and information processing. Devices that use photons

---

*Current address: Harvard University, 9 Oxford Street, Cambridge, MA 02138, USA.

rather than electrons as information carriers can benefit from higher speeds and reduced cross talk between optical channels. As a result, miniaturization of compact optical components such as resonators, waveguides and interferometers has become very desirable. At the same time, microfabrication has emerged as a powerful technology that enables the construction of sub-100 nm structures in a reproducible and controllable manner. The same technology that was driven by the continuing desire to miniaturize electronic components on silicon microchips has now evolved to a precision that allows us to control the flow of photons.

Optical nanodevices can now be constructed in standard electronic semiconductor materials, such as silicon on insulator (SOI), GaAs and InP. By combining the need for integrated photonics with the capabilities offered by high-resolution microfabrication, the field of nanophotonics has emerged. Thus, optical devices that have traditionally been constructed in glass and lithium niobate can now be scaled down using the higher refractive index contrast available in silicon, GaAs, or InP. Ultra-small optical systems can also be integrated, thus realizing for the first time the dream of large-scale multifunctional all-optical chips for information processing. Moreover, since nanophotonics devices are constructed from standard electronic materials, these can be integrated side by side with electronic components, enabling the construction of hybrid systems of higher complexity. On the other hand, nanoscale photonic structures also offer great promise for the investigation of fundamental physical laws that govern the behavior of photons. Strong coupling between light and matter, efficient control of spontaneous emission and enhanced nonlinear optics behavior are only a few examples of the many interesting phenomena that can be explored when light and matter interact at nanoscale dimensions. As an interesting example of these new opportunities, we will show the results obtained by integrating nanophotonics with microfluidics for biochemical and chemical sensing of ultra-small volumes of analyte.

2. Vertical Microcavities

The one-dimensional equivalent to photonic crystals are the dielectric Bragg mirrors, in which layers of different refractive indices, one quarter of the desired incident wavelength in thickness, are stacked to provide a maximum in reflectivity for light that is coherently back-reflected from all of the interfaces. In most cases the incident light is perpendicular to the mirror layers, and reflectivities over 99.9999% can be achieved in carefully prepared ion beam deposited Bragg reflectors. When a layer with a thickness different from a quarter of the wavelength (usually a half of a wavelength is chosen) is introduced between two Bragg mirrors, a Fabry–Perot cavity results, in which light is localized and concentrated within that layer.

Vertical cavity Fabry–Perot etalons were intensively studied in the late 1970s and 1980s as structures for performing all-optical switching and logic operations.[1,2] The main driving-force for the development of such "vertical" cavities was the potential to increase the effective pathlength of light and thus its interaction with active or nonlinear materials. Surrounding active material with high reflectivity mirrors and forming a high-finesse cavity could increase this pathlength 100–1000 times beyond the geometric cavity length. Initially, this was accomplished by vapor or sputter deposition of dielectric mirrors onto thin semiconductor slabs, which often required the bulk micromachining of GaAs or InGaAsP membranes as part of the fabrication process.[3,4] With the emergence of molecular beam epitaxy (MBE) and organo-metallic chemical vapor deposition (OMVPE) growth technologies, which offered excellent accuracy over the deposited layer thickness in the early 1980s, it became possible to construct monolithically grown etalon cavities. Monolithic growth of etalon structures by epitaxial techniques not only ensured unprecedented thickness and refractive index control, but also the possibility of integrating high-quality electronic structures within the optical cavity. Epitaxially grown Fabry–Perot devices soon became manufacturable and structurally stable on wafer scales, yielding

Fig. 1.   Vertical cavity surface emitting lasers with InGaAs quantum wells emitting at 980 nm and using GaAs/AlAs Bragg mirrors. Gold surface contacts are used for electrical pumping. The smallest laser shown is $1\,\mu$m in diameter and $6\,\mu$m tall.

optical logic switches such as electro-optically tunable spatial light modulators (SEED) devices.

Miniaturization of the lateral dimensions of Fabry–Perot logic gates, which were based on a detuning of the cavity length through changes in the carrier density within the optical cavity, led to a rapid reduction of energy requirements as well as recovery times for optical logic operations. 150 fJ switches with 50 ps recovery time were demonstrated by Jewell *et al.*[5] in one-dimensional GaAs/AlGaAs etalons with diameters of 1–2 micrometers. Faster recovery times of 20 ps, although at higher switching powers, were demonstrated when sub-micrometer diameter etalons were fabricated for AND and OR gate operations in the late 1980s, and controlling energies as low as 17 fJ with recovery times of 6 ps were predicted for these devices.[6–8] Even lower switching powers and faster response times were possible when exchanging the quantum well active material used in these early experiments with quantum dot materials. To obtain high quality electrically and optically active devices, Fabry–Perot etalons composed of hundreds of layers

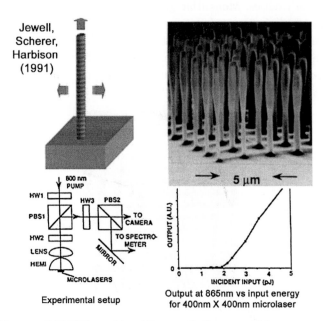

Fig. 2. Ultra-small VCSEL cavities. These optically pumped lasers were optimized as narrow single-mode etched vertical waveguides. The smallest laser measured was a 400 × 400 nm wide device, etched 8000 nm deep.

have been epitaxially deposited, and optical pathlengths could be controlled within <1% accuracies.

One of the most important semiconductor devices that emerged from this crystal growth capability of the early 1980s and the design of monolithic one-dimensional Fabry–Perot cavities was the vertical cavity surface emitting laser (VCSEL). Initially pioneered by Iga[9] in the early 1980s, this device was first MBE-grown in 1988 and miniaturized into lateral sizes as small as 400 nm in 1991.[8] Electrical pumping was accomplished by doping the AlGaAs Bragg mirrors with p- and n-dopants, permitting electro-luminescence to be excited through the electrically conducting Bragg mirrors. In these devices, the high finesse of the optical cavity radically increases the effective optical pathlength of light as well as the interaction between light emitted into the lasing mode and the thin quantum wells that provide the laser gain. When the quantum wells are positioned at an anti-node in the field distribution within the cavity, this improved interaction between emitted light and the gain material enables the dramatic reduction in the cavity length to approximately one wavelength in thickness. The vertical cavity geometry thus lends itself to miniaturization by lithographic definition of the individual laser elements, and avoids the need for forming optical cavities by cleaving the semiconductor crystals. This in turn results in the opportunity of on-chip testing of devices before lithographic processing, shaping of the laser far-field pattern, fabrication of multi-wavelength sources, and efficient mode coupling into an optical fiber. All of these attributes have made the VCSEL a very popular device as a data communications light source.

Fig. 3. VCSEL array with 700 × 700 nm laser elements. Lasers in this coherently coupled array were closely spaced to encourage photon recycling.

The first monolithic VCSELs were defined by ion etching, ion implantation, or a combination of both of these techniques. Although ion etching provided efficient waveguiding and current flow of the light, the resulting devices suffered from surface recombination of carriers at the sidewalls of the laser pillars. Similar problems with carrier recombination were observed in traps of ion implanted VCSELs. To solve these problems, VCSELs with Al_2O_3 electrical current injection apertures were designed by Deppe[10] in 1992. He employed a selective steam oxidation technique in which epitaxially grown AlAs layers were converted into insulating Al_2O_3 apertures. The resulting new GaAs VCSEL designs soon resulted in lasers with the lowest threshold currents ($<100 \mu A$) and the highest differential quantum efficiencies ($>80\%$), operating at 2.5 Gb/sec with excellent lifetimes and stability. As data communications required higher modulation speeds and longer wavelengths, VCSELs with InGaAsP active material were developed. A major difficulty in fabricating these devices results from the modest refractive index contrast available in quaternary materials lattice matched to InP when compared to index contrast of GaAs/AlAs. This low-index-contrast requires much thicker grown mirror stacks to obtain the 99% reflectivities needed, and wafer bonding of GaAs/AlAs mirrors onto InGaAs/InGaAsP active material has emerged as an appealing alternative to monolithic growth. With the development of new gain materials lattice matched to GaAs, such as InGaAs quantum dots, InGaAsN and antimonides, monolithically grown 1.55μm VCSELs are now within reach and may soon be commercialized. An interesting alternative to the growth of Bragg mirrors for the definition of microcavities by using epitaxial techniques, however, lies in the lithographic construction of mirrors and the definition of 2D and 3D photonic "crystals".

3. Photonic Crystals

The photonic crystal[11–14] (PC) is one of the platforms that can enable the miniaturization of photonic devices and their large-scale integration. These microfabricated periodic nanostructures can be designed to form frequency bands (photonic bandgaps) within which the propagation of electromagnetic waves is forbidden irrespective of the propagation direction. Depending on the dimensionality of spatial periodicity, we distinguish between several different classes of photonic crystals. One-dimensional photonic crystals are well-known dielectric stacks. For example, in Fig. 4(a) we show a vertical cavity surface emitting laser that utilizes Bragg mirrors to

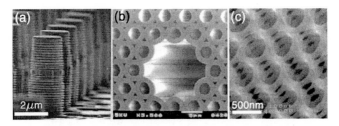

Fig. 4. Photonic crystals of different dimensionality (a) 1D: Bragg mirror,[8] (b) 2D: Microstructured fiber[15] (c) 3D: Yablonovite structure.[16]

achieve light localization in the vertical direction. If the periodicity is two-dimensional (2D), we talk about 2D photonic crystals. Strictly speaking, these structures are assumed to be infinitely long in the direction perpendicular to the plane in which 2D periodicity exists. One of the most interesting applications of 2D PCs is photonic crystal fiber, shown in Fig. 4(b).[15] In such a structure, light is confined to the core by the photonic band gap (PBG) and propagates down the fiber, along the PC holes. By introducing spatial periodicity in all three dimensions, real three-dimensional (3D) photonic crystals[16] can be realized [Fig. 4(c)]. 3D PCs can have complete bandgap, and therefore can control propagation of light in all directions. These structures can be realized using standard top-down etching techniques, multiple thin-film deposition techniques, self-assembly, micromanipulation, etc.

Fabrication of 3D PC structures is still a difficult process, and a more appealing approach is based on the use of lower-dimensional photonic crystals. A structure that has recently attracted a lot of attention is a semiconductor slab perforated with a 2D lattice of holes (Fig. 5).[17–21] The big advantage of planar photonic crystals (PPC) is their fabrication procedure, which is compatible with standard planar technology used to construct microelectronic devices. Lithographic control is another great advantage of PPCs — for example, lasers that operate at different wavelengths can be monolithically integrated within the same semiconductor slab, and the lasing wavelengths can be tuned by changing the periodicity of the structure.[22] The size of nanophotonic devices based on the PPC concept is comparable to the wavelength of the light in the material and therefore it is possible to realize functional photonics components that are at least an order of magnitude smaller than conventional devices. Therefore, planar photonic crystal devices offer the promise to replace conventional photonics in highly integrated optical systems. PPCs show even greater promise for

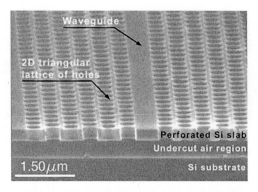

Fig. 5. Planar photonic crystal waveguide fabricated in silicon membrane suspended in the air.

realization of devices that cannot be made using a conventional approach. Highly dispersive super-prisms[23] and self-collimators are some examples of using the unusual dispersion characteristics of photonic crystals. However, one of the most attractive planar photonic crystal devices is a compact and efficient nanocavity. This is due to extraordinary feature of PPCs to localize high electromagnetic fields into very small volumes for long periods of time.

3.1. The Origin of the Photonic Bandgap

As it is well known, the periodicity of electronic potential in semiconductor materials, due to the regular arrangement of atoms in a crystal lattice, results in the formation of forbidden energy bands for electrons, i.e. the electronic bandgap. Similarly, the periodicity created in the refractive index by micro-patterning a dielectric material to form a dielectric lattice of photonic crystals results in a photonic bandgap. We will explain the existence of the photonic bandgap in photonic crystals by using the example of a 2D photonic crystal fabricated in bulk Si perforated with a square lattice of air holes [Fig. 6(a)].

The dispersion relation of the unpatterned Si block, shown in Fig. 6(b), can be expressed as

$$\omega = kc_0/n_{\mathrm{Si}}$$

where ω is the frequency of light, k propagation constant, c_0 speed of light in air and n_{Si} refractive index of silicon. Assuming an artificial lattice with period a in the silicon block, an equation can be written in the normalized

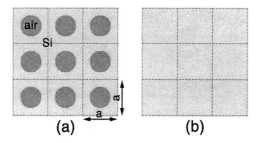

Fig. 6.   (a) Square lattice photonic crystal. (b) Si block, without air holes. Unit cell is represented with dashed lines.

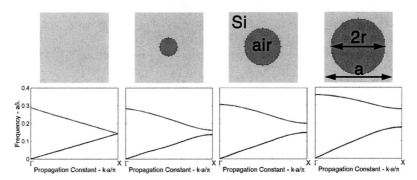

Fig. 7.   Dispersion diagrams for light propagating along the x-axis direction for various sizes of air holes in a Si block (only unit cell of the lattice is shown). Normalized frequency in units $a/\lambda = \omega a/2\pi c$ is shown on the y-axis and normalized propagation constant on the x-axis.

form common for photonic crystals:

$$\omega a/(2\pi c_0) = ka/(2n_{Si}\pi).$$

This dispersion relation is plotted in the first panel in Fig. 7. The light line is shown in the reduced scheme, and is artificially folded back into the first Brillouin zone. Addition of a periodic lattice of holes into this Si block justifies the representation in the reduced scheme, and results in the opening of a bandgap at the edge of the Brillouin zone. The evolution of the dispersion diagram, obtained by using finite-difference time-domain calculations[24] (FDTD), is shown in Fig. 7. As the holes become larger, the bandgap becomes wider and shifts towards higher frequencies. The latter can be attributed to the increased overlap between light and low-dielectric constant material (air) as the holes grow. Field intensity profiles were calculated for two modes at the edge of the Brillouin zone ($ka/\pi = 1$) and are

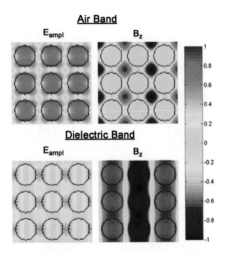

Fig. 8.   The field profiles of dielectric and air band. It can be seen that light is concentrated in Si in the case of then dielectric band, and in the air hole region in the case of air band.

shown in Fig. 8. It can be seen that the two modes have the same symmetry, but localize their energy in materials with different dielectric constants (air or dielectric). As a result, these modes have different eigen-frequencies and a bandgap opens. From these observations, we conclude that a photonic bandgap arises due to the periodicity and symmetry of the photonic crystal lattice.

So far, we have only concerned ourselves with light propagating along the x-axis direction. For a complete analysis of the optical properties of uniform, un-patterned material (e.g. a Si block), it is sufficient to study light propagation along one spatial direction, since all directions are equivalent. This is not the case when studying multidimensional periodic dielectric lattices. The introduction of a periodic lattice reduces the symmetry of the system, and it becomes necessary to study light propagation along various directions in order to describe the optical properties of 2D and 3D patterned materials. Fortunately, as in the case of electronic bandgap, it is sufficient to study light propagation along high symmetry directions of the photonic crystal lattice.[25] In the case of a square lattice, these directions are labeled ΓX, XM and ΓM. In Fig. 9, we show a dispersion diagram for modes propagating in a 2D square photonic crystal lattice. In this figure it can be seen that a complete bandgap exists between the M point in the

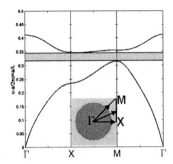

Fig. 9.   Band diagram calculated along high symmetry directions in 2D photonic crystal lattice of square symmetry. The position of the first-order bandgap is indicated. The first Brillouin zone is shown in the inset.

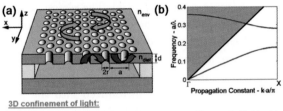

Fig. 10.   (a) Schematic of planar photonic crystal. (b) Light-line (black solid line) and light cone (gray region) overlaid on a dispersion diagram of the square lattice PPC along the ΓX direction.

dielectric band and the X point in the air band, and that this bandgap is narrower than the stop-band calculated only along the ΓX direction.

3.2. *Photonic Crystal Slabs*

While the theoretical treatment of 2D photonic crystals is relatively easy, their fabrication is not. To ease the fabrication requirements, the concept of a planar photonic crystal was originally proposed. A planar photonic crystal is essentially a 2D photonic crystal with a finite third dimension. One typical example is the perforated membrane shown in Fig. 10. An optically thin semiconductor slab (of thickness $\approx \lambda/2n_{\mathrm{diel}}$) is surrounded with a lower refractive index material (usually air) and patterned with a 2D lattice of holes. In such a structure, the localization of light in all three dimensions is made possible by combining two mechanisms: in the vertical direction, light is confined to the slab by means of total internal reflection (TIR) resulting from the high-index-contrast between the patterned dielectric slab and the low-index surrounding, while in the lateral direction light is controlled by distributed Bragg reflection resulting from the presence of a periodic 2D lattice of holes. The confinement in vertical direction can also be achieved using low index contrast semiconductor waveguide structures.[26,27]

In a planar photonic crystal shown in Fig. 10, the third dimension is neither periodic nor infinite. Therefore, photons incident to the interface between the semiconductor and air can escape from the slab provided that their angle exceeds the critical angle for TIR. These photons can then couple into the continuum of radiation modes, and therefore represent a loss mechanism from the planar photonic crystal. In order to take these losses into account, we introduce the notion of a light cone (light line) into the analysis of planar photonic crystals (PPCs). The region above the lightline, where leaky modes exist, is represented as the gray region in Fig. 10(b). Since radiative modes exist at all frequencies, including the bandgap region, these close the bandgap, and a complete bandgap is no longer found in such PPCs. The forbidden frequency range remains, however, for those modes guided within the slab. Any defects introduced into the photonic crystal lattice can now couple propagating guided modes to radiative modes and scatter light from the slab. These defects can either be intentional, in the form of missing holes for example, or unintentional, resulting from fabrication imperfections. Both of these scatterers enhance mode coupling to leaky modes and increase the photonic crystal losses. Clearly, care needs to be taken when designing and fabricating photonic crystals to minimize undesired losses.

To completely understand the operation of a planar photonic crystal, a full three-dimensional analysis is required. Waveguides in thin slabs require vertical confinement of light, and in contrast to the infinite 2D case, dielectric slabs can support modes with higher-order vertical oscillations. If a slab is made too thick, the presence of these modes can again result in closing the bandgap. Therefore, the thickness of the slab is a critical parameter in a PPC and needs to be modeled. 2D analysis of vertically extended structures (infinitely thick slabs) result in band diagrams that are shifted toward lower frequencies. The frequency shift arises as guided modes are not completely confined in the slab, but also extend into the air. This can be taken into account by using an effective refractive index.[28] Although an effective-index method can give good predictions for position of the bandgap, it does not correctly predict the shape of the bands in dispersion diagrams. Therefore, we have proposed an alternative 2D method for the analysis of 3D PPC structures, based on the use of frequency offsets.[29] Due to their simple implementation and relaxed computational requirements, these "effective" 2D methods are very appealing and are employed by many groups. However, they are limited only to the phenomenological analysis of the properties of photonic crystals and cannot be used in the analysis of losses in waveguides and cavities.

The properties of planar photonic crystals, such as position and width of the bandgap, depend on several important parameters: the type of lattice (e.g. triangular, square, honeycomb), the thickness of the slab (d), the refractive index of both the slab (n_{slab}) and the environment surrounding the slab (n_{env}), the periodicity of the lattice (a) and the size of the holes (r). In the next two sections we will investigate the influence of these parameters on the properties of photonic crystal slabs.

3.3. *Comparison between Square and Triangular Photonic Crystal Lattices*

The structure that we first analyze is a silicon slab, suspended in air and patterned with a square or triangular lattice of holes[30,31] (Fig. 11). A 3D FDTD model was used to calculate the band diagrams. In both cases, we analyzed a lattice unit cell by applying appropriate boundary conditions to its sides: Bloch boundary conditions are used for all four sides perpendicular to the plane of the slab, and Mur's Absorbing Boundary Conditions[32] (ABCs) for the top boundary. A mirror boundary condition is implemented at the middle of the slab to reduce the computation time by analyzing only

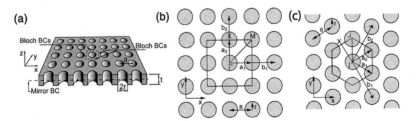

Fig. 11. (a) Si slab patterned with 2D square lattice of holes. Unit cell and boundary conditions used in the 3D FDTD calculation are also indicated. (b) Square lattice PPC. Without loss of generality, the lattice constant a in the figure is chosen to be $a = \pi/a$ (dimensionless units), in order to conveniently show reciprocal lattice vectors. The square represents the first Brillouin zone. (c) Triangular lattice PPC. This time we chose $a = 4\pi/3a$. The hexagon represents the first Brillouin zone.

one half of the unit cell. By changing the type of mirror symmetry we can select even (TE-like) or odd (TM-like) eigenmodes of the photonic crystal structure. The spatial resolution that we used to discretize our structures is 20 computational points per lattice constant ($a = 20$).

In the case of square PPC, the lattice vectors are $a_1 = a \cdot (1, 0, 0)$ and $a_2 = a \cdot (0, 1, 0)$, and the reciprocal lattice vectors are $b_1 = 2\pi/a(1, 0, 0)$ and $b_2 = 2\pi/a \cdot (0, 1, 0)$. The square in Fig. 11(b) represents the first Brillouin zone, and Γ, X and M are the high symmetry points with coordinates (in reciprocal space): Γ $= (0, 0, 0)$, X $= (0, \pi/a, 0)$, M $= (\pi/a, \pi/a, 0)$. In the case of a triangular lattice PPC, lattice vectors in real space can be expressed as $a_1 = a(\sqrt{3}/2, -1/2, 0)$ and $a_2 = a \cdot (\sqrt{3}/2, 1/2, 0)$, and in reciprocal space as $b_1 = 4\pi/a\sqrt{3} \cdot (1/2, -\sqrt{3}/2, 0)$ and $b_2 = 4\pi/a\sqrt{3} \cdot (1/2, \sqrt{3}/2, 0)$. The hexagon in Fig. 11(c) represents the first Brillouin zone, and the coordinates of high symmetry points are Γ $= (0, 0, 0)$, X $= 2\pi/a\sqrt{3} \cdot (1/2, -\sqrt{3}/2, 0)$ and J $= 4\pi/3a \cdot (0, 1, 0)$.

Band diagrams for the vertically even (TE-like) and odd (TM-like) eigenmodes of the PPC structure with square and triangular symmetry are shown in Fig. 12. As it can be seen, in both structures the first-order bandgap is open for the guided modes of the slab (modes below the light line) that have an even symmetry (TE-like). The triangular lattice planar photonic crystal has a much wider bandgap than the square lattice, a result of the greater symmetry and the smoother Brillouin zone in that geometry.

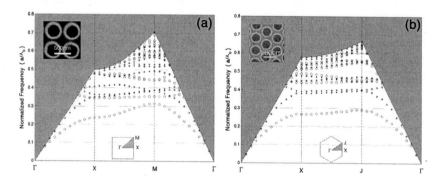

Fig. 12.   Band diagrams of eigenmodes of Si slab perforated with 2D lattice of holes of (a) square and (b) triangular symmetry. The x-axis represents different directions in the reciprocal lattice, and the y-axis normalized frequency in the units a/λ. The gray region represents the light cone. Even modes (TE-like) are represented with circles and odd (TM-like) with stars. Insets show the SEM micrographs of the fabricated photonic crystal structure, and the first Brillouin zones.

Table 1.   Design parameters of the photonic crystal with bandgap around $\lambda = 1.5\,\mu$m.

Lattice	PBG width	Midgap (f_0)	Lattice constant $a = f_0 \cdot \lambda$ [nm]	r [nm]	d [nm]
square	0.031	0.330	496	198	272
triangular	0.148	0.365	547	219	301

Therefore, the triangular lattice is in many ways a more robust candidate for the realization of wide bandgap planar photonic crystal mirrors. However, in many cases it is not necessary, and sometimes is not even beneficial, to work with photonic crystals that have a wide bandgap and enable strong light localization. This is most notable in the case of PPC cavities, where high quality factor cavity designs were realized even in square lattice PPC.[33] In Table 1, we specify geometries of both square and triangular planar photonic crystals designed to operate at wavelength $\lambda = 1.5\,\mu$m.

3.4. *Geometry-Dependent Properties of Photonic Crystals*

Next, we describe the influence that hole size, thickness of the slab, refractive index of both the slab and environment has on the properties of a PPC. A triangular lattice PPC is studied, but results can be generalized to the square lattice PPC as well.

3.4.1. *Influence of the Hole Size*

In Fig. 13, we show band edges of the air and dielectric bands calculated for various slab thicknesses and hole sizes. In panel (a) it can be seen that the bandgap becomes wider as the holes become larger. Also, due to increased overlap with low-dielectric material (air), the band edges are shifted towards higher frequencies as the hole sizes are increased. The air-band modes that localize their energy in the air-holes are more sensitive to changes in the hole size, and they experience a larger blue-shift than the dielectric-band modes, when holes are enlarged. This results in widening of the photonic bandgap. When holes are too large ($r/a > 0.5$), the bandgap for TE-like modes can be closed and a bandgap for TM-like modes can open.[13,30] On the other hand, when holes are too small, the bandgap can also be closed. The slab thickness, in the range that we have explored, does not have strong influence on the width of the bandgap [Fig. 13(b)], and it only affects the position of the bandgap.

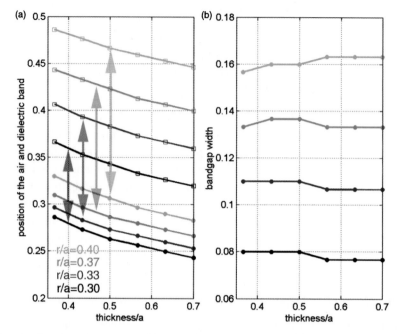

Fig. 13. Bandgap position as a function of relative hole size (r/a) and relative thickness of the slab. (a) Positions of the band edges are shown: top of the dielectric band with full circles, and bottom of the air band with open squares. (b) The width of the bandgap for various geometries.

3.4.2. *Influence of the Refractive Index of the Environment*

In most common applications, planar photonic crystals are in the form of a free standing membrane suspended in air. However, it is also of interest to explore the PPC properties when the air is replaced with material with a refractive index bigger than 1. For example, we may use PPC lasers as chemical sensors, and the operation of PPC membranes depends on the refractive indices of the surrounding.[34] Also, nanocavities based on PPCs can be backfilled with electro-optic or nonlinear polymer, and their resonant frequencies can be changed by applying strong electromagnetic fields.

In Fig. 14, we show the dependence of the dielectric band and air band edges when a PPC is immersed in materials with various refractive indices. The normalized thickness of the PPC is $d/a = 0.75$ and its refractive index is $n_{slab} = 3.4$. The normalized hole sizes (r/a) are 0.3 and 0.4. As expected, the band edges experience red-shift when the PPC slab is immersed in a material with higher refractive index. Also, the width of the bandgap

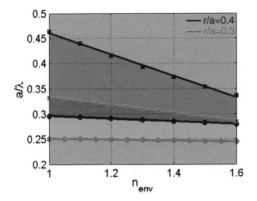

Fig. 14. Air band and dielectric band edge dependence on the refractive index of environment (n_{env}) surrounding PPC slab for $r/a = 0.3$ (red) and $r/a = 0.4$ (blue).

decreases as n_{env} increases. The reason is similar to the decrease in the bandgap when the holes of PPC are made smaller. The bottom of the air band is more affected with increased n_{env}, since air-band modes have a larger spatial overlap with the environment. From Fig. 14 we deduce that the bandgap will close when $n_{env} \approx 2$ in both structures. In the case of a crystal with $r/a = 0.4$, the dependence of the band edges on n_{env} can be well approximated with

$$(a/\lambda)_{diel} = -0.0289 n_{env} + 0.3255$$
$$(a/\lambda)_{air} = -0.2139 n_{env} + 0.6748.$$

When $r/a = 0.3$, the band edges can be expressed as

$$(a/\lambda)_{diel} = -0.0104 n_{env} + 0.2621$$
$$(a/\lambda)_{air} = -0.0812 n_{env} + 0.4144.$$

As expected, a structure with larger holes is more sensitive to the changes in the refractive index of the environment. From these equations, we find that the sensitivity of the air-band edge is over $\Delta\lambda \sim 700\Delta n$ (in nanometers), when $r/a = 0.4$. This suggests that very small changes in the refractive index of the ambient can be detected by monitoring shifts in the emission wavelength of a band-edge laser. In a later section we will explore a similar possibility of using confined cavity modes instead of extended band-edge modes for sensing.

3.4.3. *Influence of the Refractive Index of the Slab*

For some applications, it is necessary to make PPCs in dielectric materials that have a small refractive index. For example, if we want to make PPC devices that operate in the visible spectrum, we cannot use many of the semiconductor materials, due to their absorption in visible range. One promising candidate for PPCs that operate in the visible is silicon-nitride[35,36] (Si_3N_4), with a refractive index of $n_{SiN} \approx 2.02$. In Fig. 15, we show band diagrams for $d/a = 0.55$, and $r/a = 0.3$ and $r/a = 0.4$. As expected, due to the small refractive index contrast between core and environment, the bandgap is narrow in these cases, especially in the case of smaller holes.

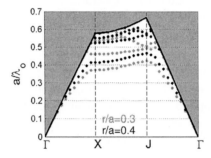

Fig. 15.   Band diagrams for PPC realized in Si_3N_4 slab for two different hole sizes. Only TE-like modes are shown.

4. Propagation of Light through Photonic Crystal Slabs

In previous sections, we described the existence of a bandgap in photonic crystals, and we investigated the influence of various design parameters on the position and width of the bandgap. This was an important task since photonic crystals are used as mirrors to confine light in space in a majority of applications. However, by exciting the modes of photonic crystal with frequencies outside the photonic bandgap, light can propagate in the PPC. Then, due to the strong dispersion of planar photonic crystals, phenomena associated with anomalous refraction of light such as a super-prism or self-collimation can take place.[29,37–47] In order to understand light propagation in patterned slabs, it is not sufficient to study band diagrams along high-symmetry directions only, as done in previous sections. Instead, a complete band diagram for all possible directions in the first Brillouin zone (BZ) needs to be constructed.

The planar photonic crystal that we study here is a silicon slab ($n_{\text{Si}} = 3.5$) of thickness $d = 0.57a$, patterned with a 2D square or triangular lattice of holes of radius $r = 0.3a$, where a is the periodicity of the lattice. The slab is surrounded by air on both sides. 3D FDTD is used to calculate dispersion diagrams with discretization of 30 computation points per lattice constant ($a = 30$). Better discretization in time and space in FDTD algorithm would be needed in order to correctly predict the position and shape of the allowed bands of a complete photonic crystal. Here, we analyze only one half of the structure in the vertical direction and apply even mirror symmetry at the center of the slab to study TE-like (vertically even) modes.

Figure 16(a) shows a complete dispersion diagram for the first two bands for all k vectors in the first BZ. The light cone is represented as the unshaded mesh. From Fig. 16(a), we conclude that the first band is outside the light cone (and guided) throughout the entire frequency range, while the second band is guided only for normalized frequencies $a/\lambda < 0.306$. Furthermore, the second band is almost flat and therefore light in that frequency range will be slowed down significantly. In Figs. 16(b) and 16(c), we plot the iso-frequency contours for the first and second bands, and the gradient of frequency change as a function of the in-plane k vector is indicated by the

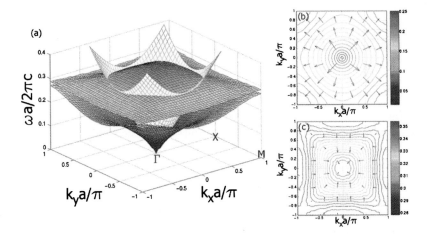

Fig. 16.   (a) The dispersion ($\omega(k)$) relation for the first two bands of the square PPC, calculated for all k vectors in the first Brillouin zone. The light cone is represented as an unshaded mesh. The iso-frequency contours for the (b) first and (c) second band. The color represents different a/λ. The vectors represent the gradient of frequency change as the function of k_x and k_y components.

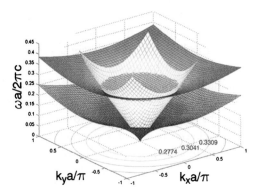

Fig. 17. The dispersion of the first two bands supported in the unpatterned Si slab. The iso-frequency contours in the case of the unpatterned slab are circles (shown for $a/\lambda = 0.2774, 0.3041, 0.3309$) since all in-plane directions are equivalent.

length of the arrows. Those vectors are oriented towards the Γ point in the second band — an indication of negative group velocity (the group velocity and phase velocity are oriented along opposite directions). Moreover, iso-frequency contours of the second band are almost perfect squares in the frequency range where light is guided. This is very different from dispersion in unpatterned Si slabs, where iso-frequency contours of guided modes are circles (Fig. 17). The modification of the dispersion relation by patterning leads to very interesting collimation and dispersion effects, which we describe below.

4.1. *Self-Collimation in Square Lattice PPCs*

In Fig. 18 we again show iso-frequency contours for the second band, but this time only for frequencies that lie outside of the light cone ($0.273 < a/\lambda < 0.306$). The light cone for $a/\lambda = 0.306$ is represented with a dashed circle. It can be seen that the iso-frequency contours can be approximated by squares in the range of $0.295 < a/\lambda < 0.306$. The energy of the excited mode will propagate with a group velocity $\boldsymbol{v_g} = \boldsymbol{\nabla_k}\omega$, and the direction of propagation (in real space) will be perpendicular to the iso-frequency contour for that frequency (in the k-space). Therefore, if the input light has a range of k_x vectors all of which are, for example, between $-k_0$ and k_0 in Fig. 18, then light in the PPC will propagate along the y-axis direction (ΓX), as indicated in gray in Fig. 18. In other words, the light beam in the PPC can be self-collimated.

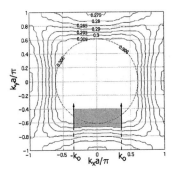

Fig. 18. The iso-frequency contours of the Band 2 of square lattice PPC. Only the region outside the light cone is shown. The light cone, for $\lambda = 0.306$, is represented by the dashed circle. The light of frequency $a/\lambda = 0.3$ is self-collimated and propagates in the direction (in the real space) indicated by the gray color.

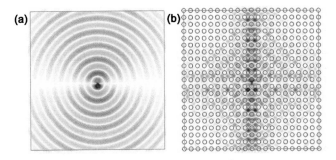

Fig. 19. B_z component of the field in the (a) unpatterned slab and (b) slab patterned with square lattice PPC. Structures were excited with dipole sources (with E_x component) at frequencies (a) $a/\lambda = 0.19$ and (b) $a/\lambda = 0.295$. Self-collimation can be observed in (b) as predicted.

In order to verify the prediction that self-collimation is possible in a square lattice PPC, we have performed 3D FDTD calculations for such a structure. The results presented in Fig. 19 show the field evolution (B_z component) in the un-patterned (a) and patterned (b) Si slab. The structures were excited using a dipole source placed at the center of the slab. In the case of the unpatterned slab, the dipole radiation is coupled into the slab mode and light propagates in all directions. However, when the PPC structure is excited with the dipole source with frequency $a/\lambda = 0.295$ (in the frequency region of the second band where the iso-frequency contours are squares) light is well collimated and radiates predominantly in

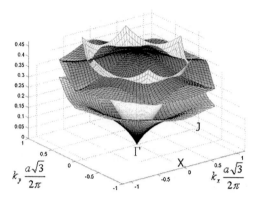

Fig. 20. The complete band-diagram of the first three bands of a triangular planar photonic crystal.

the ΓX directions. Radiation in y-axis direction is stronger because the structure was excited with E_x field only.

4.2. *Complete Band-Diagram of Triangular Lattice PPC*

The complete band-diagram for a triangular lattice was also calculated and is shown in Fig. 20. The complete band-diagram of the triangular lattice photonic crystal has more structure than the square lattice dispersion diagram, as a result of the higher symmetry of the triangular lattice. Iso-frequency contours of the third band now have a characteristic star-shape that can be used to achieve self-focusing, spot-size conversion, super-prism effects, etc.[40,43–47]

5. Planar Photonic Crystal Cavities and Lasers

One of the most promising applications of planar photonic crystals is the realization of a compact and efficient optical nanocavities with high quality factors (Q) and small mode volumes (V_{mode}). Moreover, PC cavities can be engineered to concentrate light in the air, and thus are natural candidates for the investigation of interaction between light and matter on a nanoscale level. Some of the most interesting applications of planar photonic crystal nanocavities are:

- *Low-threshold lasers*
  Q/V_{mode} ratio can be large in photonic crystal cavities resulting in low-threshold powers of photonic crystal nanolasers.

- *Cavity QED experiments*

  Strong coupling between neutral Cs atoms, or single quantum dots, placed within the PC cavity and interacting with the cavity field has been predicted. Integration of several cavities will lead towards realization of quantum-optical networks.
- *Bio-chemical sensing/single molecule detection*

  Ultra-small quantities of biochemical reagents can be placed in the air region where field intensity is the strongest and their (strong) influence on the optical signature of the resonator can be monitored. This can lead towards realization of integrated spectroscopy systems (e.g. on-chip Raman spectroscopy).
- *Channel drop filters for telecom applications*

  PC nanocavities can have high Q factors (>10,000) and can be highly integrated (less than $5\,\mu$m apart), which makes them promising candidates for realization of channel drop filters in dense wavelength-division multiplex (WDM) systems.

Photonic crystal cavities can be formed by modifying one or more holes in the photonic crystal lattice. By changing the size of one of the holes (Fig. 21), we form a potential well for modes supported in a bulk planar photonic crystal. This is similar to the formation of potential wells in a periodic electrostatic potential typically studied in quantum mechanics. By enlarging one of the holes, we locally increase the amount of low-dielectric constant material (air) and therefore increase the energy of the

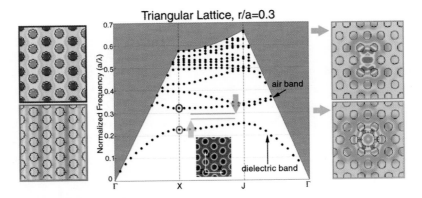

Fig. 21. Dispersion diagram for the modes supported in the triangular lattice planar photonic crystal ($r/a = 0.3$ this time). Mode profiles for one component of the E field in the dielectric (red) and air (blue) band, taken at the X point, are also shown.

modes supported in the bulk photonic crystal. Modes that were originally confined within the dielectric material (dielectric band modes) will now be pulled up into the bandgap, and are trapped in the energy well formed by the larger hole. This bound state exists close to the dielectric band and in its nature is similar to an acceptor level in semiconductors. Thus, modes created by modification of the dielectric band are often called acceptor modes. Similarly, by reducing the size of one of the holes, we form bound states close to the air band — donor modes.[48] Acceptor and donor modes are shown in Fig. 21.

In case of 2D photonic crystal, with infinitely many crystalline layers surrounding a defect, light can be completely trapped at the defect. However, in optical cavities defined within photonic crystal slabs, modes also suffer from radiation losses through coupling into the continuum of radiation modes that exist within the light cone. Since the cavity mode is localized in real space, it is extended in reciprocal space as governed by the uncertainty principle. It therefore consists of k-vector components that are positioned within the light cone and these components contribute to the out-of-plane losses of the cavity, as shown in Fig. 22. At the same time, light can leak laterally due to the limited number of the photonic crystal layers surrounding the cavity, contributing to the in-plane losses of the resonator. The efficiency of a resonator, described by a quality factor (Q), can be expressed as the ratio of energy stored in the cavity and energy lost (emitted) from the cavity in one period

$$Q = 2\pi W_{\text{stored}}/W_{\text{lost}}.$$

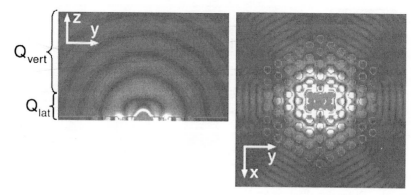

Fig. 22.   One of the dipole modes supported in a single defect cavity. In-plane and out-of-plane energy leaks can be observed.

This quality factor can now be broken into the inverse sum of the lateral quality factor (Q_{lat}) and vertical quality factor (Q_{vert}) that take into account in-plane and out-of-plane losses, respectively:

$$1/Q = 1/Q_{\text{lat}} + 1/Q_{\text{vert}}.$$

By lithographically adding more photonic crystal layers around the cavity, lateral leakage can be completely suppressed and we can assume that Q_{lat} can be arbitrary high.[28] Therefore, the cavity Q is typically limited by Q_{vert}. Unfortunately, the simplest photonic crystal cavity, formed by reducing the size of one of the holes, known in literature as a single defect cavity, suffers from large radiative losses (small Q_{vert}). Therefore, the problem of high Q cavity design has attracted much research attention and several designs were proposed and characterized experimentally.[33,49−59] Photonic crystal lasers constructed according to some of these cavity designs have also been reported by several groups.[18,33,59−74]

5.1. *High Q Cavity Designs*

One of the high Q cavity geometry that we have developed in the past is based on the introduction of a fractional edge dislocation together with a single defect into a triangular lattice photonic crystal.[49] Our planar photonic crystal is based on a free-standing membrane consisting of a high dielectric constant slab (refractive index $n = 3.4$), perforated with a 2D lattice of holes and suspended in air. The cavity consists of a defect hole (radius r_{def}) that is smaller than the surrounding holes (radius r) which define the photonic crystal mirror. The row that contains the defect is elongated by moving the two photonic crystal half-planes a fraction of a lattice constant apart in the ΓX direction (Fig. 23). Each half-plane is moved by $p/2$, yielding a total dislocation of p.

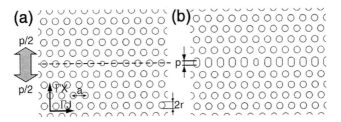

Fig. 23. (a) Conventional single defect cavity ($p = 0$). When structure is cut along the dashed line, and two PPC half-planes are dislocated along ΓX direction by $p/2$, (b) high Q cavity can be formed ($p = 0.25a$).

It was shown that in such a photonic crystal cavity, with $r/a = 0.275$, $r_{\text{def}}/a = 0.2$ and $d/a = 0.75$ (d is thickness of the slab), it is possible to achieve Q factors as high as 11,000 by optimizing the dislocation parameter p. The cavity Q was maximized when $p/a = 10\%$. These high Q values were obtained while maintaining a very small mode volume of $V_{\text{mode}} \sim 0.1 \cdot (\lambda/2)^3$. These cavities were originally designed for cavity QED experiments, where strong coupling between atoms emitting in the high field region of the cavity and light trapped in the cavity was to be investigated.[49] It is clear that the presence of a hole at the point of maximum field intensity is not desirable for the construction of low-threshold laser designs, since the overlap with the gain region provided by quantum wells is decreased. Therefore, we have to compromise the overlap between the optical field and the gain material with the cavity Q, and optimize the defect hole size (r_{def}) for an optimal Q factor of the cavity.

In order to improve the lateral confinement of light in the cavity, we decided to analyze photonic crystals with slightly larger holes ($r/a = 0.3$). This results in a more compact cavity, at the expense of increased scattering of light in the vertical direction from the more porous photonic crystal mirrors and decreased Q factors. As a first step, we calculated the band-diagram of the bulk photonic crystal with parameters $r/a = 0.3$, $d/a = 0.75$, and $n_{\text{slab}} = 3.4$, and found that a bandgap exists for vertically even modes (TE-like) for the normalized frequencies in the range of $0.2508 < a/\lambda < 0.3329$. 3D FDTD was used to calculate this dispersion diagram as described previously. A single defect donor cavity in a triangular lattice photonic crystal without the fractional edge dislocation is known to support two doubly-degenerate, linearly polarized, dipole modes.[28] However, as we stretch the photonic crystal lattice by introducing a fractional dislocation, these modes start to interact and the degeneracy between them is lifted. In Fig. 24 we show the results of 3D FDTD analysis of a structure with $p/a = 10\%$, $r_{\text{def}} = 0.2a$, $r = 0.3a$ and $d = 0.75a$. Two dipole modes, labeled LQ and HQ, are found to exist in the cavity. Additional modes are found close to the air band, as well. Those modes are not localized to the defect hole, but are instead attributed to the waveguide modes of the elongated central row. The mode at longer wavelength can have an order of magnitude better Q factor value and therefore is called HQ (high Q) mode.

In Fig. 25(a) we show the dependence of the eigen-frequency and Q of the HQ dipole mode on the stretching parameter of the central row (p/a) and on the size of the defect hole (r_{def}/a). It can be seen that by increasing the dislocation, the splitting between the two dipole modes also

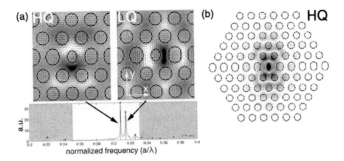

Fig. 24. Defect modes of the cavity with $p/a = 10\%$ and $r_{\text{def}} = 0.2a$. (a) The cavity supports two dipole modes, and their profiles are shown (B_z component and vector of the **E** field). The spectrum of the modes supported in the cavity, obtained using 3D FDTD, is also shown. The bandgap is shown in white. (b) The amplitude of the E field is shown. It can be seen that light is localized in the small defect hole.

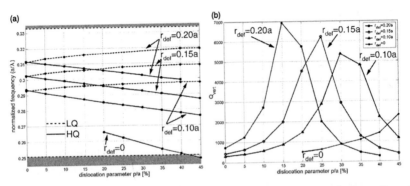

Fig. 25. Dependence of (a) the position of the two dipole modes of the cavity and (b) vertical quality factor of the HQ mode on the amount of dislocation introduced (p), and the size of the defect hole (r_{def}).

increases. As the defect hole becomes larger, the modes shift towards higher frequencies. This shift occurs since a larger hole in the center of the cavity leads to an increased overlap between the optical field and air. One more interesting feature is that splitting between LQ and HQ modes does not depend strongly on the size of the central hole, and is mostly dependent on the stretching parameter (p) introduced. The mode of interest for laser applications is the HQ mode, since it has an order of magnitude of higher Q and therefore will reach threshold first. The quality factors of LQ modes are limited to several hundreds and are typically not of practical importance. In Fig. 25(b) we show the dependence of the vertical (Q_{vert}) quality factors of the HQ mode on the stretching parameter (p) for various sizes of central defect hole (r_{def}). In our calculations, the cavity was surrounded with five layers of photonic crystal.

Figure 25(b) tells us that the best Q that we can hope to achieve in the modeled cavity geometry is around 7,000. In comparison, we were able to achieve Qs as high as 11,000 when $r/a = 0.275$. As expected, due to the increased size of the bulk photonic crystal holes ($r/a = 0.3$), light scattering in the vertical direction increases, and the cavity Q is reduced. The optimal design (Q factor maximized) requires more dislocation as r_{def} decreases. This can be understood by looking at spatial frequencies that exist in the Fourier spectrum of the HQ mode. In order to increase the Q of the cavity, components that lie within the light cone need to be minimized. This can be achieved by changing the size of the air-region in the cavity in order to balance the energy that exists in each lobe of the mode, thus minimizing its DC component.[75] The change of the area occupied by the defect hole, induced by stretching of the central row is $\Delta A = 2r_{def}p$. From this equation it follows that in order to achieve the same influence on the mode, larger p's are needed when r_{def} is reduced. In other words, for a big defect hole, a small change in p has larger influence on the mode since most of the light is located in the hole.

Another important figure of merit of any laser design is the gain provided by the laser cavity. As the defect hole diameter is decreased and the amount of dislocation is increased, we expect a better overlap between the optical cavity mode and the quantum wells, and therefore a reduced laser threshold. However, it is important in our application that the central defect hole is as large as possible so that we can achieve strong interaction between light emitted from the laser and the material (nanoparticles, single molecules, chemical fluids, gasses, etc.) placed into the optical field of the laser. Therefore, we have chosen $r_{def} = 0.15a$ and $p = 0.25a$ as a good compromise for our initial laser sensor design.

By introducing different material into the photonic crystal holes, or by immersing our lasers in various chemicals (liquids, gasses), the refractive index of the surrounding is changed. Therefore, it is of interest to determine the change in Q and the eigen-frequency of the cavity resonance as the cavity is back-filled with reagents. In Fig. 26(a), we observe that the highest Q that we could hope to achieve in the modeled cavity design occurs at an ambient refractive index $n_{env} = 1$ (air), and this value deteriorates as the refractive index of the ambient surrounding the photonic crystal cavity is increased. This reduction in Q is a direct result from the weaker vertical confinement of light within the perforated slab and can be compensated for by increasing its thickness.

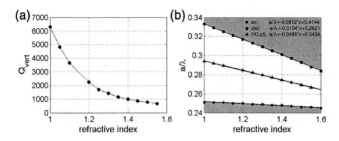

Fig. 26. Dependence of (a) Q factor and (b) eigen-frequency of cavity resonance on the refractive index of analyte introduced in the cavity with $p/a = 25\%$.

It is also interesting to note that the frequency of the resonant mode, as well as the band edges of the photonic crystal mirror, depend linearly on the refractive index of the environment (n_{env}) [Fig. 26(b)]. From linear fits of the dependence of the resonant frequency on n_{env}, we estimate the sensitivity of the cavity as approximately $\Delta\lambda \approx 266\Delta n_{env}$ (in nanometers) where Δn_{env} is the change in refractive index, and $\Delta\lambda$ wavelength shift in nanometers. The simplest method of optically sensing ambient material uses wavelength shifts in a laser when the laser is immersed into a solution or exposed to a material to measure its refractive index. In this method, the sensitivity of the sensor depends on the smallest change in refractive index that can be optically detected. In passive devices, this is related to the width of the cavity resonance peak, which in turn is determined by the cavity quality Q. If we assume that our cavity is embedded in a typical polymer ($n_{env} = 1.4$) a wavelength shift that is still observable from such a cavity with $Q = 1,000$ is $\Delta\lambda = 1.55\,\text{nm}$, which corresponds to a change in refractive index of $\Delta n \approx 0.0056$. On the other hand, once we introduce optical gain into the cavity, as in the case of the photonic crystal laser, the emission linewidth is significantly narrowed (in our case $\Delta\lambda \approx 0.12\,\text{nm}$), and sensitivities of $\Delta n < 0.0001$ can be measured even in cavities with modest Q factors. Figure 26 suggests that the edge of the air band is even more sensitive to changes in the refractive index than the cavity mode itself. Therefore, band edge lasers[76-78] might prove to be even better choice for applications where high sensitivity to changes in refractive index is needed. However, band edge lasers operate at extended bulk photonic crystal modes with large mode volumes, and therefore are not suitable for applications where high spatial sensitivity is required (e.g. for single molecule detection).

The high Q cavity described so far can be used in the applications listed at the beginning of this section. The material of choice, wavelength of

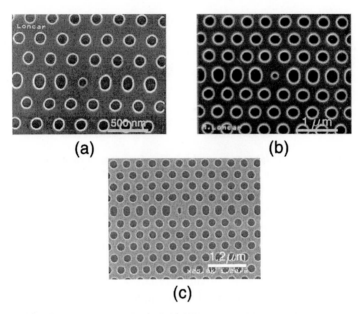

Fig. 27.   High Q cavities fabricated in (a) AlGaAs ($\lambda = 852\,$nm), (b) silicon on insulator ($\lambda = 1550\,$nm) and (c) InGaAsP ($\lambda = 1550\,$nm).

interest and fabricated structures are shown in Fig. 27. In next sections we study in detail photonic crystal nanolasers fabricated in InGaAsP material.

5.2. *Low-Threshold Planar Photonic Crystal Nanolasers*

In this section we explain the fabrication and experimental characterization of lasers realized in InGaAsP material system. InGaAsP quantum well material was grown on InP substrate using metal-organic chemical vapor deposition (MOCVD). Optical gain is provided by four compressively strained quantum wells, each 9 nm thick, with an electronic bandgap at $\lambda_{\mathrm{bg}} = 1.55\,\mu$m. The quantum wells are separated by 20 nm thick InGaAsP barriers ($\lambda_{\mathrm{bg}} = 1.22\,\mu$m). Because of the compressive strain, the coupling is the strongest to the TE polarized modes of the slab. This is desirable since in triangular lattice PPC the bandgap is larger for TE-polarized light. This active material is placed in the center of a 330 nm thick InGaAsP slab ($\lambda_{\mathrm{bg}} = 1.22\,\mu$m), with a $1\,\mu$m thick sacrificial InP layer underneath the slab. An InGaAs etch stop is introduced above the InP substrate, and the active quaternary material is designed to operate at $\lambda = 1.55\,\mu$m. The wafer structure is shown in Table 2.

Table 2. InGaAsP wafer structure.

Composition	Layer description	λ_{bandgap} [μm]	Strain	Thickness [nm]
InP	cap			50
InGaAsP	confinement	1.22		117
InGaAsP	QW	1.55	0.85% comp.	9
InGaAsP	confinement	1.22		20
InGaAsP	QW	1.55	0.85% comp.	9
InGaAsP	confinement	1.22		20
InGaAsP	QW	1.55	0.85% comp.	9
InGaAsP	confinement	1.22		20
InGaAsP	QW	1.55	0.85% comp.	9
InGaAsP	confinement	1.22		117
InP	sacrificial			1,000
InGaAs	etch stop			20
InP	substrate			300,000

5.3. *Fabrication Procedure*

The fabrication procedure consists of electron-beam lithography, followed by several dry- and wet-etching steps. Ideally, only one mask layer is needed to define patterns in the InGaAsP material of interest. However, due to poor etching selectivity between e-beam resist and InGaAsP we have to use a mask amplification method. A total of three mask layers is used before the patterns are defined in InGaAsP.

The fabrication procedure (Fig. 28) begins by RF sputtering of 100–150 nm of Si_3N_4 and is followed by thermal evaporation of 40 nm of Au. Then, 120 nm of polymethyl methacrylate (PMMA), electron-beam (e-beam) sensitive resist, is deposited on top of the Au layer by spin coating. By varying the electron dose and the periodicity of the lattice during lithography, we could span the entire design space of the PPC nanocavities by changing a, r, r_{def} and p/a independently. Upon completing e-beam lithography, PMMA was developed. Ion milling is used to transfer the mask from the PMMA into the Au. After milling, we undertake reactive ion etching (RIE) to transfer the mask from Au into Si_3N_4. Finally, the patterns are transferred into the InGaAsP using inductive-coupled plasma (ICP) RIE etching. We used the mixture of Ar and Cl to perform this final etch step. Following the ICP-RIE, the remaining Si_3N_4 mask is removed in HF acid and the InGaAsP membrane is released from the substrate by wet etching in 4 : 1 HCl : water solution at 4°C. HCl goes into the photonic crystal holes, and selectively attacks InP sacrificial layer, leaving InGaAsP

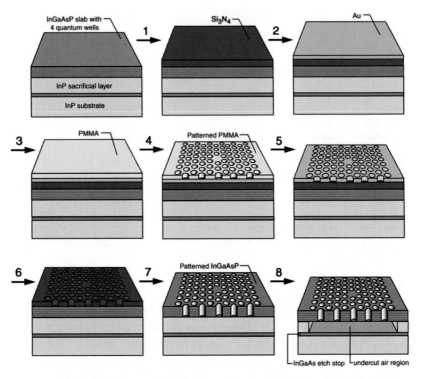

Fig. 28.   Fabrication procedure for PPC lasers made in InGaAsP material. RF sputtering (1) is used to deposit dielectric and is followed by deposition of Au mask (2).
E-beam lithography consists of PMMA deposition (3), e-beam writing (3) and PMMA
developing (4). Patterns are transferred from PMMA into Au using IBE (5) and then
into Si₃N₄ using RIE (6). Finally, cavities are etched into InGaAsP using ICP-RIE (7)
and the membrane is released from the substrate in HCl (8). Final structure is a free
standing InGaAsP membrane with four quantum wells (thin red layers).

membrane and InGaAs etch-stop layer intact. The processing sequence is
summarized in Fig. 28, and the conditions used are listed in Table 3.

In Fig. 29 we show some of the fabricated structures. Each structure
consists of six different cavities that have received the same electron-dose
during the e-beam lithography step, and therefore should have similar hole
sizes (r) and lattice constants (a). The only difference between the cavities
in one structure is the value of the dislocation parameter p which assumes
values in the range $p/a = 0$ to $p/a = 0.25$. Fifty different structures are
fabricated on the same sample, and these are arranged in the matrix with
5 rows and 10 columns, yielding the total of $5 \times 10 \times 6 = 300$ cavities. In this

Table 3.   Process parameters for fabrication of the first generation samples.

IBE	RIE	ICP RIE
$V_{\text{beam}} = 500\,\text{V}$	$P_{\text{RF}} = 70\,\text{W}$	$P_{\text{ICP}} = 800\,\text{W}$
$V_{\text{acc}} = 100\,\text{V}$	$U_{\text{DC}} = 320\,\text{V}$	$P_{\text{RF}} = 155\,\text{W}$
$I_{\text{beam}} = 10\,\text{mA}$	$P_{\text{process}} = 35\,\text{mTorr}$	$P_{\text{process}} = 1\,\text{mTorr}$
$P_{\text{process}} = 0.2\,\text{mTorr}$	C_2F_6 flow $= 25\,\text{sccm}$	Cl flow $= 15\,\text{sccm}$
Ar flow $= 1.7\,\text{sccm}$	Ar flow $= 5\,\text{sccm}$	Ar flow $= 10\,\text{sccm}$
etch time $= 3\,\text{min}$	etch time $= 4\,\text{min}$	etch time $= 15\,\text{sec}$

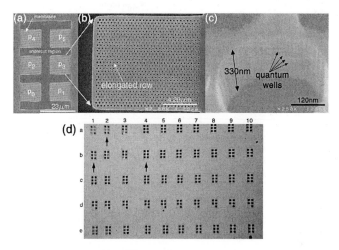

Fig. 29.   SEM micrograph of the structure $b4$. Each structure consists of (a) six different cavities with different elongation parameters: $p_0 = 0$, $p_1 = 0.05a$, $p_2 = 0.1a$, $p_3 = 0.15a$, $p_4 = 0.2a$ and $p_5 = 0.25a$. (b) Blow-up of p_3 cavity, and (c) of a single hole (tilted). Quantum wells and undercut air region can be seen. (d) Optical image of all 50 fabricated structures. Structures for which we show experimental data are indicated by arrows.

work we show results for three structures (total of 18 resonators) a_2, b_1 and b_4, according to their position within 5×10 matrix of fabricated structures.

A scanning electron (SEM) micrograph of a typical laser cavity characterized in the following section is shown in Fig. 30. The distribution of the hole sizes in that laser cavity is also shown. In Table 4 we summarize the geometry for four lasers that we study in the next section. In case of the b_4 structure, for which we show the most experimental data, the average hole size is found to be $r \approx 140\,\text{nm}$, and average periodicity of the photonic crystal lattice is $a \approx 435\,\text{nm}$. This geometry yields relative thickness

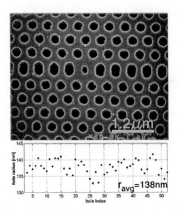

Fig. 30. SEM micrographs of b_4p_5 cavity along with the distribution of hole sizes in the photonic crystal mirrors surrounding the cavity.

Table 4. Geometry of fabricated devices.

Structure	r [nm]	r/a	r_{def} [nm]	$r_{def/a}$
a_2p_4	126	0.290	50	0.115
b_1p_5	125	0.290	71	0.162
b_4p_5	139	0.320	73	0.168
b_4p_4	138	0.317	73	0.168

of $d/a \approx 0.76$ and relative hole size $r/a \approx 0.32$, and thus is a slightly different geometry than the one analyzed in the previous section. Due to the increased hole size, the Q factors of our fabricated structures are expected to be smaller than those reported in the previous section. Moreover, the Q factors are expected to be even further decreased due to increased scattering of light at the rough hole walls.

5.4. Characterization of High Q Cavities

Fabricated structures were tested at room temperature using micro-photo-luminescence (μPL). The experimental setup is shown in Fig. 31. A diode laser emitting at $\lambda = 830$ nm is used as a pump source. In most cases the lasers were pumped with 1% duty cycles, using 10 ns pulses with 1 μs periodicity, or 30 ns with 3 μs periodicity. The pump beam was focused through a 100× objective lens onto the sample surface to obtain a spot size of $\sim$2.5 × 1.5 μm^2. The emission from the cavities was collected through the same lens, and the spectrum of the emitted signal was analyzed with an optical spectrum analyzer (OSA). Flip-up mirrors were used to obtain the optical images of the excitation pump-spot as well as the cavity modes.

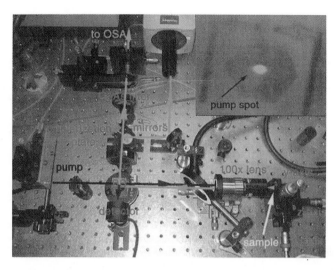

Fig. 31. Experimental setup. Pump beam ($\lambda = 830$ nm) is focused on the sample using high-power NIR lens (100×). The Si detector behind beam splitter (BS) is used to monitor the pump power. Emission from the sample is collected using the same lens, and is analyzed with an optical spectrum analyzer (OSA). The emission from photonic crystal resonators can also be monitored with an infrared camera (IR) using a pair of flip-up mirrors. Inset shows the size of the pump beam.

First, the emission from the unprocessed InGaAsP material was measured to obtain a gain spectrum of the active material, emitting between 1300 nm and 1650 nm, with a maximum at 1550 nm [Fig. 32(a)]. Assuming a photonic crystal lattice constant of $a = 436$ nm, this wavelength range corresponds to normalized frequencies in the range $0.264 < a/\lambda < 0.335$, well within the bandgap of the photonic crystal mirrors surrounding the cavity. Emission from the quantum wells is modified when the slab is suspended in air, even without presence of a PPC lattice, and Fabry–Perot resonances can be observed in slab luminescence spectra [Fig. 32(b)] due to the weak cavity formed between the etched facets at the edge of the membrane.

We tested all six cavities in the b_4 laser set in order to characterize their resonant modes. Two prominent resonant peaks are found in all cavities, both positioned well within the bandgap of the photonic crystal mirror. The peak linewidths are resolution bandwidth limited in Fig. 33, and the sub-threshold cavity peaks are limited by the cavity Q when carefully analyzed. We observe that the positions of these resonances depend strongly on the value of the elongation parameter p. This is the signature of the HQ and

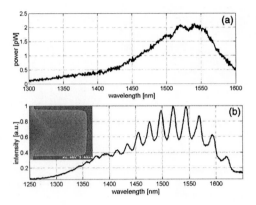

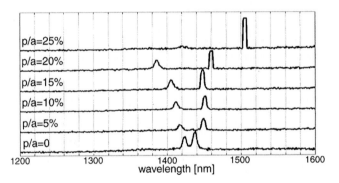

Fig. 32.   (a) Emission from unprocessed InGaAsP material and (b) from rectangular membrane suspended in air. Fabry–Perot resonances, due to rectangular cavity formed between etched facets, can be seen. Structures were pumped under CW conditions.

Fig. 33.   Structure b_4. Position of resonant modes detected in cavities p_0 div p_5 as a function of the elongation parameter p.

LQ modes of our cavity, as predicted by the theoretical treatment presented in the previous section. Originally, double degenerate dipole modes of a single defect cavity (p_0) are separated in frequency as the dislocation is introduced, and this splitting between the two orthogonal modes increases as the length of the dislocation increases. The low Q modes positioned at higher frequencies are shifted towards even shorter wavelengths, whereas the high Q modes are shifted towards longer wavelengths. The positions of HQ and LQ resonances are also strongly dependent on the defect hole size. Furthermore, the LQ and HQ modes are typically split even when no

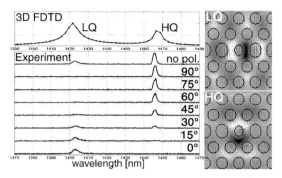

Fig. 34. Structure b_4p_3. Polarization dependence of the resonant modes. 0° corresponds to the y-axis direction. The figure also shows mode profiles (B_z component), polarization (E) and spectrum calculated using 3D FDTD analysis.

deliberate anisotropy is introduced (when $p = 0$), which we attribute to the breaking of the cavity symmetry due to fabrication imperfections.

The two resonances are linearly polarized with orthogonal polarization (Fig. 34). Moreover, the mode at longer wavelength (lower frequencies) is polarized along the x-axis, whereas the mode at shorter wavelength is polarized along the y-axis, as expected from our FDTD calculations. We also find that our numerical predictions of the positions of the modes are in excellent agreement with our experimental results. Based on the polarization properties of the observed resonances, their strong dependence on the amount of stretching of the central row, their position within the bandgap of photonic crystal mirror as well as the excellent agreement with theoretical predictions, we conclude that the resonances that we observe are localized dipole modes. According to theoretical predictions, for $r_{\text{def}} \approx 0.15a$ (the case of fabricated structures), Q is maximized when $p = 0.25a$, and we expect that the cavity p_5 has the highest Q factors of all the fabricated structures and is the best candidate for the realization of a low-threshold laser.

5.5. *Room Temperature Lasers*

Since our cavities have high Q factors and small mode volumes, it is expected that lasing can occur at room temperature. In Fig. 35, we show the dependence of the detected peak output power as a function of peak input optical power — a light–light (L–L) curve — for the high Q resonant mode of the p_5 cavity in structure b_4. The peak input power is the amplitude of the pump beam at the sample surface. The reflection from the sample is not taken into account, and the actual power that "pumps" the quantum

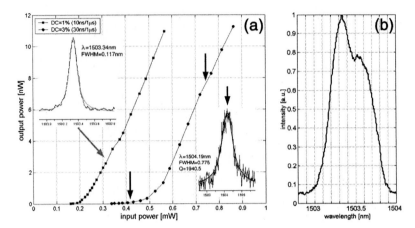

Fig. 35.   Cavity b_4, p_5. L–L curve for two different duty cycles (DC). The pulse period-icity was $1\,\mu s$ in both cases. Spectra taken above threshold for DC = 1% (red) and below threshold for DC = 3% (black arrow) are shown as insets. (b) Wavelength chirp in case of DC = 3%.

wells is significantly lower. When structure p_5 is pumped with a 1% duty cycle, a characteristic lasing curve with a threshold power of $P_{th} = 214\,\mu W$ is measured. In the same figure, we also show the laser spectrum above threshold, and note that the linewidth is then measured as $0.117\,nm$, and is limited by the resolution of our spectrometer.

When the duty cycle is increased to 3%, the threshold power increases to $520\,\mu W$. We have attributed this increase in threshold power to poor heat dissipation from our nanolasers and increased operating temperatures. The quality factor in the case of a p_5 structure was estimated from below threshold luminescence measurements to be approximately 2,000. However, it should be noted that by estimating Q factors from sub-threshold spectra, the actual Q value is typically underestimated. The measurement error is a result of additional cavity losses from reabsorption of light in the mirrors, which contain unpumped quantum well material. We expect the actual Q values to be higher and in the range of 6,000 as predicted from our numerical calculations. In Fig. 35(b) we show a spectrum for the lasing mode above threshold at a duty cycle of 3%. Wavelength chirping is apparent, and most likely a result of overheating of the cavity when pumping with high power, or due to the variation of the refractive index due to the carrier dynamics.

In Fig. 36, we also show an L–L curve for the p_4 cavity. It can be seen that the threshold is higher in this case, around $P_{th} = 950\,\mu W$. There

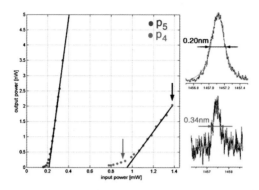

Fig. 36. Cavity b_4. L–L curves for structures p_5 and p_4 pumped with DC = 1%. The reduced differential quantum efficiency in the case of p_4 structure is due to the smaller Q factor. Spectra for p_4 structure, at and above threshold, are also shown.

are several reasons for this increased threshold. The gain provided from quantum wells is smaller at this wavelength ($\lambda = 1457.5$ nm) than in the case of p_5 cavity ($\lambda = 1504$ nm). Moreover, for $r_{\text{def}} \approx 0.15$, the Q factor of a p_4 cavity is smaller than that of a corresponding p_5 cavity, the p_5 structure has more dielectric material in the cavity region (longer dislocation), and a better mode-overlap with the gain material. In the inset of Fig. 36, we show the spectrum of the HQ peak of the p_4 cavity (set b_4), taken at the lasing threshold. A full-width half-maximum value of the resonance peak is measured to be FWHM = 0.34 nm, which corresponds to a Q factor of 4,300. This result is in good agreement with our theoretical predictions of $Q_{p4} = 4,000$. However, since the spectrum is taken at threshold, it is possible that the line-width is narrowed due to the gain provided by quantum wells. Therefore, this high measured Q value should be taken with caution, and more detailed analysis needs to be conducted in order to obtain more reliable estimates for such Q factors. For example, structures defined within passive materials (e.g. SOI) must be characterized in order to measure unambiguous "cold-cavity" Q factors.

Based on the experimental results presented so far, we can conclude that in spite of the unusual design of our nanocavities, with an air hole pierced through the center of the laser where the cavity field is the strongest, our structures lase at low-threshold powers. We have attributed this to high-quality factors and small mode volumes of our cavities (large Q/V_{mode}). In order to even further reduce threshold powers, it is beneficial to increase the Q of the cavities. This can be accomplished by decreasing the size of photonic crystal holes and thereby reduce the out-of-plane scattering.

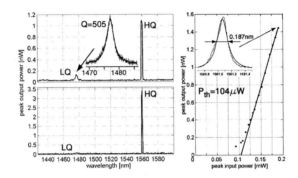

Fig. 37.   Structure b_1, cavity p_5. Photoluminescence spectra taken at two different pumping levels. Linewidth narrowing for HQ mode can be observed. The Q factor of LQ mode is $Q_{\rm LQ} = 505$. The L–L curve of HQ mode is also shown (DC = 1%).

In addition, the defect hole should be reduced in order to increase the optical mode overlap with the gain in the cavity. As can be seen in Table 4, cavity p_5 from set b_1 satisfies these conditions, and we expect to observe a lower threshold power in this cavity. In Fig. 37 we show results for the cavity b_1 p_5. Quality factor of the LQ mode was estimated to be $Q \approx 505$. The spectra taken for two different pumping levels show evidence of the spectral line-narrowing in the case of HQ mode. The HQ mode is positioned at $\lambda = 1560$ nm, almost perfectly matching the position of the maximum gain provided by the quantum wells. L–L curve for the HQ mode shown in Fig. 37 indicates reduced threshold power in this structure, as expected.

In Fig. 38, we show L–L curves for the a_2p_4 cavity, as well as tuning characteristics of the structures in laser set a_2. These cavities also support two modes, and like before, the one at longer wavelength (HQ) lases. Since both holes in the bulk photonic crystal mirror surrounding the cavity and the central defect hole are smaller in this case (Table 4), all of the resonances are shifted towards longer wavelengths. A photoluminescence spectrum taken above threshold, as well as the mode images from an infrared detector array are shown. The size of the emitted light spot from the nano laser is measured as $\approx 3.9\,\mu m^2$, a strong indication that this laser has a small mode volume. When the pump beam is moved from the center of the cavity by less than $1\,\mu m$, the strong signal shown in Fig. 38 disappears. In order to obtain quantitative estimation of the mode size as well as an image of the mode profile near field scanning optical microscopy (NSOM) can be used.[79]

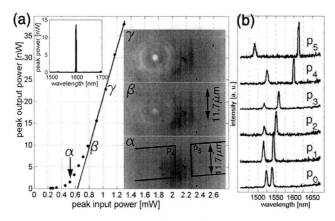

Fig. 38. Structure a_2p_4. (a) L–L curve. Lasing action occurs from HQ mode positioned at $\lambda = 1598$ nm. Insets show spectrum above threshold, and mode profiles of the lasing mode for several pump levels. The boundaries of the structure can also be seen. The mode is very well localized to the center of the cavity. (b) Tuning properties of a_2 structure, as a function of the elongation parameter p.

6. Photonic Crystal Lasers as Chemical Sensors

The construction of compact spectroscopic tools for the optical analysis of ultra-small ($<10^{-15}$ liter) sample volumes remains an important goal in the development of integrated microfluidics systems. Miniaturization of appropriate light sources and detectors can enable very compact and versatile "laboratory on a chip" devices, in which many analytical functions can be monolithically combined. One of the device integration platforms which is ideally suited to enable such integration of ultra-small and efficient optical components is the membrane-based planar photonic crystal. Here, we propose an application of planar photonic crystal cavities in the development of chemical sensors, with high spectral resolution and excellent sensitivity to changes in the absorption or refractive index of their surrounding. By combining our unconventional cavity geometry with optical gain at 1550 nm, we have defined ultra-small sensor elements which emit a very narrow spectral line. Our porous cavity design permits the introduction of analyte directly into the highest optical field of this laser cavity and, due to the ultra-small mode volume of our lasers, permits the sensing of optical changes within femtoliter volumes (Fig. 39). The introduction of absorbing or fluorescing molecules into such cavities is also expected to have a large influence on the optical signature, and the high fields obtained in the cavity can be used for spectroscopy of the cavity contents

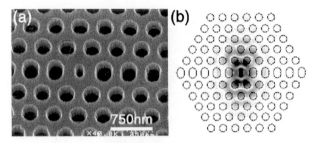

Fig. 39.   (a) Scanning electron micrograph and (b) calculated field distribution of a photonic nanocavity laser sensor.

(e.g. Raman or absorption spectroscopy). This can enable the sensing and analysis of individual organic molecules or self-assembled quantum dots, and offers a unique opportunity to achieve strong interaction between light and molecules on a nanoscale level.

6.1. *Fabrication of Photonic Crystal Biochemical Sensors*

The cavity geometry that we have chosen for our chemical sensors is based on the fractional edge dislocations introduced in previous sections. To demonstrate our biochemical sensors, photonic crystal nanolasers were fabricated in the same InGaAsP quantum well material as the one described in previous section (Table 5). This time, we controlled the precise emission wavelength either by scaling the lattice parameter or by changing the size of the defect hole introduced into the lattice to form the cavity. The structures were tested using a micro-photoluminescence approach and were optically pumped at room temperature with 30 ns pulses of $3\,\mu$s periodicity at $\lambda_{\text{pump}} = 830$ nm. In this work we used the same photo-luminescence setup as the one described in the previous section.

The fabrication procedure was slightly modified in order to improve the quality of fabricated structures. In this case, the RIE process was modified by replacing C_2F_6 gas with CHF_3, which has good etch selectivity between Si_3N_4 and PMMA. Therefore, we were able to etch Si_3N_4 using PMMA as the only mask. Patterns were again defined in PMMA by means of e-beam lithography and then were transferred into Si_3N_4 using RIE. 20 sccm of CHF_3 reactive gas was used and chamber pressure was kept at 16 mTorr. RF power used in this RIE step was 90 W and built-up DC voltage was $U_{\text{DC}} = 480$ V. The etch was done for 3 min. Finally, patterns are transferred from Si_3N_4 into InGaAsP using an ICP reactive ion etching process with the same conditions as in the previous section. The membrane was again released from the substrate using HCl.

As shown in the previous sections, the properties of photonic crystal devices depend strongly on the periodicity of the lattice as well as on the size and shape of all the holes in the photonic crystal lattice. Therefore, it is necessary to control these parameters and to minimize fabrication related disorders. In order to quantitatively characterize the quality of our fabrication process we have developed a simple pattern recognition technique. The procedure is based on detection of the edges of photonic crystal holes in the SEM image of the structure (Fig. 40). Panel (b) in Fig. 40 shows the holes detected in SEM micrograph shown in (a), and panel (c) shows distribution of the hole radii. The radius of each hole is estimated by calculating the area of the hole, assuming a circular shape of the hole. The average lattice constant is estimated as the average periodicity in each row of holes shown. The resolution of this procedure depends on the magnification and resolution of the SEM image. In case of Fig. 40 this resolution was 4 nm. As it can be seen fluctuation of the hole radius (±4 nm) is within the resolution of our technique. The quasi-periodic fluctuation of the hole size observed within each row (holes in the middle of the row are slightly larger than the holes at the ends of the row) can be attributed to the proximity effect during e-beam lithography — the holes in the center receive bigger effective electron dose. This can be corrected by reducing the size of the holes or the dose that holes in the center receive. By comparing the quality

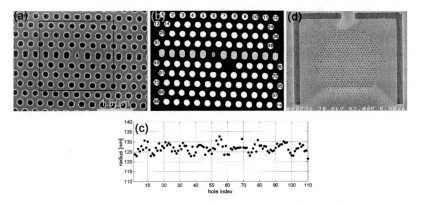

Fig. 40. Pattern recognition. (a) SEM micrograph of one of the structures. (b) Detected holes are shown in white and their area is calculated and used to estimate (c) the radius of each hole. (d) Zoomed-out view of the structure, showing the undercut membrane supported at two sides.

and uniformity of these fabricated structures with those described in the previous section (Fig. 30) we conclude that new fabrication procedure has much better uniformity and reproducibility.

Our hole-detection procedure can be used to compare theoretical and experimental results for our high Q cavities. For example, we used the information from Fig. 40(b) to generate an input structure for a 3D FDTD algorithm and to numerically model real structure, with all imperfections taken into account. Then, the only discrepancy between experimental and numerical results comes from the discretization of FDTD algorithm and the assumption that the walls in our etched structures are straight.

6.2. *Chemical Sensing*

To test the influence of a change in ambient refractive index on the laser spectrum of a cavity, we have backfilled our photonic crystal lasers with isopropyl alcohol and methanol. Figure 41(a) shows position of the resonances from six different lasers after immersion in air, isopropyl alcohol (IPA) and methanol. It can be seen that wavelength shifts of up to 67 nm can be observed when a cavity is immersed in IPA. This red-shift corresponds to a change in refractive index from 1.0 to 1.377, and yields roughly 1 nm spectral shift for a 0.0056 change in refractive index. When IPA is replaced with methanol ($n = 1.328$), the laser resonance experiences a blue shift of 9 nm (Fig. 26). We have also investigated the dependence of the cavity resonance wavelength on the lithographic laser geometry, particularly the lattice constant (a) and the dislocation (p/a) in the photonic crystal cavity. The cavity resonances experience red-shifts of 80 nm when the periodicity is changed from $a = 446$ nm (dashed lines) to $a = 460$ nm (solid lines). This lithographic tuning of the emission wavelength can be used to ensure an overlap of the cavity resonance peak with the InGaAsP quantum well emission gain curve even when the cavities are immersed in a reagent. Laser threshold curves before and after immersion into alcohol are also presented in Fig. 41(c). After immersion, the laser threshold power for the cavity measured was reduced since the emission wavelength was shifted to match the maximum gain of the quantum wells. However, the differential quantum efficiency of the immersed cavity is slightly lower, which may reflect the lower laser cavity Q after immersion.

To perform detailed analysis of the influence of the ambient refractive index on laser spectra, we have backfilled our photonic crystal lasers

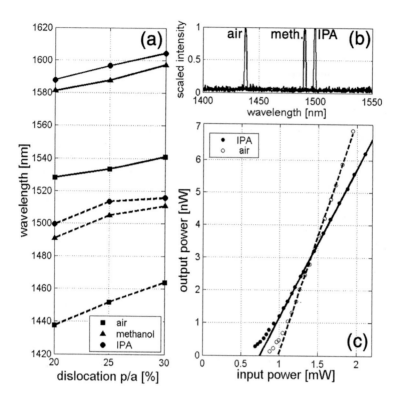

Fig. 41. (a) Sensor response measured from six nanocavities with a lattice parameter of $a = 446$ nm (dashed lines) and $a = 460$ nm (solid lines) as a function of p/a dislocation parameter. The three curves correspond to the laser wavelength with air, methanol and IPA backfilled into the cavity. (b) Spectrum change and (c) threshold curve change of laser before and after filling with IPA.

with refractive index calibration fluids (based on mixtures between perfluorocarbon and chlorofluorocarbon), with refractive indices in the range $n_{\mathrm{env}} \in [1.295, 1.335]$ with step $\Delta n_{\mathrm{env}} = 0.005$. To facilitate the testing the fabricated photonic crystal lasers were embedded into polydimethylsiloxane (PDMS) microfluidic flow channels. The PDMS chip is fabricated using replication molding soft lithography.[80–82] The microfluidic flow channels are aligned on top of the PPC lasers so that one of the flow channels passes over several photonic crystal cavities. For applications in which separate flow channels are not required on each PPC nanolaser, a large fluid cell encompassing the entire semiconductor chip with all PPC nanolasers was fabricated through replication molding. We have filled such reservoirs

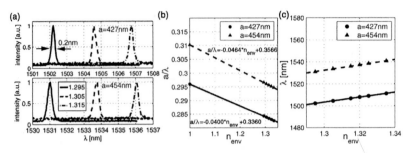

Fig. 42. (a) HQ modes of two cavities experience red-shifts when n_{env} increases. The linewidth of the laser is $\Delta\lambda = 0.2$ nm. Dependence of (b) a/λ and (c) λ of HQ mode on refractive index of the environment in the case of $p/a = 20\%$ cavities with two different periodicities.

defined in PDMS (depth of $100\,\mu$m) with refractive index fluids to perform experimental characterization of PPC nanolasers.

We have analyzed several different cavities and present typical results from two cavities with $p/a = 20\%$ and different lattice constants: (i) $a = 427$ nm and (ii) $a = 454$ nm. In Fig. 42(a) we plot the spectra of HQ modes in these two cavities for various n_{env}, and in (b) we show the dependence of the frequency of the resonant HQ mode as a function of n_{env}. It can be seen that frequency shifts depend linearly on n_{env}. The experimentally obtained sensitivities (slopes) are in good agreement with numerical model predictions [Fig. 26(b)].

While a relation between a/λ and n_{env} is useful for comparison between theory and experiments, a relation between λ and n_{env} is of more practical importance. That dependence is shown in Fig. 42(c) and we can see that the emission wavelength depends almost linearly on the refractive index of environment. The sensitivity of the laser emission wavelength on the changes in n_{env} is $\Delta\lambda = 243 \cdot \Delta n_{env}$ (in nanometers) in case of structure with $a = 454$ nm, and this is in good agreement with sensitivity obtained using 3D FDTD model ($\Delta\lambda = 266 \cdot \Delta n_{env}$). The smallest index change of index calibration fluids that we used was $\Delta n_{env} = 0.005$ and we measured wavelength shifts of $\Delta\lambda = 1.2$ nm in that case. It can be seen in Fig. 42(a) that the typical full-width half-maximum linewidths of lasers immersed in fluids are better than $\Delta\lambda_{FWHM} = 0.2$ nm. Therefore, our nanolasers can detect refractive index change as small as $\Delta n_{env} = 8.2 \cdot 10^{-4}$, within volume of 50 femtoliter (limited by V_{mode}). This resolution can be further improved by introducing more gain in the cavity.

6.3. *Dense Integration of Laser Sensors*

For many spectroscopy applications, it is desirable to construct dense arrays of ultra-small laser cavities, tuned to different wavelengths. We have explored the integration of photonic crystal lasers into such multiwavelength sources with lithographically predetermined spectra. These devices can be used as compact light sources for monitoring of several analytes at the same time, for example. Individual reactions can be observed in laser cavities which have predetermined spectral signatures, and can be optically read by observing changes in the collective spectrum of a multiwavelength laser array. In Fig. 43 we show both the structure and accompanying spectra of three optical cavities fabricated within a common photonic crystal slab with $a = 446$ nm and $r = 134$ nm. The sizes of the defect holes which define the optical cavities were varied from $r_{small} = 74$ nm, $r_{mid} = 85$ nm and

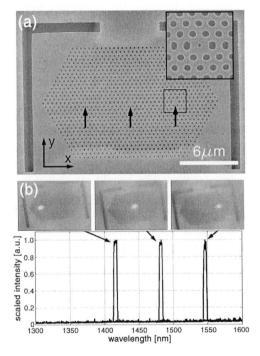

Fig. 43. Fabricated structure consists of three cavities integrated within the same photonic crystal mirror. (a) Defect holes are indicated by arrows, and their size increases from right to left. (b) Resonances detected in each cavity (normalized spectra). Mode experience blue-shift as the size of defect hole increases. Positions of pump-beam are shown.

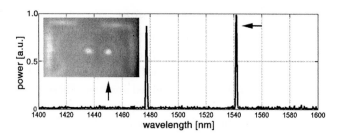

Fig. 44.   Simultaneous emission from two adjacent nanolasers pumped simultaneously using big pump spot. Insets show near-field image of the light emitted from nanolasers.

r_{big} = 97 nm, and a detailed view of one of the cavities is shown in the inset. The distance between the cavities is 10 lattice periods or ∼4.5 μm. This distance can be even further reduced (to about 2 μm) since the radiation of the HQ dipole mode of the cavity is predominantly along y-axis direction [Fig. 39(b)] and therefore the cross-talk between two adjacent cavities is minimal.

When our laser cavities are pumped individually, well-confined spectra are obtained from each of these nanocavities. The lasing wavelength of the cavities can be tuned from $\lambda = 1420$ nm (for r_{big}) to $\lambda = 1550$ nm (for r_{small}). It is important to emphasize that the emission from each laser could be observed only when the pump beam was positioned exactly on top of the nanocavity. Even slight variations in the position of the pump beam resulted in turning off the laser. We were also able to achieve simultaneous emission from two adjacent nano-lasers by defocusing the pump beam so that the pump spot covers two cavities. The results are shown in Fig. 44. Emission from cavities with defect hole radius r_{small} and r_{mid} was detected. It can be seen that by lithographic tuning of the nanocavities and by the choice of pump position, it is possible to achieve simultaneous lasing at two different wavelengths with comparable output powers. By adjusting the position of the pump beam, we were also able to choose between the two lasers, and enhance the signal from one laser while suppressing the other. The inset shows the corresponding near field image of the emitted laser light obtained using our IR camera. This time, the pump beam is filtered using a GaAs wafer, and the light detected with camera is only emitted luminescence.

6.4. *Future Directions*

In this section we briefly review possible methods to improve the performance of our sensors.[58] In particular we are interested in improvements of the cavity design (higher quality factors) and improvements in the fabrication procedure (smooth and vertical side walls in order to minimize unwanted light scattering).

In previous sections we showed that high-quality factors could be achieved in the single defect PPC cavity by tuning the cavity length by elongating one row of holes. An intuitive way to understand the origin of the high Q in this cavity is to consider that by introducing dislocation we alter the phase matching condition of the two-dimensional Fabry–Perot resonator. An alternative way to tune the cavity length is shown in Fig. 45. Two holes along x-axis, closest to the defect hole, are cut in half, and the halves facing the defect are turned into ellipses. The equation of the ellipse that we use is $x^2/(r - \Delta/2)^2 + y^2/r^2 = 1$, and therefore the length of the cavity can be expressed as $L = a(2 - 2r/a + \Delta/a)$. We show in Fig. 45(b) that by changing the ellipticity of these ellipses the y-dipole mode can reach $Q = 21,000$ when $\Delta/a = 40\%$. When such a cavity is backfilled with fluids with refractive index $n_{env} = 1.4$, Q factor drops to 3,100. This quality factor is three times higher than Q obtained in cavities with fractional edge dislocations. Also, a quality factor as high as 1,800 is achieved when $n_{env} = 1.6$. This suggests that it is possible to achieve room temperature lasing when the cavity is surrounded by liquid crystals, for example.

Our novel cavities are fabricated in the same InGaAsP quantum well material as the lasers described so far. An etch mask consists of 170 nm SiO_2 and 200 nm PMMA. Electron-beam lithography is performed using a Leica EBPG 5000 + e-beam writer, and the structures are developed in 3 : 1 solution of IPA and MIBK. Reactive ion etching with CHF_3 reactive gas is used to transfer patterns from PMMA into SiO_2. Inductively coupled plasma RIE, using hydrogen-iodine based gas chemistry[83,84]

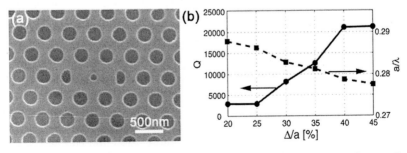

Fig. 45. (a) Improved cavity design. (b) Dependence of Q factor and normalized frequency (a/λ) on the ellipticity of two modified holes (Δ/a).

 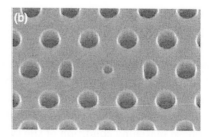

Fig. 46.   (a) Cross-section of one of the holes after ICP-RIE. Vertical side-walls can be observed at position of 330 nm thick InGaAsP active region. (b) 20° view of the final undercut structure.

is used to transfer the pattern from SiO_2 into InGaAsP. This gas chemistry ($HI:H2:Ar = 30:10:3$ sccm) provides smooth and straight side walls in quaternary material. The chamber pressure is $P = 10$ mTorr and ICP power and RF power are 850 W and 100 W, respectively. The DC bias is 320 V. The etch rate of InGaAsP/InP is ≈ 500 nm/min, and the selectivity between InGaAsP and SiO_2 is better than 8 : 1. The remaining mask is removed with HF and the InGaAsP membrane is subsequently released from the substrate by wet etching in a 4 : 1 HCl water solution at 4°C. The final structure is a free-standing membrane supported on one side. In Fig. 46(a) we show the cross-section of the sample after the ICP RIE step, and in Fig. 46(b) the final undercut structure viewed at 20° angle. It can be seen that straight walls are achieved at the position of InGaAsP slab that contains quantum wells, while some overcut is observed in the InP sacrificial layer. We have backfilled our lasers with liquid crystals and by applying the external gate voltage we were able to tune the emission wavelength of the laser.[85] In spite of the high refractive index of liquid crystals ($n_{env} > 1.5$) room temperature lasing action was observed. We have attributed this to high Q in our cavities when they are backfilled with liquid crystals.

7. Conclusions and Comparison between Types of Optical Cavities

In this chapter, we have described the evolution of optical cavities from one-dimensional Fabry–Perot etalons to two-dimensional photonic crystal geometries. We have presented the design, fabrication and characterization of such high Q photonic nanocavities, with particular attention to the dispersion characteristics of the photonic crystals. 3D FDTD methods, which we use as cavity design tools, permit us to develop cavities in various material systems (InGaAsP, AlGaAs, and silicon on insulator) for many applications. Nanocavities made from InGaAsP active material can be used to define room temperature optically pumped laser sources with ultra-small mode volumes. Light emission can be observed from modes with different directions in the same cavity, and photonic crystal cavities with several laser emission lines can be designed. High Q nanocavities, with Q values exceeding 20,000 and mode volumes of 50 femtoliters (for $\lambda = 1.55\,\mu m$) can be fabricated by using photonic crystals as two-dimensional mirrors, and these are expected to be important for strong interaction of light with matter. One of the most unusual capabilities of photonic crystals is their ability to concentrate light in regions where it would not be expected in classical devices. Thus, we could design a photonic crystal with a hole defined in the highest optical field region of the cavity. In spite of the unusual design of such lasers, which have a hole etched through the center of the cavity and therefore reduced overlap with gain material, we observe low lasing thresholds even in such devices. We have attributed this to both the small mode volume and the high Q factors inherent to the device design. The polarization and lithographic tuning properties of the dipole modes supported in the cavity are in excellent agreement with theoretical FDTD predictions, and the mode profile taken by our IR camera shows that the lasing resonance is well localized to the center of our cavity. In the nanolasers that we have described here, the highest intensity of the optical field is localized in air. This is fundamentally different from the majority of semiconductor lasers, in which light is typically confined in the high-index material.

As we have shown in detail, the precise geometry of microfabricated optical cavities determines the spectrum, polarization and direction of the lasing mode. This geometry in turn can be lithographically defined. One key advantage of lithographic definition of mirrors particular to quarternary materials is that it provides a good alternative to the difficult task of bonding or growing high-reflectivity mirrors epitaxially. Moreover, a more profound advantage of photonic crystal cavities over vertical cavities as found in VCSELs is the ease with which these can be coupled together. Lithographic alignment of optical cavities with one another enables the definition of "circuits" of cavities, and may provide a route towards defining cascadeable optical logic. The opportunity of miniaturizing the optical cavities by orders of magnitude also results in increases the spontaneous emission rates and the efficiency and speed of optical components. Finally, it is possible to align a high-finesse optical cavity with a quantum-mechanical source, such as a quantum dot or an atom, resulting in a strongly coupled system to experimentally explore the limits of quantum mechanics.

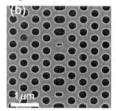

- mirrors are defined by lithography and etching
- only one lasing mode can be supported in the cavity
- lasers can easily be coupled together

- mirrors are defined by growth
- the cavity can support many lasing modes
- devices are difficult to couple together

Acknowledgments

We would like to thank Dr. Yueming Qiu, Dr. Mark L. Adams, Tomoyuki Yoshie, Michael Hochberg, Tom Baehr-Jones, Will Green, Brett Maune, Prof. Jelena Vuckovic and Prof. Stephen Quake for their help. We acknowledge generous support from the National Science Foundation under grant ECS-9912039, the Air Force Office of Scientific Research under contract F49620-01-1-0497 and DARPA under contract MDA972-00-1-0019.

References

1. H. M. Gibbs, S. L. McCall, T. N. C. Venkatesan, A. C. Gossard, A. Passner and W. Wiegmann, *Appl. Phys. Lett.* **35**, 451, 1979.
2. D. A. B. Miller, S. D. Smith and A. Johnston, *Appl. Phys. Lett.* **35**, 658, 1979.
3. H. Soda, K. Iga, C. Kitahara and Y. Suematsu, *Japan. J. Appl. Phys.* **18**, 2329, 1979.
4. S. Uchiyama and K. Iga, *El. Lett.* **21**, 162, 1985.
5. J. L. Jewell, S. L. McCall, Y. H. Lee, A. Scherer, A. C. Gossard and J. H. English, *Apl. Opt.* **29**, 5050, 1990.
6. J. L. Jewell, S. L. McCall, A. Scherer, H. H. Houh, N. A. Whitaker, A. C. Gossard and J. H. English, *Appl. Phys. Lett.* **55**, 22, 1989.
7. J. L. Jewell, A. Scherer, S. L. McCall, A. C. Gossard and J. H. English, *Appl. Phys. Lett.* **51**, 94, 1987.
8. J. L. Jewell, J. P. Harbison, A. Scherer, Y. H. Lee and L. T. Florez, *IEEE J. Quant. El.* **27**, 1332, 1991; J. Jewell, J. Harbison and A. Scherer, *Sci. Amer.* **86**, 56, 1991.
9. K. Iga, F. Koyama and S. Kinoshita, *IEEE J. Quant. El.* **24**, 1845, 1988.
10. D. L. Huffaker, J. Shin and D. G. Deppe, *El. Lett.* **30**, 1946, 1994.

11. E. Yablonovitch, *Phys. Rev. Lett.* **58**, 2059, 1987.
12. S. John, *Phys. Rev. Lett.* **58**, 2486, 1987.
13. J. D. Joannopoulos, R. D. Meade and J. N. Winn, *Photonic Crystal*, Princeton, NJ: Princeton University Press, 1995.
14. K. Sakoda, *Optical Properties of Photonic Crystals*, Springer, Berlin: Springer, 2001.
15. R. F. Cregan, B. J. Mangan, J. C. Knight, T. A. Birks, P. S. J. Russell, P. J. Roberts and D. C. Allan, *Science* **285**, 1537, 1999.
16. C. Cheng and A. Scherer, *J. Vac. Sci. Tech.* **B13**, 2696, 1995.
17. T. Krauss, R. DeLaRue and S. Brand, *Nature* **383**, 699, 1996.
18. J. O. Brien, J. O. Painter, R. K. Lee, C. C. Cheng, A. Yariv and A. Scherer, *El. Lett.* **32**, 2243, 1996.
19. A. Scherer, O. Painter, B. D'Urso, R. K. Lee and A. Yariv, *J. Vac. Sci. Tech.* **B16**, 3906, 1998.
20. B. D'Urso, O. Painter, J. O'Brien, T. Tombrello, A. Yariv and A. Scherer, *JOSA* **B5**, 1155, 1998.
21. P. Villeneuve, S. Fan, S. Johnson and J. Joannopoulos, *IEE Proc. Optoel.* **145**, 384, 1998.
22. O. Painter, A. Husain, A. Scherer, P. Lee, I. Kim, J. O'Brien and P. Dapkus, *IEEE Phot. Tech. Lett.* **12**, 1126, 2000.
23. H. Kosaka, T. Kawashima, A. Tomita, M. Notomi, T. Tamamura, T. Sato and S. Kawakami, *J. Lightwave Tech.* **17**, 2032, 1999.
24. A. Taflove, *Computational Electrodynamics — The Finite-Difference*, Time-Domain Method, Norwood, Massachussetts: Artech House, 1995.
25. N. Ashkroft and N. D. Mermin, *Solid State Physics*, Saunders College Publishing, 1976.
26. H. Benisty, C. Weisbuch, D. Labilloy, M. Rattier, C. Smith, T. Krauss, R. De la Rue, R. Houdre, U. Oesterle, C. Jouanin and D. Cassagne, *J. Lightwave Tech.* **17**, 2063, 1999.
27. C. Weisbuch, H. Benisty, S. Olivier, M. Rattier, C. J. M. Smith and T. F. Krauss, *Phys. Stat. Sol.* **B221**, 93, 2000.
28. O. Painter, J. Vučković and A. Scherer, *J. Opt. Soc. Am.* **B16**, 275, 1999.
29. J. Witzens, M. Lončar and A. Scherer, *IEEE J. Sel. Top. Quant. El.* **8**, 1246, 2002.
30. S. Johnson, S. Fan, P. Villeneuve, J. Joannopoulos and L. Kolodziejski, *Phys. Rev.* **B60**, 5751, 1999.
31. M. Lončar, T. Doll, J. Vučković and A. Scherer, *J. Lightwave Tech.* **18**, 1402, 2000.
32. G. Mur, *IEEE Trans. Electromagnet. Comput.* **23**, 377, 1981.
33. H. Ryu, S. Kim, H. Park, J. Hwang, Y. Lee and J. Kim, *Appl. Phys. Lett.* **80**, 3883, 2002.
34. M. Lončar, A. Scherer and Y. Qiu, *Appl. Phys. Lett.* **82**, 4648, 2003.
35. M. D. B. Charlton, S. W. Roberts and G. J. Parker, *Mat. Sci. Eng.* **B49**, 155, 1997.
36. M. E. Zoorob, M. D. B. Charlton, G. J. Parker, J. J. Baumberg and M. C. Netti, *Nature* **404**, 740, 2000.

37. H. Kosaka, T. Kawashima, A. Tomita, M. Notomi, T. Tamamura, T. Sato and S. Kawakami, *Phys. Rev.* **B58**, R10096, 1998.
38. H. Kosaka, T. Kawashima, A. Tomita, M. Notomi, T. Tamamura, T. Sato and S. Kawakami, *Appl. Phys. Lett.* **74**, 1212, 1999.
39. E. Cubukcu, K. Aydin, E. Ozbay, S. Foteinopoulou and C. M. Soukoulis, *Nature* **423**, 604, 2003.
40. M. Notomi, *Phys. Rev.* **B62**, 10696, 2000.
41. E. Silvestre, J. Pottage, P. Russell and P. Roberts, *Appl. Phys. Lett.* **77**, 942, 2000.
42. B. Gralak, S. Enoch and G. Tayeb, *JOSA* **B17**, 1012, 2000.
43. C. Luo, S. Johnson and J. Joannopoulos, *Appl. Phys. Lett.* **81**, 2352, 2002.
44. T. Baba and T. Matsumoto, *Appl. Phys. Lett.* **81**, 2325, 2002.
45. T. Baba and M. Nakamura, *IEEE J. Quant. El.* **38**, 909, 2002.
46. L. Wu, M. Mazilu, T. Karle and T. Krauss, *IEEE J. Quant. El.* **38**, 915, 2002.
47. C. H. Chen, A. Sharkawy, D. M. Pustai, S. Y. Shi and D. W. Prather, *Opt. Exp.* **11**, 3153, 2003.
48. E. Yablonovitch, T. Gmitter, R. Meade, A. Rappe, K. Brommer and J. Joannopoulos, *Phys. Rev. Lett.* **67**, 3380, 1991.
49. J. Vučković, M. Lončar, H. Mabuchi and A. Scherer, *Phys. Rev.* **E6501**, 016608, 2002.
50. K. Srinivasan and O. Painter, *Opt. Exp.* **10**, 670, 2002.
51. E. Miyai and K. Sakoda, *Opt. Lett.* **26**, 740, 2001.
52. S. Johnson, S. Fan, A. Mekis and J. Joannopoulos, *Appl. Phys. Lett.* **78**, 3388, 2001.
53. J. Vučković and Y. Yamamoto, *Appl. Phys. Lett.* **82**, 2374, 2003.
54. Y. Akahane, T. Asano, B. S. Song and S. Noda, *Nature* **425**, 944, 2003.
55. H. Park, J. Hwang, J. Huh, H. Ryu, Y. Lee and J. Kim, *Appl. Phys. Lett.* **79**, 3032, 2001.
56. K. Inoshita and T. Baba, *El. Lett.* **39**, 844, 2003.
57. T. Yoshie, J. Vučković, A. Scherer, H. Chen and D. Deppe, *Appl. Phys. Lett.* **79**, 4289, 2001.
58. M. Lončar, M. Hochberg, A. Scherer and Y. Qiu, to appear in *Opt. Lett.* 2003.
59. K. Srinivasan, P. E. Barclay, O. Painter, J. X. Chen, A. Y. Cho and C. Gmachl, *Appl. Phys. Lett.* **83**, 1915, 2003.
60. M. Lončar, T. Yoshie, A. Scherer, P. Gogna and Y. Qiu, *Appl. Phys. Lett.*, **81**, 2680, 2002.
61. T. Yoshie, O. Shchekin, H. Chen, D. Deppe and A. Scherer, *Elect. Lett.* **38**, 967, 2002.
62. P. Pottier, C. Seassal, X. Letartre, J. Leclercq, P. Viktorovitch, D. Cassagne and C. Jouanin, *J. Lightwave Tech.* **17**, 2058, 1999.
63. M. Meier, A. Mekis, A. Dodabalapur, A. Timko, R. Slusher, J. Joannopoulos and O. Nalamasu, *Appl. Phys. Lett.* **74**, 7, 1999.
64. K. Inoue, M. Sasada, J. Kawamata, K. Sakoda and J. Haus, *Jap. J. Appl. Phys. 2* **38**, L157, 1999.

65. R. Lee, O. Painter, B. Kitzke, A. Scherer and A. Yariv, *Elect. Lett.* **35**, 569, 1999.
66. O. Painter, R. Lee, A. Scherer, A. Yariv, J. O'Brien, P. Dapkus and I. Kim, *Science*, **284**, 1819, 1999.
67. A. Sugitatsu and S. Noda, *Elect. Lett.* **39**, 213, 2003.
68. J. Hwang, H. Ryu, D. Song, I. Han, H. Song, H. Park, Y. Lee and D. Jang, *Appl. Phys. Lett.* **76**, 2982, 2000.
69. M. Notomi, H. Suzuki and T. Tamamura, *Appl. Phys. Lett.* **78**, 1325, 2001.
70. T. Happ, A. Markard, M. Kamp, J. Gentner and A. Forchel, *Elect. Lett.* **37**, 428, 2001.
71. C. Monat, C. Seassal, X. Letartre, P. Viktorovitch, P. Regreny, M. Gendry, P. Rojo-Romeo, G. Hollinger, E. Jalaguier, S. Pocas and B. Aspar, *Elect. Lett.* **37**, 764, 2001.
72. P. Lee, J. Cao, S. Choi, Z. Wei, J. O'Brien and P. Dapkus, *IEEE Phot. Tech. Lett.* **14**, 435, 2002.
73. S. Kim, H. Ryu, H. Park, G. Kim, Y. Choi, Y. Lee and J. Kim, *Appl. Phys. Lett.* **81**, 2499, 2002.
74. C. Monat, C. Seassal, X. Letartre, R. Regreny, P. Rojo-Romeo, P. Viktorovitch, M. d'Yerville, D. Cassagne, J. Albert, E. Jalaguier, S. Pocas and B. Aspar, *Appl. Phys. Lett.* **81**, 5102, 2002.
75. J. Vučković, M. Lončar, H. Mabuchi and A. Scherer, *IEEE J. Quant. El.* **38**, 850, 2002.
76. H. Ryu, S. Kwon, Y. Lee, Y. Lee and J. Kim, *Appl. Phys. Lett.* **80**, 3476, 2002.
77. M. Imada, A. Chutinan, S. Noda and M. Mochizuki, *Phys. Rev.* **B65**, 195306, 2002.
78. S. Noda, M. Yokoyama, M. Imada, A. Chutinan and M. Mochizuki, *Science* **293**, 1123, 2001.
79. K. Okamoto, M. Lončar, T. Yoshie, A. Scherer, Y. Qiu and P. Gogna, *Appl. Phys. Lett.* **82**, 1676, 2003.
80. M. Unger, H. Chou, T. Thorsen, A. Scherer and S. Quake, *Science* **288**, 113, 2000.
81. Y. Xia and G. Whitesides, *Angew. Chem. Int. Ed. Engl.* **37**, 551, 1998.
82. M. L. Adams, Ph.D. Thesis, California Institute of Technology, Pasadena, California, 2003.
83. S. J. Pearton, U. K. Chakrabarti, A. Katz, F. Ren and T. R. Fullowan, *Appl. Phys. Lett.* **60**, 838, 1992.
84. D. C. Flanders, L. D. Pressman and G. Pinelli, *J. Vac. Sci. Tech.* **B8**, 1990, 1990.
85. B. Maune, M. Lončar, J. Witzens, M. Hochberg, T. Baehr-Jones, D. Psaltis, Y. Qiu and A. Scherer, submitted to *Appl. Phys. Lett.*, December 2003.

CHAPTER 3

SEMICONDUCTOR LASERS FOR TELECOMMUNICATIONS

Thomas L. Koch

Departments of Electrical and Computer Engineering, and Physics
205 Sinclair Laboratory, 7 Asa Drive
Lehigh University, Bethlehem, Pennsylvania
tlkoch@lehigh.edu

This chapter reviews the fundamental elements of semiconductor lasers for telecommunications, including materials, device structures and dynamical properties. Basic laser properties such as laser linewidth, modulation response and chirp characteristics are discussed. This chapter also includes the role of the cavity Q, and the use of grating-based microresonator cavities in distributed feedback lasers and other integrated structures that combat dispersive transmission limitations or add tunable functionality.

1. Introduction

Semiconductor lasers hold a special place in a book on microresonators as perhaps the most pervasive and mature application of microresonator technology. In fact, semiconductor lasers have also been the first widespread application of integrated optics, and they represent a marvelous confluence of decades of basic research in quantum electronics and laser physics, solid state and semiconductor device physics, epitaxial crystal growth techniques, and a wealth of planar and nonplanar integrated circuit microfabrication technologies.

The semiconductor laser was simultaneously invented at three different industrial labs, IBM,[1] General Electric,[2,3] and MIT Lincoln Laboratory.[4] This resulted from a period of intensive investigation of new laser media following the introduction of the laser concept in 1958 by Townes and Schawlow,[5] and Maiman's ruby laser demonstration[6] in 1960. The first CW low-threshold heterostructure semiconductor laser was realized by Hayashi and Panish[7] in 1970. In the context of microresonators, Kogelnik and

Shank[8] introduced a breakthrough in 1971 with their Distributed Feedback (DFB) laser by replacing the cleaved facet mirrors of waveguide resonators with a corrugated-layer grating. This provided a highly wavelength-selective resonator critical to today's wavelength-division-multiplexed (WDM) communications systems, and was also key to integration since DFB lasers did not rely on chip boundaries for feedback. Some of the enabling breakthroughs in this era have been well recognized for their extraordinary ingenuity, including the Nobel Prize winning work of Herbert Kroemer[9] and Zhores I. Alferov[10] in the early 1960s conceiving the semiconductor heterostructures that are so critical for today's high-performance lasers.

The years since then have witnessed remarkable advances in the materials growth of the underlying III-V compound semiconductors, understanding materials defects and laser failure modes, and defining the sophisticated device structures and engineering properties that are essential for use in optical communication systems and other applications of today. Today's lasers range in wavelength from the UV to the IR, are reliable enough for undersea and space applications, can be modulated at rates in excess of 10 Gbit/sec with wavelengths manufactured with precision to one part in ten thousand, and in most cases the chips themselves are cheap enough to be nearly inconsequential in cost even for consumer electronics.

In the context of this book, there are several noteworthy and remarkable observations to be made at the outset that distinguish semiconductor lasers from passive microresonators. These devices have provided an incredibly rich arena for device and materials engineering, in that they simultaneously perform as (1) highly sophisticated and often dynamically functional microresonators, (2) electrical P-N junction diodes, replete with requirements of energetic confinement of electrically injected carriers within the resonator, and sometimes integration with other junction devices, capacitors and resistors, and finally (3) these functions are usually implemented in highly sophisticated morphological structures with dozens of dissimilar material sectors, all formed together as *one seamless lattice-matched crystal*. Designs must be simultaneously optimized for optical, thermal and often RF performance. To achieve low cost, they must also be optimized with minimum fabrication and testing steps, and high reproducibility and reliability.

1.1. *Applications*

Semiconductor lasers have seen widespread use in consumer electronics, optical communications, and printing. They range from the ubitquitous CD lasers at 780 nm wavelength to the shorter 635–650 nm wavelength lasers used in (now ubiquitous) DVD players, laser pointers, etc., to 850 nm lasers used in short-reach data communications, to longer wavelength 1300 nm and 1550 nm lasers for intermediate and long reach communications. The optical storage and printing applications are expected to benefit from advances in GaN-based blue and UV semiconductor lasers in the 400 nm range, while wavelengths up to 2000 nm and beyond are finding applications in a variety of sensors.

The majority of these lasers are *single-spatial-mode*, and for the more advanced optical communication applications, these lasers are also *single-longitudinal-mode*. The nature of the optical microresonators used in semiconductor lasers play a key role in determining these characteristics, as well as the static and dynamic properties such as threshold current, quantum efficiency, modulation response, and characteristics such as intensity noise or laser linewidth. For a number of very high power applications, *multi-spatial-mode* edge emitting designs are also used, where even in a single 50 μm width stripe, output powers in excess of 5 Watts are practical. Much higher powers are still achievable in multistripe bars or even stacks of multistripe bars which produce multi-kW powers on a pulsed basis. Some applications for these lasers include precision delivery of heat for materials processing and soldering, as well as pump sources for other solid state lasers. In the communications area, one example is cladding-pumped fiber lasers or amplifiers, where multi-spatial-mode laser bar emission is injected into the multimode cladding of a single-mode fiber. The pump light is then absorbed into the single-mode core by dopants in the core (Yb or Er, for example) that can provide an effective gain medium for high power lasers or amplifiers.

This chapter will focus exclusively on *edge-emitting* semiconductor lasers, where the resonator is formed from an optical waveguide in the plane of the semiconductor substrate, and in particular on lasers for optical fiber telecommunications. Edge-emitting lasers provide ample gain at practical current levels, allowing output powers in single-spatial-mode designs as high as 1 Watt. Over the past decade vertical-cavity surface-emitting lasers (VCSELs) have also become commercially significant. In a VCSEL the resonator is typically formed between multilayer reflector mirror stacks and light propagates normal to the wafer surface. The major uses for VCSELs have been in short-reach data communication applications, usually at 850 nm, and have predominantly been *multi-spatial-mode* for use in multimode optical fiber. These devices have introduced a new regime of microresonator engineering, both due to their small net mode volume and also to the high-reflectivity mirrors required, and are the subject of Chapter 9.

1.2. *Brief Materials Background*[11]

Commercial semiconductor lasers are almost exclusively fabricated in III-V compound semiconductors due to their direct bandgap and resulting strong coupling to the radiation field for high absorption and gain characteristics. These semiconductors grow in the *zinc-blende* structure, essentially forming two face-centered-cubic lattices for the groups III and V compounds, displaced along a cubic diagonal by one quarter the diagonal, and where each atom of one group is tetrahedrally bonded to atoms of the other group. The most common substrate materials are GaAs and InP, with active layers made from alloys of Ga, Al and In for the group III, and P, Al, As, and more recently N for the group V. For example, in an alloy of $In_{1-x}Ga_xAs_{1-y}P_y$ the group III site is randomly occupied by either In or Ga, with average occupancy of $1 - x$ and x, and similar for As and P on the group V site.

The bandgap and lattice constants typically depend on the exact alloy compositions. In the case of $In_{1-x}Ga_xAs_{1-y}P_y$ to lattice match to the InP substrate as required for thicker, dislocation-free layers, these values must be chosen to satisfy[12,13] $y = 1 - 2.13x$ according to Vegard's law, with the 300 K bandgap energy given by $E_g = 0.74 + 0.61y$.

For example, to lattice match to InP and emit at 1300 nm, the values must be $x \approx 0.30$ and $y \approx 0.35$, whereas 1550 nm emission requires $x \approx 0.42$ and $y \approx 0.10$. In the early days of laser development, it was considered essential that all epitaxial layers be grown perfectly lattice matched to the substrate and to each other and, hence, free of strain. Beginning in mid 1980s, however, it was pointed out by Yablonivich and Kane[14] and Adams[15] that strain could induce beneficial modifications of the bandstructure for lasers, and in the ensuing years extensive additional experiments and analysis have validated these benefits. As a result, nearly all high-performance modern lasers are designed with strain incorporated in layers thin enough to accommodate the strain without inducing dislocations and reliability problems. Most often this is compressive strain, which was shown to allow transparency and lasing threshold at lower currents, and increase differential gain for improved high-speed and chirp characteristics. The reasons for these latter two improvements will become clear later in this chapter.

It has already been noted that the optical and electrical functions required for a laser are engineered in complex geometries that are realized in a single lattice-matched crystal. The devices must thus be grown as sequential layers of crystals using epitaxy processes, as opposed to more common amorphous thin-film technologies using evaporation and chemical vapor deposition. The three historically most significant epitaxy processes are Liquid Phase Epitaxy (LPE), Molecular Beam Epitaxy (MBE) and Metal-Organic Vapor Phase Epitaxy (MOVPE).

The first growth technology for lasers that was studied in great depth was LPE in the $Al_xGa_{1-x}As$ system, lattice matched to GaAs substrates. In this technique, liquid melts of Ga metal, containing small amounts of dissolved Al, GaAs, and semiconductor p- and n-type dopants such as Si, Ge, Te, or Sn are used to achieve the desired composition and conductivity. This materials system proved particularly attractive because the lattice constant of $Al_xGa_{1-x}As$ system is nearly independent of the value of x, allowing relatively thick waveguide cladding layers to be grown without strain-induced dislocations. The liquid melts are prepared by precise weighing of the constituents and loading them into a sequence of wells in a graphite strip (one well per melt composition) that can slide over the crystal substrate. The assembly is heated in a precision furnace to the 750°C range in an inert or reducing ambient such as He or H_2. After baking the entire apparatus for an extended period, the melts are cooled slightly until they become supersaturated, and then the crystal substrate is brought into contact with each melt in turn for a specified time to precipitate out the desired semiconductor crystal layers. The $Al_xGa_{1-x}As$ lasers covered the 700–900 nm range, and a host of other compounds and wavelengths were investigated by LPE as well. In particular, for the longer wavelength telecommunication applications it was found that the InGaAsP alloys can also be grown effectively by LPE on InP substrates and this formed the basis of the early commercial deployments of 1300 nm and 1550 nm systems.

In 1971 the MBE technique was introduced for growing AlGaAs epitaxial by Al Cho and coworkers at Bell Labs.[16] This new technique made it possible to grow highly-precise and uniform single crystal GaAs and AlGaAs layers of only a few atomic layers thickness. In MBE the III-V substrate is mounted in an ultrahigh vacuum system (10^{-10} torr) and heated to a temperature around 700°C. The system then directs molecular beams evaporated from effusion cells onto the substrate to grow a series of epitaxial layers forming the desired device structure; for example, beams of Ga atoms, Al atoms and As_2 molecules are employed to grow the AlGaAs alloy.

Motivated by the capabilities of MBE, in 1974 Dingle, Weigmann and Henry at Bell Labs conducted optical spectroscopy on structures where the absorption layer was less than 30 nm and could display quantum size effects.[17] Subsequent theory and experiments soon revealed that the use of these structures as active layers could provide tunability of bandgaps, and also that such a "quantum well" (QW) laser could have dramatically higher gain, lower threshold, and lower losses than previously demonstrated. These beneficial properties result from quantization in one dimension, with the result that more of the carriers are concentrated into energy states that can participate in the laser's stimulated emission. It has since been found that the best performance is usually obtained when the active region of the laser provides higher gain by using multiple quantum wells (MQW's), and these form the basis of most of today's lasers.

For the InGaAsP system of primary interest to optical fiber telecommunications, MBE proved difficult to implement for manufacturing, principally due to difficulties with handling phosphorus. The epitaxial technique of choice for fabricating MQW lasers in this material system has proven to be MOVPE,[18] often called MOCVD for *Metal-Organic Chemical Vapor Deposition*, a name which belies its true epitaxy character. In MOVPE technique, the substrate wafer is placed in a flowing gas ambient that is bubbled through liquid chemical precursors suitable for the growth of the desired alloy. The metal group III components are supplied as metalorganics, for example, trimethylgallium and trimethylindium, and the group V components are usually supplied as hydrides, for example, arsine (AsH_3) and phosphine (PH_3). At a sufficiently high substrate temperature (e.g. 630C, for the case of InGaAsP) these compounds decompose on the substrate surface enabling the lattice matched growth of the desired alloys.

MOVPE has proven to broadly have many of the same characteristics of MBE, including the ability to grow highly uniform quantum well and MQW structures on large area substrates with atomically abrupt interfaces. An additional key attribute of MOVPE growth is the nearly conformal coverage of etched features, such as the laser active mesa stripes, waveguides and corrugated-layer gratings, including the ability to fill-in regions that would be shadowed in MBE. This makes possible the complex structures to be discussed in the next section. Another feature that has become commercially significant is selective area growth (SAG), especially for its ability to fabricate complex integrated structures.[19,20] By covering adjacent regions on a wafer with growth-inhibiting SiO_2 masks, the relative growth rate in unmasked gaps can be locally enhanced relative to regions with no masking at all. This allows the thickness to be varied along the length of a layer that will form a waveguide core. Due to the quantum size effect, if this thickness variation is in the quantum wells, it translates to continuously controllable bandgap variations, making possible the integration of gain regions, passive waveguide regions, and electroabsorption modulator regions, all formed continuously along the length of a chip. This process has been used extensively to make integrated photonic circuits such as a DFB laser monolithically integrated with an electroabsorption modulator for ultra-high-speed telecommunications, preserving the spectral purity of the DFB source even under modulation.

2. Basic Semiconductor Laser Structures

The material advances discussed in the introduction have provided a rich foundation for semiconductor laser designs of ever-increasing sophistication and complexity. The resulting improvements in laser performance have been instrumental to the rapid advances in both span lengths and raw capacity that can be achieved with optical fiber transmission. This section will examine the evolution in laser designs that have become critical enablers for many new regimes of optical fiber communications system design.

Early research and development in laser designs focused primarily on improvements in basic laser operating parameters that were essential to prove the viability of semiconductor lasers as a communications light source. These improvements included reductions in threshold current I_{th}, improvements in differential quantum efficiency η_d, and closely related improvements in device reliability. These laser characteristics were clearly paced by the rapid improvements in material quality, but were also governed by elements of design that rapidly matured into highly sophisticated, topologically rich laser structures. It is this three-dimensional processing and design environment which most markedly distinguishes the manufacture of semiconductor lasers from that of other microelectronic components and circuits.

Fundamental to efficient laser operation are the simultaneous confinement of light in a low-loss resonator, and the confinement of a population inversion inside the resonator to provide optical gain. For *edge-emitting* semiconductor lasers, in contrast to VCSELs, the most basic and ubiquitous optical resonator consists of a low-loss dielectric optical waveguide terminated by cleaved crystal facets serving as feedback mirrors as shown in Fig. 1.

The population inversion is generated by injection of a high-density nonequilibrium electron-hole plasma using a *p-n* junction in forward bias. The basic elements that are needed to understand the performance of this resonator are thus simply (1) the loss of the waveguide, and (2) the static and dynamic characteristics of the current-injection-induced gain.

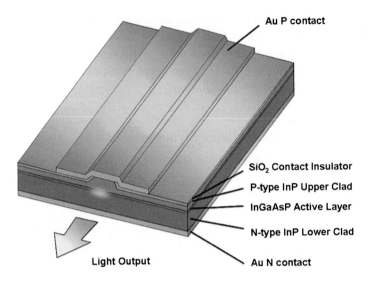

Fig. 1.   Typical Fabry–Perot semiconductor laser displaying the use of a double-heterostructure active region; in this example a ridge-waveguide is used for lateral confinement.

Critical to the performance of all modern lasers is the concept of the double heterostructure, and its relationship to optical waveguide design. Optical waveguides in semiconductor lasers typically consist of a waveguide *core* comprising epitaxially grown layers with a composition chosen to have an average index of refraction that is higher than the surrounding epitaxially grown layers above and below the core that comprise the waveguide *cladding*. It was a fundamental observation by Kroemer[9] and Kazarinov and Alferov[10] as early as 1963 that a *heterostructure* sandwich formed by placing a narrow bandgap layer between higher bandgap layers provided not just the aforementioned optical waveguide, but also served to confine the electron-hole population inversion in the narrow bandgap waveguide core, exactly as required for efficient laser operation. Such a structure is shown in Fig. 2 for an InGaAsP/InP bulk heterostructure at 1.55 mm, which also shows qualitatively the electron and hole wavefunctions in the conduction and valence bands, along with the optical mode profile of the slab waveguide formed by the active layer and InP cladding. This optical and electrical confinement, while operative only in one dimension, was nevertheless key to the demonstration of room-temperature *continuous wave* (CW) operation.[7] This one-dimensional confinement was readily modified to stripe geometry and gain-guided designs simply by limiting the lateral

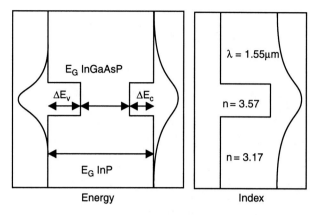

Fig. 2.   Double heterostructure active layer concept, where a narrow-bandgap material is sandwiched between two higher-bandgap layers, providing both energetic confinement of the inversion, as well as a high-index core for an ideally superposed optical waveguide.

extent of the electrical excitation by limiting current injection with narrow electrical contacts (typically 5–20 μm) or by proton implantation to render all but a protected central stripe region highly resistive. Typical chip lengths were in the range of 250–600 μm between the cleaved facet mirrors, and AlGaAs lasers of this early generation were used in the first network deployment of optical fiber communications on the Eastern coast of the US in 1981 with 0.82 μm lasers in multimode fiber operating at 45 Mb/s.

Major advances through the 1970s focused on the refinement of designs and fabrication techniques that extended this optical and electrical confinement more efficiently to the lateral dimension as well. This required defining a full low-loss two-dimensional *rib* waveguide, and also providing for the confinement of the highly mobile injected minority carriers efficiently inside the micron-scale core of this waveguide. Figure 3 shows a modern example of a *buried heterostructure* laser, a key generic design that first emerged[21] in 1974 and achieved the required optical and electrical lateral confinement by the epitaxial regrowth of lateral cladding layers around a mesa etched through the active layer stack. This design, and numerous variants resulting from a proliferation of etching and crystal growth techniques, also provides for reverse-biased junctions in the lateral cladding to block current leakage paths and to force the current into the excitation-confining active region. As a result, this design and its descendents have provided for many of the most efficient and highest-performance InGaAsP lasers, routinely achieving threshold currents of $I_{\text{th}} < 10$ mA and efficiencies of $\eta_d \sim 0.4$ W/A.

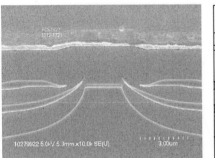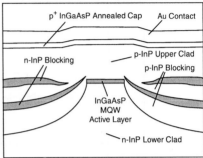

Fig. 3.   Example of a modern buried heterostructure semiconductor laser design.[22] (Left) Scanning electron micrograph of laser waveguide cross-section, "stained" using a chemical etchant to reveal layers of different composition and doping. (Right) Drawing identifying the layers seen in the SEM, illustrating the use of alternating p- and n-doped layers for blocking current except in the region of the active layer.

Another laser design that continues to enjoy widespread use, especially in high-power InGaAs/AlGaAs lasers required for pumping optical amplifiers, is the *ridge-waveguide* laser as was shown in Fig. 1. Here the active layer is not laterally restricted, but the upper cladding is etched to form a ridge-shaped dielectric loading above and along the length of the active core, thereby providing weak stripe optical waveguiding. The ridge upper cladding also serves to laterally restrict current injection into the active layer beneath the ridge. The tight layer thickness control and uniformity required to achieve good performance at high yield using this design were practical once advanced MBE and MOCVD crystal growth became prevalent, and this design does offer a simplified fabrication and epitaxial growth sequence. Both the buried heterostructure and ridge waveguide lasers have clearly demonstrated reliability exceeding the requirements for deployment even in undersea applications.

3. Laser Resonator Static and Dynamic Characteristics

The laser resonator plays a key role in determining many of the laser's key characteristics, including output coupling and efficiency, dynamic performance, and especially spectral characteristics. To understand this role in a simple fashion, we will use a standard rate equation analysis governing the interplay between population inversion, or free-carrier density in a semiconductor active layer, and the optical field intensity or photon density in the cavity. The simple form we will use for discussion is:[23]

$$\dot{S} = av_g\Gamma(N - N_t) \cdot S \cdot (1 - \varepsilon S) - \frac{S}{\tau_{\text{ph}}} + \beta\frac{\Gamma N}{\tau_{\text{sp}}} \qquad (1)$$

$$\dot{N} = \frac{I}{eV_{\text{act}}} - av_g(N - N_t) \cdot S \cdot (1 - \varepsilon S) + \frac{N}{\tau_{\text{total}}} \qquad (2)$$

$$\dot{\phi} = -\frac{\alpha}{2}av_g\Gamma(N - N_t) \cdot S \qquad (3)$$

where $N(t)$ is the carrier density in the active layer, ϕ is the phase of the optical field, and we define the photon density $S(t)$ in terms of the number of photons in the cavity mode $N_P(t)$ or the optical power from one of the two (assumed equally transmissive) facets $P(t)$ as

$$S(t) \equiv \Gamma N_P(t)/V_{\text{act}} = \Gamma 2\tau_{\text{ph}}P(t)/h\upsilon\eta V_{\text{act}} \qquad (4)$$

Here V_{act} is the active layer volume, Γ the usual modal confinement factor, τ_{sp} the spontaneous lifetime, τ_{total} the total recombination lifetime including radiative and non-radiative events, τ_{ph} the photon lifetime, and η the total differential quantum efficiency assuming no leakage currents, β is the fraction of spontaneous emission into the lasing mode given by $\beta = \Gamma\lambda^4/4\pi^2 n_g^3 V_{\text{act}}\Delta\lambda_{\text{sp}}$ where $\Delta\lambda_{\text{sp}}$ is the FHWM of the spontaneous emission peak.[24] This expression arises simply by calculating the solid angle subtended by the laser mode compared to 4π steridian of emission, and additionally the frequency span $\Delta\lambda_{\text{FP}} = \lambda^2/2n_g L$ between modes compared to the total emission band $\Delta\lambda_{\text{sp}}$.

The gain has been modeled linearly in excitation as $g(N) = a(N - N_t) \cdot (1 - \varepsilon S)$ with the average emission cross-section $a \sim 2.5 \times 10^{-16}\,\text{cm}^2$ and $N_t \sim 1 \times 10^{18}\,\text{cm}^{-3}$ being the carrier density required to reach transparency (equal absorption and stimulated emission) at the lasing wavelength, and v_g is the modal group velocity. Here we have also included a small contribution of optical nonlinearity in the gain as can arise from spectral hole-burning of optically-induced carrier heating, with the inclusion of the factor $(1 - \varepsilon S)$. The equation for the phase ϕ is a fairly accurate linearization of the gain medium rate of differential change of real index versus imaginary index with carrier density, using the alpha factor defined as $\alpha \equiv (\partial n_r/\partial N)/(\partial n_i/\partial N)$, and noting that the gain is given by $g = 4\pi n_i/\lambda$.

While these equations appear to be a vast phenomenological simplification, they can be more rigorously related to first-principles calculations where fruitful.

Of interest here is that nearly all the resonator characteristics have been captured in two parameters, namely τ_{ph} and β. The photon lifetime τ_{ph} is the decay rate of optical excitation in the "cold cavity", i.e. with no gain or absorption from the active transition — it does include absorption from all other means, and certainly decay due to output coupling from the cavity.

In the usual fashion, we can average the mirror facet losses (assumed equal for now) along the length of the cavity to contribute effectively a continuous loss of $\gamma_m = -(1/L)\ln(R)$ and thus if we assume a waveguide loss per unit length of γ_{wg}, the decay rate from this is

$$\tau_{\mathrm{ph}} = [v_g(\gamma_{wg} + \gamma_m)]^{-1} = Q/\omega \tag{5}$$

where Q is the usual definition of the cavity "Q" at optical frequency ν of the resonator, $Q \equiv \nu/\Delta\nu$ where $\Delta\nu$ is the full-width-half-maximum of the resonator response, or alternatively from above, the number of angular cycles in one exponential decay time of the energy in the resonator. Only the power coupled from the facets is accessible, and thus the differential quantum efficiency η is given by $\eta = \gamma_m/(\gamma_m + \gamma_{wg})$. This is true because all additional excitation goes into stimulated emission over threshold. This is evident by solving the equations in steady state which allows the simple and intuitive observation that the output power is given by the total noise source divided by the difference in loss and gain,

$$P = \frac{\eta}{2}\left(\frac{h\nu N V_{\mathrm{act}}}{\tau_{\mathrm{sp}}}\right)\left(\frac{\beta}{1 - v_g\Gamma g(N)\tau_{\mathrm{ph}}}\right) \tag{6}$$

and hence the requirement that for reasonable output powers, the gain is "clamped" at the threshold value where the denominator is nearly zero.

Typical values of mirror reflectivities are ~30%, but of course coatings are often used to modify this, with typical values of ~70–90% on the back facet, and 10–20% on the front facet to improve power efficiency from the launch end of the device. Waveguide losses for InGaAsP bulk lasers are typically ~25 cm^{-1} primarily resulting from dopant and excitation-induced free-carrier and intervalence-band absorption, with holes producing approximately 20 cm^{-1} alone for each 10^{18} cm^{-3} of density. Together these lead to photon lifetimes on the order of $\tau_{\mathrm{ph}} \sim 2\,\mathrm{psec}$, illustrating that edge-emitting semiconductor lasers are usually operated in a relatively high-gain, high-loss regime.

To a remarkable extent this simple picture captures most of the critical ingredients required to understand the basic properties of semiconductor lasers. However more formal approaches can certainly be used to discuss the resonator-related characteristics. For example, if we rigorously solve the electromagnetic Green's function for a simple Fabry–Perot resonator as discussed above, and derive power outputs based upon intracavity spontaneous dipoles using this Green's function, the output will have the familiar resonance denominator

$$D_{\text{cold}}(\omega) = 1 - R \cdot \exp[-i2n_r\omega L/c - i\gamma_{wg}L] \qquad (7)$$

If we then imagine an impulse excitation, for example, this will oscillate and decay based on the poles of the Green's function, or the zeros of this denominator. This provides exactly the same answer as the heuristic and physical description of Eq. (6), with $\omega_{\text{pole}} = \omega_{\text{resonance}} - i/2\tau_{\text{ph}}$ where the factor of 2 is derived from a field representation rather than intensity. However, such an approach is useful in more complex resonators that may be spatially inhomogeneous, etc. Looking at the above resonance denominator also reveals the multilongitudinal mode nature of the Fabry–Perot resonator. If, in addition to the "cold cavity" loss above, we can insert the optical modal gain to get the "hot cavity" resonance denominator

$$D_{\text{hot}}(\omega) = 1 - R \cdot \exp[-2in_r\omega L/c + (\Gamma g(N) - \gamma_{wg})L] \qquad (8)$$

which provides insight into a number of interesting things. One of these is the multimode character. The rate equations can easily be extended to multimode situations by having a separate photon density equation for each mode, and the carrier density driven by the sum of all modes. In a Fabry–Perot laser, the multimode character is governed by the curvature of the gain with wavelength. The gain is usually approximated as a parabola near its peak as $g(N, \lambda) = a(N - N_t)[1 - b(\lambda - \lambda_{\text{peak}})^2]$, and the relative mode outputs are easily understood through Eq. (8). As in the earlier discussion, there are subtleties that arise from optical nonlinearities of the gain medium that can affect the relative gain of one mode versus another, including spectral hole-burning and four-wave mixing effects that arise from the carrier population being driven by intensity beat notes of the various modes.

For optical communication purposes, resonators are often modified to encourage one mode to have significantly lower loss than others, leading to a situation where threshold is reached for this mode while other modes are still negligibly small. In general, it is easy to show for the case of one dominant mode with power P_1 that the ratio of output powers $\chi = P_1/P_2$

for two modes is given by[25]

$$\chi = \frac{2P_1}{h\nu v_g n_{\rm sp}(\gamma_m + \gamma_{\rm wg})} \left[\frac{\Delta\gamma L}{\gamma_m L}\right] \tag{9}$$

where $\Delta\gamma$ is the difference in net loss minus gain for the two modes, and $n_{\rm sp}$ is determined by the gain medium near threshold as the ratio of *actual* downward transitions and *net* downward transitions (i.e. stimulated emission minus absorption for the levels), and is called the "spontaneous emission factor"[26] with a typical value of $n_{\rm sp} \approx 2$.

This ratio is true for average powers, but for communication application assumptions based on average power may not be adequate, as temporal fluctuations may occur on the time scale of a digital bit period and cause errors in the system. For example, in dispersive systems, a bit produced at a different wavelength will propagate at a different group velocity and leave its expected time slot. It can be shown that,[23] to maintain a probability below 10^{-9} for a statistical fluctuation of 50% of the main mode power, the ratio above must be $\chi > 40$, and the time scale of the fluctuations is inversely proportional to χ. This is a severe event, and a better rule of thumb is that $\chi > 100$, which is consistent with the experimental rule of thumb that "single longitudinal mode" operation requires at least 20 dB side mode suppression. Commercial sources require at least 30 dB in most cases to insure adequate margin.

Other properties such as phase noise, or linewidth, can also be understood in a basic sense using these simple considerations. The linewidth of a passive resonator is given by $\Delta\nu = 1/(2\pi\tau_{\rm ph})$ with the "cold cavity" Q given by

$$Q_{\rm cold} \equiv \nu/\Delta\nu = \omega\tau_{\rm ph} = \omega \cdot [v_g(\gamma_{wg} + \gamma_{\rm facet})]^{-1} \tag{10}$$

The "hot cavity" Q is given by substituting in the *net loss* (i.e. including gain) as in our discussion of the hot cavity resonance denominator,

$$Q_{\rm hot} \equiv \omega\tau_{\rm ph} \cdot [1 - v_g\Gamma g(N_{\rm th})\tau_{\rm ph}]^{-1} \tag{11}$$

which leads immediately to the expectation, using the earlier expression for output power, that the hot cavity linewidth would be given by the expression

$$\Delta\nu_{\rm hot} = \frac{1}{2\pi\tau_{\rm ph}} \frac{\beta \cdot (h\nu N_{\rm th}V_{\rm act})/\tau_{\rm sp}}{(2P/\eta)} \tag{12}$$

This illustrates a reasonably intuitive result that the linewidth is the cold cavity linewidth multiplied by the ratio of spontaneous emission power feeding into the lasing mode to the intracavity power of the mode (i.e. the cold cavity linewidth divided by the spontaneous emission amplification factor for the lasing mode).

This replacement of Q_{cold} with Q_{hot} as done above is essentially the basis for the early Schawlow–Townes linewidth formula,[5] and while it does capture the basic phenomenon, it omits two significant effects. The first is that many gain media, such as a semiconductor, can dynamically adjust the gain level to maintain the "gain clamping" action noted earlier, and essentially actively cancel out intensity fluctuations. This works at least for frequencies where the medium can respond, which we will discuss briefly below. Since the noise captured by the expression above is half "in phase" (intensity) noise, and half "in quadrature" or phase noise, the expression above must be reduced by multiplying with a factor of $1/2$, which then makes it the so-called "modified" Schawlow–Townes formula.[27] It should thus be remembered that the laser linewidth is essential phase noise.

Early measurements of semiconductor laser linewidths revealed serious discrepancies from simple "hot cavity" modified Schawlow–Townes linewidth formula. Henry was the first to realize the impact on laser dynamics that results from the fundamental difference of band-to-band gain in a semiconductor compared to typical isolated atomic or molecular laser transitions.[28] In the semiconductor, small increases in gain with increasing excitation are inherently accompanied by *reductions in absorption* at shorter wavelengths that are actually substantially larger in magnitude than the gain increase at the gain peak. The peak in the *differential* gain change is thus inherently shifted to shorter wavelength than the gain peak, which is most often the lasing wavelength. Examination of the famous causal Kramers–Kronig relations between real and imaginary index of refraction then requires that a decrease in the real index of refraction occur at the gain peak when gain is increased. This negative change is further enhanced by contributions resulting from the mobile carrier "plasma" index of refraction, both producing negative changes in index with increases in gain.

This effect was captured earlier with our phase equation and the α-factor, which is also often called the "linewidth enhancement factor". Using the formalism of Lax,[29] Henry showed that the dynamic gain fluctuations noted above are converted to phase fluctuations as the medium "clamps out" the intensity fluctuations. The Lax formalism provided an additional increase or enhancement of the linewidth by a factor of $(1+\alpha^2)$, and Henry showed that the semiconductor gain medium is inherently "detuned" with α-factor values as large as $\alpha \sim 5$ for bulk active layers, yielding a huge increase in linewidth by a factor of 25 or more. The corrected expression that Henry used for linewidth was

$$\Delta\nu_{\text{linewidth}} = \frac{1}{8\pi} \frac{v_g^2 h\nu n_{\text{sp}} \gamma_m \gamma_{\text{total}} \cdot (1+\alpha^2)}{P} \tag{13}$$

While this may seem quite different from the earlier expression based on the rate equations, we leave it to the reader to show that they are reconciled entirely if we identify the average emission cross-section "a" in our gain formula earlier as $a \sim (\lambda/2\pi n_g)^2(1/\tau_{\text{sp}}\Delta\nu_{\text{sp}})$ which is a basic equation for simple optical transitions,[30] and the spontaneous emission factor as $n_{\text{sp}} \sim N_{\text{th}}/(N_{\text{th}} - N_t)$ which is consistent with its definition as well.

Vahala and Yariv[31] and Henry[32] enhanced Eq. (13) to include more completely the dynamic effects of the laser response, producing for example, additional structure on the linewidth due to gain medium relaxation oscillations. The importance of Henry's α factor permeates nearly all aspects of semiconductor laser dynamics and noise since it represents a fundamental amplitude-phase coupling in the gain medium. Since changes in feedback into lasers alter the threshold gain requirement and thus the lasing frequency through α, the dynamics of feedback instabilities and injection locking are also governed by α. We will see shortly that it also governs frequency chirping under modulation.

3.1. Distributed Feedback Lasers

The lasers described above are termed *Fabry–Perot* (FP) lasers since the longitudinal optical resonator structure comprises a waveguide terminated on each end by cleaved facet mirrors, similar to the Fabry–Perot etalon. The resonances of such an optical cavity are equally spaced in frequency ν by $\Delta\nu = c/2n_g L$ where c is the speed of light, n_g is the group index of refraction of the waveguide (typically $\sim$3.8) and L is the chip length between facets (typically $\sim$300 μm). Since the optical gain spectrum provided by the electrical excitation is quite broad (typically $>$30 nm), the output of a typical FP laser, especially when the device is kept from reaching equilibrium by direct modulation, consists of a small number of longitudinal modes spaced in wavelength by $\sim$1 nm. At a wavelength of 1.3 μm, near the chromatic dispersion zero of conventional single-mode fiber, even this spectral width can permit transmission to distances of 40 km at speeds of 1.7 Gb/s without excessively restrictive tolerances on the center wavelength of laser operation. However, speeds of 2.5 Gb/s and higher make the wavelength tolerance around the fiber dispersion zero prohibitive, both due to variance in fiber dispersion zero and variance in laser manufacture. It provided an incentive to design lasers that restrict their operation to a single longitudinal mode of the laser cavity, providing a dramatic reduction in laser spectral width.

A more important incentive came from the significantly lower fiber loss at 1.5 μm, with values below 0.2 dB/km allowing loss-limited spans of 100 km or more, and the resulting savings from increasing the span length between regenerators beyond 40 km. It was also attractive to provide increased capacity by offering coarse wavelength division multiplexing (WDM) with channels at both 1.3 μm and 1.5 μm simultaneously on one fiber. However, the high dispersion value of $D = 17$ ps/nm-km for conventional fiber at 1.5 μm prohibited the use of FP lasers.

While a number of structures were examined in the early 1980s for achieving single longitudinal mode operation in semiconductor lasers, the *distributed feedback* (DFB) laser emerged as the clear choice for widespread manufacture and deployment. First demonstrated by Kogelnik and Shank in dye lasers,[8] this laser design replaced the FP cavity with optical feedback from a corrugated waveguide grating as illustrated in Fig. 4. Instead of relying on a discrete mirror reflection with no spectral selectivity, this corrugation provides a multitude of tiny sub-reflections from each corrugation period that are phased properly for a large cumulative net reflection only near the Bragg wavelength λ_B given by $\lambda_B = 2n\Lambda_B$ where n is the

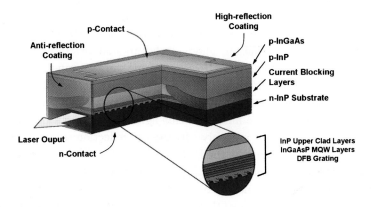

Fig. 4.   Example of a modern buried heterostructure distributed feedback laser design.

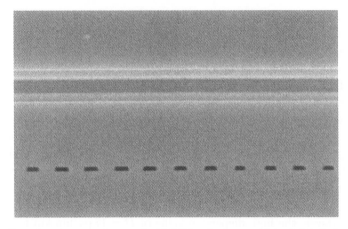

Fig. 5.   Scanning electron micrograph of MQW active layer with buried InGaAsp grating
sections for distributed feedback.

phase refractive index of the waveguide mode, and Λ_B is the spatial period
of the corrugation. Typical values of Λ_B for $1.5\,\mu$m operation are $0.23\,\mu$m,
requiring ultra-violet laser interference to lithographically pattern the mask
features for etching the waveguide corrugation. In today's DFB lasers, the
corrugation is typically etched into a surface with buried quantum wells and
then planarized with a burying epitaxial growth to form buried rectangular
islands as shown in Fig. 5.

Gratings and periodic structures play a key role in the development of
optical microresonators. One of the basic facts to be grappled with is that
light propagating at the Bragg wavelength can actually not be resonant in
a continuous DFB grating structure. Light is highly reflected by a Bragg
reflector at its Bragg wavelength, but this does not constitute a resonance.
Figure 6 depicts this situation, where all the cumulative sub-reflections add
up to give a large net reflection. However, if one were to stand in the middle
of such a structure and trace light that propagates in one direction, gets
reflected, propagates back in the other direction, and gets reflected again to
the original direction, one finds that this light is exactly out of phase with
the original light. This is precisely why the net reflection is high — there is
no transmission resonance at the Bragg wavelength, and lasing takes place
at the transmission resonances of a resonator structure.

In order for a resonance to occur, the wavelength needs to slip away
from the Bragg wavelength so that it can get some reflection from the ends
of a span of grating and have some phase slip as it crosses the span, in

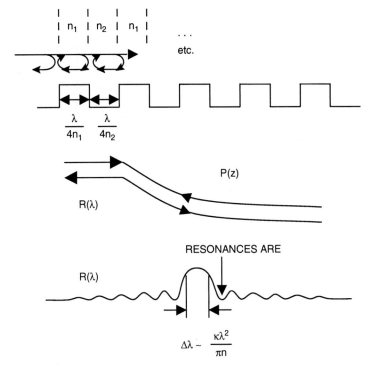

Fig. 6. Schematic showing sub-reflections from the periods of a waveguide grating structure. Light is *antiresonant* in the "forbidden gap" wavelength band near the Bragg wavelength.

order to get a cumulative constructive interference. This is shown in Fig. 7, where the resonances occur just outside of the grating "stop band" that has a width of $\Delta\lambda \sim \kappa\lambda^2/\pi n_g$ where κ is the grating coupling constant given by $\kappa = \pi\delta n/\lambda$. Here δn is the average amplitude (i.e. half peak-to-peak) of the modal index perturbation seen by the waveguide mode.

These structures are readily analyzed with coupled-mode equations, which we will not reproduce here.[33] The situation is entirely analogous to electron waves propagating in the periodic potential of a crystal lattice, where the bandgap represents a forbidden zone. In actuality, states can exist in the forbidden gap, but they can do so only when there is some discontinuity to be allowed for the appropriate matching of boundary conditions. That is, there can be a localized state in a periodic structure if a "defect" is introduced, with the exponential-like decay shown above trailing off from the "defect". In a crystal, such a defect can be an impurity atom

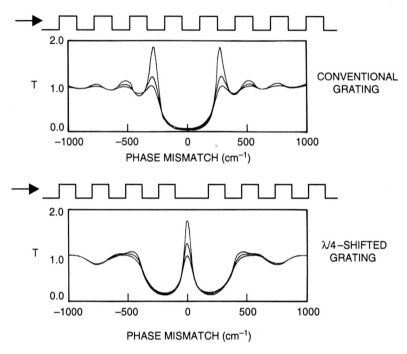

Fig. 7. Transmission spectra of corrugated-waveguide structures for increasing gain levels, showing resonances in the case of a standard continuous grating (top) and a quarter-wave-shifted grating (below).

or a crystalline defect, or simply the surface of the crystal. For periodic optical microresonators, a defect can similarly be an intentional deviation from the periodicity, or a "surface" such as an end of the device.

The ability for facets to act in this regard was illustrated by Streifer *et al.*,[34] while Haus and Shank[35] generalized this with the introduction of "Antisymmetric Tapers", the simplest of which was the introduction of a quarter-wave phase shift in the middle of the grating. This shift, of half a grating period or a quarter of the optical wavelength, can be understood based on the discussion above to introduce just the additional phase required for a cumulative round trip at this position to add up in phase. This allows for a transmission resonance, as shown in Fig. 7, displaying the transmission with successively increasing gain levels for both a standard grating and a $\lambda/4$-shifted grating. As can be seen in this figure, the conventional grating has two degenerate modes that are equally high Q, and this is undesirable for single-longitudinal-mode applications.

The linewidth formula Eq. (13) can be applied to such a structure provided the mirror loss, previously $\gamma_m = -(1/L)\ln(R)$ for a FP resonator, is replaced by the appropriate value for a grating resonator. For an arbitrary phase shift θ in the center of a grating, the value of γ_m is derived from solving the transcendental equation

$$\kappa^2 e^{-2i\theta} = [\Delta\beta + i\gamma_m/2 - iS \coth(SL/2)]^2 \tag{14}$$

where $S \equiv [\kappa^2 + (\gamma_m/2 - i\Delta\beta)^2]^{1/2}$ and $\Delta\beta \equiv \beta - 2\pi/\Lambda_B$. Here a $\lambda/4$-shifted grating is a phase shift $\theta = \pi/2$ and it can be shown[36] that for large values of κL, γ_m is roughly given by $\pi^2/\kappa^2 L^3$.

To avoid FP laser operation from the facet reflections, the output facet of any DFB laser is typically *anti-reflection* (AR) coated while it is desirable for the other end to be *high-reflection* (HR) coated to avoid wasting power from the back. This back-facet HR coating provides a very strong "surface defect" for the cavity, and while the exact phase of this facet is never controlled for manufacture, the majority of the lasers have ample selectivity for one mode compared to any others due to this high-reflectivity facet; one can think of this facet as providing a "folded" phase-shifted cavity. For this reason, the $\lambda/4$-shifted grating DFB has not seen widespread application; to perform ideally it requires AR coatings on both facets, at which point half the optical power is usually then sent off in the wrong direction for system use. However, there have been some integrated chip architectures where the precise a *priori* positioning of different wavelength lasers make the $\lambda/4$-shifted DFB the design of choice. It should be noted that the addition of some periodic loss or gain will also lead to a preference of one of the two degenerate DFB modes in a conventional nonshifted grating, which can be used for deterministic wavelength placement.

DFB lasers routinely provide highly single-longitudinal mode operation, with other modes rejected by values of 30 dB or more, and the advent of $1.5\,\mu m$ DFB lasers quickly led to demonstrations of transmission rates as high as 4 Gb/s over distances of 100 km[37] even in fibers with standard dispersion of $D \sim 17\,\mathrm{ps/nm \cdot km}$. DFB lasers have also been critical to the optical transmission of analog signals in the cable television industry, where the linearity of the light-current relationship and reduced dispersive distortion are extremely critical to signal fidelity. Here, highly optimized laser designs have also been implemented where analog distortion metrics, such as composite second and third order distortions, can be kept at maximum values of $-65\,\mathrm{dB}$ below the carrier[38] in 60-channel NTSC systems.

3.2. Direct Modulation

A critical feature of semiconductor lasers for communications is their ability to directly modulate the intensity of the light output simply by modulating the drive current. This can be analyzed rather easily from the rate equations provided earlier by performing a small-signal analysis about the steady-state CW values of drive current I, photon density S, carrier density N, and phase ϕ, as in $I = I_0 + \delta I e^{i\omega t}$, and $S = S_0 + \delta S e^{i\omega t}$, etc. The result of this simple exercise is Eq. (15) below, which compares the response of the photon density or optical output power to a modulation of the drive current at a frequency ω in this case normalized to its DC response for convenience.

$$\frac{(\delta S(\omega)/\delta I(\omega))}{(\delta S(0)/\delta I(0))} = \frac{\left(1 + \frac{\Gamma_d^2}{\omega_{RO}^2}\right)}{\left(1 - \frac{\omega}{\omega_{RO}} + i\frac{\Gamma_d}{\omega_{RO}}\right) \cdot \left(1 + \frac{\omega}{\omega_{RO}} - i\frac{\Gamma_d}{\omega_{RO}}\right)} \tag{15}$$

Here we have defined the relaxation oscillation angular frequency $\omega_{RO} \equiv 2\pi f_{RO}$ as the real part of the pole of this expression (rather than the "peak amplitude") which is given by

$$\omega_{RO} = \sqrt{\frac{v_g g' \cdot S}{\tau_{\mathrm{ph}}} - \Gamma_d^2} \tag{16}$$

where g' is the "differential gain" given by the rate of change of optical gain (in cm^{-1}) per unit change in excited carrier density (the cross-section "a" in our linearized model), and the damping or imaginary part of the pole Γ_d is given by

$$\Gamma_d = \frac{1}{2}\left(\frac{1}{\tau_{\mathrm{total}}} + v_g g' \cdot S + \frac{\varepsilon S}{\tau_{\mathrm{ph}}} + \cdots\right) \tag{17}$$

where in both expressions we have ignored small terms with β and terms of $\vartheta(\varepsilon^2)$. The omission of the β terms would not be justified in some microcavity configurations where β can become larger. As f_{RO} is approached, the response peaks up and then decays rapidly with a second-order low-pass filter characteristic at 40 dB per decade. Since the output power of a laser is converted to current by a photodetector we have drive current mapping to detector current, and when the frequency response is measured by an RF network analyzer apparatus one usually plots RF power response (i.e. current squared) vs. frequency. Such a plot is shown in Fig. 8, where we have used typical laser parameters and $\varepsilon = 1.5 \times 10^{-17}\,\mathrm{cm}^{-3}$.

As communication lasers were first put into manufacture, explicit engineering speeds approaching $\sim$1 GHz were readily achieved and allowed early laser designs to be deployed for systems operating at speeds up to 622 Mb/s.

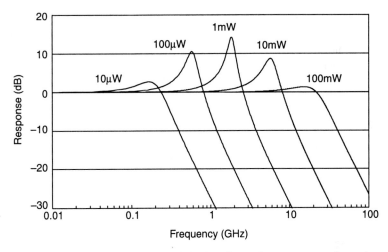

Fig. 8. Frequency response of a typical semiconductor laser structure from the small signal analysis of rate equations. Note that the inclusion of realistic parasitic capacitance and inductance would lead to additional filtering and roll-off. Also, other damping mechanisms can contribute to more suppression of the relaxation oscillation resonances.

Throughout the 1980s work in both the AlGaAs system[7] and the InGaAsP system[8] showed that speed is typically limited by the electrical parasitic capacitance, series resistance and bonding wire inductances. Chief among these was the capacitance of the reverse-biased lateral current blocking layers in buried heterostructure lasers.

Designs were introduced employing thick, low-capacitance semiinsulating Fe-doped InP layers in the lateral blocking structure to reduce this capacitance, often accompanied by longitudinal trenches to electrically isolate a $\sim 10\,\mu$m wide region containing the buried active stripe. With the addition of thick dielectrics under the bonding pads sharp reductions in the capacitance down to the range of several pF have proven practical, allowing for commercial devices to be modulated at speeds up to 10 Gb/s, with the fastest small-signal research results — often achieved in ridge laser designs with very few compromises in capacitances — extending out to 30 GHz for 1550 nm devices.[39]

Examining the factors determining f_{RO} in Eq. (16) above reveals that achieving high speed strictly by increasing the photon density S often requires undesirable or impractically high output powers. For example, it has the disadvantage of the resulting need for excessively high modulation current swings to turn the laser on and off, placing difficult demands on the driver ICs and overall power consumption employed in transmitters.

Increases in differential gain that result from strained multiple-quantum-well active layers have been shown to provide for high speeds (10 Gb/s) at reasonable drive currents (<100 mA), a fact that is very important for low-cost directly modulated OC-192 data links.

A number of applications for lasers are extremely cost sensitive and some reductions in performance can be traded against cost, operational simplicity and power consumption. Since the *thermoelectric coolers* (TECs) used in traditional long-haul laser modules consume substantial power and size, and add to cost; uncooled lasers where the laser is free to swing in temperature with the ambient are desirable. This requires high performance over a typical range of −40 to 85°C, a daunting requirement when laser thresholds typically increase by a factor of two for every 40°C increase in temperature. This has required great care in optimizing both blocking structures to avoid increases in leakage currents, and especially the detailed doping and layer sequence of the MQW active region to provide ample gain at the highest operating temperature.[40] Some of the best results have been obtained using InGaAlAs MQW active layers, due to the higher conduction-band offset and resulting reduction in laser leakage currents past the active layer at higher temperatures as the laser is pumped. Today's uncooled 1300 nm lasers can achieve lasing in the laboratory to temperatures as high as 130°C and commercially meet customer requirements up to module case temperatures of 85°C.

In the long haul digital applications, further increases in transmission data rates and span lengths using directly modulated DFB lasers were limited by both loss and remaining dispersion impairments related to laser *chirp*. Laser chirp refers to the dynamic spectral broadening that occurs during direct modulation, even for a single longitudinal mode DFB laser. The reasons for this chirp could be understood largely due to the theoretical work of Henry on laser linewidth mentioned earlier and has been captured here in our Eq. (3) for the laser phase. Directly from these equations, one can show that frequency excursions inherently accompany optical power excursions $P(t)$ in the simple relation[41]

$$\Delta v_{\text{chirp}} = \frac{-\alpha}{2\pi} \left\{ \frac{d}{dt} \ln P(t) + \mu \cdot P(t) \right\} \qquad (18)$$

where μ is a coefficient related to nonlinear gain saturation and its relaxation oscillation damping effect based on the parameter ε introduced earlier.

Equation (18) reveals that the very act of driving a laser from one power level, for example a low state for a digital "0", to a higher state for a digital "1" requires that the gain depart from equilibrium to establish this ramping of optical power. During this ramping, the gain (imaginary index) is higher than in CW operation, and due to the linewidth enhancement factor α, this also introduces a change in the real index or optical path of the cavity. This introduces a frequency shift of the laser output, a shift that is actually imposed on the photons in the cavity by the dynamically ramping index of the cavity (i.e. it does not require "old photons" to decay and "new photons" to be generated at the new frequency). The major impact of this effect is that the leading edge of a pulse will be blue-shifted ($d \ln P(t)/dt$ positive), and the trailing edge ($d \ln P(t)/dt$ negative) will be red-shifted. Furthermore, for lasers where the relaxation oscillation is excited, there will be rapid oscillatory shifting of the frequency from red to blue.

All of these effects play havoc with dispersive transmission, where the different frequency-shifted components of the pulse propagate at different group velocities. These effects were observed experimentally, and using computer implementation of the analysis discussed here, implemented numerically to reproduce with remarkable accuracy the experimental results.[22]

A very approximate rule of thumb is that the dispersive transmission limitations that would otherwise be imposed strictly from the inherent bandwidth of the information in the signal are more severe by a factor of $\sim (1 + \alpha^2)^{1/2}$.

Since the α-factor is defined as the differential real index change per unit carrier density, divided by the differential imaginary index (gain) change per unit carrier density, materials and structures with high differential gain will also have small α factors. Typical bulk active layers have values of $\alpha \sim 6$, but MQW structures have values as small as $\alpha \sim 2$. This can be further reduced by forcing the laser to operate, using the DFB corrugation for example, at wavelengths that are blue-shifted relative to the gain peak, further increasing the differential gain. By using highly optimized MQW designs with very low a values, today's commercial DFB lasers can achieve transmission in conventional fiber over distances in excess of 200 km at 2.5 Gb/s with negligible dispersion penalty.

4. Integration and Higher Functionality Modules

The extended transmission spans of the 1990s were achieved with the addition of erbium fiber amplifiers, which have been responsible for a revolution in optical transmission system design. In particular, it became practical to cascade amplified spans to provide for unregenerated transmission over dispersion limited distances of 600 km and more in conventional fiber at 1.5 μm. However, this required spectrally pure sources with chirp still smaller than achievable with a directly modulated DFB laser, and will be discussed below as higher-functionality modules.

The obvious answer to this was the use of *external modulation*, where the laser is operated as CW and its output is gated on and off through external means such as a LiNbO$_3$ traveling wave Mach–Zehnder modulator. However, it was observed in the late 1980s that *electroabsorption* (EA) modulators have design features very similar to semiconductor lasers, and thus offered the potential to form a *photonic integrated circuit* (PIC) where both the laser and EA modulators are simultaneously fabricated on one InP substrate. This offered the promise of reductions in cost and power consumption compared with LiNbO$_3$ solutions, and its reduced size also allowed nominally the same footprint as the prevalent DFB packages of the day.

Electroabsorption modulators operate on a quantum tunneling principle, termed the Franz–Keldysh effect in bulk materials and the quantum-confined Stark effect in quantum wells. Here light is propagated through a layer selected such that the light has a photon energy lower than the bandgap, or onset of absorption, of the layer. The application of a large enough electric field across this layer results in a voltage drop, generated in the very short distances accessible by quantum mechanically tunneling, that is sufficient to effectively reduce the transition energy and allow absorption for the lower energy lasing photons. Thus the application of a field across the modulator causes the light to be extinguished and converted to photocurrent that is drawn from the modulator.

The integration of a DFB laser and an EA modulator was first demonstrated by locally etching away the laser gain layers and regrowing a new, higher bandgap waveguide layer for the modulator.[42] The electric field is generated by reverse biasing the p-n junction that is fabricated in the same steps that are used for the forward biased laser p-n junction. However, the development of the advanced epitaxial techniques such as selective area growth described earlier, allowed for a simpler process. Here, suitable masking during epitaxy allows for a change in thickness of the waveguide core quantum wells along the length of the device. As shown below in Fig. 9, this can produce a structure where the thinner quantum wells in the modulator region have a higher effective bandgap due to the shift in the quantum ground state from the thinner quantum well.

This technique has been used for high-volume manufacture of the *electroabsorption modulated laser* (EML) for deployment in long-haul, optically

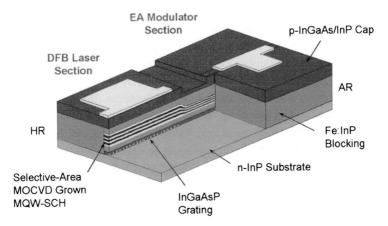

Fig. 9. Integrated electroabsorption-modulated laser (EML) integrating a DFB laser monolithically with an electroabsorption modulator formed using selective area growth (SAG).[43]

amplified systems.[43] These devices routinely provide 2.5 Gb/s sources with peak wavelength excursions of ∼0.01 nm, or frequency excursions of ∼1 GHz, resulting in only a small frequency-modulation contribution to the inherent bandwidth of the digital intensity encoding of the optical signal. For this reason, these sources can transmit over optically amplified distances in excess of 600 km in conventional fiber, close to the fundamental limits imposed by dispersion of a pure intensity encoded waveform. Achieving this level of spectral purity also required great care to eliminate electrical cross-talk between the modulator drive and the laser bias, as well as extraordinary suppression of output facet reflections that would provide time-varying, destabilizing feedback into the laser.

In addition to optically amplified spans, the 1990s also witnessed the rampant deployment of dense WDM (DWDM) transmission systems, providing cost-effective, upgradable capacity while maintaining the benefits of optical amplifiers capable of boosting an entire wavelength channel set in one device. In the mid-1990s, the *International Telecommunications Union* (ITU) accelerated the acceptance of DWDM by providing for a set of standardized channel wavelengths evenly spaced in frequency above and below 193.1 THz in 0.1 THz (100 GHz) increments.

The succession from initial deployments of eight channels at 200 GHz channel spacing (1.6 nm spacing at 1.5 μm wavelength), to 16 channels at 100 GHz, to 80 or even 160 channels at 50 GHz, has required sources with extreme wavelength stability, in addition to all the spectral attributes described above. It remains a remarkable feature of DFB lasers that the "gain clamping" at the threshold value also fixes the operating index of refraction of the excited optical waveguide, and hence operating wavelength of the laser. This made the DFB laser, or the DFB-based EML discussed above, the universal ideal source for WDM systems. Reliability screening similar to that already used for standard operation readily yielded lasers that could maintain their operating wavelength within ±0.1 nm over system life, provided that the laser temperature drifts were kept below a few tenths of a degree centigrade using TECs in the laser modules. This latter requirement stems from the useful fact that the operating lasing wavelength increases in a DFB laser at a rate of approximately 0.1 nm/°C.

As the channel spacing in systems has narrowed to current values of 50 GHz (0.4 nm) and less, system designers have increasingly resorted to external optical wavelength references to specify the desired ITU wavelength channel. These references, typically etalons, narrow-band thin-film filters, or fiber Bragg grating filters, are used with photodetectors in electrical servo loops, adjusting laser wavelength with temperature using the TEC in the laser module. Very recent work has seen the inclusion of the reference inside the laser module, reinforcing the trend towards ever-increasing functionality from the same module through monolithic and hybrid integration. Using etalons with cyclically repeating transmission resonances, such modules are capable of locking on a number of different successive ITU channels, forming the basis for the first reliable multichannel WDM modules that offer channel selection to the end user based on laser temperature settings.

This *wavelength selectable laser* (WSL) feature is particularly interesting to WDM network operators who must inventory *optical terminal units* (OTUs) for populated channels in the system. In addition to sparing for failures in the field, this requires difficult vendor supply logistics to assure timely delivery and deployment of required channels in specified geographic areas. A simpler solution would be universal OTUs that can be assigned any ITU channel dynamically under software control. Ultimately this functionality may be used for dynamic routing and bandwidth allocation in flexible add-drop elements in a WDM network.

Considerable work has gone into tunable lasers for this WSL functionality. While the laser temperature tuning is a well-accepted and reliable method, the dynamic range is limited to ~3 nm for reasonable temperature swings. Alternative structures include arrays of DFB lasers, and a variety

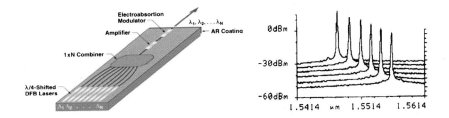

Fig. 10. (Left) Schematic drawing of an array source where a particular laser is activated for addressing a WDM channel, with boosting and modulation imposed by a monolithically integrated amplifier and electroabsorption modulator. (Right) Sequential spectra from an example integrating 6 λ/4-shifted DFB lasers.

of tunable *distributed Bragg reflector* (DBR) lasers. In the array sources, a DFB array is fabricated with wavelengths spanning the desired tuning range.

To access a particular ITU channel, the DFB with its wavelength closest to that channel is activated and temperature tuned to the exact value. This is accompanied by a servo loop similar to that described above. An example of such a chip architecture and the accompanying spectra for an experimental six-laser chip with integrated booster amplifier and modulator is shown in Fig. 10, and was capable of extended-reach transmission with negligible penalty.[44]

The tunable DBR laser uses a single laser that includes a tunable Bragg filter to change wavelengths as shown in Fig. 11. In contrast to the DFB laser where the grating feedback is continuously located along the gain medium, the DBR laser is fundamentally a two-mirror laser cavity. One mirror is the cleaved facet, while the other mirror is a transparent (higher bandgap) Bragg reflector waveguide containing the corrugated grating. This provides a wavelength selective narrow-band mirror that selects out a single longitudinal mode for operation, rejecting adjacent modes to levels of 40 dB or better as in the DFB laser. However, current injection into this higher bandgap Bragg reflector section changes its index of refraction and thus moves the central wavelength of this narrow-band mirror to shorter wavelengths. This in turn selects longitudinal modes at shorter wavelengths, thereby providing a WSL function in a single resonator. As shown in Fig. 11, such lasers have also been fabricated incorporating both integrated EA modulators for low-chirp information encoding, and integrated semiconductor optical amplifiers as power boosters for improved transmission. Wavelength coverage of ~8 nm is readily achievable with this design, providing

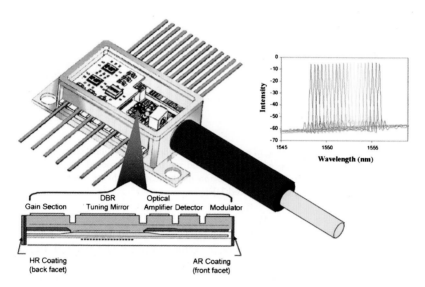

Fig. 11. Schematic drawing of a tunable DBR source with boosting, monitoring and modulation imposed by a monolithically integrated amplifier, waveguide detector and elctroabsorption modulator. (Inset) Sequential spectra showing 20 wavelengths selected from the ITU grid at 50 GHz spacing.[45]

20 ITU channels at 50 GHz spacing from one module incorporating suitable reference etalon and servo control for the module.

Advanced designs of tunable lasers have also been demonstrated that, like the tunable DBR laser, incorporate a tunable filter inside the resonator. The goal of these designs is to address a larger tuning range, and therefore utilize a broadly tunable filter. A number of such filters have been developed that are not limited, as in the case of the DBR reflector, to a percentage of the operating wavelength equal to the percentage change of modal index. Two examples of such devices are sampled-grating distributed Bragg reflector (SGDBR) lasers[46−48] and vertical grating coupler lasers.[49]

The SGDBR laser replaces the DBR mirror at one end of the resonator with a *sampled grating* at each end of the laser. In a sampled grating, the corrugated teeth of the grating are periodically blanked out, leaving the corrugation in "patches" that repeat periodically with a period much coarser than the teeth themselves — typically the blanked regions with no teeth have a higher duty cycle in each period. The result is that the reflectivity of such a sampled grating produces a comb of reflection peaks. The width of each peak is just as narrow as if the whole structure was a weak grating, while the range or spread of the comb is governed by how short

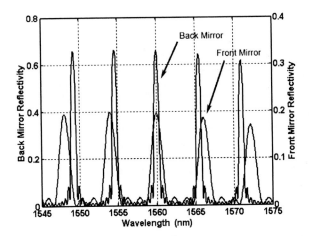

Fig. 12. Calculated reflectivity spectra for the two end reflectors in a sampled-grating DBR laser, each illustrating the comb of narrow reflection peaks that result from the periodically sampled grating.[47] Each end has a slightly different sampling period, and lasting can occur when the two reflectivity profiles labeled Front and Back Mirror have a reflection peak at the same wavelength, in this case at $\lambda = 1.56\,\mu$m.

each patch is — i.e. a short patch provides relatively little specificity of the corrugation period, and thus can support a broad comb of reflection peaks. The tuning principle behind the SGDBR laser is illustrated in Fig. 12.

At each end of the laser, the periodicity of the sampling or blanking is different as illustrated in Fig. 13 below. Thus, while each end provides a broad comb of narrow reflection peaks, the period of the comb is different in wavelength for each reflector. Since the laser needs two peaks to line up to get reasonable threshold lasing, tuning is accomplished by shifting one comb relative to the other. This is typically done with free-carrier injection into a higher-bandgap waveguide, containing the sampled grating, that is nominally transparent just as in the case of the tunable DBR laser. With a small shift of one comb relative to the other, a large shift in operating wavelength can be obtained since the next peak that lines up can be some distance away. This is similar to a vernier effect, or "stroboscopic" effect. SGDBR lasers have been developed commercially with tuning ranges of 40 nm, and have also been integrated with amplifiers and electroabsorption modulators[48] as shown in Fig. 13.

4.1. *Recent Advances*

Advances in semiconductor lasers have clearly been paced by the maturation of the underlying materials and processing technologies. In fields

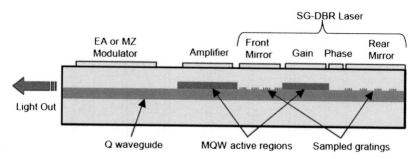

Fig. 13. Configuration for a tunable SGDBR laser with integrated amplifier and modulator.

outside of telecommunications, the 1990s saw exciting basic material advances in the GaN/InGaN system that resulted in practical lasers emitting in the *blue* and *ultraviolet* region of the spectrum. For long-wavelength materials, there has been continued research and commercialization of the InGaAlAs/InP system, where advantageous band-offsets have shown improvements both in high-temperature laser performance and quantum-well modulator performance. The remarkable achievements in the infrared quantum cascade lasers have shown how artificially engineered optical transitions can form the basis for fundamentally new semiconductor laser designs, and explorations are beginning to assess the potentiality of these structures in the 1.5 μm band. Research continues on the potential merits of reduced dimensionality quantum confinement in *quantum-wire* or *quantum-dot* lasers, and performance from these materials is now reaching a level where it warrants serious consideration as the heir to commercial quantum-well active layer devices.[50]

Critically important material advances continue to stem from improved epitaxial reactor design leading to larger wafers and higher levels of uniformity. As we begin to see a migration to larger 3″ and 4″ InGaAsP/InP reactors, this should also result in more routine thickness uniformity at the 0.5% level and ~1 nm-level photoluminescence (composition) uniformity. These advances will enable designers to implement high-performance designs with high yield, and ultimately eliminate much of the testing and yielding that takes place between manufacturing steps.

Process advances include the development of new automation equipment that is specific to optoelectronic chip handling, including automated cleaving, automated AR/HR coating processes, and fully automated chip testing and sorting operations. Other advances are expected in the area of process modeling for etching, doping, specialized epitaxial growth steps, and associated wafer-level characterization tools. Examples of this have already been heavily utilized to optimize integration using selective area growth techniques.

Since laser packaging usually represents the largest component in final module cost, significant work continues on implementation of laser designs with low-divergence output beams that may permit passively aligned module assembly, or at least simplification of the packaging optics. Such lasers are termed expanded beam lasers (XBL) or spot-size converted (SSC) lasers, and were shown in 1990 to be realizable based on tapered waveguide extensions that *reduce mode confinement* monolithically added to the InP chip.[51] Today XBL laser designs have been realized with performance nearly equivalent to standard laser designs, but with alignment tolerances reduced by factors of ~3. Such designs may become enablers for cost-reduced automated passively-aligned packaging assembly by keeping the complex processes to the batch, or wafer-level, compared to complexity at the individual module assembly level.

In the area of higher performance, EML modules have emerged with speeds reaching 40 Gb/s.[52] There are now indications that a *return-to-zero* (RZ) or pulse transmission format may emerge, in addition to today's common *non-return-to-zero* (NRZ) format, suggesting that new integrated modulator configurations may become desirable. For pump lasers, we expect powers reaching to the 1-Watt level becoming commercially practical. We expect continued pressure in the area of WDM, with advances in manufacturing allowing on-demand fulfillment of transmitter orders at arbitrary ITU channels. This is likely to promote the proliferation of WSL transmitter assemblies with wavelengths that can be locked to any of 80 or more ITU channels.

The broader application of WDM, the potential of ultrafast OTDM, the drive to higher density and levels of integration, and continued pressure on advanced low-cost packaging technologies all provide a huge range of challenges and exciting research for the coming decade. The potential for impact in these areas suggest that advances in semiconductor lasers and related components are likely to sustain their role as defining elements both in the progress of optical communications and in the applications of microresonator technology.

Acknowledgments

The author would like to gratefully acknowledge his support from the Daniel E. '39 and Patricia M. Smith Endowed Chair of Director, Center for Optical Technologies at Lehigh University.

Glossary of Terms

WDM	Wavelength Division Multiplexed
DWDM	Dense Wavelength Division Multiplexed
VCSEL	Vertical Cavity Surface Emitting Laser
LPE	Liquid Phase Epitaxy
MBE	Molecular Beam Epitaxy
MOVPE	Metal-Organic Vapor Phase Epitaxy
MOCVE	Metal-Organic Chemical Vapor Deposition
QW	Quantum Well
MQW	Multi-Quantum Well
SAG	Selective Area Growth
Q_{cold}	Q-factor $(=\omega\tau_{\text{ph}})$ of "cold" resonator, i.e., with no gain
Q_{hot}	Effective Q-factor of "hot" resonator, including gain
η	Differential quantum efficiency of laser
v_g	Group velocity of guided-wave mode
β	Fraction of Spontaneous Emission into the lasing mode
α	Linewidth Enhancement Factor
a	Average gain cross section for injected carriers
N_t	Carrier density to achieve transparency near the lasing λ
Γ	Optical mode active layer confinement factor
τ_{ph}	Photon lifetime of cold resonator
ε	Phenomenological nonlinear gain compression parameter
ω_{RO}	Relaxation oscillation frequency of laser
Chirp	Angle modulation, or variation of carrier frequency, usually undesired
DFB	Distributed Feedback
Bragg Reflector	Waveguide reflector formed by periodic index variation
κ	Coupling constant of Bragg reflector grating
HR	High-Reflection coatings for resonator mirrors
AR	Anti-Reflection coatings for facets to avoid reflection

EA Modulator	Electroabsorption modulator, using field-induced absorption edge shifts
EML	Electroabsorption modulated laser
PIC	Photonic integrated circuit, combining multiple components on a chip
ITU	International Telecommunications Union, a standards body
WSL	Wavelength Selectable Laser
DBR	Distributed Bragg reflector, a mirror comprising a Bragg reflector
SGDBR	Sampled-Grating DBR, coarse periodically "blanked" periodic index variation
XBL	Expanded-Beam Laser, also SSC (Spot-Size Converted Laser)
RZ, NRZ	Return-to-Zero or Non-Return-to-Zero digital modulation formats

References

1. M. I. Nathan, W. P. Dunke, G. Burns, F. H. Dill, Jr. and G. Lasher, *Appl. Phys. Lett.* **1**, 62, 1962.
2. R. N. Hall, G. E. Fenner, J. D. Kingsley, T. J. Soltys and R. O. Carlson, *Phys Rev. Lett.* **9**, 366, 1962.
3. N. Holonyak, Jr. and S. F. Bevacquat, *Appl. Phys. Lett.* **1**, 82, 1962.
4. T. M. Quist, R. H. Rediker, R. J. Keyes, W. E. Krag, B. Lax, A. L. McWhorter and H. J. Zeiger, *Appl. Phys. Lett.* **1**, 91, 1962.
5. A. L. Schawlow and C. H. Townes, *Phys. Rev.* **112**, 1940, 1958.
6. T. H. Maiman, *Nature* **187**, 493, 1960.
7. I. Hayashi, M. B. Panish, P. W. Foy and S. Sumski, *Appl. Phys. Lett.* **17**, 109, 1970.
8. H. Kogelnik and C. V. Shank, *Appl. Phys. Lett.* **18**, 152, 1971.
9. H. Kroemer, *Proc. IEEE* **51**, 1782, 1963.
10. Zh. I. Alferov and R. F. Kazarinov, Patent 181737 (U.S.S.R.), 1963.
11. Portions of this section's text have been edited from: W. F. Brinkman, T. L. Koch, D. V. Lang and D. P. Wilt, *Bell Labs Tech. J.* **5**(1), 150, 2000, including in particular contributions from D. P. Wilt.
12. R. E. Nahory, M. A. Pollack, W. D. Johnston, Jr. and R. L. Burns, *Appl. Phys. Lett.* **33**, 659, 1978.
13. K. Nakajima, A. Yamaguchi, K. Akita and T. Kotani, *J. Electrochem. Soc.* **125**, 123, 1978.
14. E. Yabliovitch and E. O. Kane, *IEEE J. Lightwave Tech.* **LT-4**, 504 and 961, 1986.
15. A. R. Adams, *Electron. Lett.* **22**, 249, 1986.

16. A. Y. Cho, *J. Vac. Sci. Technol.* **8**, S31, 1971; *Appl. Phys. Lett.* **19**, 467, 1971.
17. R. Dingle, W. Weigmann and C. H. Henry, *Phys. Rev. Lett.* **33**, 827, 1974.
18. R. D. Dupuis and P. D. Dapkus, *Appl. Phys. Lett.* **31**, 466, 1977.
19. T. Kato, T. Sasaki, N. Kida, K. Komatsu and I. Mito, *Proc. ECOC/ ICOC '91*, **Pt. II**, WE-B7, 429, 1991.
20. M. Aoki, M. Suzuki, M. Takahashi, H. Sano, T. Ido, T. Kawano and A. Takai, *Electron. Lett.* **28**, 1157, 1992.
21. T. Tsukada, *J. Appl. Phys.* **45**, 4899, 1974.
22. J. L. Zilko, L. Ketelsen, Y. Twu, D. P. Wilt, S. G. Napholtz, J. P. Blaha, K. E. Strege, V. G. Riggs and D. L. Van Haren, *IEEE J. Quant. Electron.* **25**, 2091, 1989. SEM complements of D.-N. Wang, ref: D.-N. Wang, D. Venables, D. Waltemyer and J. Lentz, *Proc. IEEE IPRM 2000 Conf.* **MA2.5**, 60, 2000.
23. P. J. Corvini and T. L. Koch, *IEEE J. Lightwave Tech.* **LT-5**, 1591, 1987.
24. Y. Suematsu and K. Furuya, *Trans. IECE Japan* **E-60**, 467, 1977.
25. C. H. Henry, P. S. Henry and M. Lax, *J. Lightwave Tech.* **LT-2**, 209, 1984.
26. See for ex., C. H. Henry, *Semiconductors and Semimetals*, W. T. Tsang, Ed., Academic Press, NY, **Vol. 22, Part B., Ch. 3**, 1985.
27. M. Lax, *Proc. 4th Int. Quantum Electronics Conf.*, eds. P. L. Kelley, B. Lax and P. E. Tannenwald, McGraw-Hill, NY, 1966.
28. C. H. Henry, *IEEE J. Quant. Electron.* **QE-18**, 259, 1982.
29. M. Lax, *Phys. Rev.* **160**, 290, 1967.
30. See for ex., A. Yariv, *Quant. Electron.*, 3rd, John Wiley, NY, 170, 1989.
31. K. Vahala and A. Yariv, *IEEE J. Quant. Electron.* **QE-19**, 1096 and 1102, 1983.
32. C. H. Henry, *IEEE J. Quant. Electron.* **QE-19**, 1391, 1983.
33. H. Kogelnik and C. V. Shank, *J. Appl. Phys.* **43**, 2327, 1972.
34. W. Streifer, R. D. Burnham and D. R. Scifres, *IEEE J. Quant. Electron.* **QE-11**, 154, 1975.
35. H. A. Haus and C. V. Shank, *IEEE J. Quant. Electron.* **QE-12**, 532, 1976.
36. K. Kojima and K. Kyuma, *Electron. Lett.* **20**, 869, 1984.
37. A. H. Gnauck, B. L. Kasper, R. A. Linke, R. W. Dawson, T. L. Koch, T. J. Bridges, E. G. Burkhardt, R. T. Yen, D. P. Wilt, J. C. Campbell, K. C. Nelson and L. G. Cohen, *IEEE J. Lightwave Tech.* **LT-3**, 1032, 1985.
38. H. Yonetani, I. Ushijima, T. Takada and K. Shima, *IEEE J. Lightwave Tech.* **LT-11**, 147, 1993.
39. O. Kjebon, R. Schatz, S. Lourdudoss, S. Nilsson, B. Stalnacke and L. Backbom, *Electron. Lett.* **33**, 488, 1997.
40. K. Kojima, *Tech. Digest of Optical Fiber Comm. Conf. (OFC '95)* **ThG3**, 253, 1995.
41. T. L. Koch and R. A. Linke, *Appl. Phys. Lett.* **48**, 613, 1986; see also T. L. Koch and J. E. Bowers, *Electron. Lett.* **20**, 1038, 1984.
42. See for ex., H. Soda, M. Furutsu, K. Sato, M. Matsuda and H. Ishakawa, *Electron. Lett.* **25**, 334, 1989.
43. J. E. Johnson, P. A. Morton, T. V. Nguyen, O. Mizuhara, S. N. G. Chu, G. Nykolak, T. Tanbun-Ek, W. T. Tsang, T. R. Fullowan, P. F. Sciortino,

A. M. Sergent, K. W. Wecht and R. D. Yadvish, *Tech. Dig. OFC '95* **TuF2**, 21, 1999.

44. M. G. Young, U. Koren, B. I. Miller, M. Chien, T. L. Koch, D. M. Tennant, K. Feder, K. Dreyer and G. Raybon, *Electron. Lett.* **31**, 1835, 1995.

45. J. E. Johnson, L. J.-P. Ketelsen, D. A. Ackerman, L. Zhang, M. S. Hybertsen, K. G. Glogovsky, C. W. Lentz, W. A. Asous, C. L. Reynolds, J. M. Geary, K. K. Kamath, C. W. Ebert, M. Park, G. J. Przybylek, R. E. Leibenguth, S. L. Broutin, J. W. Stayt Jr., K. F. Dreyer, L. J. Peticolas, R. L. Hartman and T. L. Koch, *IEEE J. Sel. Topics Quant. Electron.* **7**, 168, 2001.

46. V. Jayaraman, A. Mathur, L. A. Coldren and P. D. Dapkus, *IEEE Phot. Tech. Lett.* **5**, 489, 1993.

47. H. Ishii, H. Tanobe, F. Kano, Y. Tohmori, Y. Kondo and Y. Yoshikuni, *IEEE J. Quant. Electron.* **32**, 433, 1996.

48. Y. A. Akulova, G. A. Fish, P. C. Koh, C. Schow, P. Kozodoy, A. Dahl, S. Nakagawa, M. Larson, M. Mack, T. Strand, C. Coldren, E. Hegblom, S. Penniman, T. Wipiejewski and L. A. Coldren, *IEEE J. Sel. Topics Quant. Electron.* **8**, 1349, 2002.

49. B. Broberg, P.-J. Rigole, S. Nilsson, L. Andersson and M. Renlund, *Proc. OFC/IOOC 1999.* **2**, 137, 1999.

50. O. B. Shchekin and D. G. Deppe, *Appl. Phys. Lett.* **80**, 3277, 2002.

51. T. L. Koch, U. Koren, G. Eisenstein, M. G. Young, M. Oron, C. R. Giles and B. I. Miller, *IEEE Phot. Tech. Lett.* **2**, 88, 1990.

52. H. Kawanishi, Y. Yamauchi, N. Mineo, Y. Shibuya, H. Murai, K. Yamada and H. Wada, *Tech. Dig. OFC '01* **MJ3**, 2001.

CHAPTER 4

CAVITY-ENHANCED SINGLE PHOTONS FROM A QUANTUM DOT

Jelena Vučković[*], Charles Santori[†], David Fattal, Matthew Pelton[‡],
Glenn S. Solomon[§] and Yoshihisa Yamamoto[¶]

*Quantum Entanglement Project, ICORP, JST
Ginzton Laboratory, Stanford University, Stanford CA 94305, USA*

Single-photon sources rarely emit two or more photons in the same pulse, compared to a Poisson-distributed source of the same intensity, and have numerous applications in quantum information processing. The quality of such a source is evaluated based on three criteria: high efficiency, small multiphoton probability, and quantum indistinguishability. We have demonstrated a single-photon source based on a quantum dot in a micropost microcavity that exhibits a large Purcell factor together with a small multiphoton probability. For a quantum dot on resonance with the cavity, the spontaneous emission rate has been increased by a factor of five, while the probability to emit two or more photons in the same pulse has been reduced to 2% compared to a Poisson-distributed source of the same intensity. The indistinguishability of emitted single photons from one of our devices has been tested through a Hong–Ou–Mandel-type two-photon interference experiment; consecutive photons emitted from such a source have been largely indistinguishable, with a mean wave-packet overlap as large as 0.81. We have also developed new cavity designs that could lead to single photon sources of even higher quality.

1. Introduction and Overview

Generation of single photons at a well-defined timing or clock is crucial for practical implementation of quantum key distribution (QKD),[1]

---

[*]Also at: Department of Electrical Engineering, Stanford University, Stanford, CA 94305; jela@stanford.edu.
[†]Also at: IIS, University of Tokyo, Tokyo, Japan.
[‡]Now at: University of Chicago, IL.
[§]Also at: SSPL, Stanford University, Stanford, CA 94305.
[¶]Also at: NTT Basic Research Labs, Atsugishi, Japan.

as well as for quantum computation[2] and networking based on photonic qubits.[3,4] Single-photon sources have recently been demonstrated using a variety of devices, including molecules,[5–7] mesoscopic quantum wells,[8] color centers,[9] trapped ions,[10] and semiconductor quantum dots.[11–15] These sources rarely emit two or more photons in the same pulse, compared to a Poisson-distributed source of the same intensity. Three different criteria are taken into account when evaluating the quality of a single-photon source: high efficiency, small multiphoton probability (measured by the second-order coherence function $g^{(2)}(0)$), and quantum indistinguishability. For example, high efficiency and small $g^{(2)}(0)$ are required, but quantum indistinguishability is not necessary for BB84 QKD.[16] On the other hand, for almost all other applications in the field of quantum information, including linear-optical quantum computation,[17] we need photons that are indistinguishable and thus produce multiphoton interference. For a source based on a single quantum emitter, the emitter must therefore be excited in a rapid or deterministic way, and interact little with its surrounding environment.

One of the popular approaches to generation of single photons is based on a pulsed excitation of a semiconductor quantum dot (QD) combined with spectral filtering.[11–15] Although a single quantum dot by itself can be used to generate single photons,[12] the efficiency of such a system is poor, as the majority of emitted photons are lost in the substrate. Moreover, emitted photons are unlikely to be indistinguishable, with coherence lengths shorter than the radiative limit (Fourier transform limit), and the single-photon generation rate is low, as determined by the long excitonic lifetime. Microcavities can help in correcting all of these deficiencies[11,15,18–20] by reducing the radiative lifetime of an emitter on resonance with the cavity (i.e. by enhancing the spontaneous emission rate).

Spontaneous emission is not an intrinsic property of an isolated emitter (e.g. atom or exciton), but is rather a property of an emitter coupled to its electromagnetic vacuum environment. The spontaneous emission rate is directly proportional to the density of electromagnetic states, and can be modified with respect to its value in free space by placing the emitter in a cavity.[21] Experimental demonstrations of the inhibition and enhancement of spontaneous emission rate were carried out since mid-70's,[22–27] with atoms coupled to single mirrors, planar cavities, or spherical Fabry–Perot resonators. Advances in microfabrication techniques enabled the construction of high-quality semiconductor micropost and microdisk microcavities in the late 80's and early 90's, and ignited interest in solid-state cavity quantum electrodynamics (QED) experiments.[28-30] In 1987, photonic-crystal structures were proposed as promising candidates for strong spontaneous emission modification,[31,32] but the first experimental results on photonic crystal microcavities were reported a decade later.[33,34]

Advantageous feature of solid-state microcavities is that a single narrow-linewidth emitter (quantum dot) can be embedded in them during the growth process, enabling cavity field interaction with an "artificial atom".[35] The observation of cavity QED phenomena in semiconductor systems relies on the construction of optical microcavities with high quality factors (Q) and small mode volumes (V). Most of the semiconductor cavity QED research to date has been focused on distributed-Bragg-reflector (DBR) microposts[18,35] or microdisk microcavities,[36,37] with a recent exception of preliminary photonic crystal microcavity work.[38] Microposts were also used in all experiments with microcavities presented in this chapter. The advantages of microposts relative to other microcavities are that the light escapes normal to the sample, in a single-lobed Gaussian-like pattern, and that it is relatively straightforward to isolate a single quantum dot in them. It is also possible to optimize the micropost designs to enable the observation of the single QD-cavity QED in the strong coupling regime.[39] However, stringent fabrication requirements of these structures motivated us to investigate planar photonic crystal microcavities as a more robust and manufacturable solution.

The first successful optical characterizations of photonic-crystal microcavities with quantum dots were performed recently.[40-42] Q factors as large as 2800 were reported, together with mode volumes as small as $0.5(\lambda/n)^3$.[43] The possibility of improving the quality factor by more than an order of magnitude, while preserving such a small mode volume makes these structures good candidates for cavity QED, both with neutral atoms[44-46] and quantum dots.[47] This has not yet been demonstrated experimentally, however.

In Sec. 2, we describe assembly and spectroscopy of a single semiconductor quantum dot without a microcavity, mostly covering material from Refs. 12 and 48. Section 3 focuses on results from Refs. 39 and 47; it describes designs of microposts that were used in experiments discussed in Secs. 4 and 5, and designs of two-dimensional photonic crystal microcavities that will be used in our future experiments. In Secs. 4 and 5 we present our experimental work on solid-state cavity QED and sources of nonclassical light.[19,20]

2. Single Quantum Dots: Assembly and Spectroscopy

Semiconductor quantum dots (QDs) are nanoscale inclusions of a low-bandgap semiconductor inside a semiconductor with a larger bandgap. The bandgap difference acts as a potential barrier for carriers, confining them inside the dot. Moreover, the dots are small enough that the confined carriers can only occupy discrete energy levels, and the transitions between such levels in the conduction band and the valence band involve the absorption or emission of photons at near-optical frequencies. Quantum dots can be formed spontaneously during epitaxial growth of lattice-mismatched materials, and such dots are called self-assembled.[49] For example, when InAs is deposited on GaAs, a strained planar layer, known as a wetting layer, is initially formed. The strain energy that builds up in this layer is eventually partially relieved by the formation of nanometer-scale islands on the surface, which can subsequently be covered with a capping layer of GaAs. InAs/GaAs quantum dots used in our experiments were self-assembled during molecular beam epitaxy (MBE), under conditions that give relatively sparse dots, with a surface density of 11–$75\,\mu\mathrm{m}^{-2}$. Figure 1 shows an atomic-force microscope image of dots similar to those used in our experiments, except for the absence of a GaAs capping layer. Single dots can be spatially isolated by etching mesas in the MBE-grown sample. In the experiments described in this section, the mesas were about $120\,\mathrm{nm}$ tall

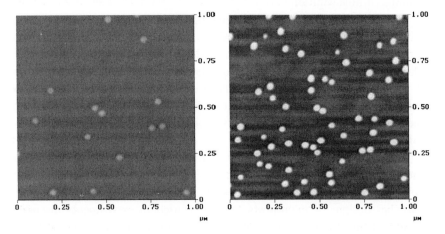

Fig. 1. Atomic Force Microscope images of quantum dot arrays grown by Molecular Beam Epitaxy. The size and density of dots can be moderated by controlling the growth rate, the substrate temperature, the ratio of As to In impinging on the surface, and the amount of material deposited.

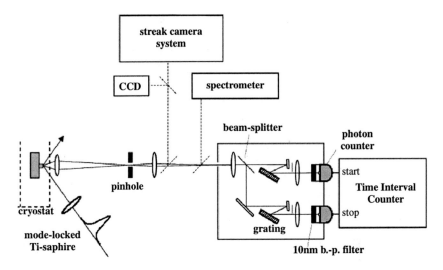

Fig. 2. Experimental setup used to measure luminescence from single quantum dots.

and 200 nm wide, spaced by 50 μm, and were fabricated by a combination of electron-beam lithography and dry etching. The single quantum dots were probed by photoluminescence (PL), and Fig. 2 shows the experimental apparatus used. The sample is placed in a liquid He cryostat, and mesas containing quantum dots are excited from a steep angle by Ti:sapphire laser pulses, 3 ps in duration, with a 76 MHz repetition rate. The emission from the dot is collected normal to the sample surface with an aspheric lens that has a numerical aperature of 0.5, and is focused onto a pinhole that effectively selects a 5 μm region of the sample for collection. The collected emission is then directed towards a streak camera preceded by a spectrometer, for time-resolved photoluminescence measurements. The spectral resolution of the system is 0.1 nm, together with the time resolution of 25 ps. For photon correlation measurements, the collected emission is first spectrally filtered and then directed towards a Hanbury Brown and Twiss-type (HBT) setup. In the HBT setup, photon counters are placed at both outputs of a nonpolarizing beamsplitter for detection. The electronic signals from the counters are sent to a time-to-amplitude converter followed by a multichannel analyzer computer card, which generates a histogram of the relative delay time $\tau = t_2 - t_1$ between a photon detection at one counter (t_2) and the other (t_1).

The excitation frequency of the Ti:sapphire laser is either above the bandgap of the GaAs surrounding InAs QDs, or is resonant with higher

level confined states of a QD. The former type of the excitation is called
above-band excitation, and the latter is referred to as *resonant excitation*.
In above-band excitation, electron-hole pairs are created in the surrounding semiconductor (GaAs) matrix, after which they diffuse towards the dot,
where they relax to the lowest confined states. The created carriers recombine in a radiative cascade, leading to the generation of several photons
for each laser pulse. All of these photons have slightly different frequencies,
resulting from the Coulomb interaction among carriers. The last emitted
photon for each pulse has a unique frequency, and can be spectrally isolated. With resonant excitation, electron-hole pairs are created within the
dot, and the favored absorption of a single electron-hole pair and formation of one-exciton state is expected. This is illustrated in Fig. 3. With
continuous-wave (CW) excitation above the GaAs bandgap, the emission
spectrum displays several lines. When the laser is tuned to an absorption
resonance at 857.5 nm of a higher-level confined state of a QD, thus creating
excitons directly inside the dot, emission peaks 3 and 4 almost disappear.
We therefore believe that they represent emission from charged states of
the dot. We identify peak 1 as ground-excitonic state emission after the
capture of a single exciton, and peak 2 as "biexcitonic" emission after the
capture of two excitons. A biexcitonic energy shift of 1.7 meV is due to
electrostatic interactions among carriers. Assignment of the different peaks

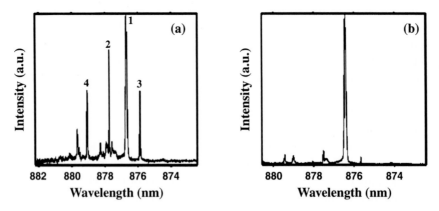

Fig. 3.   Photoluminescence from a single InAs/GaAs quantum dot under above-band
excitation (left) and resonant excitation (right). With resonant excitation, the favored
absorption of a single electron-hole pair and formation of one-exciton state is expected.

is supported by the dependence of the emission line intensities on pump power: under above-band excitation peak 1 grows linearly, while peak 2 grows quadraticaly in the weak-pump limit, as expected for excitons and biexcitons, respectively.

Further support for the peak identification comes from time-dependent spectra, as collected by the streak camera.[48] The camera produces two-dimensional images of intensity versus wavelength and time after exciting QD with a laser pulse. By integrating intensity within frequency windows corresponding to the peaks shown in Fig. 3, time-dependent intensities are obtained for the different lines. Under weak, above-band excitation, the single-exciton line (line 1) appears quickly after the excitation pulse, and then decays exponentially. This decay time has been measured accurately under resonant excitation to be 0.47 ns. Under higher excitation power, however, line 1 reaches its maximum only after a long delay. Most of the emission immediately after the excitation pulse now comes from line 2. A simple explanation for this behavior is that, since the laser pulse now initially creates several electron-hole pairs on average, some time is required before the population of the dot reduces to one electron-hole pair, and only then can the single-exciton emission occur. In an even higher excitation power, the biexcitonic emission is delayed and new peak (multiexcitonic emission at even longer wavelengths than the biexciton line) appears immediately after the excitation pulse.

As described previously, the last photon emitted for each excitation pulse corresponds to a single-exciton state and has a unique frequency corresponding to peak 1 in Fig. 3. If this photon (i.e. peak 1) is spectrally filtered at the output of a quantum dot source, then a single-photon emission on demand is achieved, since only one photon is present at the output of the source for each laser pulse.[12] Besides narrow-bandwidth spectral filtering, it is also necessary that the laser excitation pulses be much shorter than the lifetimes of carriers in the dot, thereby suppressing the re-excitation of the dot in the same pulse. A pulse stream with reduced multiphoton probability compared to the Poissonian case is said to be *antibunched*, and can only be described quantum mechanically. Mathematically, such nonclassical photon statistics can be described using the second-order coherence function $g^{(2)}(\tau)$, defined as follows:

$$g^{(2)}(\tau) = \frac{\langle \hat{a}^\dagger(t)\hat{a}^\dagger(t+\tau)\hat{a}(t+\tau)\hat{a}(t) \rangle}{\langle \hat{a}^\dagger \hat{a} \rangle^2} , \tag{1}$$

where $\hat{a}^\dagger$ and $\hat{a}$ are the photon creation and annihilation operators, respectively. A pulsed source with Poissonian statistics will have a $g^{(2)}(\tau)$ function consisting of a series of peaks with unit area, when normalized by the pulse repetition period.

To characterize our single-photon source, we employ the previously described HBT setup. The HBT setup can be used to measure the photon correlation histogram which corresponds to the second-order coherence function $(g^{(2)}(\tau))$ in the limit of low collection and detection efficiency;[50] the results of our experiment are shown in Fig. 4. In order to select only the photon corresponding to the single-exciton state, spectral filtering with the resolution of 2 nm was used. Of special interest is the central peak in $g^{(2)}(\tau)$

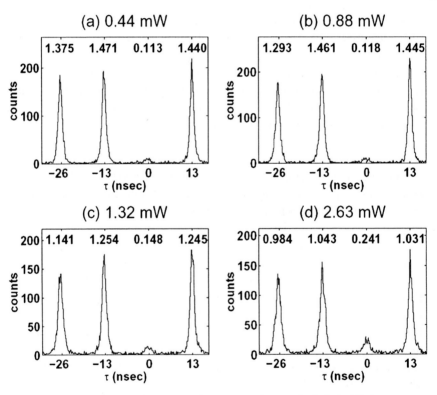

Fig. 4. Photon correlation histograms for emission from a single InAs/GaAs quantum dot under pulsed, resonant excitation. The numbers above the peaks correspond to the normalized peak areas calculated using a 5.6 ns integration window. The width of the peaks is determined by the resolution of the photon counters (0.3 ns) and the excitonic radiative lifetime (0.7 ns).

at $\tau = 0$, which gives an upper bound on the probability to generate two or more photons in a given pulse[12]:

$$P(n \geq 2) \leq \frac{1}{2}\bar{n}^2 g^{(2)}(0), \qquad (2)$$

where $\bar{n}$ is the mean photon number per pulse. A large reduction in the size of the central peaks (at $\tau = 0$) in Fig. 4 indicates a strong antibunching, i.e. a large suppression of multiphoton pulses. The numbers printed above the peaks indicate the peak areas, properly normalized by dividing the histogram areas by both singles rates, the repetition period and the measurement time. For the numbers shown, the only background counts subtracted were those due to the known dark count rates of the photon counters ($130\,\mathrm{s}^{-1}$ and $180\,\mathrm{s}^{-1}$), almost negligible compared to the singles rates, $19800\,\mathrm{s}^{-1}$ and $18000\,\mathrm{s}^{-1}$ for the two counters at $0.88\,\mathrm{mW}$ pump power. From the ratio of the central peak to the side peaks integrated over the $5.6\,\mathrm{ns}$ window, we conclude that the probability to generate two or more photons for the same laser pulse is reduced to 11% in this single-photon source relative to the Poisson-distributed source of the same intensity, for the pump power equal to $0.44\,\mathrm{mW}$.

A single quantum dot by itself can therefore be used to generate single photons, but the efficiency of such a system is poor, as the majority of emitted photons are lost in the substrate. In addition, the radiative lifetime can be as long as 1ns, which is greater than the dephasing time estimated to be $\sim 0.9\,\mathrm{ns}$ at $4\,\mathrm{K}$.[51] Emitted photons are unlikely to be indistinguishable, with coherence lengths shorter than the radiative limit (Fourier transform limit). Finally, the single-photon generation rate is low, as determined by the long excitonic lifetime. Microcavities can help in correcting all of these deficiencies.[11,15,18-20] The radiative lifetime of an emitter on resonance with the cavity can be decreased significantly below the dephasing time, bringing the emitted photon pulses closer to the Fourier transform limit. Moreover, the spontaneous emission rate can be enhanced, and a large fraction of spontaneously emitted photons can be coupled into a single cavity mode, thereby increasing the outcoupling efficiency. In the following sections, we will discuss design and construction of optical microcavities for interaction with quantum dots, and present our experimental work on an improved source of single photons based on a quantum dot in a microcavity.

3. Microcavities for Interaction with Semiconductor Quantum Dots

3.1. *Motivation for Maximizing the Ratio of Quality Factor to Mode Volume*

Let us assume that a single quantum dot is isolated in a microcavity, and is coupled to the cavity field. The isolation of a single quantum dot prevents inhomogeneous broadening of the emission spectrum. We also assume that the transition frequency ω from the one-exciton state $|e\rangle$ to the zero-exciton state $|g\rangle$ is on resonance or nearly on resonance with the fundamental optical cavity mode frequency. Under these conditions, the excitation of other cavity modes can be neglected, and the system can be modeled as a single two-level atom coupled to a single cavity mode. This coupled system can be described by the Jaynes–Cummings Hamiltonian:

$$H = H_E + H_F + H_I, \tag{3}$$

where

$$H_E = \hbar\omega\sigma_z, \tag{4}$$

$$H_F = \hbar\omega\left(a^\dagger a + \frac{1}{2}\right), \tag{5}$$

and

$$H_I = i\hbar(g(\vec{r}_E)a^\dagger\sigma_- - g^*(\vec{r}_E)\sigma_+ a). \tag{6}$$

The three terms of the Jaynes–Cummings Hamiltonian are the excitonic Hamiltonian (H_E), the field Hamiltonian (H_F), and the exciton-field interaction Hamiltonian (H_I). a and $a^\dagger$ are the photon annihilation and creation operators, respectively, $\sigma_- = |g\rangle\langle e|$ and $\sigma_+ = |e\rangle\langle g|$ are the quantum dot lowering and raising operators, respectively, while $\sigma_z = \frac{1}{2}(|e\rangle\langle e| - |g\rangle\langle g|)$ is the population operator. The coupling parameter $g(\vec{r})$ is the product of the Rabi frequency g_0, a position dependent part $\psi(\vec{r})$, and a polarization dependent part $\cos(\xi)$:

$$g(\vec{r}) = g_0\psi(\vec{r})\cos(\xi) \tag{7}$$

where

$$g_0 = \frac{\mu}{\hbar}\sqrt{\frac{\hbar\omega}{2\epsilon_M V}}, \tag{8}$$

$$\psi(\vec{r}) = \frac{E(\vec{r})}{|E(\vec{r}_M)|}, \tag{9}$$

and

$$\cos(\xi) = \frac{\vec{\mu} \cdot \hat{e}}{\mu}, \tag{10}$$

where $\vec{r}_M$ denotes the point where the field intensity $\epsilon(\vec{r})|E(\vec{r})|^2$ is maximum and ϵ_M is the dielectric constant at this point ($\epsilon_M = \epsilon(\vec{r}_M)$). Electric field orientation at the location $\vec{r}$ is denoted as $\hat{e}$, $\vec{\mu}$ is the dipole moment matrix element between the states $|e\rangle$ and $|g\rangle$, and $\mu = |\vec{\mu}|$. $\vec{\mu}$ is defined as $q\langle e|\vec{d}|g\rangle$, where $\vec{d}$ is the coordinate operator and q is a unit charge. $g(\vec{r}_E)$ denotes the value of the coupling parameter at the exciton location $\vec{r}_E$, and $|g(\vec{r}_E)|$ reaches its maximum value of $|g_0|$ when the exciton is located at the point $\vec{r}_M$ where the field intensity is maximum, and when its dipole moment is aligned with the electric field (i.e. when $\psi(\vec{r}_E) = 1$ and $\cos(\xi) = 1$). $E(\vec{r})$ is the electric field magnitude, and V is the cavity mode volume, defined as

$$V = \frac{\iiint \epsilon(\vec{r})|E(\vec{r})|^2 d^3\vec{r}}{\epsilon_M |E(\vec{r}_M)|^2}. \tag{11}$$

H_I can be derived from the interaction Hamiltonian in the dipole approximation ($H_I = -\vec{\mu} \cdot \vec{E}$), after the expansion and quantization of electric field in terms of the cavity modes.

The losses of the system can be described in terms of the *cavity field decay rate* κ, equal to $(\omega/2Q)$, and the *excitonic dipole decay rate* γ. κ is the decay rate of the resonant cavity mode, while γ includes losses to modes other than the cavity mode and to nonradiative decay routes. Depending on the ratio of the coupling parameter $|g(\vec{r}_E)|$ to the decay rates κ and γ, we can distinguish two regimes of coupling between the exciton and the cavity field: the *strong-coupling regime*, for $|g(\vec{r}_E)| > \kappa, \gamma$, and the *weak-coupling regime*, for $|g(\vec{r}_E)| < \kappa, \gamma$.[52] In the strong-coupling case, the time scale of coherent coupling between the exciton and the cavity field is shorter than that of the irreversible decay into various radiative and noradiative routes. *Rabi oscillation* occurs in this case, and the time evolution of the system can be described by oscillation at frequency $2|g(\vec{r}_E)|$ between the states $|e, 0\rangle$ and $|g, 1\rangle$, where $|e, 0\rangle$ corresponding to one exciton is in the quantum dot and no photons in the cavity, and $|g, 1\rangle$ corresponds to zero excitons in the quantum dot and one photon in the cavity. We will assume that losses into modes other than the resonant cavity mode and into nonradiative routes are not modified in the presence of a cavity, and estimate γ from the homogeneous linewidth of the exciton without a cavity (γ_h).[53] The strong coupling regime in the solid state is not only of academic interest; it can also be used in construction of sources of Fourier-transform limited single photons and components of quantum networks.[54]

On the other hand, in the weak-coupling case, the irreversible decay rates dominate over the coherent coupling rate; in other words, the exciton-cavity field system does not have enough time to couple coherently before dissipation occurs. The spontaneous emission decay rate Γ for this case is[55]

$$\Gamma = |g(\vec{r}_E)|^2 \frac{4Q}{\omega}. \tag{12}$$

For an exciton coupled to free-space electromagnetic fields, the spontaneous emission decay rate is given by

$$\Gamma_0 = \frac{\omega^3 \mu^2}{3\pi\epsilon_0 \hbar c^3}. \tag{13}$$

Γ_0 is known as Einstein's A coefficient. The ratio of the two decay rates given above is called the *Purcell factor*.[21] For an exciton positioned at the maximum of the field intensity, and aligned with the electric field, the Purcell factor is therefore equal to

$$F_0 = \frac{\Gamma}{\Gamma_0} = \frac{3Q\lambda^3 \epsilon_0}{4\pi^2 V \epsilon_M}. \tag{14}$$

If the Purcell factor is much greater than one, the exciton can radiate much faster in a cavity than in free space. The radiative-rate enhancement is proportional to the ratio of the quality factor to the volume of the cavity mode, according to Eq. (14). The Purcell factor increases with Q/V only to the point where the coupling parameter $|g(\vec{r}_E)|$ becomes larger than the decay rates of the system (κ and γ). At that point, the coupled exciton-cavity system enters the strong-coupling regime and the spontaneous emission becomes reversible. We usually define the Purcell factor F as the spontaneous emission rate enhancement relative to the bulk material. The spontaneous emission rate in the bulk material with refractive index n_h is enhanced n_h times with respect to its value in free space,[56] which implies that $F = F_0/n_h$. When the excitonic emission wavelength λ is off-resonance with the cavity mode wavelength λ_c, the spontaneous emission rate modification (with respect to the bulk material) as a function of detuning from the cavity resonance ($\lambda - \lambda_c$) follows a Lorentizan lineshape. The Lorentzian is centered at the cavity resonance λ_c where it reaches the maximum value equal to the Purcell factor, and has the linewidth equal to the cavity resonance linewidth $\Delta\lambda_c$, where $\Delta\lambda_c = \lambda_c/Q$ (Q is the quality factor of the cavity mode):

$$\frac{\Gamma}{n_h\Gamma_0} = \frac{3Q\lambda_c^3}{4\pi^2 n_h^3 V}\frac{\Delta\lambda_c^2}{\Delta\lambda_c^2 + 4(\lambda-\lambda_c)^2} + f. \tag{15}$$

Parameter f in the previous equation corresponds to the spontaneous emission rate modification far from the cavity resonance. For example, three-dimensional photonic crystal cavities would feature a complete spontaneous emission suppression off-resonance with the cavity mode, and f would be equal to zero. However, this is not the case in micropost microcavities studied in this chapter, where f is determined by the coupling to leaky modes and is approximately equal to one for large micropost diameters.[18,35]

Increasing Q/V can also lead to a reduction in laser threshold. The fraction of the light emitted by an exciton that is coupled into one particular cavity mode is known as the spontaneous emission coupling factor β, and is approximately related to the Purcell factor via the following expression[57]

$$\beta = \frac{F}{1+F}. \tag{16}$$

Therefore, if the emission rate of an exciton is strongly enhanced by its interaction with a cavity mode, most of the light emitted by this exciton will be channelled into that mode. This, in turn, means that the fraction of spontaneous emission into all other modes $(1-\beta)$ is reduced. The fraction of spontaneous emission going into nonlasing modes is one of the fundamental losses in a laser, and by decreasing it, we can lower the laser threshold. Finally, we point out that by combining Eqs. (8) and (13), the Rabi frequency g_0 of a system on resonance can be expressed as

$$g_0 = \frac{\Gamma_0}{2}\sqrt{\frac{V_0}{V}}, \tag{17}$$

where $V_0 = (3c\lambda^2\epsilon_0)/(2\pi\Gamma_0\epsilon_M)$.

3.2. *Micropost Microcavities*

Micropost microcavities consist of a high-refractive-index region (spacer) sandwiched between two dielectric mirrors, as shown in Fig. 5. Confinement of light in these structures is achieved by the combined action of distributed Bragg reflection (DBR) in the longitudinal direction (along the post axis), and total internal reflection (TIR) in the transverse direction (along the post cross-section). The microposts analyzed in this chapter are rotationally symmetric around the vertical axis. The DBR mirrors can be viewed as one-dimensional (1D) photonic crystals generated by stacking high- and

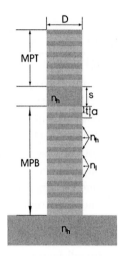

Fig. 5.   Parameters for a micropost microcavity. The micropost is rotationally symmetric around the vertical axis.

low-refractive-index disks on top of each other. The microcavity is formed by introducing a defect into this periodic structure. The periodicity of the photonic crystal is denoted as a, the thickness of the low-refractive-index disks is t, the diameter of the disks is D, and the refractive indices of the low- and high-refractive-index regions are n_l and n_h, respectively. The defect is formed by increasing the thickness of a single high-refractive-index disk from $(a - t)$ to s, as shown in Fig. 5. The number of photonic crystal periods above and below the defect region (i.e. the number of DBR pairs) is labeled as MPT and MPB, respectively.

The mode of interest to us is the doubly-degenerate fundamental (HE_{11}) mode, whose field pattern is shown in Fig. 6. The component of the electric field parallel to the DBR mirrors is dominant in this mode, and has an antinode in the center of the spacer. Furthermore, in this central plane, the electric field is practically linearly polarized at the vertical axis of the micropost, while there is a small deviation from the linear polarization at larger distances from this axis. All theoretical analyses presented in this chapter are performed by the Finite-Difference Time-Domain (FDTD) method; for the detailed description of this method, please refer to our earlier publications.[58,59]

The rule of thumb generally used for designing microposts is to make mirror layers one-quarter wavelength thick, and to choose the optical thickness of the spacer equal to the target wavelength. In the case of a planar

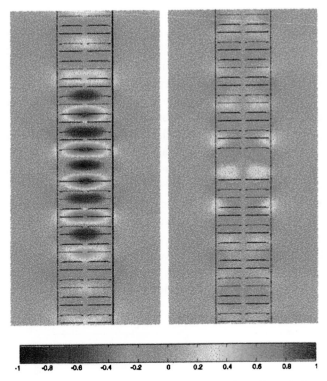

Fig. 6. Electric-field components for the fundamental (HE$_{11}$) mode in a micropost microcavity. The left figure illustrates the electric-field component parallel to the distributed Bragg reflectors (DBR's), while the figure on the right represents the electric field component perpendicular to the DBR's.

DBR cavity (with $D \to \infty$), this choice of parameters leads to the maximum reflectivities of the mirrors and the maximum Q-factor of the cavity mode: the cavity operates at the Bragg wavelength, for which the partial reflections from all high- and low-refractive-index interfaces add up exactly in phase. However, the strength of the cavity QED phenomena is proportional to the ratio of the cavity Q-factor to the mode volume V, as discussed previously in the text, and we will try to design microposts in such a way that this ratio is maximized.

In our earlier work,[58] we analyzed the Q-factor of the HE$_{11}$ mode in a GaAs/AlAs micropost as the cavity diameter was tuned between 0.5 μm and 2 μm. The remaining cavity parameters were chosen according to the large-cavity rule of thumb, i.e. in such a way that the cavity would operate at the Bragg wavelength for $D \to \infty$. When the cavity diameter was decreased from $2\,\mu$m to $0.5\,\mu$m, the mode volume decreased by a factor of almost ten, from $19.2(\lambda/n_h)^3$ to $2(\lambda/n_h)^3$, while the cavity Q dropped by only a factor of two, from 11500 to 5000. Thus, in order to maximize the ratio of the quality factor Q to the mode volume V, we need to explore structures with small diameters D, and try to improve their Q-factors.

The reduction in Q with a decrease in D is due to the combination of two loss mechanisms: *longitudinal loss* through DBR mirrors, and *transverse loss* due to imperfect TIR confinement in the transverse direction. Let us address the *longitudinal loss* first. The decrease in the post diameter D implies a change in the dispersion relation of the 1D photonic crystal, and the size and position of its bandgap, as illustrated in Fig. 7. In this figure, it is assumed that the high- and low-refractive-index regions of the photonic crystal consist of GaAs and AlAs, with refractive indices of $n_h = 3.57$ and $n_l = 2.94$, and thicknesses of 70 nm and 85 nm, respectively, or that they consist of GaAs and $\text{Al}_x\text{Ga}_{1-x}\text{As}$, with refractive indices of $n_h = 3.57$ and $n_l = 3.125$, and thicknesses of 70 nm and 80 nm, respectively. When the diameter D decreases, the frequencies of the bandgap edges increase, and the size of

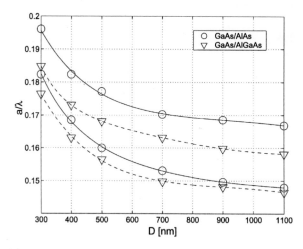

Fig. 7. Bandgap edges, calculated using the FDTD method (points), of the fundamental (HE_{11}) mode in a cylindrical one-dimensional photonic crystal in the GaAs/AlAs or GaAs/$\text{Al}_x\text{Ga}_{1-x}\text{As}$ material systems. The lines are guides to the eye. The GaAs/AlAs photonic crystal has the following parameters: $n_h = 3.57$, $n_l = 2.94$, $t = 85$ nm, and $a = 155$ nm. The GaAs/$\text{Al}_x\text{Ga}_{1-x}\text{As}$ photonic crystal has the following parameters: $n_h = 3.57$, $n_l = 3.125$, $t = 80$ nm, and $a = 150$ nm. (See Fig. 5 for definition of parameters.) The bandgap edges for $D \rightarrow \infty$ are positioned at a/λ equal to 0.1445 and 0.1634 for the GaAs/AlAs photonic crystal, and at a/λ equal to 0.1431 and 0.1565 for the GaAs/$\text{Al}_x\text{Ga}_{1-x}\text{As}$ photonic crystal.

the bandgap decreases. For structure diameters larger than $2\,\mu$m, bandgap edges can be approximated by their values at $D \to \infty$. Therefore, as D decreases, the blue shift of the cavity mode wavelength λ increases relative to the target wavelength at which the 1D cavity operates.[58] Simultaneously, the size of the photonic bandgap decreases, implying that the cavity mode is less confined in the longitudinal direction than in the planar cavity case.

The cavity mode is strongly localized in real space, and consequently delocalized in Fourier space (k-space), meaning that it consists of a wide range of wavevector components. Some of these components are not confined in the post by TIR; i.e. they are positioned above the light line, where they can couple to radiative modes, leading to *transverse loss*. A cavity mode which is strongly confined in the longitudinal direction by high-reflectivity mirrors is delocalized in Fourier space and suffers large transverse loss. Similarly, a mode that is delocalized in the longitudinal direction is more localized in Fourier space and suffers less transverse loss. Therefore, when optimizing the quality factor of three-dimensional microposts, there is a tradeoff between these two loss mechanisms.

In the middle of a large bandgap, the longitudinal confinement is strongest, but the Q-factor is limited by transverse loss. By shifting the resonant wavelength away from the mid-gap (e.g. by tuning the thickness of the cavity spacer) one can delocalize the mode in real space, localizing it more strongly in Fourier space, reducing the contribution of wavevector components above the light line, and thereby decreasing the transverse radiation loss. Eventually, as the mode wavelength approaches the bandgap edges, the loss of longitudinal confinement starts to dominate and Q drops. Therefore, in the microposts with high reflectivity mirrors and finite diameter, it is expected that the maximum Q will be located away from the mid-gap position. Moreover, since the mode wavelength can be tuned from the mid-gap towards any of the two bandgap edges, two local maxima of Q (i.e. a double peak behavior in Q versus mode wavelength) are expected. Besides detuning the mode wavelength from the mid-gap, we can also suppress the transverse loss by relaxing the mode slightly in the longitudinal direction, i.e. by reducing the reflectivities of photonic crystal mirrors and decreasing the bandgap size. This can be achieved by shrinking the cavity diameter, or by changing the photonic crystal parameters (e.g. by reducing the refractive-index contrast).

In this section, we study both of these approaches to Q optimization: tuning the mode wavelength away from the mid-gap by changing the spacer thickness, and tuning the mirror reflectivities by changing photonic crystal parameters or cavity diameter. We also show that the employment of very high reflectivity mirrors cannot lead to high-Q cavities with small diameters, as the transverse radiation loss is high, resulting from very strong mode localization in the longitudinal direction. Finally, we discuss prospects for achieving the strong coupling regime with a single quantum dot embedded in a micropost, and we comment on the degradation of the micropost quality factor as a result of fabrication imperfections.

3.2.1. GaAs/AlAs Cavities

Let us first study the same set of micropost parameters as in our earlier work,[58] chosen in such a way that the cavity would operate at the Bragg wavelength for $D \to \infty$: $a = 155\,nm$, $t = 85\,nm$, $s = 280\,nm$, $n_h = 3.57$ and $n_l = 2.94$, $MPT = 15$ and $MPB = 30$. In an attempt to maximize the ratio of the Q-factor cavity to the mode volume, we analyze here only the behavior of the HE_{11} mode for the cavity diameter D below $0.5\,\mu m$.

In order to tune the mode frequency within the bandgap, we tune the spacer thickness s. Results for λ, Q, V and Q/V are shown in Figs. 8 and 9. From Fig. 7, we see that the bandgaps in these structures extend from 875 nm to 969 nm, from 850 nm to 920 nm, and from 790 nm to 850 nm, for structure diameters of $0.5\,\mu m$, $0.4\,\mu m$ and $0.3\,\mu m$, respectively. As we have noted previously, when D decreases, the bandgap edges shift towards lower wavelengths, and the size of the bandgap decreases. The cavity mode wavelength is blue-shifted in this process, as can be seen in Fig. 8.

The mode volume V is minimized when the mode wavelength is located near the middle of the bandgap, as shown in Fig. 9. For structures with D equal to $0.4\,\mu m$ and $0.3\,\mu m$, the maximum Q-factor also occurs close

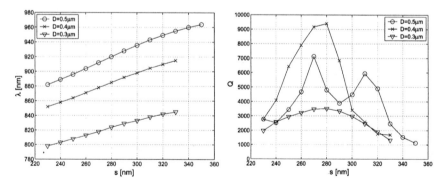

Fig. 8.   Wavelength λ and quality factor Q of the fundamental mode in a micropost with $a = 155\,nm$, $t = 85\,nm$, $n_h = 3.57$, $n_l = 2.94$, $MPT = 15$ and $MPB = 30$. The cavity diameter D and the spacer thickness s are tuned.

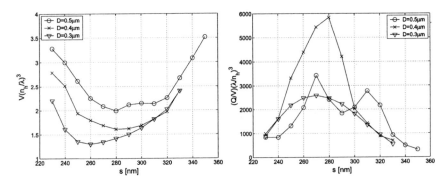

Fig. 9. Mode volume V and ratio of quality factor Q to V for the HE_{11} mode in a micropost with $a = 155\,nm$, $t = 85\,nm$, $n_h = 3.57$, $n_l = 2.94$, $MPT = 15$ and $MPB = 30$. The cavity diameter D and the spacer thickness s are tuned.

to the mid-gap. Different behavior is seen for the structure with D equal to $0.5\,\mu m$, which has a local minimum of Q at mid-gap and exhibits a double-peak behavior.

The double-peak behavior was already introduced previously in the text. In the middle of the bandgap, where the longitudinal mode confinement is strongest and the mode volume is minimum, the radiation loss in the transverse direction is high, and the Q-factor is degraded. By shifting the resonant wavelength away from the mid-gap, the mode is delocalized in real space, leading to a reduction in the transverse radiation loss (e.g. at the positions of the two peaks in Q). Eventually, as the mode wavelength approaches the bandgap edges, the loss of longitudinal confinement starts to dominate, Q drops, and the mode volume increases.

To support this explanation, we analyze the same structure, with $D = 0.5\,\mu m$, but with the number of mirror pairs on top (MPT) increased from 15 to 25. As expected, at mid-gap, Q does not increase significantly with MPT. The mode there is already strongly confined in the longitudinal direction, and the addition of extra pairs does not change the longitudinal loss. The modal Q-factor is determined by the radiation loss in the transverse direction, which is independent of MPT. On the other hand, the Q's at the two peaks increase with MPT. At these points, the mode is not confined as well in the longitudinal direction, and longitudinal loss can be reduced by adding more mirror pairs.

As an even stronger demonstration of our explanation for the double-peak behavior, we separate the radiation loss into the loss above the top micropost surface (L_a), and the loss below it (L_b). The total Q is a combination of two newly introduced quality factors, Q_a and Q_b, which are inversely proportional to L_a and L_b, respectively:

$$\frac{1}{Q} = \frac{1}{Q_a} + \frac{1}{Q_b}. \tag{18}$$

It follows from their definition that Q_a and Q_b are measures of the longitudinal and transverse loss, respectively. We analyze two sets of structure parameters, corresponding to the local maximum or minimum in Q. For $s = 270\,\text{nm}$ and $D = 0.5\,\mu\text{m}$ (local maximum), we calculate $Q_a \approx 14500$ and $Q_b \approx 13910$, while, for $s = 290\,\text{nm}$ and $D = 0.5\,\mu\text{m}$ (local minimum), we calculate $Q_a \approx 16000$ and $Q_b \approx 5100$. These results show that the local minimum in Q is due to an increase in the transverse loss, manifested as a drop in Q_b.

Let us now address the single-peak behavior of Q as a function of cavity spacer thickness, when D is equal to $0.4\,\mu\text{m}$ or $0.3\,\mu\text{m}$. Structures with smaller diameters have smaller bandgaps, as illustrated in Fig. 7, and the cavity modes are more delocalized in the longitudinal direction, relative to the structure with $D = 0.5\,\mu\text{m}$. The defect modes must therefore be more localized in Fourier space, and will thus suffer less radiation loss in the transverse direction. This implies that the Q-factors are determined mostly by the longitudinal loss. They reach their maxima at the mid-gap, where the mode volume is minimum, and the longitudinal confinement is strongest.

The maximum Q/V ratio of almost 6000 (where V is measured in cubic wavelengths in the high-refractive-index material) is achieved for the structure with $D = 0.4\,\mu\text{m}$, as shown in Fig. 9. For this structure, the Q-factor is close to 9500, and the mode volume is $1.6(\lambda/n_h)^3$. For $D = 0.4\,\mu\text{m}$, a variation in the thicknesses of the mirror layers allows us to achieve a small increase in the Q-factor, to 10500, and in the Q/V ratio, to 6500. This result is obtained for $a = 155\,\text{nm}$, $t = 75\,\text{nm}$, and $s = 290\,\text{nm}$.

3.2.2. $GaAs/Al_x Ga_{1-x} As$ Cavities

Previously in this chapter, we stated that a potential route to maximizing Q for small micropost diameters is the construction of a photonic crystal with a small-refractive-index perturbation. As the perturbation gets smaller, the cavity mode becomes more delocalized in real space, and consequently more localized in Fourier space. This, in turn, leads to reduction in the transverse radiation loss. Furthermore, the cavity resonance can be located at lower frequencies, where the density of free-space radiation modes is smaller.

In order to compensate for the increased longitudinal loss, we need to put more mirror pairs on top of these structures.

We will now analyze a micropost with the following parameters: $a = 150\,\text{nm}$, $t = 80\,\text{nm}$, $MPT = 25$, $MPB = 30$, $n_h = 3.57$, and $n_l = 3.125$. This choice of refractive indices corresponds to GaAs/Al$_x$Ga$_{1-x}$As layers. Both the cavity diameter D and the spacer thickness s are tuned. The positions of the bandgap edges as a function of D are illustrated in Fig. 7. In comparison to the positions of the bandgap edges for the GaAs/AlAs system, we confirm that the bandgap in the GaAs/Al$_x$Ga$_{1-x}$As system is shifted to lower frequencies, and that its size is decreased. This affects the HE$_{11}$ mode dramatically, as can be seen in Fig. 10.

By comparing Figs. 9 and 10, we can see that the mode volume increases when the refractive-index contrast is reduced, as a result of the reduction in bandgap size. Even though Q larger than 14000 can be achieved for $D = 0.5\,\mu\text{m}$ (see Fig. 10), V also increases, and the maximum Q/V ratio is similar to that calculated for the GaAs/AlAs system. Furthermore, this Q/V ratio can be achieved in the GaAs/AlAs system with fewer top mirror pairs. Longitudinal loss dominates in the GaAs/Al$_x$Ga$_{1-x}$As system, and Q versus s plots demonstrate a single-peak behavior.

When the number of mirror pairs on top is reduced from 25 to 20, the peak Q-factor of the GaAs/Al$_x$Ga$_{1-x}$As micropost with diameter $0.5\,\mu\text{m}$ drops from 14000 to 4000, showing that the longitudinal loss is dominant in this case, and a large number of mirror pairs is necessary to achieve large Q-factors.

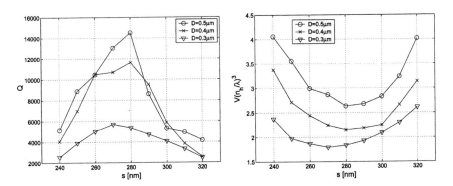

Fig. 10.   Quality factor Q and mode volume V of the HE$_{11}$ mode in a micropost with $a = 150\,\text{nm}$, $t = 80\,\text{nm}$, $n_h = 3.57$, $n_l = 3.125$, $MPT = 25$ and $MPB = 30$. The cavity diameter D and the spacer thickness s are tuned.

3.2.3. $GaAs/AlO_x$ Cavities

From the results already presented in this section, it is clear that a material system with a high-refractive-index contrast, such as $GaAs/AlO_x$, is not a good choice for high Q, small mode volume microposts. High-refractive index contrast can certainly produce larger bandgaps, and thereby provide a better longitudinal confinement of the cavity mode. However, if the contrast is increased, the mode suffers more radiation loss in the transverse direction, which limits its Q-factor. To confirm this, we analyzed a structure with $n_h = 3.57$, $n_l = 1.515$, $a = 235\,\mathrm{nm}$, $t = 165\,\mathrm{nm}$, $MPT = 15$, and $MPB = 30$, for different D. We were unable to obtain good mode localization for $D < 0.8\,\mu\mathrm{m}$, and the calculated Q-factors were under 250. For $D = 0.8\,\mu\mathrm{m}$, the mode has $Q = 600$ and $\lambda = 947\,\mathrm{nm}$. If we keep increasing D to $1.3\,\mu\mathrm{m}$, Q-factors remain below 1000.

3.2.4. Cavity Quantum Electrodynamics with Microposts

The question that we would like to address now is whether such cavity QED phenomena as strong coupling with a single quantum dot can be observed in microposts. Let us revisit our best design, with $Q \approx 10^4$, $V = 1.6(\lambda/n_h)^3$, $D = 0.4\,\mu\mathrm{m}$, $\lambda = 885\,\mathrm{nm}$, and the cavity field decay rate $\kappa = (\pi c)/(\lambda Q) = 106\,\mathrm{GHz}$. Let us assume that a quantum dot exciton without a cavity has a homogenous linewidth $\gamma_h = 2\,\mathrm{GHz}$ (measured from the intensity spectrum), and a spontaneous emission rate of $\Gamma = 2\,\mathrm{GHz}$. The free-space spontaneous emission rate is $\Gamma_0 = \Gamma/n_h = 0.56\,\mathrm{GHz}$. The Rabi frequency for our optimized cavity, calculated from Eq. (17), is equal to $g_0 = 400\Gamma_0 = 224\,\mathrm{GHz}$. If we assume that the quantum dot is located in the center of the micropost and that its dipole is aligned with the electric field, we have $g = g_0$. Strong coupling is therefore possible in this case, since $g_0 > \kappa, \pi\gamma_h$, and the minimum quality factor necessary to achieve it is approximately equal to 5000.

3.2.5. Influence of Fabrication Imperfections on the Micropost Quality Factor

In reality, imperfections in the fabrication process used to construct microposts inevitably lead to deviation from perfectly vertical micropost wall profile. Although we have improved our fabrication procedure dramatically, undercutting can still be seen in scanning electron micrographs of our microposts (see Fig. 15).

This undercut and the insufficient etch depth through the lower DBR lead to an inrease in the transverse losses and the diffraction in the lower

Fig. 11. Electric field amplitude for the fundamental mode in a micropost with realistic wall profile.

DBR, respectively, and cause a degradation of the quality factor of the fundamental mode whose field pattern is shown in Fig. 11. This phenomenon has been discussed in greater detail in our earlier work.[58] The problem of the insufficient etch depth has been eliminated in our recent samples, while the undercutting is still present, although significantly minimized. Achieving the Q-factor of 5000 in a micropost with diameter of 0.4 μm, necessary for the strong coupling regime, would thus require an immense fabrication effort.

3.3. *Photonic Crystal Microcavities*

Even though the optimized designs of DBR microposts can potentially enable the observation of the single QD-cavity QED phenomena in the strong coupling regime, as discussed above, stringent fabrication requirements of these structures motivated us to investigate planar photonic crystal microcavities as a more robust and manufacturable solution. The design and fabrication of photonic crystal microcavities that can achieve the strong coupling regime with a neutral atom trapped in one of the photonic crystal holes have been proposed recently;[44,45] these structures can have Q-factors larger than 10^4 together with mode volumes of the order of one-half of the cubic optical wavelength in material $(1/2(\lambda/n)^3)$. A Q-factor close to 3000 has been subsequently demonstrated from such a cavity where theory predicts $Q \approx 4000$,[43] and the lowest-threshold lasing at $1.5\,\mu m$ has been observed from optically pumped multi quantum wells embedded in it.[60] Moreover, by employing inversion methods, the theoretical Q-factor of this design has been increased even further, to a value around 10^5.[46] However, as these structures have been optimized for interaction with neutral atoms, they have an electric field maximum in the air region and are therefore not suitable for interaction with a single QD emitter embedded in a semiconductor. Other high-Q photonic crystal cavity designs that have recently appeared in the literature are also not tailored for semiconductor cavity QED, since they have a node of the electric field in the high-index region at the cavity center or an increased mode volume.[61-63] In this section, we present a design that could enable very high Q/V ratios, but with the electric field maximum in the high-index region at the cavity center, to facilitate its alignment to a QD and to prevent nonradiative surface recombination. Although we have been primarily motivated by a single QD-cavity QED experiment, this design can also be employed to significantly reduce the lasing threshold of standard photonic crystal lasers.

The proposed microcavity is illustrated in Fig. 12. The structure is formed in the underlying planar hexagonal photonic crystal with lattice

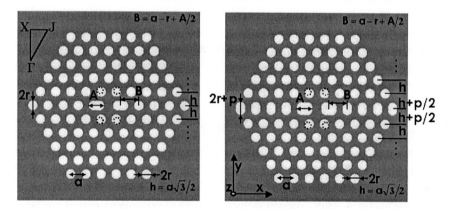

Fig. 12.   Photonic crystal microcavity for cavity quantum electrodynamics with a single quantum dot emitter embedded in semiconductor. After omitting the central hole and modifying its nearest-neighbors (left), fractional edge dislocations are inserted along the x-axis to tune the Q-factor of the x-dipole mode (right). The orientations of x-, y-, and z-axes are indicated in the plot, and the point $(0,0,0)$ is located laterally at the center of the defect and vertically in the middle of the slab.

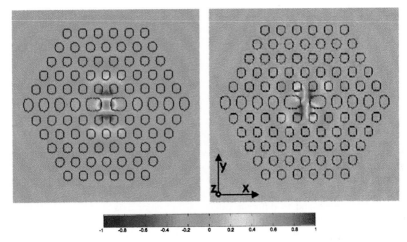

Fig. 13.   Electric field components of the x-dipole mode in the middle of the slab for the microcavity in Fig. 12. The figure on the left corresponds to the x-component, and the figure on the right to the y-component of electric field (the z-component is equal to zero at this plane). Electric field is maximum and is x-polarized at the cavity center, where an emitter should be located in order to provide a strong interaction. The parameters of this particular cavity are $r/a = 0.3$, $d/a = 0.65$, $n = 3.4$, $A/r = 3.33$ and $p/a = 0.2$, leading to the Q-factor larger than 5000 together with the mode volume of $1/2(\lambda/n)^3$ with only seven photonic crystal layers around the defect.

periodicity a and hole radius r, constructed in the slab with refractive index n and thickness d in the z-direction. As illustrated in the figure on the left, the central hole is omitted and its two nearest neighbors along the x-axis are made ellipsoidal with major axis equal to A and minor axis equal to $2r$. The edge-to-edge distance between the ellipses and their next-nearest neighbor holes along the x-axis is preserved and is equal to $a - 2r$. In order to tune the Q-factor of the x-dipole mode (whose electric field pattern is shown in Fig. 13), fractional edge dislocations of the order p are then applied along the x-axis, as illustrated on the right in Fig. 12. Fractional edge dislocations and their effect on the Q-factor improvement of the x-dipole mode have been described in more detail in our earlier publications.[44,45] Briefly, by inserting them, one can perform small adjustments to the central lobe-size of the even field components (E_x and B_y), without significantly perturbing mode volume and photonic crystal reflectivity. This, in turn, results in cancellation of the Fourier components within the light cone and consequent suppression of radiation losses when the balance between the positive and negative field lobes is reached.

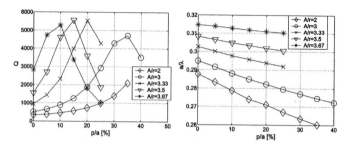

Fig. 14. $Q_\perp$-factor (left) and normalized frequency a/λ (right) of the x-dipole mode as a function of the normalized elongation parameter p/a, for the cavity design in Fig. 12. The underlying hexagonal photonic crystal parameters are $r/a = 0.3$, $d/a = 0.65$ and $n = 3.4$, leading to the bandgap for even modes extending from $a/\lambda = 0.26$ to 0.33. Various plots correspond to different values of the major axis A of the two ellipsoidal holes next to the defect region in the x-direction. For the peak-Q point, $Q \approx Q_\perp$ can be achieved with as few as seven photonic crystal layers around the defect.

Figure 14 displays Q-factor and frequency a/λ of the x-dipole mode in the structure in Fig. 12, as a function of p/a and A. The underlying hexagonal photonic crystal parameters are $r/a = 0.3$, $d/a = 0.65$ and $n = 3.4$, with a photonic bandgap for the TE-like (even) modes extending roughly from $a/\lambda = 0.26$ to 0.33. In order to minimize the computational requirements of the employed FDTD method, we have analyzed structures with only five photonic crystal layers around the defect. Q-factor can be expressed as $Q = \frac{Q_\parallel Q_\perp}{Q_\parallel + Q_\perp}$, where $Q_\perp$ corresponds to the radiation losses above the surface of the membrane, for $|z| > 3d/2$, and $Q_\parallel$ includes only the in-plane losses within the membrane, for $|z| < 3d/2$. By increasing the number of photonic crystal layers above five, the in-plane losses are suppressed, while the out-of-plane (radiation) losses remain approximately unchanged; $Q_\parallel$ becomes consequently much larger than $Q_\perp$, leading to the Q-factor being determined only by the radiation losses, $Q \approx Q_\perp$.[44,45] For structures operating away from the bandgap edges, such as the peak-Q point in Fig. 14 with electric field pattern shown in Fig. 13, the condition of $Q \approx Q_\perp$ can be achieved with as few as seven photonic crystal layers.

The Q-factor can therefore be larger than 5000 at its peak, while the mode volume remains roughly equal to one half of the cubic optical wavelength $(1/2(\lambda/n)^3)$ in the studied range of p. The significance of the two ellipsoidal holes next to the defect region is clear from Fig. 14: without them (i.e. for $A/r = 2$), the resonant mode frequency is too low, and it exits the bandgap before the out-of-plane losses are minimized by tuning p. As we have noted previously,[44] the resonant mode frequency drops as a function of the elongation parameter p; it is therefore important to start in the elongation process as close to the top of the photonic bandgap as possible, in order to minimize the out-of-plane losses within the bandgap and thus preserve a strong in-plane confinement and a small mode volume. By making the two central holes ellipsoidal and for increasing A, the resonant mode's overlap with air region increases, thereby increasing its frequency and leaving enough space for the Q-optimization within the bandgap (see Fig. 14).

Further improvement in the Q-factor can be achieved by tuning the four holes closest to the defect in the ΓJ directions, as illustrated with dotted lines in Fig. 12. The radius of the four holes is reduced to r_1, and they are simultaneously shifted by $r - r_1$ in the ΓJ directions to preserve their edge-to-edge distance from the next-nearest neighbor holes and thus keep a high photonic crystal reflectivity in these directions. The tuning of these four holes as a tool for increasing the Q-factor of the dipole mode has been explored earlier.[44,46] In this case, the mechanism behind the Q improvement is the suppression of radiation losses due to mode delocalization.[63] As expected, the mode volume of the resonant x-dipole mode slightly increases to $0.8(\lambda/n)^3$, but the Q-factor is improved to 45000; the best Q/V ratio is thus larger by an additional factor of six. This maximum is obtained in the structure with $A/r = 3.33$, $p/a = 0$, $r_1/a = 0.25$, $r/a = 0.3$, $d/a = 0.65$, $n = 3.4$ and $a/\lambda = 0.2847$.

Let us consider the cavity with the peak Q/V ratio and assume that it operates at $\lambda \approx 900$ nm, on resonance with an InAs/GaAs QD exciton whose radiative lifetime without a cavity is equal to 0.5 ns. If the exciton is initially slightly detuned from the cavity resonance, it can be easily brought to it by varying the sample temperature.[20,37] The cavity field decay rate is $\kappa = (\pi c)/(\lambda Q) = 23$ GHz, and the spontaneous emission rate of the exciton without a cavity is $\Gamma = 2$ GHz. The homogenous linewidth (γ_h) of excitons on our samples is of the order of a few GHz (e.g. 2 GHz), and therefore smaller than the cavity field decay rate κ. From Eq. (17), the Rabi frequency g_0 of the photonic crystal cavity-QD exciton system on resonance is equal to 315 GHz. For the exciton located at the electric field maximum and aligned with the field, a very strong coupling (high-Q regime) is possible, as the coupling parameter g is much larger than the decay rates of the system. Moreover, strong coupling is possible even if the ratio of g to κ is reduced by another factor of fourteen, due to the misalignment between the QD and the field maximum or the reduction in the Q-factor resulting from fabrication inaccuracies or material absorption. As an example, even if the QD is positioned at one-half of the electric field maximum, and the

Q-factor is up to seven times smaller than the theoretically predicted value of 45000, strong coupling would occur. Despite this robustness of the proposed design, we are also developing a fabrication method that would enable lithographic alignment of a microcavity to the location of a QD previously selected during the spectroscopy performed on unprocessed wafers. With such alignment, the minimum Q-factor necessary to achieve strong coupling would be approximately equal to 3000, which is in the range of already demonstrated experimental values for similar photonic crystal cavities.[43]

4. Experimental Demonstration of Solid-State Cavity Quantum Electrodynamics

Let us now describe our cavity QED experiments with a previously described InAs quantum dot embedded in the middle of a GaAs spacer in a distributed Bragg reflector (DBR) micropost microcavity. The GaAs spacer is approximately one optical wavelength thick (274 nm), and sandwiched between twelve DBR mirror pairs on top, and thirty DBR mirror pairs at the bottom. Each DBR pair consists of a 68.6 nm thick GaAs and a 81.4 nm thick AlAs layer (both layers are approximately a quarter-wavelength thick). The micropost structures (shown in Fig. 15) were constructed by a combination of molecular-beam epitaxy (MBE) and chemically assisted ion beam etching (CAIBE). MBE was used to grow a

Fig. 15.   Scanning electron micrograph showing a fabricated array of microposts.

wafer consisting of DBR mirrors and a GaAs spacer with embedded self-assembled QDs. Microposts with diameters ranging from $0.3\,\mu m$ to $5\,\mu m$ and heights of $5\,\mu m$ were fabricated in a random distribution by CAIBE, with Ar^+ ions and Cl_2 gas, and using sapphire (Al_2O_3) dust particles as etch masks. Sapphire was chosen as a mask because of its chemical stability and hardness, which enabled etch depths larger than other mask-materials. Unfortunately, these same properties impede its removal from the top of the structures, without endangering the microposts.[64] The presence of sapphire on top of the structures decreases their quality factors and consequently reduces outcoupling efficiencies. This can be confirmed theoretically by the FDTD calculation of the Q-factor of the fundamental HE_{11} mode (whose field pattern is shown in Fig. 6) in a micropost with and without sapphire on top. The FDTD unit cell size used in this case is $5\,nm$. We assume that the DBR layers of the simulated post have the same parameters as in the experimentally studied structures, that the post has perfectly straight walls, diameter of $0.5\,\mu m$, and that refractive indices of GaAs, AlAs and Al_2O_3 are 3.5, 2.9 and 1.75, respectively. Without sapphire, the Q-factor of the HE_{11} mode is 2600 and its wavelength is $882\,nm$; with a $0.5\,\mu m$ thick sapphire disk on top of the post (with diameter equal to that of the post), the Q-factor drops to 1400, while the mode wavelength remains unchanged. Due to the irregular shapes of the fabricated posts, the HE_{11} mode is typically polarization-nondegenerate, and many microposts have only one or two QDs on resonance with this fundamental mode.

We performed lifetime and $g^{(2)}$ measurements on a QD chosen for its bright emission under resonant excitation. Tuning of the sample temperature was used to tune the emission wavelength relative to the cavity resonance.[37] In the studied temperature range, the shift of the cavity resonance with temperature (due to refractive index change) is much slower than the shift in the excitonic emission wavelength, which is attributed to the changes in the bandgaps of InAs and GaAs with temperature.[37] This can be seen in Fig. 16, where streak camera images taken at $10\,K$ and $30\,K$ for the same resonantly excited dot are shown. Clearly, the emission wavelength increases with temperature; as a result of this shift, the radiative lifetime also changes. This particular dot (presented in Figs. 16 and 17) is almost exactly on resonance with the cavity at low temperature, so by heating the sample and increasing the QD emission wavelength, one also increases the detuning from the cavity resonance and the radiative lifetime. The opposite process is observed if the QD emission wavelength is initially smaller than the cavity resonance.

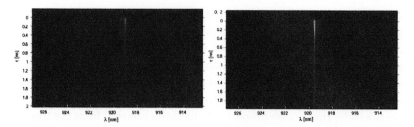

Fig. 16.   Streak camera images of a quantum dot emission taken at the temperatures of 10 K (left) and 30 K (right).

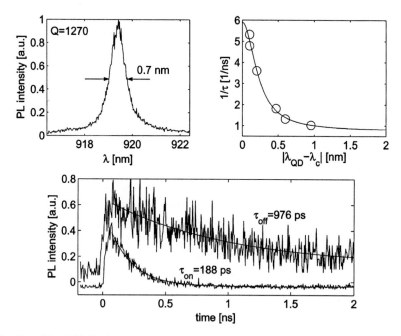

Fig. 17.   (Top-left) Background emission filtered by the cavity. (Top-right) Decay rate of the emission line as a function of the absolute value of its detuning from the cavity resonance ($|\lambda_{QD} - \lambda_c|$). The dot emission wavelength was tuned by changing the sample temperature within the 6–40 K range. (Bottom) Time-dependent photoluminescence from the emission line on-resonance with the cavity, as opposed to this same emission line off-resonance.

In addition to a good correspondence between the cavity resonance linewidth and the linewidth of Lorentzian describing decay rate of the emission line as a function of detuning from the cavity resonance (see Fig. 17, top-left and top-right), these processes indicate that cavity QED has a dominant effect on the radiative lifetime. Figure 17 (bottom) shows the time-resolved photoluminescence of an emission line on- and off- resonance with the cavity mode. The decay lifetime differs by a factor of five for these two cases. The decay rate for this emission line as a function of the absolute value of its detuning from the cavity resonance ($|\lambda_{QD} - \lambda_c|$) is plotted with circles in the top-right plot in Fig. 17. The solid line corresponds to the Lorentzian fit to the experimental data,[18] and a good match is observed between our experiment and the theoretically predicted behavior. The fitting parameters are the linewidth, maximum and minimum of the Lorentzian, while it is assumed that its central wavelength is equal to λ_c, the cavity resonance wavelength. λ_c and the quality factor ($Q = 1270$) are determined from the photoluminescence intensity taken at high pump powers (see the top-left plot in Fig. 17). The cavity resonance wavelength red-shifts by roughly 0.3 nm with increasing temperature in the studied range, and this shift is included in plotting the data. Up to five-fold spontaneous emission rate enhancement (Purcell factor F) is observed for the dot coupled to a cavity, as opposed to the dot off-resonance (top-right plot in Fig. 17).

The theoretical limit of the Q-factor of the studied microposts is 4000, which is the value calculated for a planar cavity (before etching) without any absorption losses and inaccuracies in the growth of DBR layers. The maximum Q-factor of a micropost with a finite diameter has to be below this limit, due to the additional loss mechanisms in the transverse directions (see Sec. 3.2 and Ref. 39) and the presence of sapphire dust particles, as discussed above. According to the detailed theoretical treatment of the QD micropost device,[39,58] the cavity Q-factor and Purcell factor can be much larger ($Q \sim 10000$ and $F \sim 100$) for optimized cavity designs with 15 and 30 DBR pairs on top and bottom, respectively, perfectly straight cavity walls, and a QD located at the cavity center.

5. Cavity-Enhanced Single Photons on Demand

5.1. *Generation of Single Photons with Small Multiphoton Probability and Large Purcell Factor*

In addition to the small multiphoton probability, a strong Purcell effect is important in a single-photon source for improving the photon outcoupling efficiency and the single-photon generation rate, and for bringing the emitted photon pulses closer to the Fourier transform limit. In this section, we describe experimental demonstration of a single-photon source based on a quantum dot in a micropost microcavity that exhibits a large Purcell factor together with a small multiphoton probability: for a quantum dot on resonance with the cavity, the spontaneous emission rate is increased by a factor of five, while the probability to emit two or more photons in the same pulse is reduced to 2% compared to a Poisson-distributed source of the same intensity.

A photon correlation histogram was measured for the quantum dot in Sec. 4 on resonance with the cavity, and with an excitonic radiative lifetime below 200 ps. The collected emission was first spectrally filtered with a bandwidth of 0.1 nm, and then directed towards the previously described Hanbury Brown and Twiss-type (HBT) setup; the measured histogram is shown in Fig. 18. The distance between peaks is 13 ns, corresponding to the repetition period of pulses from the Ti:sapphire laser. The decrease in the height of the side peaks as $|\tau|$ increases indicates the dot blinking behavior generally observed under resonant excitation, and can be approximated with a double-sided exponential.[12] The vanishing central peak (at $\tau = 0$) indicates a strong antibunching and a large suppression of multiphoton

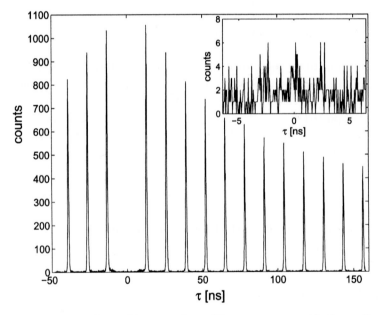

Fig. 18. Photon correlation histogram for a QD on resonance with the cavity (the lifetime of this dot is shown in Fig. 17), under pulsed, resonant excitation. The inset depicts the magnified central portion (from $\tau = -6.5$ ns to 6.5 ns) of the histogram. The missing central peak (at $\tau = 0$) indicates a large suppression of multiphoton pulses.

pulses. The probability of generating two or more photons for the same laser pulse compared to a Poisson-distributed source of the same intensity $(g^{(2)}(0))$ is estimated from the ratio of the areas of the central peak and the peaks at $|\tau| \to \infty$. Each area is calculated by integrating all the counts within the integration window centered at the peak and without subtracting any background counts. The peak at $|\tau| \to \infty$ is estimated from the decaying exponential fit to the heights of the side peaks. $g^{(2)}(0)$ is estimated to be equal to 2% for an integration window of 4 ns. The integration window is chosen so that the contribution of the peak tails outside to the peak area can be neglected (it is below 1%). The width of the histogram peaks is determined by the photon counter timing resolution (0.3 ns) and the excitonic lifetime (below 0.2 ns).[12] Owing to the strong Purcell effect, this small multiphoton probability should be preserved even for a repetition period much smaller than 13 ns (e.g. 2 ns). If the integration window is reduced to 1 ns, $g^{(2)}(0)$ drops to 1%. Depending on the application of the photon-source, a different definition of the two-photon probability may be necessary. For example, in an interference experiment using two consecutive photons emitted from a dot, the relevant parameter is the probability to emit two photons in the same pulse, as opposed to the probability to emit one photon in each of two consecutive pulses.[19] This parameter, which we denote as g, is calculated from the ratio of the area of the central peak to the area of the nearest side peak, and is equal to 0.9% for the integration window of 4 ns. The difference between g and $g^{(2)}(0)$ is a result of the blinking behavior of the dot.

5.2. *Generation of Indistinguishable Single Photons on Demand*

As we stated previously, for most applications in the field of quantum information we need photons that are quantum-mechanically indistinguishable and can thus produce multiphoton interference. We now describe a test of the indistinguishability of photons emitted by a semiconductor quantum dot in a microcavity through a Hong–Ou–Mandel-type two-photon interference experiment.[65,66] We find that consecutive photons have a mean wavepacket overlap as large as 0.81, making this source useful for a variety applications in quantum optics and quantum information. The sample used in the experiments presented here is the same sample that was used for the measurements in Secs. 4 and 5.1, but different pillars were tested here.

When single photons with identical wavepackets enter a 50–50 beam-splitter from opposite sides, quantum mechanics predicts that both photons must exit in the same direction. This effect, known as "two-photon inter-ference," can be thought of as originating from the Bose–Einstein statistics of photons. Such behavior was first observed with pairs of highly corre-lated photons produced by parametric downcoversion,[66] but until recently this behavior had never been observed with single, independently generated photons. Although such behavior was predicted for spontaneous emission from single atoms,[65] the difficulty of isolating and trapping single atoms makes such an experiment technically challenging.

The experiment described here used single semiconductor quantum dots at 4K excited by 3 ps laser pulses on resonance with transitions typically about 20–30 nm shorter in wavelength than the fundamental (single-exciton ground state) transition. After the quantum-dot emission was collected with a lens, a single polarization was selected. The emission was then spectrally filtered with a resolution of about 0.1 nm and coupled into a single-mode fiber. Three quantum dots were chosen for this study, denoted as dots 1, 2 and 3, having emission wavelengths of 931 nm, 932 nm and 937 nm, respec-tively. We first discuss three properties of the spontaneous emission from these quantum dots that are especially important for a two-photon inter-ference experiment: two-photon suppression, spontaneous-emission lifetime, and coherence length.

As discussed above, a commonly used two-photon suppression parame-ter is $g^{(2)}(0)$, the probability of generating two photons in the same pulse divided by that same probability for an equally bright Poisson-distributed source. This $g^{(2)}(0)$ is estimated from photon correlation measurements to be equal to 0.053, 0.067, and 0.071 for dots 1, 2 and 3, respectively. But for the experiment described below, the important parameter is the probability of generating two photons in the same pulse (for either of two consecutive pulses), divided by the probability of generating one photon in each of two consecutive pulses, and we denote this quantity as g. For quantum dots 1, 2 and 3, g is estimated to be equal to 0.039, 0.027 and 0.025, respectively. This g differs from $g^{(2)}(0)$ only because of the blinking in our source.

The spontaneous emission lifetime is estimated from the streak camera measurements shown in Fig. 19(left). By fitting decaying exponentials to the time-resolved photoluminescence from the quantum dots (averaged over many pulses), we estimate the spontaneous emission lifetimes τ_s of dots 1, 2 and 3 to be 89 ps, 166 ps, and 351 ps, respectively. This variation is due in part to differences in how well each quantum dot couples to its microcavity.

To estimate the time-averaged coherence length (equivalent to inverse spectral linewidth), a Michelson-type interferometer was used. The setup

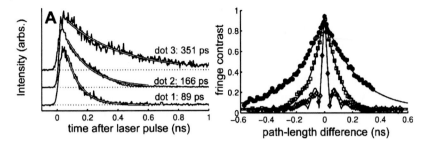

Fig. 19.   (Left) Spontaneous emission decay under resonant excitation, measured by a streak camera. Lifetimes estimated from exponential fits are indicated. (Right) Coherence length, measured using a Michelson interferometer, showing fringe contrast versus pathlength difference. The $1/e$ coherence lengths τ_c are 48, 223, and 105 ps for dots 1 (diamonds), 2 (solid circles), and 3 (squares), respectively.

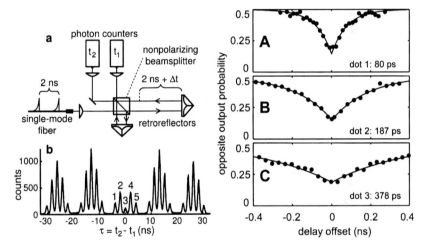

Fig. 20.   (Left-a) Two-photon interference experiment. (Left-b) An interferometric photon correlation histogram (53 ps bin size) obtained for quantum dot 2, with $\Delta t = 0$. The number of repetitions was $N = 2.3 \times 10^{10}$ (5 min.), and the combined two-photon generation and detection efficiency was $\eta^{(2)} = 2.5 \times 10^{-6}$. The small area of peak 3 demonstrates two-photon interference. (Right) For quantum dots 1 (A), 2 (B) and 3 (C), the probability that two photons collide at the beamsplitter exit in opposite directions, plotted as a function of interferometer delay offset, Δt, showing "Mandel dips". The fitted (solid lines) $1/e$ widths of these dips are indicated.

was the same as shown in Fig. 20, but with the 2 ns delay removed. The fringe contrast was measured by monitoring the intensity of one of the interferometer outputs while varying one of the arm lengths over several wavelengths using a piezoelectric transducer. The arm length was then moved over long distances by a motor stage. The measured curves, shown in

Fig. 19(right), give the magnitude of the Fourier transform of the intensity spectra. When we did not select a single polarization, we sometimes saw oscillatory behavior due to polarization splitting of the emission lines.[67] For dots 2 and 3 (with splittings of 13 meV and 17 meV), we were able to eliminate this effect by selecting a particular linear polarization. For dot 1, the 45 meV splitting could not easily be eliminated, perhaps because the quantumdot emission couples to just one cavity mode having a polarization rotated 45° relative to the splitting axis of the quantum dot. We estimate the $1/e$ coherence lengths τ_c (divided by c) for quantum dots 1, 2 and 3 to be 48 ps, 223 ps, and 105 ps, respectively. Quantum dot 2 is closest to being Fourier transform-limited, with $2\tau_s/\tau_c = 1.5$. If this ratio were equal to 1, no dephasing would be present, and perfect two-photon interference would be expected.

The measured ratios $2\tau_s/\tau_c$, especially for dot 2, are good enough that at least a partial two-photon interference effect should be possible. One can further hope that the remaining spectral broadening might be due largely to a slow, environment-induced dephasing process,[51] so that consecutive photons emitted only 2 ns apart might have identical wavepackets, even if photons emitted 1 sec apart might not.

The main elements of the two-photon interference experiment are shown in Fig. 20(left-a). The single-photon source is as described above, except that the quantum dot is excited twice every 13 ns by a pair of equally intense pulses with 2 ns separation. Two pulses, each containing zero or one photon, emerge from the single-mode fiber, and interfere with each other in a Michelson-type interferometer. The incident beam is first split into two arms by a nonpolarizing beamsplitter, with one arm ($2 \text{ ns} + \Delta t$) longer than the other. Corner-cube retroreflectors are used at the ends of the arms, so that the mode overlap is insensitive to slight angular misalignment of the optical elements. The length of the short arm can be adjusted over long distances by a 15 cm motor stage. The beams then recombine at a different place on the same beamsplitter. The fringe contrast measured using a laser with a long coherence length was 0.92, limited probably by optical surface imperfections in the beamsplitter and retroreflectors. The two outputs from the beamsplitter are collected by photon counters, and a photon correlation histogram is generated of the relative delay time $\tau = \tau_2 - \tau_1$ for two-photon coincidence events, where τ_1 and τ_2 are the times at which photons are detected at detectors 1 and 2, respectively. A histogram obtained in this way for dot 2 with $\Delta t = 0$ is shown in Fig. 20(left-b).

Five peaks appear within the central cluster, corresponding to three types of coincidence events. For peaks 1 and 5 at $\tau = \mp 4\,\text{ns}$, the first photon follows the short arm of the interferometer, the second photon follows the long arm, and one photon goes to each counter. For peaks 2 and 4 at $\tau = \mp 2\,\text{ns}$, both photons follow the same arm. For peak 3 at $\tau = 0$, the first photon follows the long arm, and the second photon follows the short arm, so that the two photons collide upon their second pass through the beamsplitter. Only in this case can two-photon interference occur, and for perfect two-photon interference, peak 3 vanishes.

The probability of successful delivery of a pair of photons depends on the internal quantum efficiency of the quantum dot, the collection efficiency, and subsequent optical losses. The total two-photon efficiency, including detector losses, for dot 2 was $\eta^{(2)} = 2.5 \times 10^{-6}$. This efficiency could have been improved somewhat by increasing the laser intensity, possibly at the expense of increased $g^{(2)}(0)$ and increased decoherence. For future experiments involving more than two photons, the efficiency will have to improve. When the source does successfully deliver a pair of photons, the two-photon state can be written as:

$$|\psi\rangle = \int ds\, x(s) \int dt\, y(t) a^\dagger(s) a^\dagger(t + 2ns)|0\rangle, \qquad (19)$$

where $a^\dagger(t)$ is the photon creation operator at time t, $x(s)$ and $y(t)$ define the photon wavepackets, and $|0\rangle$ is the vacuum state. We assume that the photon wavepackets are much shorter than 2 ns. In the limit of low collection efficiency, the mean areas of peaks 1–5 are:

$$A_1 = N\eta^{(2)} R^3 T \qquad (20)$$

$$A_2 = N\eta^{(2)}[R^3 T(1 + 2g) + RT^3] \qquad (21)$$

$$A_3 = N\eta^{(2)}[(R^3 T + RT^3)(1 + 2g) - 2(1 - \epsilon)^2 R^2 T^2 V(\Delta t)] \qquad (22)$$

$$A_4 = N\eta^{(2)}[R^3 T + RT^3(1 + 2g)] \qquad (23)$$

$$A_5 = N\eta^{(2)} RT^3 \qquad (24)$$

where N is the number of repetitions, $\eta^{(2)}$ is the combined two-photon generation and detection efficiency, and R and T are the beamsplitter intensity coefficients of reflection and transmission, respectively. As defined above, the parameter g characterizes the two-photon emission probability, with $g = 0$ for an ideal single-photon source, and $g = 1$ for a Poisson-distributed source (without blinking). The parameter $1 - \epsilon$ is the interference fringe contrast measured when an ideal monochromatic calibration source is sent into the interferometer, and accounts for optical surface imperfections. The parameter $V(\Delta t) = \langle |\int dt\, x(t)y^*(t+\Delta t)|^2 \rangle$ in the expression for peak 3 is the mean overlap between the wavepackets of the two photons for interferometer pathlength difference $(2\,\text{ns} + \Delta t)$. An ensemble average is performed over all possible two-photon states generated by the source.

The signature of two-photon interference that we observe is the small size of peak 3 in Fig. 20(Left-b), compared with peaks 2 and 4. We define the quantity $M(\Delta) = A_3/(A_2+A_4)$ in terms of the peak areas in Eqs. (20)–(24). This quantity is equal to the conditional probability, given two photons collide at the beamsplitter, that the photons exit in opposite directions, in the limit $g \approx 0$. We measured $M(\Delta t)$ while varying the interferometer pathlength offset Δt, and the results are shown in Fig. 20(Right). For all three quantum dots, we observe reductions in the coincidence probability near $\Delta t = 0$, by factors of 0.61, 0.69 and 0.62 for dots 1, 2 and 3, respectively. The remaining coincidences we see are partly due to independently measured optical imperfections in our setup, $R/T = 1.1$ and $(1 - \epsilon) = 0.92$. Without these imperfections, the coincidence reduction factors would be $V(0) = 0.72$, 0.81 and 0.74 for quantum dots 1, 2 and 3, respectively.

To analyze these data, we fit the function $M(\Delta t) = 0.5[1 - a\exp^{-|\Delta t|/\tau_m}]$, where the fitting parameters a and τ_m characterize the depth and the width of the coincidence dip, respectively. The fits, shown as solid lines in Fig. 20(Right), match the data well. For an ideal spontaneous-emission source, with instantaneous initial excitation and no decoherence, a would differ from 1 due only to imperfections in the optical setup, and τ_m would be equal to the spontaneous emission lifetime. The fitted values of τ_m we obtain are 80 ps, 187 ps, and 378 ps for quantum dots 1, 2 and 3, respectively. These values agree quite well with the spontaneous emission decay lifetimes τ_s obtained in Fig. 19(a). For quantum dots 1 and 3, this result is surprising, given the short coherence lengths τ_c listed above. One might conclude that, for quantum dots 1 and 3, the primary spectral broadening mechanism occurs on a timescale much longer than 2 ns. Such a "spectral diffusion" effect could occur due to charge fluctuations in the vicinity of the quantum dot, for example.[51]

For quantum dot 2, we calculate a mean two-photon overlap of at least 0.81. The remaining imperfection could arise from several decoherence mechanisms. When the quantum dot is first excited by a laser pulse, the generated electron-hole pair is initially in an excited state, and must relax to its lowest state through phonon emission before a photon can be emitted at the proper wavelength. The ratio of this relaxation time, which could be as long as tens of picoseconds, to the lowest-state radiative lifetime, could limit the performance of this source. Decoherence by phonons[68,69] is another possible mechanism, though we see little temperature dependence from 3–7 K. Finally, the spectral diffusion mechanism noted above could also potentially contribute to decoherence on short timescales.

The measured degree of photon indistinguishability is large enough that a few interesting quantum-optical experiments are already possible. The performance of most schemes based on two-photon interference depends on the same wavepacket overlap measured here. For example, for a single-photon implementation of a post-selective scheme to generate single pairs of polarization-entangled photons,[70] the polarization correlation would ideally be unity in the H–V basis, and 0.81 in the 45°/−45° basis, violating Bell's inequality. Other applications, such as quantum teleportation and quantum logic gates, may become feasible as the performance of single-photon sources continues to improve.

6. Conclusions and Future Prospects

Single photons on demand can be generated by combining pulsed excitation of a single semiconductor quantum dot and spectral filtering, but properties of such a source can be significantly improved by embedding a quantum dot inside an optical microcavity with a high quality factor and small mode volume. The radiative lifetime of an emitter on resonance with the cavity can be decreased significantly below the dephasing time, bringing the emitted photon pulses closer to the Fourier transform limit. Moreover, a large fraction of spontaneously emitted photons can be coupled into a single cavity mode, thereby increasing the outcoupling efficiency. Finally, the single-photon generation rate increases with a cavity-induced reduction in the excitonic lifetime. We have demonstrated a single-photon source based on a quantum dot in a micropost microcavity that exhibits a large Purcell factor together with a small multiphoton probability. For a quantum dot on resonance with the cavity, the spontaneous emission rate is increased by a factor of five (Purcell factor $F = 5$), while the probability to emit two or more photons in the same pulse is reduced to 2% compared to a Poisson-distributed source of the same intensity ($g^{(2)}(0) = 2\%$). The indistinguishability of emitted single photons from one of our devices was tested through a Hong–Ou–Mandel-type two-photon interference experiment. We found that consecutive photons are largely indistinguishable, with a mean wavepacket overlap as large as 0.81, making this source useful in a variety of experiments in the field of quantum information. We have also designed a planar photonic crystal microcavity with Q-factor as high as 45000, mode volume smaller than a cubic optical wavelength, and maximum field intensity located in the high-refractive-index region, that can be used for enhancing interaction between quantum dots and cavity field. Reaching the strong coupling regime with a single quantum dot should be possible in such a cavity, opening a possibility for novel solid-state quantum optical devices, including sources of Fourier-transform limited single photons or components of quantum networks.[54]

Acknowledgments

The authors would like to thank A. Scherer and T. Yoshie from Caltech for providing access to CAIBE and for helping with fabrication, and B. Zhang for AFM images of quantum dot arrays.

References

1. C. H. Bennet and G. Brassard, *Proc. IEEE Int. Conf. Computers, Systems and Signal Processing*, IEEE, Bangalore, India, 1984.
2. E. Knill, R. Laflamme and G. J. Milburn, *Nature* **409**, 46, 2001.
3. J. Cirac, P. Zoller, H. Kimble and H. Mabuchi, *Phys. Rev. Lett.* **78**, 3221, 1997.
4. L. Duan, M. Lukin, J. Cirac and P. Zoller, *Nature* **414**, 413, 2001.
5. F. D. Martini, G. D. Giuseppe and M. Morocco, *Phys. Rev. Lett.* **76**, 900, 1996.
6. C. Brunel, B. Lounis, P. Tamarat and M. Orrit, *Phys. Rev. Lett.* **83**, 2722, 1999.
7. B. Lounis and W. E. Moerner, *Nature* **407**, 491, 2000.
8. J. Kim, O. Benson, H. Kan and Y. Yamamoto, *Nature* **297**, 500, 1999.
9. A. Beveratos, S. Kuhn, R. Brouri, T. Gacoin, J. Poizat and P. Grangier, *Eur. Phys. J.* **D18**, 191, 2002.
10. A. Kuhn, M. Hennrich and G. Rempe, *Phys. Rev. Lett.* **89**, 067901, 2002.
11. P. Michler, A. Kiraz, C. Becher, W. V. Schoenfeld, P. M. Petroff, L. Zhang, E. Hu and A. Imamoğlu, *Science* **290**, 2282, 2000.
12. C. Santori, M. Pelton, G. Solomon, Y. Dale and Y. Yamamoto, *Phys. Rev. Lett.* **86**, 1502, 2001.
13. V. Zwiller, H. Blom, P. Jonsson, N. Panev, S. Jeppesen, T. Tsegaye, E. Goobar, M. E. Pistol, L. Samuelson and G. Bjork, *Appl. Phys. Lett.* **78**, 2476, 2001.
14. Z. Yuan, B. E. Kardynal, R. M. Stevenson, A. J. Shields, C. J. Lobo, K. Cooper, N. S. Beattie, D. A. Ritchie and M. Pepper, *Science* **295**, 102, 2002.

15. E. Moreau, I. Robert, J. M. Gérard, I. Abram, L. Manin and V. Thierry-Mieg, *Appl. Phys. Lett.* **79**, 2865, 2001.
16. E. Waks, C. Santori and Y. Yamamoto, *Phys. Rev.* **A66**, 042315, 2002.
17. E. Knill, R. Laflamme and G. J. Milburn, *Nature* **409**, 46, 2001.
18. J. M. Gérard, B. Sermage, B. Gayral, B. Legrand, E. Costard and V. Thierry-Mieg, *Phys. Rev. Lett.* **81**, 1110, 1998.
19. C. Santori, D. Fattal, J. Vučković, G. Solomon and Y. Yamamoto, *Nature* **419**, 594, 2002.
20. J. Vučković, D. Fattal, C. Santori, G. Solomon and Y. Yamamoto, *Appl. Phys. Lett.* **82**, 3596, 2003.
21. E. M. Purcell, *Phys. Rev.* **69**, 681, 1946.
22. K. H. Drexhage, *Progress in Optics*, Vol. 12E, North-Holland, New York, pp. 165–232, 1974.
23. D. Kleppner, *Phys. Rev. Lett.* **47**, 233, 1981.
24. P. Goy, J. M. Raymond, M. Gross and S. Haroche, *Phys. Rev. Lett.* **50**, 1903, 1983.
25. G. Gabrielse and H. Dehmelt, *Phys. Rev. Lett.* **55**, 67, 1985.
26. F. DeMartini, G. Innocenti, G. R. Jacobowitz and P. Mataloni, *Phys. Rev. Lett.* **59**, 2955, 1987.
27. D. J. Heinzen, J. J. Childs, J. F. Thomas and M. S. Feld, *Phys. Rev. Lett.* **58**, 1320, 1987.
28. Y. Yamamoto, S. Machida, K. Igeta and Y. Horikashi, in *Coherence and Quantum Optics*, eds. J. H. Eberly, L. Mandel and E. Wolf, Plenum, New York, 1989.
29. J. L. Jewell, J. P. Harbison, A. Scherer, Y. H. Lee and L. T. Florez, *IEEE J. Quant. Electron.* **27**, 1332, 1991.
30. S. L. McCall, A. F. J. Levi, R. E. Slusher, S. J. Pearton and R. A. Logan, *Appl. Phys. Lett.* **60**, 289, 1992.
31. E. Yablonovitch, *Phys. Rev. Lett.* **58**, 2059, 1987.
32. S. John, *Phys. Rev. Lett.* **58**, 2486, 1987.
33. T. Baba, *IEEE J. Sel. Topics Quant. Electron.* **3**, 808, 1997.
34. J. S. Foresi, P. R. Villeneuve, J. Ferrera, E. R. Thoen, G. Steinmeyer, S. Fan, J. D. Joannopoulos, L. C. Kimerling, H. I. Smith and E. P. Ippen, *Nature* **390**, 143, 1997.
35. G. S. Solomon, M. Pelton and Y. Yamamoto, *Phys. Rev. Lett.* **86**, 3903, 2001.
36. B. Gayral, J. M. Gérard, A. Lematre, C. Dupuis, L. Manin and J. L. Pelouard, *Appl. Phys. Lett.* **75**, 1908, 1999.
37. A. Kiraz, P. Michler, C. Becher, B. Gayral, A. Imamoğlu, L. Zhang, E. Hu, W. V. Schoenfeld and P. M. Petroff, *Appl. Phys. Lett.* **78**, 3932, 2001.
38. T. D. Happ, I. I. Tartakovskii, V. D. Kulakovskii, J. P. Reithmaier, M. Kamp and A. Forchel, *Phys. Rev.* **B66**, 041303, 2002.
39. J. Vučković, M. Pelton, A. Scherer and Y. Yamamoto, *Phys. Rev.* **A66**, 023808, 2002.
40. C. J. M. Smith, H. Benisty, D. Labilloy, U. Oesterle, R. Houdre, T. F. Krauss, R. M. D. L. Rue and C. Weisbuch, *Electron. Lett.* **35**, 228, 1999.

41. C. Reese, C. Becher, A. Imamoglu, E. Hu, B. D. Gerardot and P. M. Petroff, *Appl. Phys. Lett.* **78**, 2279, 2001.
42. T. Yoshie, A. Scherer, H. Chen, D. Huffaker and D. Deppe, *Appl. Phys. Lett.* **79**, 114, 2001.
43. T. Yoshie, J. Vučković, A. Scherer, H. Chen and D. Deppe, *Appl. Phys. Lett.* **79**, 4289, 2001.
44. J. Vučković, M. Lončar, H. Mabuchi and A. Scherer, *Phys. Rev.* **E65**, 016608, 2002.
45. J. Vučković, M. Lončar, H. Mabuchi and A. Scherer, *IEEE J. Quant. Electron.* **38**, 850, 2002.
46. J. M. Geremia, J. Williams and H. Mabuchi, *Phys. Rev.* **E66**, 066606, 2002.
47. J. Vučković and Y. Yamamoto, *Appl. Phys. Lett.* **82**, 2374, 2003.
48. C. Santori, G. Solomon, M. Pelton and Y. Yamamoto, *Phys. Rev.* **B65**, 073310, 2002.
49. M. Tabuchi, S. Noda and A. Sasaki, *Science and Technology of Mesoscopic Structures*, Springer-Verlag, Tokyo, 1992, p. 375.
50. S. Reynaud, *Ann. Physique* **8**, 315, 1983.
51. M. Bayer and A. Forchel, *Phys. Rev.* **B65**, 041308, 2002.
52. H. J. Kimble, in *Cavity Quant. Electrodynamics*, ed. P. Berman, Academic Press, San Diego, 1994.
53. In our notation, γ and κ correspond to *field* decay rates, and g is the angular frequency. If γ_h estimated from the linewidth of the *intensity* spectrum in units of Hz (i.e. not angular, but usual frequency, and not field decay, but intensity decay), then $\gamma = \frac{1}{2} 2\pi\gamma_h = \pi\gamma_h$.
54. Mabuchi, M. Armen, B. Lev, M. Lončar, J. Vučković, H. J. Kimble, J. Preskill, M. Roukes and A. Scherer, *Quantum Information and Computation (Special issue on Implementation of Quantum Computation)* **1**, 7, 2001.
55. Y. Yamamoto and A. İmamoğlu, *Mesoscopic Quantum Optics*, John Wiley & Sons, INC., New York, 1999.
56. R. J. Glauber and M. Lewenstein, *Phys. Rev.* **A43**, 467, 1991.
57. J. M. Gérard and B. Gayral, *Physica* **E9**, 131, 2001.
58. M. Pelton, J. Vučković, G. S. Solomon, A. Scherer and Y. Yamamoto, *IEEE J. Quant. Electron.* **38**, 170, 2002.
59. O. Painter, J. Vučković and A. Scherer, *J.Opt. Soc. America* **B16**, 275, 1999.
60. M. Lončar, T. Yoshie, A. Scherer, P. Gogna and Y. Qui, *Appl. Phys. Lett.* **81**, 2680, 2002.
61. H. Y. Ryu, H. G. Park and Y. H. Lee, *IEEE J. Sel. Topics Quant. Electron.* **8**, 891, 2002.
62. H. Y. Ryu, S. H. Kim, H. G. Park, J. K. Hwang, Y. H. Lee and J. S. Kim, *Appl. Phys. Lett.* **80**, 3883, 2002.
63. S. G. Johnson, S. Fan, A. Mekis and J. D. Joannopoulos, *Appl. Phys. Lett.* **78**, 3388, 2001.
64. P. Levy, M. Bianconi and L. Correra, *J. Electrochem. Soc.* **145**, 344, 1998.
65. H. Fearn and R. Loudon, *J. Opt. Soc. America* **B6**, 917, 1989.
66. C. K. Hong, Z. Y. Ou and L. Mandel, *Phys. Rev. Lett.* **59**, 2044, 1987.

67. V. D. Kulakovskii, G. Bacher, R. Weigand, T. Kummell, A. Forchel, E. Borovitskaya, K. Leonardi and D. Hommel, *Phys. Rev. Lett.* **82**, 1780, 1999.
68. L. Besombes, K. Kheng, L. Marsal and H. Mariette, *Phys. Rev.* **B63**, 155307, 2001.
69. X. Fan, T. Takagahara, J. Cunningham and H. Wang, *Solid State Commun.* **108**, 857, 1998.
70. Y. H. Shi and C. O. Alley, *Phys. Rev. Lett.* **61**, 2921, 1988.

CHAPTER 5

FABRICATION, COUPLING AND NONLINEAR OPTICS OF ULTRA-HIGH-Q MICRO-SPHERE AND CHIP-BASED TOROID MICROCAVITIES

Tobias J. Kippenberg, Sean M. Spillane, Deniz K. Armani,
Bumki Min, Lan Yang and Kerry J. Vahala

Department of Applied Physics, California Institute of Technology
Pasadena, CA 91125, USA

Surface-tension-induced microcavities (STIMs) possess ultra-high-Q values (typically greater than 100 million) within micron-scale dimensions. Their "whispering-gallery" type optical modes are confined by continuous total internal reflection at the dielectric cavity interface. Silica microspheres are a well-known example of STIMs, which possess record Q values in excess of 10^9. In this chapter we will discuss a novel type of ultra-high-Q (UHQ) microcavity fabricated on a microelectronic silicon chip, which allows a level of integration and control previously not available in the UHQ regime. The device merges concepts used in traditional surface-tension-induced microcavity fabrication with standard microfabrication techniques to combine UHQ with wafer-scale integration. We have studied the quality-factor (Q) and modal structure of these toroidally-shaped microcavities and have observed Q-factors exceeding 100 million, similar to values achieved in microspheres. Investigations of optical coupling to UHQ microcavities using tapered, optical fibers are also described and show that highly efficient coupling can be obtained, with efficiencies in excess of 99%. ultra-high-Q combined with micronscale modal volumes and the ability to efficiently transfer optical power both to and from the whispering gallery modes, leads to extremely high circulating intensities. These intensities can easily exceed the thresholds for all common nonlinear phenomena in silica. This has allowed us to observe a variety of nonlinear oscillations at threshold levels several orders lower than in any prior work and typically in the micro-Watt regime. In particular, stimulated and cascaded Raman scattering and parametric oscillation are studied both experimentally and theoretically. The ability to dope silica with rare earths to create on-chip silica microlasers in the important 1500 nm band is also investigated. Finally, we review the regime of strong modal coupling induced by backscatter effects within the resonator, and examine how backscattering modifies

waveguide coupling properties including a change of the condition of
"critical coupling".

1. Introduction

This chapter reviews the fabrication, fiber-optic coupling and nonlinear
optical properties of UHQ STIMs, including a recently-developed chip-
based toroidal device. The pioneering work of Braginsky and Ilchenko[1]
demonstrated the unique properties of silica microsphere-STIMs to confine light for extended periods of time. Silica microspheres exhibit record Q
values due to both the low intrinsic material loss of silica in the visible and
near infrared and the nanometer-scale surface smoothness of the resonator
dielectric interface.[2] The highest Q to date observed in silica microspheres
has been 9×10^9 (and a cavity Finesse of 2.3×10^6 at a wavelength of
850 nm) which is close to the fundamental material absorption limit at
this wavelength.[3] These attributes have inspired numerous applications of
microspheres in fundamental and applied areas. The long photon storage
times are especially attractive for cavity QED experiments. The interested
reader is referred to the chapter by S. Haroche and coworkers in the previous volume of this series (Ref. 4) as well as to recent reviews.[3,5] Furthermore silica microspheres can be used to create low-threshold lasers[6,7]
and nonlinear oscillators.[8] It is also important to note that more generally, liquid droplet STIM nonlinear oscillators were pioneered by R. Chang
and coworkers[9-11] and A. Campillo and coworkers[12,13] nearly two decades
ago. More recently, microspheres have also entered the field of biochemical
sensing[14] as optical transducers with high sensitivity. Despite the advantageous properties of UHQ silica microspheres they do have several significant disadvantages. These include the fact that their physical characteristics are difficult to control during fabrication and that their mode spectrum, while well studied,[15] is highly complex and varies in a way that is
strongly dependent upon subtle variations in the eccentricity of the sphere.
Finally, and most significant, is that sphere fabrication lacks all of the
desirable features of wafer-scale processing, now the standard method of
fabrication employed for low and moderate Q microcavity applications.
These advantages include precise lithographic control and reproducibility
of dimension, as well as fabrication parallelism and the potential for integration with additional functionality such as electrical or micromechanical
devices.

In short, wafer-processing techniques offer a far more powerful approach
to fabricating microcavities. Photonic crystal defect cavities, micropost

structures, and ring geometry whispering gallery devices at moderate Q levels are now fabricated routinely using wafer-scale technology as described in the companion chapters by Little, Scherer, Steier and Yamamoto in this book. Reported Q-factor results have ranged from 13,000[16] in InGaAs microdisks, to 130,000 in polymer rings,[17] to a record value of 3 million in a silica microdisk reported by our group and described in this chapter. However, these Q-factors are several orders of magnitude lower than those found in surface-tension-induced microcavities, since the nanometer-scale surface finish required for ultra-high-Q has not been attainable using wafer-scale processing.

Despite tremendous progress in the UHQ regime application space (Q typically greater than 100 million — see Refs. 2 and 5), wafer-based UHQ devices have remained elusive. In this chapter, we also present and explore the properties of a chip-based (planar) STIM devices in the form of a microtoroid which combine for the first time the advantages of wafer-scale processing and ultra-high-Q.[18] These cavities are made by combining conventional microfabrication techniques with a laser-assisted reflow process to achieve the atomic-scale surface finish characteristic of STIMs. We have observed Quality factors of up to 500 million in these structures, similar to values reported and measured for silica microspheres. The mode structure of these microcavities exhibits a strongly-reduced mode density compared to a microsphere of the same size. A further unique property of microtoroids is the ability to reduce their optical mode volume significantly below those of microspheres of comparable diameter. This property is particularly important for nonlinear optical and cavity QED experiments.

In another focus of this chapter, we also review the coupling and transfer of optical power to both microspheres and microtoroids on-a-chip using tapered optical fibers.[19,20] Using tapers we have demonstrated the efficient excitation of whispering-gallery-modes and observed the condition of "critical coupling".[19] We also show that in the regime of ultra-high-Q the coupling properties are altered due to back-scatter-induced modal coupling of the whispering-gallery-modes.[21] The "ideality" of the taper-resonator junction is also studied and characterized using a new method.[22] The junction is shown to exhibit an extraordinarily high ideality in terms of linking a target resonator mode with the fundamental taper mode. This property has potential importance in quantum optical studies where low-loss optical fiber is proposed for transport of nonclassical states and correlations of light.[23]

The efficient transfer of optical power and accurate control of coupling strength using the tapered fiber method when combined with the ultra-high-Q and microscale nature of the whispering gallery modes enables the resonant buildup of very high circulating intensities leading to a variety of nonlinear optical processes. The regime of very-high circulating intensities — both in silica microspheres and in microtoroids — where the nonlinear interaction of light and matter becomes prominent is reviewed in Sec. 5. In studying these processes, the fiber taper serves to both couple pump power as well as to extract newly generated waves thereby allowing precise study of nonlinear optical effects. For appropriate coupling and Q conditions nonlinear oscillation at microwatt-level fiber-coupled power levels is observed.[8,24] Near theoretical efficiencies are possible due to the high-ideality of the taper-resonator junction. As such these systems offer both a well-controlled laboratory in which to study nonlinear phenomena, and also a potentially practical on-chip nonlinear optical source. One such nonlinear effect is stimulated Raman scattering, wherein the light field interacts with the material by way of a vibrational state (an optical phonon).[25] We have observed this effect and created fiber-coupled Raman microcavity lasers using both a microsphere[8] and a microtoroid on-a-chip. In addition to stimulated nonlinearities (i.e. Raman) we have also observed a second class of nonlinear interactions (parametric interactions). Whereas stimulated processes such as Raman are intrinsically phase matched, parametric oscillation requires stringent phase matching conditions causing the stimulated Raman nonlinearity to dominate microcavity nonlinear optics. However, we show that by controlling microtoroid cavity geometry a transition from stimulated Raman to parametric oscillation can be achieved, allowing for the first time the observation of the Kerr-induced parametric oscillation in a microcavity.

In Sec. 6, we review the application of both spheres and microtoroids as lasers employing rare-earth doped silica. For the same reasons described above in conjunction with nonlinear oscillators, these devices are important because of their high-differential efficiency and low threshold operation. As was also true for the toroid-based nonlinear oscillators, rare-earth doped toroids have the significant advantage of being chip-based devices that are directly coupled to optical fiber. The processing developed to create these lasers uses a novel sol-gel based approach in which erbium is coated upon the device and subsequently selectively densified at the toroid periphery. The process, itself, thereby allows a precise application of the erbium dopant where it is required.

Finally, we will conclude the chapter with an outlook section.

2. Ultra-High-Q Microcavity Fabrication and Background

2.1. *Microspheres*

The optical modes of spherical cavities have been intensively studied, starting with Mie. Here we provide only a brief review of some of the basic properties of the microsphere mode structure. The interested reader is referred to detailed reviews such as found in Refs. 15 and 26. The mode structure of microsphere resonators is well-known and follows from solving the spherical wave equation. Owing to spherical symmetry (and in analogy with the solutions of a hydrogen atom), modes are characterized by angular (ℓ), azimuthal (m), radial (n) and polarization (p) indices. Perfect spherical symmetry guarantees degeneracy with respect to the index m, however, in practice this $2\ell+1$ degeneracy is never achieved due to eccentricity-induced splitting. The resulting "spherical" mode spectrum is therefore highly complex and strongly dependent upon fabrication-induced eccentricity. Whispering gallery modes, like their acoustical analogues,[27] are a class of modes that execute "orbits" near the surface of the sphere. Their near surface trajectories make their Q strongly dependent upon interfacial blemishes and surface roughness, however, their radiation leakage or tunneling is the lowest of all modes within the sphere and hence, in cases where the surface is exceptionally smooth, these modes also exhibit the highest quality factors. Indeed, radiation tunneling is a negligible component of loss in STIM silica spheres for sphere diameters in excess of 20–$30\,\mu$m. So-called "fundamental" whispering gallery modes are ring shaped orbits and correspond to the spherical harmonic solutions $\ell = m$. For the case of the fundamental radial index they exhibit the smallest modal volume and as such are interesting in studies of cavity QED and nonlinear optics. Figure 1 shows green luminescence from a fundamental whispering gallery mode in a microsphere.

The ultimate obtainable Q-factor for a microspherical cavity has been intensively studied by several groups, with the highest reported Q to date of 9×10^9 at a wavelength of $850\,$nm. This value is close to the theoretical absorption limit of silica at this wavelength, however, for slightly longer wavelengths in the telecom band around $1550\,$nm, the theoretical limit imposed by material loss is around 100 billion.

The microspheres described in this chapter were fabricated by heating the tip of a tapered telecommunication fiber using a $10\,$Watt CO_2 laser. Silica is highly absorptive at $10.6\,\mu$m and can be readily melted by this method. The surface tension of the molten tapered glass tip causes the end

Fig. 1.   Main graph: Optical micrograph of an Erbium-doped, microsphere resonator
that is approximately 70 μm in diameter. The green luminescence originates from Erbium
up-conversion and indicates that the modal distribution of the light is in the fundamental
whispering gallery sphere mode. Visible in the bottom of the picture is the fiber stem,
which is used to position the microsphere. Micrograph *courtesy of* M. Cai.

to form a near-spherical shape, with residual eccentricities of typically a few
percent. The unmelted portion of the optical fiber taper serves as a stem
allowing convenient positioning with respect to the fiber-taper coupler as
explained in later sections.

2.2. *Fabrication of Ultra-High-Q Microtoroids On-a-Chip*

Microtoroids are silica resonators on a silicon chip and their fabrication is
described in this section. As noted in the introduction, microtoroids enable
access to the desirable ultra-high-Q performance of microspheres but in
a design that is wafer based and that takes advantage of process control
and parallelism not available with microspheres. The planar nature of these
structures also restricts azimuthal degrees of freedom, rendering their mode
spectrum far simpler than that of a sphere. Fabrication of microtoroid cavities on a chip proceeds as described in Ref. 18 in three basic steps: creation
of a silica disk by a combination of lithography and pattern transfer, isolation of the disk periphery by selective undercut of the substrate, and
finally, selective reflow of the disk material using a CO_2 laser to create the
microtoroid. The process steps are illustrated in Fig. 2.

Further details of the fabrication process are as follows. First, photolithography using standard photoresist is performed on a silicon wafer
containing a thermally-oxidized surface layer to create circular photoresist
pads. The diameter of the photoresist pads is chosen so as to be larger than

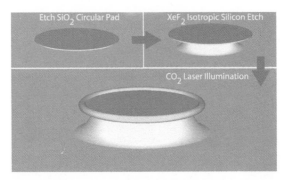

Fig. 2.   Illustration of the process sequence used to fabricate silica toroidal microcavities on-a-chip. First, silica disks are created by lithography and etching, second, the silica disks are undercut using an isotropic silicon etch, and third, the silica disks are surface-normal-irradiated using a CO_2 laser which causes the disk to melt and collapse into a toroid-shaped form.

the desired diameter of the toroid, since silica is consumed in the reflow process. After hard bake of the photoresist pads, circular disks are etched into the silica using buffered hydrofluoric acid. As the optical modes of the disk structure reside at the perimeter of the disk, the silica disks must be isolated from the high-index silicon to prevent leakage of optical energy into the substrate. This is achieved in the second step by using the silica disks as an etch mask for a silicon etch. An isotropic, selective, silicon etch employing XeF_2 gas at 3 Torr is used for this purpose. This process selectively removes the silicon, leaving circular silica disks on nearly-circular silicon support pillars. Figure 3 is a scanning electron micrograph of the resulting structure. The inset of Fig. 3 is an optical micrograph of the structure and also shows a fiber taper coupled to the resonator. The above process flow leaves lithographic blemishes at the all-important cavity-mode boundary, in many cases visible with an optical microscope.

The disks fabricated in this way exhibit whispering-gallery-type modes located near the disk periphery. Under optimal processing conditions we have measured Q-factors up to 3×10^6 in these structures,[28] believed to be limited by lithography and etching. Therefore, to obtain UHQ performance further sample processing is necessary to obtain the nanometer-r.m.s (root mean square) surface roughness typical of STIMs.[2]

In order to achieve an exceptional surface finish a processing step is introduced that reflows the undercut silica disks without affecting the underlying silicon support pillar. This is accomplished by surface-normal irradiation of the disk ("preform") with a CO_2 laser ($10.6\,\mu$m wavelength),

Fig. 3.   Scanning electron micrograph of a silica microdisk structure on a silicon wafer. The inset is a top-view optical micrograph of the structure coupled to a tapered fiber from the side.

similar to techniques once proposed for integrated-circuit planarization.[29] The laser beam intensity follows a Gaussian profile centered on the disk with a width exceeding that of the silica preform. Irradiation of the disk with laser intensities typically less than 100 Mega-Watts/cm^2 leads to melting and collapse of the silica at the disk periphery and not over the silicon post. The reason for this effect is that silicon has a far weaker absorption constant than for silica at 10.6 μm and additionally is 100 times more thermally conductive than silica. This leads to the silicon pillar staying significantly cooler (and physically unaffected) throughout the reflow process, allowing it to serve as a heat sink for optical power absorbed in the silica disk. An additional factor in the reflow process is the strong temperature dependence of the silica absorption,[30] which increases substantially with temperature so that melting occurs abruptly with increasing laser fluence through a thermal runaway process.

Several methods have been investigated to control the final toroid shape and size in the reflow process. A first method uses continuous laser irradiation. In this case, the disk collapses and toroid formation is self-quenched by a combination of thermal conduction through the silicon pillar and the reduced cross-section for absorption of CO$_2$ radiation. In this continuous-illumination formation process, the final toroid thickness (minor diameter) is determined by the preform diameter, pillar diameter and oxide thickness. Alternatively, in a second method, the reflow process can be interrupted prior to the self-quenching point by pulsed laser irradiation, which

Fig. 4.   Scanning electron microscope image of a 90 μm-diameter silica microtoroid on a silicon chip. The initial silica disk ("perform") diameter was 160 μm and 2 μm thick, with the reduction in diameter due to formation of the toroid perimeter.

produces a toroid with a final diameter in between the disk and silicon pillar diameter. The structure in Fig. 4 has been created using this method. The silica disk used in this case had a 2 μm thickness and an initial diameter of 160 μm. During irradiation the diameter of the disk was reduced to 120 μm and the consumed silica formed a 7 μm-diameter toroid-shaped perimeter. We also note that due to the long absorption length in silica at room temperature for CO_2 laser radiation (approximately 35 μm[30]), the required laser intensity to initiate disk collapse is increased significantly when using thinner oxides. Because it might not always be desirable to control the final toroid diameter by allowing complete collapse (i.e., first method) or through interruption of collapse, an alternative method of control was investigated as described below.

2.3. *Toroid Dimensional Control by Preform Design*

Increased lithographic control over the toroid geometry can be achieved by suitable design of the silica disk preform. Due to the thermal runaway effect, the thickness of the preform is an important parameter determining the required flux of CO_2 laser illumination to initiate silica reflow and toroid formation. As noted above, for thin oxide layers the required flux is strongly increased, and thus by using a variable preform thickness profile, the temperature distribution of the silica in the laser illumination process can be controlled. For example, an annular preform (i.e. thicker at the

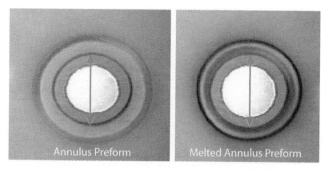

Fig. 5. An annular preform can be used to control the dimensions of the toroid. The left panel shows an annular preform prior to melting and the right panel shows the preform after irradiation. The CO_2 laser radiation selectively melts the thicker annulus perimeter, leaving the thinner interior disk unaffected.

perimeter) will be preferentially heated at the perimeter, where the oxide thickness is large. We have found that not only does this mean that the melt initiates in the annulus, but significantly it is prevented from proceeding into thinner interior. The left panel of Fig. 5 is an optical micrograph of an annular preform, featuring a thick rim at the perimeter ($2\,\mu$m), and a thinner, interior disk ($1\,\mu$m). In order to fabricate this structure two consecutive lithographic steps and buffered HF oxide etching were performed. The right panel in Fig. 5 shows the structure after pulsed laser illumination. The outer annulus region preferentially melts and surface tension causes it to form a toroid. The process is observed to self-quench when the inner toroid diameter reaches the inner annulus diameter, and, significantly, prior to reaching the silicon pillar. The advantage of this fabrication technique is that the amount of material used to form the toroid as well as the inner diameter of the toroid microcavity can be accurately controlled by lithography. In addition, the supporting disk structure can be made very thin, which thereby increases the optical confinement of the modes within the toroidal periphery.

3. Coupling to Ultra-High-Q Microcavities Using Tapered Optical Fibers

Efficient coupling of UHQ whispering gallery modes is a prerequisite for studying both microsphere and microtoroid cavities and is important in studies that either require low loss or efficient power transfer into and out of a cavity mode. To couple light into microspheres several techniques have

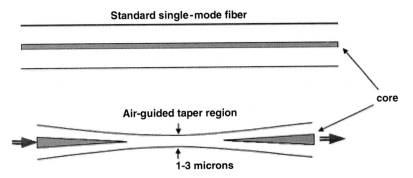

Fig. 6.   Schematic showing a standard telecommunication fiber and a tapered optical fiber.

been used, ranging from free-space illumination to evanescent coupling techniques. Since the fundamental whispering gallery modes are confined near the surface of the microsphere, evanescent coupling techniques can provide a powerful means of control of the coupling properties. A common technique used to excite microsphere resonances is prism coupling, where the evanescent field of a laser beam, which undergoes total internal reflection in a prism, is brought into overlap with the whispering gallery mode.[26,31,32] We describe here an alternate method that enables high efficiency coupling directly from optical fiber.[19] Additional details on this method which relies upon the formation of a narrow tapered segment within the optical fiber is provided in Refs. 19, 20 and 26.

Tapered optical fibers allow both efficient excitation and extraction of optical power from the same fiber, while maintaining fiber-optic compatibility. Using tapered optical fibers we have achieved very efficient excitation of whispering gallery modes and observed the condition of "critical coupling". Of even greater significance is that the fiber taper junction is highly "ideal" meaning that its behavior closely approximates a perfect two-mode coupler.[22] Indeed, it is not an exaggeration to say that the ideality and insertion loss of taper junctions vastly exceed what is considered typical for optical coupling to microresonators. As such they are potentially important junctions in systems requiring ultra-low loss coupling to optical fiber, such as in quantum optics and quantum information studies.[23,33-38] They are also inherently fiber compatible, giving them great utility in the laboratory for test and measurement. They are also useful when studying active and nonlinear microresonators, since the newly generated waves can be extracted through the same fiber that provides the pump excitation. This will be discussed further in Sec. 5.

Tapered optical fibers are made by heating a single-mode fiber (typically SMF-28) with a torch and slowly pulling the fiber apart until a waist region only a few microns in diameter is created. By maintaining an adiabatic taper profile[20,39] an initially-launched fundamental, fiber mode is converted to an air-guided, fundamental, taper mode and vice versa upon propagating through the tapered region. In principle, the loss arising from higher-order mode coupling or radiation leakage can be made arbitrarily small. Tapers used in our work typically have losses of a few tenths of a dB. We have analyzed the taper waist diameter in a scanning-electron-microscope and measured taper sizes of less than 1 μm, while maintaining loss levels of less than 0.1 dB.

We have studied taper excitation of microsphere WGMs in the optical telecommunication range. To excite the fundamental whispering gallery mode of the microsphere, the taper is brought close to an equator of the sphere, where the fundamental modes are confined. Coupling into the microsphere requires that three conditions are satisfied. First, the evanescent field of the fiber taper and the microsphere mode exhibit spatial overlap; second, the frequency of the excitation wave must be resonant with a target WGM of the sphere; and finally the propagation constants of the taper and whispering gallery mode need to be matched to some extent (i.e. phase matched). As will be described below, the last requirement can be relaxed in cases where the taper is highly single mode and where the resonator is UHQ. The overall coupling strength depends on the overlap of the taper with the sphere mode (which varies typically exponentially with the air gap between the resonator and the taper) as well as on phase-matching. Energy transfer is most efficient when the propagation constants of the fundamental taper mode and the microcavity WGM are equal. Phase-matching is accomplished in practice by matching the taper propagation constant to the WGM's propagation constant through change of the taper waist diameter. In experiments, the coupling is also controlled by varying the gap distance using a three-axis piezoelectric stage (step resolution of 20 nm) to accurately position the taper with respect to the microcavity. We found it helpful in our experiments to keep the taper under slight tension to minimize the influence of vibrations. We also discovered the particular suitability of this method to couple microtoroids on-a-chip. Since the toroidally-shaped perimeter of these cavities lies typically only a few tens of microns above the surface of the planar substrate, coupling using tapered fibers is ideally

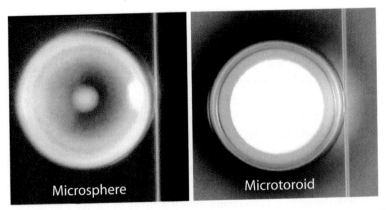

Fig. 7. An optical micrograph showing a tapered fiber that is side-coupled to a micro-toroid on-a-chip (right) and a microsphere (left). The microtoroid has a diameter of approximately 90 μm and has a supporting silicon pillar (appearing white in the image). The microsphere in the image has diameter of ca. 80 μm.

suited to this geometry, whereas bulk couplers such as prisms present addi-tional alignment difficulties. The other advantages of fiber-taper coupling will be explained in more detail in the following sections where tapered fibers are used to excite microtoroid whispering gallery modes and to study nonlinear optical effects in these structures. Figure 7 shows a microsphere and microtoroid side-coupled to a tapered fiber.

As discussed above, a tapered fiber converts the fundamental fiber mode into an air-guided taper mode. The taper section near the microcavity junc-tion can locally be described as a silica rod with guiding determined by the index of refraction of the original fiber cladding and the external environ-ment (typically air; index of refraction near unity). For the taper diameters used in this work to couple with WGM's (1.5–3 μm), a simple calculation shows that the local waveguide is multimode. Even though the adiabatic taper guarantees launch of the fundamental taper mode from the input side to the microresonator, the presence of multiple output waveguide (and radiation) modes means that, in general, a resonator mode can couple to all of the supported local radiation/waveguide modes of the taper. Any power which couples into modes other than the fundamental taper mode will be lost upon transition of the waveguide back to single-mode fiber. Nonethe-less, we are able to show that with proper choice of taper diameter, this parasitic, nonintended coupling can be reduced to a negligible level, ren-dering the taper junction highly efficient for coupling both to and from the resonant mode.[22]

To substantiate this claim, we have developed a technique to measure parasitic junction coupling losses, defined as any loss that occurs by way of unintended coupling. The junction performance is quantified in terms of the degree to which the fiber-taper/resonator junction behaves as a true single-mode/single-mode coupler, expressed as the *ideality* of the junction.

The coupling between a multimode waveguide and a resonator (illustrated in Fig. 8) can be modeled using a simple approach valid for weak coupling and small internal resonator loss. Under these conditions, the contributions of waveguide-induced cavity loss and intrinsic cavity loss are separable.[40] These conditions are met for the experimental configurations presented in this chapter. Under these assumptions, the internal cavity field ("a") is determined by accounting for all sources of cavity loss and excitation, and can be described by:

$$\frac{da}{dt} = i\Delta\omega a - \frac{1}{2}\left(\frac{1}{\tau_0} + \frac{1}{\tau_{\text{rad}}^e} + \sum_{i=0}\frac{1}{\tau_i^e}\right)a + i\sqrt{\frac{1}{\tau_0^e}}s. \qquad (1)$$

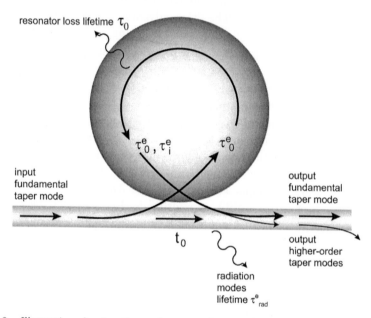

Fig. 8. Illustration showing the various, coupling mechanisms within a waveguide-resonator system. The input field (single-mode due to adiabatic launch conditions) excites the resonator with a coupling lifetime τ_0^e and transmitted amplitude t_0. The resonator mode couples into radiation modes with lifetime τ_{rad}^e, and all supported waveguide modes with lifetime τ_i^e. The intrinsic resonator lifetime is given by τ_0.

Here τ_0 is the intrinsic resonator photon lifetime, τ_{rad}^e is the lifetime for coupling/scattering to radiation modes, and τ_i^e denotes the coupling lifetime for coupling to each supported waveguide mode (τ_0^e represents the fundamental waveguide mode which is always present). The last term gives the excitation of the cavity due to only the fundamental waveguide mode with amplitude denoted by s under adiabatic tapering conditions. The transmission past the resonator consists of an interference of the transmitted optical field (i.e. that component that did not couple to the resonator) plus the field coupled out of the resonator into the fundamental taper mode (as power in all other modes is lost in the adiabatic transition), and is given by:

$$T = \left| t_0 + i\frac{a}{s}\sqrt{\frac{1}{\tau_0^e}} \right|^2 . \tag{2}$$

In steady state, the waveguide transmission on resonance can be expressed in the form

$$T = \left(\frac{1-K}{1+K}\right)^2 ; \quad K \equiv \frac{(1/\tau_0 + 1/\tau_{rad}^e + \sum_{i\neq 0} 1/\tau_i^e)^{-1}}{\tau_0^e} \equiv \frac{\tau_0'}{\tau_0^e} \tag{3}$$

where K denotes a dimensionless coupling parameter. Briefly, $K = 1$ is denoted as critical coupling, $K > 1$ as over-coupled and $K < 1$ as under-coupled.[40] The transmission expression is identical to that expected for a perfect single-mode to single-mode coupler, provided that the resonator intrinsic loss lifetime is taken as the "effective" (primed) lifetime defined above in the last equality. This interpretation is intuitive as any power not coupled into the fundamental mode of the output waveguide can be considered a resonator-specific loss when looking at the system as a whole. Thus the effective intrinsic lifetime has a contribution due to the actual resonator intrinsic lifetime (independent of coupling) as well as a lifetime associated with various components of parasitic coupling. These contributions can be separated by decomposing K as follows:

$$K^{-1} = K_I^{-1} + K_P^{-1} \tag{4}$$

where:

$$K_P \equiv \frac{(1/\tau_{rad}^e + \sum_{i\neq 0} 1/\tau_i^e)^{-1}}{\tau_0^e} \quad \text{and} \quad K_I \equiv \frac{\tau_0}{\tau_0^e}$$

and where K_I and K_P are defined as the intrinsic and parasitic coupling parameters, respectively. To describe the degree to which the coupling junction behaves as a single-mode coupler, we define the junction ideality,[22] where ideality is the ratio of power coupled into the desired fundamental waveguide mode to optical power coupled into all junction-related modes (both parasitic and fundamental modes) and is given by,

$$I \equiv \frac{1}{1 + K_P^{-1}}. \tag{5}$$

The ideality ranges from zero (no coupling at all into the desired mode) to unity (perfect single-mode coupler). Measuring the ideality of a waveguide-resonator system, thus requires one to accurately determine the parasitic coupling K_P. While, in principle, K_P can be determined from the transmission expression, coupled with knowledge of the intrinsic and coupling lifetimes, a more sensitive technique to measure K_P is to note that parasitic coupling will in general vary as the taper-resonator gap is varied. This variation will produce an observable deviation in the coupling parameter K from the ideal case that can be used to infer K_P. To see that this is true, consider the dependence of coupling strength K on gap. By inverting Eq. (3) we obtain,

$$\left(\frac{1 \pm \sqrt{T}}{1 \mp \sqrt{T}} \right) = K = \frac{(1/\tau_0 + 1/\bar{\tau}_1^e e^{\alpha_1 x})^{-1}}{\bar{\tau}_0^e e^{\alpha_0 x}} \tag{6}$$

where the upper signs are taken for transmission values in the over-coupled regime, and the lower signs in the under-coupled regime. The second equality follows from Eq. (4) and by noting that the coupling lifetimes associated with coupling to the various taper modes will vary exponentially with resonator-waveguide separation (τ-bar represents the lifetime at resonator/waveguide contact and α represents the rate of increase of lifetime with separation). In addition, it has been assumed in this equality that the parasitic coupling parameter is dominated by coupling to a single higher-order waveguide mode. This assumption is valid for the range of taper sizes typically investigated in this chapter. As demonstrated below, by plotting K on a logarithmic scale versus the gap distance, K_I and K_P can often be identified. This is possible because the spatial decay rate of the higher-order fundamental mode is distinct from that of the fundamental, causing K_P to be a line with a slope less than that of K_I. If the degree of parasitic coupling is comparable to or greater than the intrinsic resonator loss over the range of gaps measured, then the roll-off of K has asymptotes determining K_I and K_P. If, however, parasitic coupling is less than intrinsic loss, the parasitic coupling is not observable and only a lower bound on K_P (and thus ideality) can be established.

We have experimentally investigated the ideality of a fiber-taper/silica microsphere system using the above approach for WGM resonances around 1550 nm. The transmission data are obtained by normalizing the on-resonance power transmission with the power transmitted through the taper alone (i.e. in the absence of a resonator). The ideality in this measurement was determined for several taper diameters. Figure 9 shows the resultant coupling parameters K plotted versus taper-sphere gap for local taper diameters of 1.2 μm, 1.35 μm, and 1.65 μm (measured by an scanning electron microscope).

The data clearly show that taper diameter strongly influences the coupling parameter versus gap dependence. As taper diameter increases, the increased influence of higher-order mode coupling (a result of the decreased phase-mismatch between the fundamental taper mode and the next higher-order mode) leads to a strong roll-off of K with decreasing gap from the expected linear relation representing perfect ideality. A fit using Eq. (6)

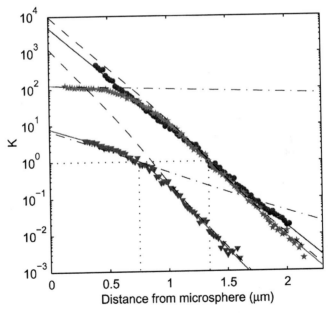

Fig. 9. Coupling parameter K versus taper-sphere gap for taper diameters of 1.2 μm (circles), 1.35 μm (stars), and 1.65 μm (triangles). The lower ideality as taper size increases is manifested as an increased roll-off of the data as gap distance is reduced. The dashed and dash-dotted lines represent the intrinsic coupling parameter K_I and the parasitic coupling parameter K_P, respectively. The dotted lines mark the critical point for each data set.

shows excellent agreement, confirming that the dominant contribution to nonideality in this measurement is from coupling to a single higher-order taper mode. Using the fit, both K_P and K_I can be extracted (dash-dotted and dotted lines in Fig. 9, respectively), which allows determination of the junction ideality versus gap distance. The $1.65\,\mu$m taper data yield a value for ideality ranging from 88% at taper-sphere contact to 13% at a gap distance of $1.5\,\mu$m. The next smaller taper size has an improved ideality value ranging from 99% at contact to 98% at the $1.5\,\mu$m gap. For the smallest taper size measured, the behavior appears completely ideal. As it is not possible to infer a dependence of K_P on taper-sphere gap, due to the finite loss of the resonator, only a lower bound on ideality at contact can be established for this case by using the value of K at taper-sphere contact. In this case we obtain a lower bound of ideality of 99.98%.

One consequence of reduced ideality is a shift of the critical coupling point due to the increased parasitic loss. This can be seen in Fig. 9 (all data sets taken using the same resonator and hence resonator intrinsic loss is unchanged from set-to-set). For perfect ideality, the critical point should occur at the gap distance where $K_I = 1$.[40] Inspection of the data for the largest taper size reveals that reduced ideality results in a shift of the critical point towards smaller gap distances. The critical point is essentially unaffected for the two smaller taper diameters, as a result of the large ideality values at critical coupling. Considering the large taper data once more, the critical point is shifted by 100 nm with respect to its position in the ideal case. An implication of this shift is that a nonideal coupler can lead to an under-estimation of the intrinsic cavity Q factor when using the measured Q at critical coupling to infer the actual intrinsic resonator Q (as is commonly done either by linewidth measurements or cavity ring-down). The data also demonstrate that the ideality of the junction actually decreases as the taper-sphere gap is increased. This, at-first-sight counter-intuitive fact, is the result of the slower evanescent decay of higher-order taper modes. This also implies, that if high ideality is to be achieved in the under-coupled or critically coupled condition, high ideality at contact is a prerequisite.

We have also investigated the absolute ideality attainable in this system by attempting to maximize the value of K through strong overcoupling to an ultra-high-Q microsphere as shown in Fig. 10.

Numerical calculations based on a modified coupled-mode theory verify that the observed slope of K is consistent with the expected ideal behavior. Using the approach described above, an extrapolated ideality

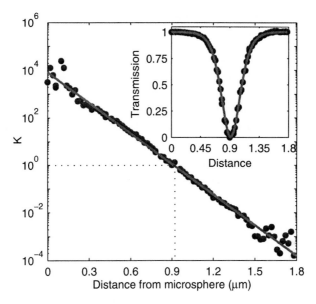

Fig. 10.   Coupling parameter K versus taper-sphere separation for a $65\,\mu$m-diameter microsphere showing highly ideal behavior. A fit assuming an ideal coupler (solid line) gives a lower bound on ideality of 99.99% in contact. The inset shows the original transmission data, with over-coupled transmission exceeding 99.95%.

value in contact exceeding 99.99% is obtained. Using a more conservative estimate based upon the actual measured data points in the vicinity of contact still results in a lower-bound value greater than 99.97%. This data includes all forms of taper insertion loss (e.g. even that associated with radiation mode scattering from juxtaposition of the taper and sphere). Thus this value provides an absolute determination of a physically realizable value. The original transmission data (inset) show that nearly complete over-coupling is obtained, with over-coupled transmission values exceeding 99.95%. The fact that fiber tapers can provide exceedingly low coupling loss to UHQ resonators allows great flexibility to study processes in microcavities where loss is highly undesirable. This includes applications in cavity QED[37] and quantum optics, and the study of active and nonlinear cavities as described below. Equally important is that this coupling is to optical fiber, the optical transport medium of choice. This last point cannot be over-emphasized as it enables experimental work to leverage the enormous investment in technology resulting from fiber communications.

3.1. *Modal Coupling in Whispering Gallery Mode Microcavities*

Whispering gallery type modes possess a natural two-fold degeneracy due to the two possible directions of propagation (clockwise and counterclockwise). Haroche *et al.*[41] have observed breaking of this degeneracy in the regime of ultra-high-Q[41] when the counter-propagating cavity modes are coupled through various mechanisms including Rayleigh scattering or scattering from surface defects at the dielectric boundary of the cavity.[42] When the mode scattering rate from such mechanisms exceeds the photon decay rates, caused either by intrinsic cavity loss or by waveguide coupling, one expects to enter a regime of "strong modal coupling" where lifting of the degeneracy (splitting of the resonant frequency into a doublet) can be observed. Curiously, many of the same mechanisms that are responsible for the intrinsic cavity loss will also tend to couple the counter-propagating modes. Hence, this mode-coupling is an effect that is a nearly inevitable consequence of entering the UHQ regime. Mode-coupling was theoretically investigated by Gorodetsky and it was predicted that in the so-called strong modal-coupling regime microresonators can act as narrowband reflectors.[32] In this section we describe experimental observations of mode splitting and in particular, the regime of strong modal coupling. We also show that in the regime of strong modal coupling a shift in the critical coupling point occurs. Understanding this shift is important in measurements of Q factor that occur in the presence of backscatter. Indeed, we provide an expression that enables the proper renormalization of Q in such circumstances. Furthermore, the waveguide-resonator coupling condition of maximum power transfer no longer coincides with the critical coupling point.

The equations of motion for counter-propagating (CCW and CW) modes that are coupled to one another as well as to a waveguide mode can be described by the couple-mode equations similar to those presented in Ref. 32:

$$\frac{da_{\mathrm{cw}}}{dt} = i \cdot \Delta\omega \cdot a_{\mathrm{cw}} - \frac{1}{2\tau} a_{\mathrm{cw}} + \frac{i}{2\gamma} \cdot a_{\mathrm{ccw}} + \kappa \cdot s$$

$$\frac{da_{\mathrm{ccw}}}{dt} = i \cdot \Delta\omega \cdot a_{\mathrm{ccw}} - \frac{1}{2\tau} a_{\mathrm{ccw}} + \frac{i}{2\gamma} \cdot a_{\mathrm{cw}}. \qquad (7)$$

Here "a" is the amplitude of the CCW and CW modes of the resonator and s denotes the input wave, which is selected to excite the CW mode. The excitation frequency is detuned by $\Delta\omega$ with respect to the resonance frequency ω_0 of the initially degenerate modes. τ is the total lifetime (i.e. which accounts for both intrinsic cavity loss as well as coupling loss to the waveguide) of photons in the resonator and is related to the quality factor $Q = \omega \cdot \tau$. The coefficient κ denotes the amplitude coupling of the input wave to the CW mode of the resonator. The relation $\kappa = \sqrt{\tau_{\text{ex}}^{-1}}$ associates the coupling coefficient with a corresponding lifetime, such that $1/\tau = 1/\tau_{\text{ex}} + 1/\tau_0$.[43] The mutual coupling of the CCW and CW mode is described by a (scattering) rate $1/\gamma$. The degenerate WGMs couple to the waveguide in opposite directions and give rise to reflected (r) and transmitted (t) signals.

$$t = s - \kappa \cdot a_{\text{cw}}, \quad r = \kappa \cdot a_{\text{ccw}}. \tag{8}$$

The scattering to all modes other than the originally degenerate pair of modes is included in the overall cavity loss, as given by the intrinsic lifetime τ_0. In the presence of modal coupling, the new cavity eigenmodes are symmetric and anti-symmetric superpositions of the original CW and CCW modes centered around new eigenfrequencies $\omega_0 \pm 1/\gamma$ (having a linewidth of $1/\tau$ as illustrated in Fig. 11). For strong modal coupling the two new eigenmodes correspond to standing waves, similar to those in a Fabry–Perot resonator. To enable a more precise definition of strong intermode

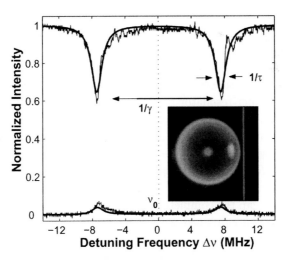

Fig. 11. Transmission and reflection spectra of a $70\,\mu$m-diameter sphere with $Q_0 = 1.2 \times 10^8$ and modal coupling of $\Gamma = 10$. The solid line represents a fit using the coupled-mode equations [Eqs. (7)]. The inset shows an optical micrograph of a microsphere coupled to a tapered fiber.

coupling, we introduce below the normalized intermode coupling parameter Γ^{21} and the normalized coupling coefficient K (which is identical to the K_I parameter introduced in the previous section).

$$\text{Modal coupling parameter: } \Gamma \equiv \left(\frac{\tau_0}{\gamma}\right), \qquad (9)$$

$$\text{Normalized coupling parameter: } K \equiv \left(\frac{\tau_0}{\tau_{\text{ex}}}\right) \qquad (10)$$

These normalized coupling coefficients describe the rate of modal coupling and waveguide-coupling with respect to the intrinsic cavity loss rate. The regime of strong modal coupling is characterized by $\Gamma \gg 1$ while, as noted earlier, the normalized coupling coefficient K describes the waveguide coupling regimes with $K > 1$ over-coupled, $K < 1$ under-coupled, and $K = 1$ critically coupled.[43]

We have observed the splitting of degenerate WGMs into doublets in both microsphere and microtoroid resonators, and have investigated the modified coupling properties using the fiber-taper coupling technique. As noted earlier, doublets are prominent in UHQ microcavities since only minute amounts of scattering either from bulk Rayleigh or from surface imperfections are sufficient to cause easily observable mode coupling.[42] In experiments, the taper contributes a negligible amount to the overall scattering (slight increase in mode splitting of less than 5% while scanning the taper from under-coupled to over-coupled), confirming that scattering centers either intrinsic to the resonator or at the resonator surface dominate. The inset to Fig. 11 is a photomicrograph of a microsphere (diameter ca. 70 μm) coupled to a tapered fiber. The coupling strength τ_{ex}^{-1} is adjusted by varying the taper-resonator gap distance in steps as small as 20 nm. A 1.55 μm tunable laser source is used to excite resonator modes. The laser is scanned repeatedly through a scan range of 60 GHz, containing a doublet structure while simultaneously recording transmission ($T = |t/s|^2$) and reflection ($R = |r/s|^2$) through the tapered fiber. A typical spectrum is shown in Fig. 11. By repeating such measurements many times for varying taper-sphere gap the complete range of waveguide coupling regimes can be observed. When combined with the coupled-mode model, the system parameters can be extracted. The data in Fig. 11 are for a 70 μm-diameter sphere (Q of $1.2 \cdot 10^8$) and the inferred mode coupling is $\Gamma = 10$.

Figure 12 contains both data and model curves for transmission and reflection versus coupling measured at a frequency corresponding to the eigenfrequency of the original degenerate modes (ω_0). We refer to this as

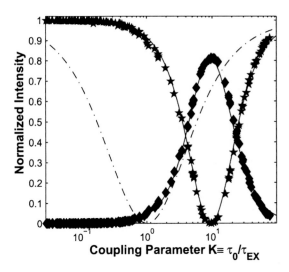

Fig. 12.   Transmission (stars) and reflection (diamonds) measured for the case of symmetric ($\Delta\omega = 0$) excitation versus K. $Q_0 = 1.2 \cdot 10^8$ and a modal coupling of $\Gamma = 10$ are inferred from the data. The solid lines provide a theoretical fit using the mode coupling model from Eq. (7). The critical point ($T = 0$) occurs at $K_{\mathrm{crit}} = \sqrt{1 + \Gamma^2}$ and is shifted with respect to the case of zero modal coupling. In addition, the shifted critical point is accompanied by a maximum reflected signal (in this case measured to be 84%). For comparison the dash-dotted curve is the theoretical coupling curve in the absence of modal coupling. The critical point occurs for $K = 1$ in this case.

"symmetric" excitation since it occurs midway between the doublets in the spectrum. In the absence of modal coupling the transmission vanishes (with zero reflection) at the conventional critical point $K = 1$ (waveguide loss and coupling loss are balanced). For comparison with data, the corresponding curve for this case is the dash-dot curve. Critical coupling also coincides in the case of no modal coupling with the point of maximum circulating power.

In the presence of modal coupling, critical coupling, as defined by vanishing waveguide transmission $T = 0$, can still be achieved, however only for symmetric excitation $\Delta\omega = 0$. The transmission and reflection for $\Delta\omega = 0$ are given by the expressions:

$$T = \left[\frac{\Gamma^2 + (1 - K)(1 + K)}{\Gamma^2 + (1 + K)^2} \right]^2, \qquad R = \left[\frac{2 \cdot \Gamma \cdot K}{\Gamma^2 + (1 + K)^2} \right]^2. \qquad (11)$$

It can be shown that the waveguide transmission vanishes identically for:

$$K \equiv \left(\frac{\tau_{\mathrm{ex}}}{\tau_0} \right)^{-1} = \sqrt{1 + \Gamma^2}. \qquad (12)$$

The critical coupling point is therefore strongly dependent on the modal coupling and occurs in what would, under normal circumstances, be considered the over-coupled regime (i.e. for $\tau_{ex} < \tau_0$). The amount of over-coupling required for zero transmission increases monotonically versus modal coupling. (As an aside, we note that the condition $K = 1$ has the special property of causing a transmission and reflection of equal magnitude regardless of inter-mode coupling strength.)

The shifted critical point also coincides with the point of maximum reflection, which is given by:

$$R_{\text{CRIT}} = \left(\frac{\Gamma}{1 + \sqrt{1 + \Gamma^2}} \right)^2. \tag{13}$$

In the limit of large Γ the reflection approaches unity, and all incoming waveguide power is reflected so that the system behaves as a "frequency-selective reflector". We have experimentally verified the above equation as shown in Fig. 13.

The inset shows measured transmission and reflection spectra at the critical point, where the microcavity behaves as a frequency selective reflector. These particular spectra were measured using a 30 μm-diameter microsphere exhibiting an inter-mode coupling parameter of $\Gamma = 30$ (the highest observed in our measurements). This large modal coupling

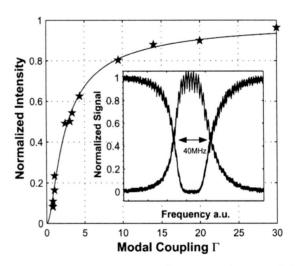

Fig. 13.   Stars are the measured maximum reflection as a function of the normalized intermode coupling parameter and the solid line is a theoretical curve using the maximum reflection formula given in the text. The inset shows measured spectral reflection and transmission properties for the case of $\Gamma = 30$.

parameter implies a strong likelihood for photon scattering into the counter-propagating mode versus loss to numerous other possible mechanisms. We believe this could in part be related to suppression of certain forms of scattering loss by the low angular mode density made possible in the whispering gallery. Only light that is scattered into an angular segment which exceeds the modal cutoff angle is lost; all the remaining scattered power is channeled back into clockwise and counterclockwise propagating WGMs, thereby inducing modal coupling. In agreement with this conjecture, large modal coupling parameters were only found in spheres typically smaller than $40\,\mu$m. The origin of strong modal coupling was also investigated by analyzing the sphere's surface with scanning electron microscopy. Spheres exhibiting strong modal coupling were found to contain small, isolated and randomly-distributed contaminants on their surface. These had dimensions of typically hundreds of nanometers. For spheres with negligible mode splitting these structures were absent. The presence of isolated sub-wavelength-scale scattering centers can also be a possible explanation for the occasional observation of asymmetric mode splitting. The standing waves that are present in the regime of strong modal coupling, probe different regions of the sphere's surface, which can lead to different loss rates when scattering is caused by highly localized structures. This, in turn, will cause an asymmetry in the line-shapes of the doublet structure.

As noted above maximum power transfer does not occur at the critical coupling point in the presence of backscattering. On the contrary, it occurs for finite transmission. However, analysis reveals that, unlike the effect backscatter induces in shifting the critical point, the point of maximum power transfer does not significantly change. Indeed, the largest shift occurs for a modal coupling parameter of $\Gamma = 1.5$ at which the maximum power transfer point shifts from $K = 1$ to $K = 1.52$. It is interesting to note that for large modal coupling ($\Gamma \gg 1$) the maximum power transfer condition again approaches the condition $K = 1$, identical to the condition with no modal coupling ($\Gamma = 0$). However, the maximum circulating power in the case of strong modal coupling is reduced by a factor of two with respect to the ideal case, due to the reflection. Figure 14 shows the reduction of circulating power, denoted by $C(\Gamma)$, with respect to the case of no modal coupling as a function Γ. The location of the point of maximum power transfer and the reduction of circulating power will become important in the subsequent sections, when analyzing threshold behavior for nonlinear optical effects.

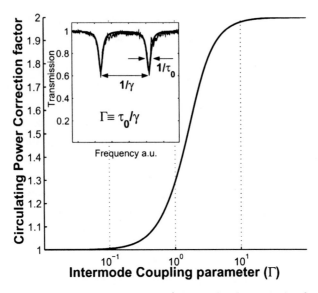

Fig. 14. Reduction of the circulating power (as given by the correction factor C) as a function of the intermode coupling parameter. In the case of strong modal coupling, the circulating power is reduced by a factor of 2 due to coupling of the resonant cavity field along both directions of the tapered fibers.

4. Mode Structure and Q-Factors of Toroid Microcavities On-a-Chip

4.1. *Q-Factors of Microtoroids*

The Q-factor of the microtoroid modes were characterized using two different techniques. First, we performed linewidth measurements by scanning a single frequency laser through the resonances in the undercoupled regime. The wavelengths measured were in the 1550 nm band. The ultra-high-Q modes typically exhibited a doublet structure and the scans were fit to the coupled-mode model of the previous section and the intrinsic lifetime inferred. The narrowest UHQ linewidths measured are in fact comparable to the short-term linewidth of the external cavity laser used in these measurements (300 kHz short-term linewidth). For this reason, as both an independent and more precise means to measure linewidth, we also measured the Q-factor by cavity ring-down.[18] In this technique, the laser was scanned successively through a frequency interval containing the resonance to be measured. The scan rate was adjusted to be slow on the timescale of the mode-lifetime so as to prevent linewidth distortion due to transient effects. The taper coupling strength was adjusted to the critical point. As the laser progressively scanned, the mode was excited until, at a point near the line center at critical coupling (measurable by setting a trigger level using the detected reflected power), a high-speed Mach–Zehnder-type amplitude modulator was used to "gate off" the input laser. With the input laser signal extinguished, the only detectable signal is due to the cavity leakage field which then decays exponentially with a time constant that gives the loaded Q factor for the mode at the critical point. Figure 15 shows a typical experimental trace in the ring-down measurement. At time $t = 0$ the laser field is gated off and an exponentially decaying field, as is expected for a single mode, can be observed. The deviation of the transmission from unity at $t = 0$ is due to the gating fall time (approximately 8 ns). The inset of Fig. 15 shows the signal on a logarithmic plot to infer the lifetime and to confirm the single-exponential behavior of the decay signal. A fit to the data in the figure yields an intrinsic lifetime of 75 ns.

As noted above, this value corresponds to a loaded cavity Q. To infer the intrinsic Q from this measurement one must take into account the waveguide loading. For an ideal traveling wave WGM the lifetime at the critical point is precisely half of the intrinsic lifetime. However, as noted earlier, in the UHQ regime the WGMs are typically split into doublets due to weak backscattering. From the previous section, the critical point

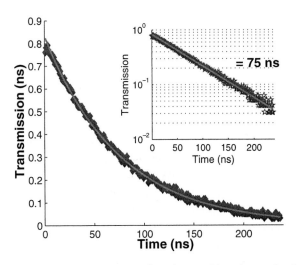

Fig. 15. Cavity ring-down measurement of a microtoroid cavity mode. At time $t = 0$ the mode is fully charged at the critical point and the laser is gated off. The inset shows the log-scale plot of the data confirming the single-exponential decay.

in the presence of intermode coupling occurs in what would (under zero-backscatter conditions) be referred to as the over-coupled regime. As a result, the actual intrinsic cavity Q is related to the critical-point lifetime by the expression:

$$Q_0 = \omega\tau_0 = \omega\tau_{\text{crit}}(1 + \sqrt{1 + \Gamma^2}). \tag{14}$$

The Γ factor can be measured as described in the previous section by measuring the normalized splitting in the under-coupled regime. For the ring-down shown in Fig. 15, a modal coupling of $\Gamma \approx 3.7$ was present yielding an intrinsic cavity Q of 4.3×10^8.

Generally, thermal effects are present in UHQ silica devices and induce distortion of the lineshape unless input power levels are adjusted to a suitably low value. These distortions tend to make the apparent linewidth larger (pull the center frequency of the mode) in one scan direction while making its appearance narrower in the opposing scan direction. These artifacts present another challenge to Q measurement based upon linewidth measurement alone.

Finally, there is an alternate method for determining the intrinsic Q factor in cases where there is strong intermode coupling induced by backscatter. This approach does not require knowledge of Γ, but rather only the doublet splitting frequency (γ^{-1}). As before, the ring-down lifetime at the critical coupling condition is measured. However, now the following expression is used to relate both this information and the measured splitting frequency to the intrinsic Q factor.

$$Q_0 = \omega\tau_0 = \omega\frac{2}{\tau_{\text{crit}}}\left(\frac{1}{\tau_{\text{crit}}^2} - \frac{1}{\gamma^2}\right)^{-1}. \tag{15}$$

This method is less sensitive to thermal effects since the splitting frequency has been observed to be nearly immune to thermal shifts (assuming that each mode is affected nearly equally by the excitation wave[41]).

In our work we have measured Q factors as high as 500 million in microtoroids and Q factors as high as 2 billion in spheres (measured in the 1550 nm wavelength band). While the intrinsic absorption of silica at 1550 nm in principle allows for Q factors of up to 10^{12}, water adsorption and surface scattering losses limit the observed Q to much lower values. Previous studies on microspheres,[44,45] have identified water adsorption on the surface as a major limiting factor of Q. Adsorption of OH molecules has been observed to occur within minutes after the fabrication procedure. The Q factors in the present work are also believed to be limited by a combination of surface adsorption of OH groups and surface scattering centers on the microcavity. Since all measurements reported here were performed in ambient conditions and after a time interval exceeding 5 min after microtoroid fabrication, water adsorption is likely to affect the Q as described in Ref. 1.

Finally, dielectric microcavities do not, in fact, provide strict confinement of modes as WGMs are always subject to tunnel leakage into the continuum (sometimes called whispering gallery loss). This leakage rate is primarily a function of the dielectric index-of-refraction as well as the diameter of the WGM orbit. Intuitively, this loss is like the bending loss in a dielectric waveguide. For microspheres the whispering gallery loss has been calculated using asymptotic expansions.[46] In silica spheres having Q values of 100 million in the 1550 nm band, this loss becomes a determining factor in degrading Q for radii less than 11 μm.[26] To investigate the influence of radiation loss on microtoroids we have measured the Q factor as a function of the principal microtoroid radius. The toroid minor radius was approximately constant in this experiment (ca. 8–10 μm). Figure 16 shows the result of this study.

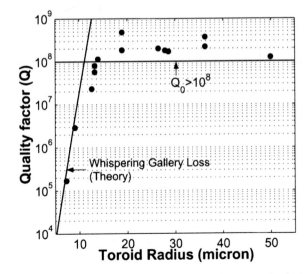

Fig. 16. Measured quality factor of microtoroids as a function of principal radii. For radii smaller than 15 μm the Q factor decreases below 10^8 due to radiation loss. The solid vertical line represents the tunnel loss for the case of spherical whispering gallery modes. The horizontal solid line is a guide to the eye set at a Q value of 100 million.

The results indicate that whispering gallery loss is negligible for principal tori-radii of more than 15 μm. Below this value the Q factors are seen to decrease as a function of radius and to approach the theoretically calculated limit of whispering gallery loss for a sphere. In addition we have varied the minor toroid diameter to a range of ca. four wavelengths ($\sim$3.8 μm) and measured Q values of more than 10^8.

4.2. *Mode Structure of a Toroid Microcavity*

Figure 17 is a typical transmission spectrum for a taper coupled to a toroidal microcavity. The observed free spectral range corresponds to the modes with successive equatorial (or angular) mode number (ℓ). Successive ℓ-modes exhibit varying amounts of coupling. Inspection of the mode structure shows that the microtoroid supports only very few transverse and radial modes. This is in contrast to microspheres, where each angular mode exhibits a $(2\ell + 1)$-fold degeneracy in the azimuthal direction. The number of transverse modes is expected to be a strong function of the toroid rim thickness (minor diameter).

Whereas the mode structure of microspheres is well known and analytical expressions for the field distribution can be derived, analytical expressions for the toroid geometry are not available as only one coordinate of

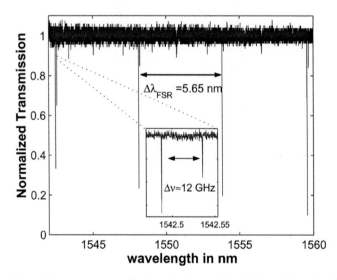

Fig. 17.   Transmission spectrum of a toroid microcavity of 94 μm-diameter. Successive modes are separated by the free spectral range of the cavity.

the wave equation separates, reducing it to a two-dimensional Helmholtz equation. In order to determine the modes of a toroidal microcavity, a finite element eigenmode solver was used to model the 2D cross-sectional geometry after accounting for the rotational symmetry. A full-vectorial method was used which allowed accurate solution of the complete field distribution, confirmed for accuracy using the resonance locations and field profiles of a spherical cavity.[47] Figure 18 shows a representative intensity profile for the fundamental TE mode of a toroid microcavity with a 6 μm minor diameter (principal diameter of 50 μm), compared with that of a microsphere of diameter 50 μm. For comparison, the first higher-order transverse modes are also shown. While degenerate in a perfect microsphere, in this toroid the closest transverse mode is now located 4.6 nm away (approximately half of the free-spectral-range for this principal diameter). This breaking of the azimuthal degeneracy can have a significant effect on the optical properties of these cavities, as discussed in more detail in Sec. 5. Further numerical

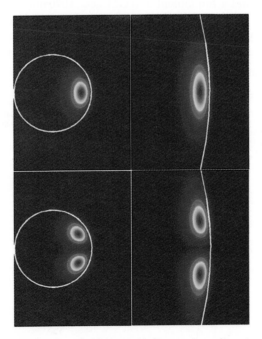

Fig. 18.   Intensity profiles for: (left) a toroid (6 μm minor diameter; 50 μm principal diameter) and (right) a microsphere with principal diameter of 50 μm. Wavelength is assumed to be near 1550 nm. Both the fundamental (top) and first higher-order transverse modes (TE polarization) are shown in each case.

investigation of toroidal modes indicates a number of additional advantages over spherical cavities, including a more regular free-spectral-range, and a decreased modal volume.

4.3. *Mode Volume of a Toroid Microcavity*

In addition to Q, mode volume has a nearly universal influence on micro-cavity performance across a wide range of applications.[5] The mode volume is commonly defined by the spatial integral over the field intensity, normalized to unity at the field maximum.[48] In the case of spherical microcavities the mode volume is readily calculated in terms of the known analytic field distributions. Fundamental-WGM mode volume on the order of $1000\,\mu m^3$ is typical for sphere diameters in the range of $100\,\mu m$ and varies to a good approximation quadratically with sphere diameter over a wide range of values. We have investigated the mode volume of toroidal-microcavities using numerical modeling as described above. Figure 19 shows the calculated mode volume versus toroid minor diameter assuming a $50\,\mu m$ toroid principal diameter and a fundamental WGM near $1570\,nm$. Note that the case of unity aspect ratio (minor diameter of $50\,\mu m$) corresponds to a spherical microcavity. Therefore the plot also provides the extent of mode volume reduction in a toroid relative to a spherical microcavity having the same (principal) diameter.

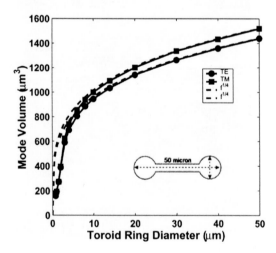

Fig. 19.   The mode volume of a fundamental toroidal WGM as a function of minor diameter for $50\,\mu m$ principal diameter. Results are presented for TE (circles) and TM (squares) modes for a mode near $1570\,nm$. The dashed lines correspond to a fit using a weak confinement approximation described in the text.

From the numerical modeling it is apparent that one can distinguish between both a slow and fast regime of mode volume reduction (or compression) with decreasing toroid minor diameter. In the slow compression regime the minor diameter is much larger than the transverse extent of the mode. A simple analysis of the wave equation reveals that in this case an effective harmonic oscillator potential (in theta) associated with the toroidal curvature determines the transverse extent of the mode in the vicinity of the equator. This mechanism is well known in a sphere.[26] In the limit of this approximation, the mode volume compression relative to the sphere should scale as the 1/4th power of the minor to major diameter. This functional form, as indicated in Fig. 19, is in excellent agreement with the exact calculation over the slow compression regime. A continued reduction then leads to the strong modal compression regime, where the toroid exerts greater influence on the mode volume as the dielectric boundary approaches and closes in upon the mode. Mode volume is reduced until the point where the toroid geometry and associated modes approach those of a step-index fiber. In analogy with a step-index fiber, further reduction of the core area leads to a point where the mode volume is minimal and additional reduction of the minor diameter is accompanied by a decreased mode confinement and increase in mode volume. As described below, we have confirmed the reduction of mode volume experimentally by using the threshold of stimulated Raman scattering to experimentally infer the cavity mode volume.

5. Observation of Nonlinear Oscillations in Ultra-High-Q Silica Microcavities

Ultra-high-Q microcavity modes are ideally suited to study stimulated and parametric optical nonlinear processes. The microcavity concentrates input power at resonant frequencies by recirculating it within a small modal volume. Also, by restricting oscillation to resonant frequencies, it can suppress the occurrence of certain nonlinearities unless specific conditions are satisfied. Nonlinear optics in liquid microdroplets has been pioneered by Campillo[13] and Chang[10] and an extensive discussion on this topic can be found in Ref. 4. Despite their unique properties (UHQ and small modal volume) microdroplets have remained a laboratory tool, due to their transient nature which makes stable and long-term study difficult. In this section we present our results concerning cavity nonlinear optics in UHQ microsphere and microtoroid resonators. These microcavity geometries allow

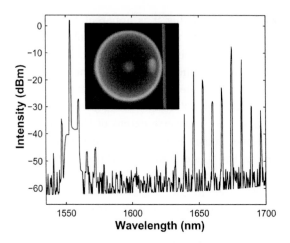

Fig. 20.   Nonlinear emission spectrum from a UHQ microsphere cavity. The pump is at 1550 nm and the peaks in the 1600 band are due to Raman oscillation. The peaks appearing symmetrically around the pump wavelength are due to Raman-assisted four-wave-mixing. The inset is a micrograph showing a fiber-taper-coupled microsphere.

stable observation of nonlinear effects and in addition are compatible with fiber-optic technology, as shown in Fig. 20.

The combination of UHQ and efficient fiber-taper coupling allows the observation of stimulated Raman scattering at unprecedented threshold powers[8] that are nearly a factor of 1000 lower than prior work using droplet microcavities. In addition to their ultra-low threshold, these devices are, from a practical standpoint, compact, fiber-compatible and exhibit high pump-to-Raman conversion efficiency (more than 45% has been measured). Due to the high conversion efficiency, the Raman mode can also build-up to appreciable intensities and exceed the threshold for higher-order Raman generation, which will be discussed in Sec. 5.2.

Whereas stimulated Raman scattering is intrinsically phase matched, parametric oscillation due to the silica Kerr nonlinearity, requires strict phase matching conditions to be satisfied. In Sec. 5.3 we show how by controlling the cavity geometry a transition from stimulated to parametric oscillation can be achieved, enabling the first demonstration of a microoptical parametric oscillator.

5.1. *Stimulated Raman Scattering in Microcavities*

Raman scattering is an inelastic light scattering mechanism[49,50] in which a photon is converted to a lower energy (Stokes) photon via the emission of an optical phonon. Raman scattering is of considerable technological and scientific importance across a range of topics including spectroscopy, biochemical sensing and quantum optics.[51,52] Also, due to the broad Raman gain bandwidth in silica[25] (extending over ca. 10 THz relative to the pump frequency with the maximum gain occurring at a Stokes shift of ca. 14.7 THz or 440 cm^{-1})[53] optical-fiber-based Raman amplifiers can provide broadband gain in almost any wavelength band[54] and can be employed to extend the wavelength range of amplifiers used in communications systems or semiconductor laser signal sources themselves.[55] There has been extensive research and progress in the demonstration of fiber-Bragg-grating and Fabry–Perot-type (e.g. those employing gas cells) Raman sources.[56] However, for silica-based devices high pump powers are typically required.[55] In this section we demonstrate the use of STIMs as efficient, compact and ultra-low threshold Raman lasers.[8] Despite the low nonlinearity of silica, these devices exhibit exceptionally low thresholds for Raman oscillation. By adjusting the taper microcavity gap we have investigated the influence of coupling on threshold, conversion efficiency and the ability to generate Raman cascades. In particular, the ability to strongly over-couple these cavities enables achievement of very high differential conversion efficiencies.

Before describing experimental results, a review of the essential physics governing Raman lasers will be presented. Raman scattering in a waveguide-coupled UHQ microcavity can be described using the coupled-mode equations shown below with nonlinear Raman terms[24] coupling the pump and subsequent Raman orders. Other nonlinear terms associated with processes such as Brillouin scattering or the Kerr nonlinearity are not considered, since their presence in a microcavity is strongly restricted as discussed at the end of this section.

$$\frac{dE_p}{dt} = -\left(\frac{1}{\tau_0} + \frac{1}{\tau_{ex}}\right)_P E_p - \left(\frac{\omega_p}{\omega_{R1}}\right) g_{R1}|E_{R1}|^2 E_p + \kappa \cdot s$$

$$\frac{dE_{R(N-1)}}{dt} = -\left(\frac{1}{\tau_0} + \frac{1}{\tau_{ex}}\right)_R E_{R(N-1)} + g_{R(N-2)}|E_{RN-2}|^2 E_{R(N-1)}$$

$$- \left(\frac{\omega_{R1}}{\omega_{R2}}\right) g_{RN}|E_{RN}|^2 E_{R(N-1)}$$

$$\frac{dE_{RN}}{dt} = -\left(\frac{1}{\tau_0} + \frac{1}{\tau_{ex}}\right)_{RN} E_{RN} + g_{RN}|E_{R(N-1)}|^2 E_{RN}.$$

(16)

In these equations, "E_i" are the amplitudes of the intra-cavity pump and Raman fields and "s" denotes the input from the waveguide, normalized such that input power is the square of the amplitude s, $P_{in} = |s|^2$. The excitation frequency of the pump mode and the resonant Raman modes are given by ω_p and ω_{RN}, respectively, with τ_0 and τ_{ex} giving the intrinsic and external photon lifetimes (i.e. decay time into the waveguide), as introduced earlier. The input field "s" is assumed to be resonant with the pump-WGM. The nonlinear Raman terms are expressed by the intracavity gain constant g_R which is related to the more commonly used bulk Raman gain coefficient g_R^{Bulk} (in units of m/W)[57] by (c is the speed of light in vacuum, and n the index of refraction):

$$g_R = C(\Gamma)^{-1}\frac{c^2}{2n^2}\left(\frac{1}{V_{\text{eff}}}\right)_R \cdot g_R^{\text{Bulk}}. \tag{17}$$

In this relation, the reduction of the cavity buildup factor in the presence of backscatter coupling is described by a circulating power correction factor $C(\Gamma)$ as introduced in Sec. 3.2. In silica the bulk maximum Raman gain coefficient is 10^{-13} m/W at a 1000 nm wavelength.[58] The spatial variation of the Raman gain across the mode profile due to the intensity dependence can be accounted for by introducing the effective mode volume.[55]

$$V_{\text{eff}} = \frac{\int |E_p|^2 dV \int |E_R|^2 dV}{\int |E_p|^2 |E_R|^2 dV}. \tag{18}$$

This definition of effective mode volume deviates from the previously introduced energy-density related definition, because it takes into account the spatial overlap of the pump and Raman modes.[55] For the microtoroid geometry, numerical modeling shows that improved modal confinement leads to a reduced effective mode volume, in comparison to a spherical microcavity.

The threshold for first-order Raman scattering can be derived by considering the steady-state solution to the coupled-mode equations. The threshold formula for first-order Raman scattering can be factorized into contributions of modal volume, coupling dependence (described by coupling parameter K) and the Q factors of the Raman and pump modes.

$$P_{\text{thresh}}^{N=1} = C(\Gamma)\frac{\pi^2 n^2}{\lambda_p \lambda_R g_R} V_{\text{eff}} \left(\frac{1}{Q_0}\right)_P \left(\frac{1}{Q_0}\right)_R \frac{(1+K_P)^2}{K_P}(1+K_R). \tag{19}$$

The threshold power has an inverse square dependence on the quality factor (assuming equal pump and Raman Q factors) and scales linearly with mode volume. Under the assumption of equal Raman and Pump quality factors, the minimum threshold occurs for $K = 1/2$ (equivalently $Q_{\text{ex}} = 2 \cdot Q_0$), which corresponds to an under-coupled pump with a waveguide transmission of $T = 1/9$ (or approximately 11%).

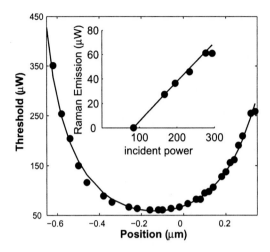

Fig. 21.   Threshold for stimulated Raman lasing in a $40\,\mu$m-diameter sphere ($Q_0 = 10^8$) as a function of taper-microcavity gap with respect to the pump critical-coupling point. The minimum threshold occurs in the under-coupled regime with 12% pump transmission in close agreement with theory. The inset shows the Pump to Raman conversion, exhibiting a bidirectional conversion efficiency of approximately 36% at minimum threshold.

We have experimentally verified the dependence of the threshold on the coupling strength. Figure 21 shows the Raman threshold measured for various taper to sphere coupling gaps using a $40\,\mu$m-diameter microsphere excited at 1550 nm. The minimum threshold for this microsphere was $62\,\mu$ Watts and occurs for a gap shifted from the critical point by $0.15\,\mu$m into the under-coupled regime (the zero position in Fig. 21 corresponds to the pump-wave critical point). The residual pump transmission at this point was measured to be 12%, in good agreement with the theoretically predicted value of 11%. The observed threshold value is in excellent agreement with the predicted value using the experimentally determined parameters $\Gamma = 3$, $R = 40\,\mu$m, and $Q = 10^8$. Figure 22 shows the Raman emission in the 1650 nm band for oscillation on a single longitudinal (i.e. equatorial or ℓ) mode. Higher-resolution inspection of the emission peak revealed that 3–5 azimuthal modes (m modes) were lasing simultaneously (upper inset of Fig. 22).

Using the coupled-mode equations, the total, first-order Raman output power (which is coupled equally into forward and backward directions of the propagating taper mode) is given by:

$$P_{\text{Raman}} = 4\frac{\omega_R}{\omega_P}\left(1 + \frac{\tau_{\text{ex}}}{\tau_0}\right)^{-2} P_{\text{thresh}}\left(\sqrt{\frac{P}{P_{\text{thresh}}}} - 1\right). \qquad (20)$$

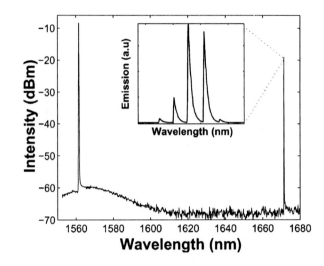

Fig. 22.   Single, longitudinal (i.e. equatorial) mode Raman lasing for a microsphere of 40 μm diameter. The threshold was measured to be 90 micro Watts. The inset provides a higher resolution scan and shows that 3–5 azimuthal WGM modes were simultaneously lasing.

It is interesting to note that the first order Raman output power follows a square-root behavior with launched pump power, as opposed to being linear. The physical origin of this behavior can be viewed as a "pumping inefficiency" resulting from the dependence of Raman gain on the internal pump wave. This effect can be illustrated by considering operation at the critical point, where the cavity leakage field interferes destructively with the transmitted pump field, causing zero transmission. However, once the onset of Raman lasing is reached, the cavity pump field is clamped at the threshold value and a subsequent increase of pump power will imbalance transmitted pump and cavity leakage fields, giving rise to a finite pump transmission. This "pumping inefficiency" causes the pump power coupled to the cavity to follow the above square-root-behavior on launched fiber power. As the Raman power scales linearly with coupled cavity power, the Raman emission also exhibits a square root behavior on pump power, resulting in the behavior of Eq. (20). We have experimentally confirmed this dependence as shown in the next section (Fig. 28). For large pump power the first-order Raman output saturates due to the presence of higher-order Raman scattering and will be discussed in the next section.

The bidirectional *external* differential pump-to-Raman conversion efficiency is given by linearizing Eq. (20) at threshold:

$$\eta_{ex} = 2 \cdot \frac{\omega_R}{\omega_P} \left(1 + \frac{\tau_{ex}}{\tau_0} \right)^{-2}. \qquad (21)$$

Figure 23 plots this efficiency as a function of coupling strength. At the point of minimum threshold ($K = 1/2$ assuming equal pump and Raman Q factors) the expected efficiency is 22%. It is 50% at the critical coupling point. The actual measured value in the experiment shown in Fig. 21 was 36%, which is slightly higher than the theoretically expected value, believed to be due to differences in pump and Raman Q, and due to the presence of modal coupling. The external differential efficiency approaches the value of $2\omega_R/\omega_P$ in the limit of strong over-coupling (i.e. $\tau_0/\tau_{ex} = \infty$). Surprisingly, this value exceeds unity, indicating that on average every added waveguide pump photon above threshold is converted to more than one Raman photon. This result might seem nonphysical, but can be explained when considering the nonlinear dependence of coupled pump power. As the loading experienced by the intra-cavity pump wave is dependent upon the Raman wave once threshold is reached (i.e. the intra-cavity pump field is clamped above threshold), a differential increase in waveguide pump will

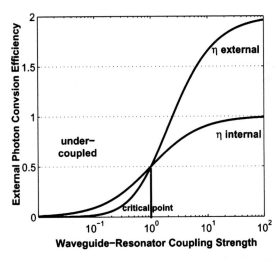

Fig. 23.   Internal and external pump-to-Raman differential quantum efficiency as a function of waveguide coupling. The internal and external efficiency are equal at the critical coupling point. In the limit of strong over-coupling the internal efficiency approaches unity, whereas the external efficiency approaches 2.

cause improved coupling to the resonator when the resonator and waveguide are over-coupled. This occurs since the additional loading induced by the, now higher-power, Raman wave shifts the pump wave closer to the critical point. As described above, the opposite effect occurs in the under-coupled regime. This nonlinear loading effect leads to the interesting phenomenon, that the external differential quantum efficiency can exceed unity. In contrast, when analyzing the *internal* pump-to-Raman differential conversion efficiency (i.e. actual-coupled pump to Raman efficiency) one obtains the expression:

$$\eta_{\text{int}} = \frac{\omega_R}{\omega_P}\left(\frac{\tau_0}{\tau_{\text{ex}}}\right)\left(1 + \frac{\tau_0}{\tau_{\text{ex}}}\right)^{-1}.\tag{22}$$

As expected, the internal conversion efficiency approaches the value $(\omega_R/\omega_P < 1)$ in the limit of strong over-coupling. This quantity is also plotted versus waveguide-resonator coupling in Fig. 23.

As noted above, the threshold scales linearly with the modal volume. The mode volume of a silica microsphere WGM scales approximately quadratically with radius. As an aside, for smaller spheres the mode volume deviates from this behavior and ultimately, for very small diameters, increases due to weakening of the whispering gallery confinement.[46] The minimum mode volume occurs for a diameter of 6.9 μm (for $l = m = 34$)[46] for 1550 nm wavelength (mode volume is $V = 173.1\,\mu m^3$). However, this size is not optimum for stimulated Raman scattering as the additional benefit of reduced mode volume is more than offset by the significant decrease in Q factor to 10^8 (Note: threshold power scales like $\propto Q^2/V$).[26]

We have investigated the dependence of the Raman threshold on mode volume by varying the microsphere radius. To enable comparison of microspheres with varying quality factor Q, modal coupling parameter Γ, Raman and pump wavelengths, we have normalized the threshold values using the dependences appearing in the threshold formula. Furthermore, the Q factors of pump and Raman waves are assumed equal. The result of this study is presented in Fig. 24. In microspheres of less than ca. 25 μm diameter, stimulated Raman scattering was not observed, since thermal drifting effects make pumping of the modes unstable and the whispering gallery loss becomes a significant loss mechanism.[26] The normalized threshold indeed exhibits a quadratic dependence (actual inferred exponent was 1.93) on microsphere radius and is in excellent agreement with the theoretical predictions ($\propto R^{1.83}$) for microspheres in this size range.[46]

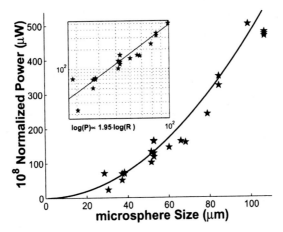

Fig. 24. Normalized Raman threshold versus sphere diameter. The inset shows a linear fit to a double logarithmic plot to infer the exponent (i.e. $V \propto R^n$), which is $n = 1.93$ and in good agreement with the theoretically expected value $n = 1.83$ for this range of sphere sizes.

In previous work on microdroplets it has been found that the Raman gain coefficient inside a microcavity experiences a further enhancement due to the presence of cavity-QED effects.[12,13,59] For the size ranges measured in this work, our experimental data do not reveal such an effect, and the bulk of Raman gain yielded good experimental agreement. This is also confirmed by recent theoretical work.[60]

Other nonlinear effects such as stimulated Brillouin scattering and four-wave mixing exhibit gain constants that exceed the Raman gain.[57] However, in the above experiments stimulated Brillioun scattering was not observed and four-wave mixing occurred only in certain cases (as a Raman assisted process). Despite the larger nonlinear gain coefficient of Brillouin[54] scattering (by a factor of 500 compared to Raman), the narrow Brillouin gain bandwidth (ca. 100 MHz) poses strong restrictions on the occurrence of this process in a microcavity due to the necessity of doubly resonant cavity modes (pump and Raman). The process is particularly restricted in toroid microcavities due to the low density of modes. Likewise, the additional condition of phase matching imposes strong restrictions on the observation of parametric oscillations due to the Kerr nonlinearity. We will turn to the observation of this interesting nonlinear process in Sec. 5.4.[61] Finally, we note that the power levels for catastrophic self focusing[62,63] are far beyond the intensity levels used in the work presented here.

5.2. *Cascaded Stimulated Raman Scattering in Microcavities*

The high internal Raman conversion efficiency within the microcavity allows the Raman mode to reach power levels sufficient for pumping a subsequent Raman laser within the same microresonator (i.e. cascaded oscillation). This higher-order process is easily observable as shown in Fig. 25, which shows a second-order cascade.

This cascade process can be described by extending the previously introduced model to include higher order Raman terms. The equations can be solved iteratively in steady state, where we have made again the simplifying assumption of equal pump and Raman Q as well as equal pump and Raman coupling properties. The corresponding pump threshold power given below takes on two distinct forms depending upon whether the Raman order is even ($N = 2m$) or odd ($N = 2m + 1$).

$$P_t^{N=2m} = \frac{fC(\Gamma)\pi^2 n^2}{\lambda_p \lambda_R g_R} V_{\text{eff}} \left(\frac{1}{Q_0}\right)^2 \cdot \left(\sum_{i=0}^{m-1} (c_i)^i\right) \left(\sum_{i=0}^{m} (c_i)^i\right)^2$$

$$P_t^{N=2m+1} = \frac{fC(\Gamma)\pi^2 n^2}{\lambda_p \lambda_R g_R} V_{\text{eff}} \left(\frac{1}{Q_0}\right)^2 \cdot \left(\sum_{i=0}^{m} (c_i)^i\right)^3 .$$

$$(23)$$

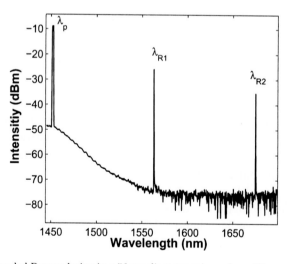

Fig. 25.   Cascaded Raman lasing in a 56 μm diameter microsphere. The pump is located at 1450 nm and the first- and second-order Stokes waves are present around 1580 and 1660 nm.

The coefficient c_i is defined as:

$$c_i \equiv \frac{\omega_i}{\omega_{i+1}} \cdot \frac{g_{i+1}}{g_{i+2}} = \frac{\omega_i}{\omega_{i+2}} \cdot \frac{V_{\text{eff}}(\lambda_{i+2})}{V_{\text{eff}}(\lambda_i)} \tag{24}$$

where the Raman gain coefficient[57] dependence has been used in the second equality. For stimulated Raman scattering in silica at optical frequencies the Raman shift is generally small compared to the frequency of the light so that the coefficient $c_i \approx 1$. In addition it has been assumed that the mode volume is independent of wavelength.

Analysis also shows that even- and odd-ordered Raman modes have different pump-to-Raman conversion characteristics. The corresponding Raman output powers for even or odd highest orders are given by the expressions:

$$P^{N=2m+1} = \eta_{\text{ex}}^N 2\left(\sqrt{P_t^N P} - P_t^N \right)$$

$$P^{N=2m} = \eta_{\text{ex}}^N (P - P_t^N). \tag{25}$$

Furthermore, when the highest-order Raman line is of odd-order all even-order modes are clamped and vice versa. Therefore, for odd, highest order, all odd-ordered Stokes modes exhibit a square-root dependence on pump power with even-orders clamped; whereas for even, highest order, all odd-ordered modes are clamped and even-orders increase linearly with pump power. This characteristic is shown graphically in Fig. 26. The corresponding external differential conversion efficiencies are given by the expressions:

$$\eta_{\text{ex}}^{N=2m+1} = \frac{\lambda_p}{\lambda_{RN}} (1 + K^{-1})^{-2} \frac{8}{(N+1)^2}$$

$$\eta_{\text{ex}}^{N=2m} = \frac{\lambda_p}{\lambda_{RN}} (1 + K^{-1})^{-2} \frac{16}{(N+2)^2}. \tag{26}$$

The conversion efficiencies decrease monotonically versus Stokes order and for large values of N follow a $1/N^2$ dependence.

From expression (23) it can be inferred that for higher orders the Raman threshold exhibits an approximate cubic scaling with order ($\propto N^3$):

$$P_t^{N=2m} = \frac{f C(\Gamma) \pi^2 n^2}{\lambda_p \lambda_R g_R} V_{\text{eff}} \left(\frac{1}{Q_0} \right)^2 \cdot \frac{(1+K)^3}{K} \cdot \frac{(N+1)^3}{8}$$

$$P_t^{N=2m+1} = \frac{f C(\Gamma) \pi^2 n^2}{\lambda_p \lambda_R g_R} V_{\text{eff}} \left(\frac{1}{Q_0} \right)^2 \cdot \frac{(1+K)^3}{K} \cdot \frac{N(N+2)^2}{8}. \tag{27}$$

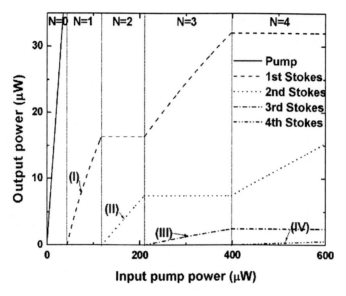

Fig. 26.   Pump-to-Raman conversion characteristic for first- and higher-order cascaded Raman modes. The highest Stokes order is denoted at the top of each region. When not clamped, even-order lines increase linearly and odd-order lines increase as the square root with launched pump power.

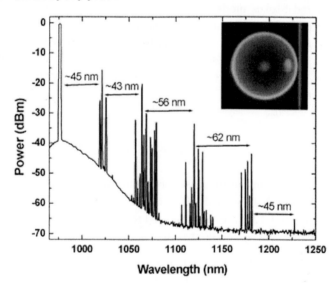

Fig. 27.   Emission spectrum of a microsphere cascaded Raman laser. The pump wavelength is set to 976 nm and approximate Stokes shifts are specified. (Inset) Micrograph of a microsphere-taper system used in the experiment.

To verify these theoretically predicted behaviors we have measured stimulated Raman scattering in microspheres up to fifth-order.[24] The experiments were carried out using pump wavelengths in the 980 nm and 1550 nm bands. Figure 27 contains a spectrum taken using a 980 nm-pump wave. The spectrum contains five distinct orders, each of which contains several oscillating modes. Figure 28 shows the Raman output power for the highest order Stokes wave for cases in which up to four orders are present.[24] The predicted linear and square root behavior of Stokes power with pump power is evident. In the upper left panel of this figure, the apparent flattening of the power is a result of the onset of oscillation in a next, higher order Stokes wave, which causes a clamping of the field.

Theoretical predictions concerning the dependence of threshold and differential conversion efficiency on Stokes order are also in agreement with the measured values as can be seen in Fig. 29.

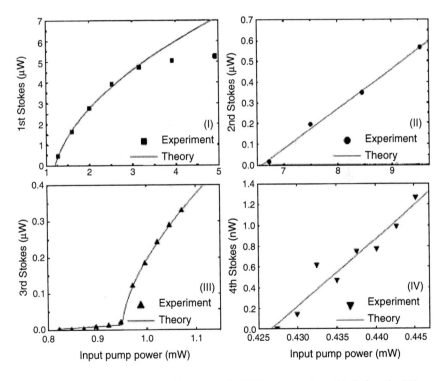

Fig. 28. Stokes lasing powers (first, second: 1550 nm pumping, third, fourth: 980 nm pumping) versus input pump power. Each data set corresponds to measurements taken using a different microsphere.

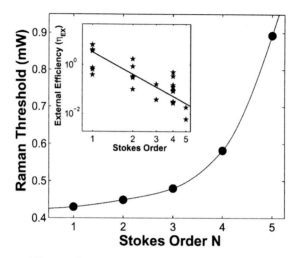

Fig. 29.   Measured Raman threshold as a function of Stokes order exhibiting an approximately cubic dependence on Raman emission order. The inset shows the measured conversion efficiency as a function of order, which decreases as predicted by theory (solid line).

5.3. *Stimulated Raman Scattering in Microtoroids*

The experimental results in the previous section concerned stimulated Raman scattering in spheres. The microsphere geometry does not, however, allow for single-mode Raman lasing, since the mode structure in the azimuthal direction is nearly degenerate and densely populated with modes (i.e. the m-index modes). As noted earlier, chip-based microtoroids possess a number of significant and practical advantages when compared to microspheres, which include a reduced mode density due to severe exclusion of azimuthal degrees of freedom. Furthermore, their chip-based nature offers enormous fabrication advantages in terms of process parallelism and the ability to integrate with electrical or optical functions. In this section we use the ultra-high-Q microtoroids to investigate the performance of the first chip-based micro-Raman lasers. Particular attention is paid to the use of the toroid geometry to achieve reduction of mode volume relative to a microsphere and hence a corresponding reduction in Raman threshold.

Figure 30 shows measured fiber-coupled Raman power versus fiber-launched pump power for a microtoroid-based Raman laser. The device exhibits a differential conversion efficiency of 45%, close to the theoretical prediction for operation at the critical point. Threshold powers as low as $70\,\mu$Watts have been measured using these devices. In contrast to

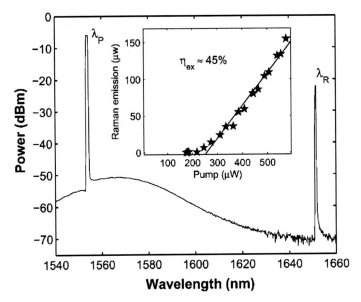

Fig. 30.   Raman laser based upon a microtoroid on-a-chip. The observed emission is in single transverse mode. Inset shows Raman emission versus launched pump power. The conversion efficiency was approximately 45% at the critical point, close to the theoretical optimum.

microspheres, Raman oscillation in microtoroids was consistently observed to be single mode, due to the highly reduced mode spectrum, as shown in the inset of Fig. 30. This degree of modal purity constitutes a significant improvement over previous reports of nonlinear optical processes in ultra-high-Q microcavities and is important for both applied and fundamental applications.

In comparison to microspheres, the toroid cavity geometry allows a higher degree of control over the mode volume, as was explained in Sec. 4.3. In a microsphere mode volume can be controlled only by variation of the principal radius, whereas in a toroid the mode volume can be controlled independently using the principal diameter and minor diameter, as shown in Fig. 31. This has potential advantages in certain applications, such as in nonlinear optical studies, where Q and mode volume must be optimal (i.e. the ratio Q^2/V being maximum). In a microsphere, the onset of whispering gallery loss imposes a lower bound on the mode volume for nonlinear wave generation. For Q values in the range of 100 million, this limit corresponds to a sphere diameter of ca. 22 μm at 1550 nm wavelength[26] (at which point Q becomes radiation loss limited to 100 million). Below this radius any

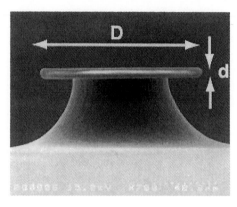

Fig. 31.   SEM side profile of a toroid microcavity showing the principal and minor toroid diameters.

reduction of Raman threshold due to the decreased mode volume is compensated for by increased whispering gallery loss. However, in a toroid of similar size a further reduction of mode volume is allowed by the extra degree of freedom given by the minor toroid diameter. As a result, the performance of a Raman laser on a chip should allow lower thresholds when compared to a comparably sized spherical microcavity and for a size range that is not whispering gallery loss limited.

To investigate the effect of toroid mode volume on Raman threshold, microtoroids were fabricated with approximately constant principal diameter (in the range of 50–65 μm) and varying minor diameter. For a principal diameter of 55 μm, the onset of the strong mode volume compression regime (see Sec. 4.3) is predicted to occur near an 8 μm minor diameter. To isolate the effect of mode volume on threshold power, a normalized threshold power was computed in a manner similar to that used for Fig. 24 in the case of spherical cavity. As before, Q factor (as well as backscatter splitting) was measured in the pump band and the assumption of identical pump and Raman band Q factor was applied. Q factors were measured by cavity ringdown. Because of the substantial difference in frequency of neighboring toroid modes, significant variations in Raman gain are possible and were corrected-for by measurement of the pump and Raman wavelengths for each data point and application of the known Raman spectra of silica.[25] The slight variation in the principal toroid diameter was also taken into account by applying a normalization factor resulting from a numerical model. To infer the effective minor diameter we used a scanning electron microscope (SEM) and recorded the cavity side profile. Figure 31 shows a

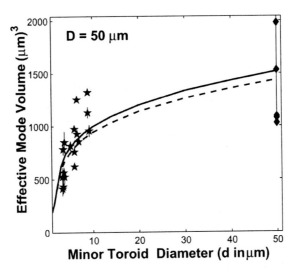

Fig. 32.   The measured effective mode volume for microtoroids (stars) and microspheres (diamonds) as a function of toroid minor diameter. The curves provide the theoretical dependence. The solid line refers to the TM polarization of a fundamental toroidal WGM, while the dashed line designates the TE polarization case.

typical side profile of a microtoroid used in this study. The toroids in most cases exhibit a circular profile, as was confirmed by fitting the SEM profile. However, in some cases a deviation to a more parabolic profile was observed and the effective toroid minor diameter was inferred by curve fitting the equatorial curvature.

A plot of the experimentally determined effective volumes is given in Fig. 32. For comparison the figure also contains data obtained from Raman threshold measurements obtained using a silica sphere of comparable diameter[8] (i.e. equal principal and minor diameter) as well as the numerically predicted dependences of the mode volume for TE and TM polarizations. Variation in the measurement results is most likely due to lack of knowledge of the actual Raman Q factors, however, a good qualitative agreement with the theoretical prediction is obtained.

5.4. *Optical Parametric Oscillations in Toroid Microcavities*

Optical parametric oscillators (OPOs) rely on energy and momentum conserving processes to generate light at new frequencies. In contrast to oscillation based on stimulated gain — e.g. Raman — optical parametric oscillation does not involve coupling to a dissipative reservoir. The

lack of such dissipation makes it uniquely suited for fundamental studies, such as the generation of nonclassical states[64,65] for quantum information research[33] as well as in numerous applied areas (e.g. photonics, spectroscopy, sensing). However, oscillation based on parametric gain requires
stringent phase matching of the involved optical fields.[53] Combined with
the necessity of high field intensity, this poses severe challenges to attaining
microscale optical parametric oscillators. In fact, whereas microscale stimulated nonlinear oscillators have been demonstrated,[4,8,10,13,66-69] microscale
optical parametric equivalents have never been realized. In this section we
demonstrate a microscale optical parametric oscillator (μOPO) using toroid
microcavites.

While ultra-high-Q ensures high circulating field intensities within the
resonator,[70] causing a variety of nonlinear optical effects,[4,8,10,13,66,71] it is
not a sufficient condition to ensure parametric oscillation. Due to inversion
symmetry, the lowest order nonlinearity in silica is the third-order nonlinearity so that the elemental parametric interaction converts two pump
photons (ω_p) into signal (ω_s) and idler (ω_I) photons.[53,65] In order for parametric oscillations to efficiently occur, both energy and momentum must be
conserved in this process.[53,61] For WGM-type modes momentum is intrinsically conserved when signal and idler angular mode numbers are symmetrically located with respect to the pump mode (i.e. $l_{i,s} = l_p \pm N$). Energy
conservation, on the other hand, is not expected to be satisfied *a priori*,
since the resonant frequencies are, in general, irregularly spaced due to both
cavity and material dispersion. As a result, the parametric gain is a function
of the frequency detuning $\Delta\omega = 2\omega_P - \omega_I - \omega_S$, which effectively gives the
degree to which the cavity modes violate strict energy conservation. It can
be shown that the existence of parametric gain requires that this detuning
be less than the parametric gain bandwidth[61]

$$\Omega = 4 \cdot \frac{c}{n}\gamma P, \quad \text{and} \quad \gamma = \frac{\omega}{c} \cdot \frac{n_2}{A_{\text{eff}}} \tag{28}$$

where is γ the effective nonlinearity, n_2 is the Kerr nonlinearity ($n_2 \approx$
$2.2 \times 10^{-20}\,\text{m}^2/\text{W}$ for silica[53]) and P is the circulating power within the
microcavity. To achieve a cavity detuning within the parametric gain band-
width i.e. $0 < \Delta\omega < \Omega$ a reduction of the toroidal cross-sectional area will
produce a two-fold benefit. First, it increases the parametric bandwidth Ω
through its dependence on γ,[61] and second, it reduces $\Delta\omega$. The increase of Ω
occurs, since a decrease in toroid cross-sectional area will reduce the modal
effective area A_{eff} as described in Sec. 4.3. The reduction of $\Delta\omega$ occurs
because of increased modal overlap with the surrounding dielectric medium

(air) and hence flattening of the modal dispersion. Thus, the desired transition can be induced with toroidal geometries of high principal-to-minor toroid diameter (high aspect ratio).

By equating parametric gain and microcavity loss (as given by the loaded cavity Q factor), the necessary launched pump power in the fiber for parametric oscillation threshold is obtained:

$$P_t^{\text{PARAM}} = \frac{\omega_0^2 Q_0^{-2}(1+K)^2 + (\Delta\omega/2)^2}{\gamma\Delta\omega \cdot c/n_{\text{eff}}} \cdot \left(\frac{\pi^2 R n_{\text{eff}}}{2C(\Gamma)\lambda_0} \frac{(K+1)^2}{K} \cdot \frac{1}{Q_0} \right). \quad (29)$$

Here K is again the coupling parameter (as defined in Eq. (10)), R is the principal cavity radius, n_{eff} is the effective index and $C(\Gamma)$ is the modal coupling correction factor as shown in Fig. 14 in Sec. 3.1. Figure 33 shows both parametric and Raman oscillation regimes as a function of the detuning frequency $\Delta\omega$ and the coupling parameter K. The threshold pump power for

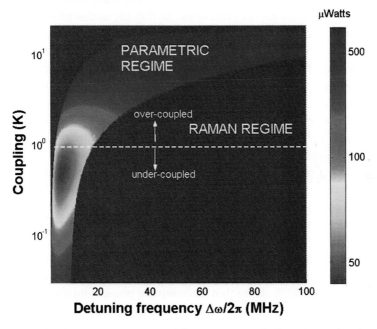

Fig. 33. Nonlinear processes in a toroidal microcavity with $D = 50\,\mu\text{m}$, $d = 4\,\mu\text{m}$ and $Q_0 = 10^8$. The vertical axis denotes coupling strength K of the waveguide-resonator system while the horizontal axis denotes frequency detuning (i.e. $\Delta\omega = 2\omega_P - \omega_I - \omega_S$). The dotted line corresponds to the critical coupling point. The dark blue region denotes areas where Raman oscillation occurs. The color-coded region corresponds to the parametric oscillation regime (where the parametric threshold is indicated by color in micro Watts).

parametric oscillation is color-coded as indicated. As stimulated Raman scattering does not depend on the detuning frequency (i.e. it is intrinsically phase matched; compare the threshold Eq. (19) in Sec. 5.1), it is the dominant nonlinear mechanism by which light is generated for large detuning values. With decreasing $\Delta\omega$ (or equivalently increasing toroidal aspect ratio D/d), a transition from stimulated to parametric regimes occurs when the threshold for parametric oscillation falls below that for Raman. Also note that for increased waveguide loading (and hence correspondingly higher threshold pump powers) the transition can be made to occur for detuning values that are progressively larger.

In order to confirm this prediction, the nonlinear optical properties of toroid microcavities with an approximately constant principal diameter ($50\,\mu\text{m} \leq D \leq 70\,\mu\text{m}$) and varying minor diameter (d) were measured. Due to the ultra-high-Q of the toroidal whispering gallery modes (WGMs), ultra-low-threshold, stimulated Raman scattering was consistently observed for toroids having an aspect ratio of ca. $D/d < 15$, as described in Sec. 5.3.

For microtoroids having an aspect ratio (D/d) in excess of ca. 15 a transition (and a subsequent quenching of Raman[50,72]) to parametric oscillation was observed. Figure 34 shows a parametric oscillation emission spectra for a microtoroid with $d = 3.9\,\mu\text{m}$, $D = 67\,\mu\text{m}$ and $Q_0 = 0.5 \times 10^8$. In this measurement a single fundamental WGM of a microtoroid is pumped in the telecommunication window near 1550 nm using tapered optical fiber waveguides.[19,20] The parametric interaction in the microcavity causes emission of copropagating signal and idler modes, which are coupled into the forward direction of the tapered fiber. Some residual signal and idler reflection were detected in the backward direction due to the presence of modal coupling, induced by backscattering.[21] The generated signal and idler modes had identical oscillation threshold, within the experimental resolution set primarily by taper coupling variations (ca. $\pm 5\%$).

Figure 35 shows the parametric oscillation threshold as a function of taper-toroid coupling gap for the toroid microcavity in Fig. 34. Analysis of the threshold Eq. (29) shows that the coupling point of minimum threshold is a function of the detuning frequency. At the optimum frequency detuning (i.e. maximum parametric gain, for $\Delta\omega_{\text{opt}} = 3 \cdot \omega \cdot Q^{-1}$), the minimum threshold occurs under-coupled for $K = 0.5$ with finite pump transmission ($T = 1/9$), as was also the case for stimulated Raman scattering considered in Sec. 5.1 (compare with Eq. (19) for the SRS case). For larger detuning the minimum threshold point shifts towards being over-coupled (compare with Fig. 33). The measured minimum threshold in the figure was $339\,\mu\text{Watts}$

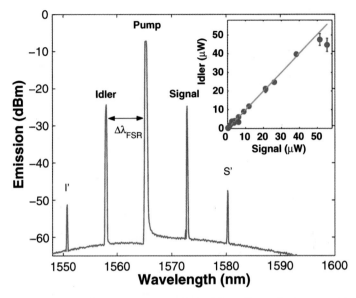

Fig. 34.   Parametric-oscillation spectrum measured for a 67 μm-diameter toroidal micro-cavity. The pump is located at 1565 nm and power levels are far above threshold. The signal and idler are modes with successive angular mode numbers and are spaced by twice the free spectral range (2 × 7.6 nm). The subsidiary peaks (denoted I', S') only appeared at high pump powers and are due to a combination of nonlinear effects, such as parametric oscillation (of signal and idler) as well as four-wave-mixing involving the idler, pump and signal. (Inset) Idler emission power plotted versus the signal emission power, recorded for different pump powers. The signal-to-idler power ratio is 0.97 ± 0.03. For higher pump powers deviation is observed due to appearance of secondary oscillation peaks (I', S') (compare with main figure).

and occurred for the taper displaced by 0.04 μm into the under-coupled regime. The corresponding pump transmission was $T \approx 4\%$ ($K = 0.7$), indicating that the frequency detuning is close to being optimum.

Above threshold, the signal and idler fields increase approximately linearly with pump power (for high pump power P, the emission scales $\propto \left(\sqrt{P/P_t - 1} \right)$ which are due to clamping of the intra-cavity pump field, as described in Sec. 5.1). The inset of Fig. 35 shows a pump-to-idler conversion characteristic at the point of minimum threshold. The corresponding external differential conversion efficiency (compare with Sec. 5.1) was 17% pump-to-idler while the total differential conversion efficiency of pump to both signal and idler fields was 34%. Comparison of the conversion efficiency to theory $2(1+K^{-1})^{-2}$ is consistent with the minimum threshold occurring under-coupled, as the inferred coupling point is $K \approx 0.7$ (corresponding to

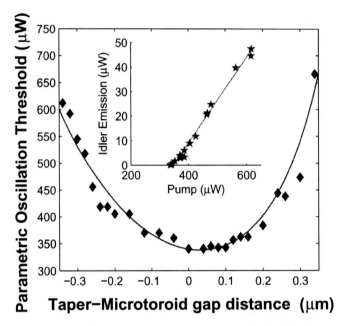

Fig. 35. The coupling-gap-dependence of the parametric threshold with respect to the critical coupling point measured using a 67 μm-diameter toroid microcavity. The minimum threshold occurs with the tapered optical fiber 0.04 μm under-coupled (with finite transmission of ca. 4%). The solid line is a theoretical fit using the threshold equation. (Inset) Idler emission versus pump power. The differential conversion efficiency from pump-to-idler was ~17% (and correspondingly 34% for pump-to-signal and idler).

$T = 4\%$ in agreement with the above measured value). Using the minimum threshold data and the cavity ringdown measurements, a detuning frequency of $\Delta\omega/2\pi \approx 24$ MHz is inferred from the threshold equation, compared with $\Delta\omega_{opt}/2\pi = 11.6$ MHz for optimum detuning frequency at the measured Q value. The lowest measured parametric oscillation threshold for the microtoroids in this study was 170 micro-Watts of launched power in the fiber (for a microcavity with $D/d = 16$, $d = 4\,\mu m$, $Q_0 = 1.25 \times 10^8$ and $\Delta\omega/2\pi \approx 18$ MHz) and is a factor of 200 lower than for fiber OPOs[73] that utilize the remarkable dispersion control provided by photonic crystal fiber.[74]

As a further independent confirmation that indeed the emission can be attributed to parametric oscillation, both signal and idler emissions were recorded simultaneously for varying pump power. From theory, a signal-to-idler photon creation ratio of unity is expected for parametric oscillation.[53] The inset of Fig. 34 shows the measured idler emission power plotted versus signal emission power through the optical fiber taper. The measurements were corrected for modal coupling[21] by measuring the reflected power for all three resonances at the critical point, using the procedure described in Sec. 3.1. After correcting for modal coupling, the ratio of signal-to-idler conversion was 0.97 ± 0.03. The observation of near-unity signal-idler emission ratio, combined with the observed identical threshold for signal and idler modes, demonstrates that the observed emission bands can solely be attributed to Kerr-induced microcavity parametric oscillation. This observation constitutes the first report of optical parametric oscillation in a microcavity.

In addition to the highly advantageous practical aspects of on-chip microcavity nonlinear oscillators, such as wafer-scale integration and control, compact form factor and possible integration with other functions, these oscillators exhibit important properties due to the nature of the underlying nonlinear process within the microcavity. Specifically, a phase-sensitive amplification process that can exclude competing Brillouin or Raman processes, as demonstrated here, can provide an excellent candidate system for the generation of nonclassical states of light[36,64,65,75-80] on a chip. Whereas the work presented here has used the third-order nonlinearity of silica itself, it should also be possible to induce second-order nonlinear interaction (such as parametric down-conversion), by using ultraviolet[81] or thermal-electric[82] glass poling techniques. This would be important in quantum information[33] and quantum optical studies[65,83-85] as well as for novel bioimaging schemes based on entanglement.[86]

6. Surface Functionalization

In contrast to the intrinsic nonlinear optical properties of silica, the surface of silica STIMs can also be modified to obtain extended optical functionality. This can be achieved in many ways, including ion-beam implantation or spin-coating techniques. In this section a novel sol-gel coating technique is described that enables functionalization of both microtoroid and microsphere cavities with layers of erbium doped silica. When pumped, the resulting devices are efficient lasers (a microtoroid is illustrated in Fig. 36 and oscillates in the $1.55\,\mu$m-telecom band).

The surface functionalization process begins by dip-coating[87] a chip (or sphere) containing silica microtoroids using erbium-doped sol-gel. The process details are contained in Refs. 87 and 88. Subsequently, the microtoroids are selectively annealed using a CO_2 laser, causing the sol-gel films to reflow and densify selectively at the toroid periphery. Owing to the high

Fig. 36. Illustration showing a silica microtoroid which has been surface functionalized with a thin Erbium sol-gel film. Also shown is a tapered optical fiber for both pump and laser emission extraction.

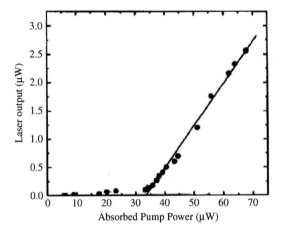

Fig. 37. Measured unidirectional laser output power versus absorbed power for an 80 μm-diameter Erbium sol-gel coated microtoroid.

thermal conductivity of silicon, sol-gel deposited in regions away from the toroid periphery remains undensified during this irradiation step, and can be selectively removed in a subsequent hydrofluoric etch step (due to the higher etch rate of nondensified sol-gel films).

The microtoroids functionalized this way will lase in the 1500 nm band when pumped at any of the standard Erbium pump bands.[88] Since high-Q factors can be maintained in this process, low-threshold lasing is obtained.

Figure 37 shows the laser emission in the 1500 nm band as a function of pump power (980 nm wavelength). Extrapolation of the data reveals a threshold of 34 μWatts.[88] Moreover, the emission spectrum was single mode with a differential quantum efficiency of 11%.

Future work will address alternative routes to optical functionalization using ion-beam implantation.

7. Outlook and Summary

In this chapter we have reviewed the properties of UHQ STIMs including their direct coupling to optical fiber via tapers. This included a review of the coupling regimes as well the concept of junction ideality. Taper junctions have been shown to be exceptionally low loss and ideal and as such are uniquely well suited for certain applications that require ultra-low-loss such as quantum optical studies. The ability to efficiently access UHQ microcavity modes using fiber-tapers allows the study of nonlinear optical mechanisms such as the Raman or Kerr nonlinearities and we have described compact, fiber-compatible Raman lasers and parametric oscillators exhibiting ultra-low threshold. In addition, we have analyzed the modified coupling properties in the UHQ regime where even weak scattering centers are sufficient to cause significant splitting of cavity modes into doublets. Modal coupling results in both finite reflection and a shift of the critical coupling point. Knowledge of this behavior is essential for proper calibration of Q and power build-up in UHQ devices.

Finally, as part of this review, a novel microcavity in the form of a microtoroid on-a-chip, which combines ultra-high-Q and wafer-scale processing, has been described. By merging the technology of wafer-scale fabrication with UHQ for the first time these devices address many of the drawbacks of previous STIMs and provide opportunities for integration of UHQ devices with other functions. In addition, they have inherent advantages over spherical cavities (such as a reduced cavity mode density and a reduced mode volume), which may facilitate the use and consideration of UHQ microresonators in many areas.[89] One important outcome described in this chapter has been the realization of the first, microcavity parametric oscillator. Other exciting avenues in which a UHQ chip-based microcavity is particularly suitable is the study of light-matter interaction with Bose–Einstein condensates on a microelectronic chip[90] or for quantum information studies based on strong-coupling. This new integration capability could result in an entirely new class of "lab on-a-chip" experiments designed to study fundamental processes. In addition the possibility of increasing functionality through use of microelectronic techniques such as ion-beam implantation can enable additional novel devices[91] and functions.

Acknowledgments

The work presented in this chapter has been supported by DARPA, the Caltech Lee Center and the National Science Foundation.

References

1. V. B. Braginskii and V. S. Ilchenko, "The properties of the optical dielectric microresonators," *Doklady Akademii Nauk Sssr* **293**(6), 1358–1361, 1987.
2. D. W. Vernooy, V. S. Ilchenko, H. Mabuchi, E. W. Streed and H. J. Kimble, "High-Q measurements of fused-silica microspheres in the near infrared," *Opt. Lett.* **23**(4), 247–249, 1998.
3. D. W. Vernooy, A. Furusawa, N. P. Georgiades, V. S. Ilchenko and H. J. Kimble, "Cavity QED with high-Q whispering gallery modes," *Phys. Rev.* **A57**(4), R2293–R2296, 1998.
4. R. K. Chang and A. J. Campillo, *Optical Processes in Microcavities,* Advanced series in applied physics, Vol. 3, World Scientific, Singapore, 1996.
5. K. J. Vahala, "Optical microcavities," *Nature* **424**(6950), 839–846, 2003.
6. V. Sandoghdar, F. Treussart, J. Hare, V. LefevreSeguin, J. M. Raimond and S. Haroche, "Very low threshold whispering-gallery-mode microsphere laser," *Phys. Rev.* **A54**(3), R1777–R1780, 1996.
7. M. Cai and K. Vahala, "Highly efficient hybrid fiber taper coupled microsphere laser," *Opt. Lett.* **26**(12), 884–886, 2001.
8. S. M. Spillane, T. J. Kippenberg and K. J. Vahala, "Ultralow-threshold Raman laser using a spherical dielectric microcavity," *Nature* **415**(6872), 621–623, 2002.
9. S. X. Qian, J. B. Snow, H. M. Tzeng and R. K. Chang, "Lasing droplets — highlighting the liquid–air interface by laser-emission," *Science* **231**(4737), 486–488, 1986.
10. S. X. Qian and R. K. Chang, "Multiorder stokes emission from micrometer-size droplets," *Phys. Rev. Lett.* **56**(9), 926–929, 1986.
11. J. Z. Zhang and R. K. Chang, "Generation and suppression of stimulated Brillouin-scattering in single liquid droplets," *J. Opt. Soc. America B — Opt. Phys.* **6**(2), 151–153, 1989.
12. H. B. Lin and A. J. Campillo, "Cw nonlinear optics in droplet microcavities displaying enhanced gain," *Phys. Rev. Lett.* **73**(18), 2440–2443, 1994.
13. A. J. Campillo, J. D. Eversole and H. B. Lin, "Cavity quantum electro-dynamic enhancement of stimulated-emission in microdroplets," *Phys. Rev. Lett.* **67**(4), 437–440, 1991.
14. F. Vollmer, D. Braun, A. Libchaber, M. Khoshsima, I. Teraoka and S. Arnold, "Protein detection by optical shift of a resonant microcavity," *Appl. Phys. Lett.* **80**(21), 4057–4059, 2002.
15. A. N. Oraevsky, "Whispering-gallery waves," *Quant. Electron.* **32**(5), 377–400, 2002.
16. P. Michler, A. Kiraz, L. D. Zhang, C. Becher, E. Hu and A. Imamoglu, "Laser emission from quantum dots in microdisk structures," *Appl. Phys. Lett.* **77**(2), 184–186, 2000.

17. P. Rabiei, W. H. Steier, C. Zhang and L. R. Dalton, "Polymer micro-ring filters and modulators," *J. Lightwave Technol.* **20**(11), 1968–1975, 2002.
18. D. K. Armani, T. J. Kippenberg, S. M. Spillane and K. J. Vahala, "Ultra-high-Q toroid microcavity on a chip," Nature **421**(6926), 925–928, 2003.
19. M. Cai, O. Painter and K. J. Vahala, "Observation of critical coupling in a fiber taper to a silica-microsphere whispering-gallery mode system," *Phys. Rev. Lett.* **85**(1), 74–77, 2000.
20. J. C. Knight, G. Cheung, F. Jacques and T. A. Birks, "Phase-matched excitation of whispering-gallery-mode resonances by a fiber taper," *Opt. Lett.* **22**(15), 1129–1131, 1997.
21. T. J. Kippenberg, S. M. Spillane and K. J. Vahala, "Modal coupling in traveling-wave resonators," *Opt. Lett.* **27**(19), 1669–1671, 2002.
22. S. M. Spillane, T. J. Kippenberg, O. J. Painter and K. J. Vahala, "Ideality in a fiber-taper-coupled microresonator system for application to cavity quantum electrodynamics," Phys. Rev. Lett. **91**(4), 043902, 2003.
23. H. J. Kimble, "Strong interactions of single atoms and photons in cavity QED," *Physica Scripta* **T76**, 127–137, 1998.
24. B. K. Min, T. J. Kippenberg and K. J. Vahala, "Compact, fiber-compatible, cascaded Raman laser," *Opt. Lett.* **28**(17), 1507–1509, 2003.
25. R. H. Stolen and E. P. Ippen, "Raman gain in glass optical waveguides," *Appl. Phys. Lett.* **22**(6), 276–278, 1973.
26. B. E. Little, J. P. Laine and H. A. Haus, "Analytic theory of coupling from tapered fibers and half-blocks into microsphere resonators," *J. Lightwave Technol.* **17**(4), 704–715, 1999.
27. L. Rayleigh, *Scientific Papers*, Cambridge University, Cambridge, pp. 617–620.
28. T. J. Kippenberg, S. M. Spillane, D. K. Armani and K. J. Vahala, "Fabrication and coupling to planar high-Q silica disk microcavities," *Appl. Phys. Lett.* **83**(4), 797–799, 2003.
29. M. Delfino and T. A. Reifsteck, "Laser activated flow of phosphosilicate glass in integrated-circuit devices," *Electron Device Lett.* **3**(5), 116–118, 1982.
30. A. D. McLachlan and F. P. Meyer, "Temperature-dependence of the extinction coefficient of fused-silica for CO_2-Laser wavelengths," *Appl. Opt.* **26**(9), 1728–1731, 1987.
31. A. Serpenguzel, S. Arnold and G. Griffel, "Excitation of resonances of microspheres on an optical-fiber," *Opt. Lett.* **20**(7), 654–656, 1995.
32. M. L. Gorodetsky and V. S. Ilchenko, "Optical microsphere resonators: optimal coupling to high-Q whispering-gallery modes," *J. Opt. Soc. America B — Opt. Phys.* **16**(1), 147–154, 1999.
33. D. Bouwmeester, A. Ekert and A. Zeilinger, *The Physics of Quantum Information*, Springer, Heidelberg, 2000.
34. S. J. van Enk, J. I. Cirac and P. Zoller, "Photonic channels for quantum communication," *Science* **279**(5348), 205–208, 1998.
35. L. M. Duan, M. D. Lukin, J. I. Cirac and P. Zoller, "Long-distance quantum communication with atomic ensembles and linear optics," *Nature* **414**(6862), 413–418, 2001.

36. M. Fiorentino, P. L. Voss, J. E. Sharping and P. Kumar, "All-fiber photon-pair source for quantum communications," *IEEE Photon. Technol. Lett.* **14**(7), 983–985, 2002.

37. H. Mabuchi and A. C. Doherty, "Cavity quantum electrodynamics: Coherence in context," *Science* **298**(5597), 1372–1377, 2002.

38. S. F. Pereira, Z. Y. Ou and H. J. Kimble, "Quantum communication with correlated nonclassical states," *Phys. Rev.* **A6204**(4), 042311, 2000.

39. A. W. Snyder and J. D. Love, *Optical Waveguide Theory*, Kluwer Academic Publisher, 2000.

40. H. A. Haus, *Waves and Fields in Optoelectronics*, Prentice-Hall, Englewood Cliffs, 1984.

41. D. S. Weiss, V. Sandoghdar, J. Hare, V. Lefevreseguin, J. M. Raimond and S. Haroche, "Splitting of high-Q Mie modes induced by light backscattering in silica microspheres," *Opt. Lett.* **20**(18), 1835–1837, 1995.

42. M. L. Gorodetsky, A. D. Pryamikov and V. S. Ilchenko, "Rayleigh scattering in high-Q microspheres," *J. Opt. Soc. America B — Opt. Phys.* **17**(6), 1051–1057, 2000.

43. H. A. Haus, *Electromagnetic Fields and Energy*, Prentice Hall, Englewood Cliffs, 1989.

44. M. L. Gorodetsky, A. A. Savchenkov and V. S. Ilchenko, "Ultimate Q of optical microsphere resonators," *Opt. Lett.* **21**(7), 453–455, 1996.

45. V. B. Braginskii, V. S. Ilchenko and M. L. Gorodetskii, "Optical microresonators with the modes of the whispering gallery type," *Uspekhi Fizicheskikh Nauk* **160**(1), 157–159, 1990.

46. J. R. Buck and H. J. Kimble, "Optimal sizes of dielectric microspheres for cavity QED with strong coupling," Phys. Rev. **A67**(3), 033806, 2003.

47. S. Schiller, "Asymptotic-expansion of morphological resonance frequencies in Mie scattering," *Appl. Opt.* **32**(12), 2181–2185, 1993.

48. L. Collot, V. Lefevreseguin, M. Brune, J. M. Raimond and S. Haroche, "Very high-Q whispering-gallery mode resonances observed on fused-silica microspheres," *Europhys. Lett.* **23**(5), 327–334, 1993.

49. C. V. Raman, K. K. S., "A new type of secondary radiation," *Nature* **121**, 501, 1928.

50. N. Bloembergen and Y. R. Shen, "Coupling between vibrations + light waves in raman laser media," *Phys. Rev. Lett.* **12**(18), 504, 1964.

51. U. Gaubatz, P. Rudecki, M. Becker, S. Schiemann, M. Kulz and K. Bergmann, "Population switching between vibrational levels in molecular-beams," *Chem. Phys. Lett.* **149**(5-6), 463–468, 1988.

52. K. Bergmann, H. Theuer and B. W. Shore, "Coherent population transfer among quantum states of atoms and molecules," *Rev. Mod. Phys.* **70**(3), 1003–1025, 1998.

53. R. W. Boyd, "Nonlinear Optics, Academic Press, Boston, 1992.

54. E. P. Ippen and R. H. Stolen, "Stimulated Brillouin-scattering in optical fibers," *Appl. Phys. Lett.* **21**(11), 539, 1972.

55. R. H. Stolen, "Fiber Raman lasers," *Fiber Integr. Opt.* **3**(1), 21–52, 1980.

56. J. K. Brasseur, K. S. Repasky and J. L. Carlsten, "Continuous-wave Raman laser in H-2," *Opt. Lett.* **23**(5), 367–369, 1998.

57. G. P. Agrawal, *Nonlinear Fiber Optics*, Academic Press, Boston, 1989.

58. R. H. Stolen, A. R. Tynes and E. P. Ippen, "Raman oscillation in glass optical waveguide," *Appl. Phys. Lett.* **20**(2), 62, 1972.

59. H. B. Lin and A. J. Campillo, "Microcavity enhanced Raman gain," *Opt. Commun.* **133**(1–6), 287–292, 1997.

60. A. B. Matsko, A. A. Savchenkov, R. J. Letargat, V. S. Ilchenko and L. Maleki, "On cavity modification of stimulated Raman scattering," *J. Optics B—Quant. Semiclass. Opt.* **5**(3), 272–278, 2003.

61. R. H. Stolen and J. E. Bjorkholm, "Parametric amplification and frequency-conversion in optical fibers," *IEEE J. Quant. Electron.* **18**(7), 1062–1072, 1982.

62. A. L. Gaeta, "Catastrophic collapse of ultrashort pulses," *Phys. Rev. Lett.* **84**(16), 3582–3585, 2000.

63. F. W. Dabby and J. R. Whinnery, "Thermal self-focusing of laser beams in lead glasses," *Appl. Phys. Lett.* **13**(8), 284, 1968.

64. D. F. Walls and G. J. Milburn, *Quant. Optics*, Springer, New York, 1994.

65. M. O. Scully and M. S. Zubairy, *Quant. Optics*, Cambridge, 1996.

66. S. Uetake, M. Katsuragawa, M. Suzuki and K. Hakuta, "Stimulated Raman scattering in a liquid-hydrogen droplet," Phys. Rev. **A6001**(1), 011803, 2000.

67. R. M. Stevenson, V. N. Astratov, M. S. Skolnick, D. M. Whittaker, M. Emam-Ismail, A. I. Tartakovskii, P. G. Savvidis, J. J. Baumberg and J. S. Roberts, "Continuous wave observation of massive polariton redistribution by stimulated scattering in semiconductor microcavities," *Phys. Rev. Lett.* **85**(17), 3680–3683, 2000.

68. A. L. Tartakovskii, M. S. Skolnick, D. N. Krizhanovskii, V. D. Kulakovskii, R. M. Stevenson, R. Butte, J. J. Baumberg, D. M. Whittaker and J. S. Roberts, "Stimulated polariton scattering in semiconductor microcavities: New physics and potential applications," *Advanced Materials* **13**(22), 1725–1730, 2001.

69. R. Butte, M. S. Skolnick, D. M. Whittaker, D. Bajoni and J. S. Roberts, "Dependence of stimulated scattering in semiconductor microcavities on pump power, angle, and energy," *Phys. Rev.* **B68**(11), 115325, 2003.

70. V. B. Braginsky, M. L. Gorodetsky and V. S. Ilchenko, "Quality-factor and nonlinear properties of optical whispering-gallery modes," *Phys. Lett.* **A137**(7–8), 393–397, 1989.

71. F. Treussart, V. S. Ilchenko, J. F. Roch, J. Hare, V. Lefevre-Seguin, J. M. Raimond and S. Haroche, "Evidence for intrinsic Kerr bistability of high-Q microsphere resonators in superfluid helium," *European Phys. J.* **D1**(3), 235–238, 1998.

72. E. Golovchenko, P. V. Mamyshev, A. N. Pilipetskii and E. M. Dianov, "Mutual influence of the parametric effects and stimulated Raman-scattering in optical fibers," *IEEE J. Quant. Electron.* **26**(10), 1815–1820, 1990.

73. J. E. Sharping, M. Fiorentino, P. Kumar and R. S. Windeler, "Optical parametric oscillator based on four-wave mixing in microstructure fiber," *Opt. Lett.* **27**(19), 1675–1677, 2002.

74. P. Russell, "Photonic crystal fibers," *Science* **299**(5605), 358–362, 2003.
75. R. E. Slusher, L. W. Hollberg, B. Yurke, J. C. Mertz and J. F. Valley, "Observation of squeezed states generated by 4-wave mixing in an optical cavity," *Phys. Rev. Lett.* **55**(22), 2409–2412, 1985.
76. R. M. Shelby, M. D. Levenson, S. H. Perlmutter, R. G. Devoe and D. F. Walls, "Broad-band parametric deamplification of quantum noise in an optical fiber," *Phys. Rev. Lett.* **57**(6), 691–694, 1986.
77. M. D. Levenson, R. M. Shelby, A. Aspect, M. Reid and D. F. Walls, "Generation and detection of squeezed states of light by nondegenerate 4-wave mixing in an optical fiber," *Phys. Rev.* **A32**(3), 1550–1562, 1985.
78. C. Silberhorn, P. K. Lam, O. Weiss, F. Konig, N. Korolkova and G. Leuchs, "Generation of continuous variable Einstein–Podolsky–Rosen entanglement via the Kerr nonlinearity in an optical fiber," *Phys. Rev. Lett.* **86**(19), 4267–4270, 2001.
79. A. Sizmann and G. Leuchs, "The optical Kerr effect and quantum optics in fibers," in *Progress in Optics*, Vol Xxxix. pp. 373–469, 1999.
80. L. J. Wang, C. K. Hong and S. R. Friberg, "Generation of correlated photons via four-wave mixing in optical fibres," *J. Optics B-Quant. Semiclass. Opt.* **3**(5), 346–352, 2001.
81. T. Fujiwara, D. Wong, Y. Zhao, S. Fleming, S. Poole and M. Sceats, "Electrooptic modulation in Germanosilicate fiber with Uv- excited poling," *Electron. Lett.* **31**(7), 573–575, 1995.
82. R. A. Myers, N. Mukherjee and S. R. J. Brueck, "Large 2nd-order nonlinearity in poled fused-silica," *Opt. Lett.* **16**(22), 1732–1734, 1991.
83. E. S. Polzik, J. Carri and H. J. Kimble, "Spectroscopy with squeezed light," *Phys. Rev. Lett.* **68**(20), 3020–3023, 1992.
84. A. Heidmann, R. J. Horowicz, S. Reynaud, E. Giacobino, C. Fabre and G. Camy, "Observation of quantum noise-reduction on twin laser-beams," *Phys. Rev. Lett.* **59**(22), 2555–2557, 1987.
85. K. S. Zhang, T. Coudreau, M. Martinelli, A. Maitre and C. Fabre, "Generation of bright squeezed light at 1.06 mu m using cascaded nonlinearities in a triply resonant cw periodically-poled lithium niobate optical parametric oscillator," *Phys. Rev.* **A6403**(3), 033815, 2001.
86. B. E. A. Saleh, B. M. Jost, H. B. Fei and M. C. Teich, "Entangled-photon virtual-state spectroscopy," *Phys. Rev. Lett.* **80**(16), 3483–3486, 1998.
87. L. Yang and K. J. Vahala, "Gain functionalization of silica microresonators," *Opt. Lett.* **28**(8), 592–594, 2003.
88. L. Yang, D. K. A. and K. J. Vahala, "Fiber-coupled erbium microlasers on a chip," *Appl. Phys. Lett.* **83**, 825, 2003.
89. J. M. Gerard, "Quantum optics — Boosting photon storage," *Nature Materials* **2**(3), 140–141, 2003.
90. W. Hansel, P. Hommelhoff, T. W. Hansch and J. Reichel, "Bose–Einstein condensation on a microelectronic chip," *Nature* **413**(6855), 498–501, 2001.
91. A. Polman, B. Min, J. Kalkman, T. J. Kippenberg and K. J. Vahala, "Ultralow-threshold erbium implanted toroidal microlaser on silicon," *Appl. Phys. Lett.* **84**(7), 1037, 2004.

CHAPTER 6

NONLINEAR OPTICAL PROPERTIES OF SEMICONDUCTOR QUANTUM WELLS INSIDE MICROCAVITIES

T. Meier, C. Sieh and S. W. Koch

Department of Physics and Material Sciences Center, Philipps University
Renthof 5, D-35032 Marburg, Germany

Y.-S. Lee

Department of Physics, Oregon State University
Corvallis, OR 97331-6507, USA

T. B. Norris

Center for Ultrafast Optical Science, The University of Michigan
Ann Arbor, MI 48109-2099, USA

F. Jahnke

Institute for Theoretical Physics, University of Bremen
P.O. Box 330 440, D-28334 Bremen, Germany

G. Khitrova and H. M. Gibbs

Optical Sciences Center, University of Arizona
Tucson, AZ 85721, USA

Experimental results on the nonlinear optical properties of semiconductors are compared with microscopic calculations which include Coulomb many-body correlations at different levels. One aim of this chapter is to show that microscopic theories, which have been developed for bare semiconductor heterostructures, are also able to describe semiconductor microcavities very well. Therefore, there is no need to phenomenologically introduce polariton–polariton interactions and parametric scattering of cavity polaritons to describe microcavity experiments, but instead a fully microscopic theory based on a Fermionic electron-hole Hamiltonian can be used.

The treatment of many-body correlations using the second-order Born approximation and the dynamics-controlled truncation scheme are introduced and analyzed for bare heterostructures. These approaches are able to successfully explain a number of important experimental results which originate from the dynamics of many-body correlations.

Then measurements of the nonlinear optical properties of a quantum-well microcavity are described. In these experiments the spectrally- and temporally-resolved nonlinear optical response is studied in detail using a pump-probe geometry. In particular, the polarization and intensity dependencies of the spectral probe reflection changes and their temporal evolution are analyzed. The prominent features of these experiments are well accounted for by the microscopic many-body theory.

1. Introduction

In this chapter, the nonlinear optical properties of semiconductors inside microcavities are analyzed. This is done by discussing experimental results and comparing them with microscopic calculations which include Coulomb many-body correlations at different levels.

We start in Sec. 2 by giving an introduction to the nonlinear optical properties of semiconductor heterostructures which are *not* inside a microcavity. Here, it is, for example, shown that the absorption spectra can be described by Elliott's formula which is the solution of the exciton Wannier equation and the bulk absorption is compared with that of quantum wells and wires. More explicitly, we review the microscopic theoretical description of the nonlinear optical response of semiconductors. Of particular importance are the possible treatments of many-body Coulomb correlations. For each level of theory, examples are presented in order to show which effects can be described by the respective approach. Furthermore, the results of numerical calculations are compared to experiments.

In Sec. 3, we focus on the nonlinear optical properties of semiconductors inside microcavities. Here, we first establish the normal-mode picture, which describes the resonances of the coupled light-matter system in the regime of nonperturbative coupling. We also review some results on the density-dependent saturation and broadening of the normal-mode splitting which eventually collapses for large carrier densities, and on biexcitonic signatures in the normal-mode spectra. The main part of this section is devoted to the detailed analysis of spectrally- and temporally-resolved nonlinear optical response of a quantum-well microcavity measured in pump-probe geometry. The polarization- and intensity-dependence of the reflection changes and their dynamical evolution are studied after selective excitation of either one or both normal modes with circularly polarized light fields. The experimental results are discussed in terms of microscopic theory in which Coulombic many-particle correlations are included at different levels.

In Sec. 4, our main results are summarized. The analysis presented in this chapter demonstrates, in particular, that microscopic theories, which have been developed for bare semiconductor heterostructures, are also able to describe the nonlinear optical response of semiconductor microcavities very well.

2. Optical Properties of Semiconductor Nanostructures

2.1. *Introduction*

A comprehensive analysis of the optical properties of semiconductors encompasses a descriptions of the light field and material excitation dynamics as well as their interaction properties.[1] At the semiclassical level the light field is treated as a classical field whose dynamics is governed by Maxwell's equations, whereas the material excitations are treated quantum mechanically. The interaction between light and semiconductor enters the electromagnetic field equation via the *optical polarization* of the medium. Therefore, the polarization is the key quantity in the description of the material excitations.

Electromagnetic field and optical polarization have to be computed self-consistently. The optical field excites the material system. In dipole approximation, the corresponding coupling is given by the scalar product of the electric field and the optical polarization. The resulting optical polarization of the material then appears as a source term for the electromagnetic field in Maxwell's wave equations. Hence, the polarization generates fields which depend on the state of the material system and therefore yield information about the medium excitations.

For interband excitation conditions the optical polarization of a semiconductor can be described as superposition of coherences between valence- and conduction-band states. The linear optical absorption spectrum of an unexcited semiconductor is determined by the sum of all interband transitions between the completely filled valence bands and the empty conduction bands. Already in the linear regime one finds characteristic signatures of the Coulomb interaction.[1,2] In particular, the Coulomb attraction between an optically excited electron and a hole leads to strong absorption resonances that appear spectrally below the bandgap. These lines are due to bound electron-hole-pair states, denoted as excitons, which often show a hydrogenic Rydberg series. Already the linear absorption shows that the

Coulomb interaction has to be treated properly if one wants to achieve a meaningful theory for the optical response of semiconductors.

If the semiconductor is not in its ground state or if the exciting light fields are not in the low-intensity limit, one has to go beyond the linear optical regime. In this case, additional quantities besides the interband polarization are required to consistently describe the nonlinear optical response. Since the Coulomb interaction introduces a many-body problem, a fully exact treatment of the nonlinear response is usually not possible. During the last decades, however, a number of approximation schemes have been developed and applied successfully to different regimes, see, e.g. Refs. 3–9.

The simplest approximation to treat a many-body system is to approximate it by an effective single-particle system with renormalized potentials. This is what is done effectively when the optical response of semiconductors is analyzed on the time-dependent Hartree–Fock (HF) level. At this level, the dynamics of the material is completely described in terms of the diagonal and off-diagonal elements of the reduced single-particle density matrix.[1] While the off-diagonal terms are given by interband transitions which determine the optical polarization, the diagonal terms are the carrier (electron and hole) occupation probabilities in the different bands. The dynamical evolution of the optically induced excitations is from solutions of the coupled equations of motion for the density-matrix elements, which are the Hartree–Fock semiconductor Bloch equations (HF–SBE). These equations include Coulombic effects via bandgap and field renormalization[1,4] and are well suited to analyze a number of important effects, especially in the coherent nonlinear optical response of semiconductors.

As usual in many-particle physics,[10] all aspects of the nonlinear optical semiconductor response that cannot be described on the basis of the HF approximation are denoted by *correlation effects*. Such correlation effects always exist and lead in certain situations to significant experimentally observable consequences. To theoretically describe these many-body correlations different approaches have been developed. One of them is the so-called dynamics-controlled truncation (DCT) scheme[9,11–13] which is based on an expansion of the nonlinear optical response in powers of the optical field.[14] Often the DCT is used together with the assumption of a fully coherent system, which strongly reduces the number of dynamic variables, since in this case only single- and multi-exciton coherences have to be treated. An approach which is able to analyze correlations in the regime of strong fields is the second Born approximation (SBA).[4–8] Density-dependent absorption spectra which show exciton saturation and broadening and gain spectra of semiconductor lasers are examples for effects where the SBA results are in good agreement with experiment.[15,16]

At the fully quantum mechanical level, the dominant light-matter correlations are described by coupled semiconductor Bloch and luminescence equations.[17-21] These equations are obtained by treating not only the material excitations but also the light field quantum mechanically. Since we are dealing in this chapter only with semiclassical aspects, we do not discuss this approach in any detail and refer the interested reader to the literature, e.g. Refs. 17–21.

Quite recently, it has been shown that one can obtain the SBA and the DCT as limiting cases of a unified theoretical approach. In Ref. 22 the light-matter and Coulombic many-body correlations were treated on the basis of a cluster expansion. For classical light fields the light-matter correlations vanish. If the analysis is furthermore restricted to a finite order in the interaction with the field and the system is fully coherent, the coherent DCT equations can be obtained from the cluster expansion. To obtain the SBA one has to treat the four-particle correlations which appear in the cluster expansion, e.g. biexciton transitions, exciton occupations, and fourth-order intraband correlations, approximately by factorizing their sources into products of polarizations and electron and hole occupations. Coupled equations which contain the DCT and the SBA as limiting cases have also been obtained using an approach based on nonequilibrium Green's functions.[23,24]

We start in this section by introducing the fundamental concepts required for the description of the optical semiconductor response in the semiclassical approach. In Sec. 2.2, it is shown that within the semiclassical theory the optical polarization of the medium appears as an inhomogeneity in the wave equation of the electric field. As described above, due to the Coulomb interaction among the carriers, one has to solve a many-body problem. In Sec. 2.3, the derivation of the semiconductor Bloch equations on the Hartree–Fock level is discussed in detail. These equations can be used to analyze the excitonic absorption for semiconductor nanostructures of different dimensionality as shown in Sec. 2.3.1. The theoretical treatment of many-body correlation effects and their influence on experiments is described in Sec. 2.4. First, the second Born approximation is introduced in Sec. 2.4.1. As shown in Sec. 2.4.2, this approach allows one to analyze the nonlinear optical response semiconductor nanostructures in various situations. As examples we present results on optical absorption spectra under different excitation conditions including gain spectra of semiconductor lasers. Then, in Sec. 2.4.3, the dynamics-controlled truncation scheme is described. The equations up to third-order in the field are derived for a coherent system, i.e. the coherent $\chi^{(3)}$-limit, where biexcitonic four-particle correlations are treated explicitly. Some experimentally observable consequences of these four-particle correlations are presented in Sec. 2.4.4. In Sec. 2.4.5 is outlined how one can extend the dynamics-controlled truncation scheme to include processes that are of fifth-order in the field, i.e. the coherent $\chi^{(5)}$-limit. Finally, some fifth-order signatures of biexcitonic correlations are presented in Sec. 2.4.6.

2.2. Semiclassical Theory

The computation of the optical properties of a material system requires the solution of Maxwell's equations. For the simple example of a one-dimensional propagation the wave equation for the electric field is given by

$$\left[\frac{\partial^2}{\partial z^2} + \frac{n^2(z)}{c^2} \frac{\partial^2}{\partial t^2} \right] E(z,t) = -d_o^{eh} \frac{\partial^2}{\partial t^2} P(z,t), \tag{1}$$

where E is the electric field, z is the space coordinate, t the time, and d_o^{eh} is a prefactor which depends on the system of units. Both E and P are two-dimensional vectors in the x-y–plane, denoting the polarization direction. In Eq. (1), the material response has been separated into two parts. The resonant part of the material excitations is included dynamically in terms of the macroscopic optical polarization P, whereas the nonresonant part is treated via the background refractive index n. Equation (1) shows that the polarization P is the key quantity which needs to be determined from the description of the material response. Below we introduce a few concepts that demonstrate how one can obtain P from microscopic theory.

2.3. Hartree–Fock Approximation

For ordered semiconductor systems it is usually convenient to expand the macroscopic polarization into a Bloch basis[1]

$$\mathbf{P} = \sum_{\mathbf{k},e,h} (\mathbf{d}^{eh})^* P_{\mathbf{k}}^{eh} + \text{c.c.}, \tag{2}$$

where $\mathbf{d}^{eh}$ is the electron-hole interband dipole matrix element between conduction band e and valence band h. Using second quantization, i.e. electron ($a_{e,\mathbf{k}}^\dagger, a_{e,\mathbf{k}}$) and hole ($b_{h,\mathbf{k}}^\dagger, b_{h,\mathbf{k}}$) creation and annihilation operators, the microscopic polarization $P_{\mathbf{k}}^{eh}$ and the carrier occupation probabilities $f_{\mathbf{k}}^{ee,hh}$ constitute the diagonal and off-diagonal elements of the reduced single-particle density matrix:

$$
\begin{pmatrix}
\langle a_{e,\mathbf{k}}^{\dagger} a_{e,\mathbf{k}} \rangle & \langle b_{h,-\mathbf{k}} a_{e,\mathbf{k}} \rangle \\
\langle a_{e,\mathbf{k}}^{\dagger} b_{h,-\mathbf{k}}^{\dagger} \rangle & \langle b_{h,-\mathbf{k}}^{\dagger} b_{h,-\mathbf{k}} \rangle
\end{pmatrix}
=
\begin{pmatrix}
f_{\mathbf{k}}^{ee} & P_{\mathbf{k}}^{eh} \\
(P_{\mathbf{k}}^{eh})^{*} & f_{\mathbf{k}}^{hh}
\end{pmatrix}.
\tag{3}
$$

The equations of motion for the components of the density matrix are given by the Heisenberg equation, which for an arbitrary operator $\mathcal{O}$ reads

$$
i\hbar \frac{\partial}{\partial t} \mathcal{O} = [\mathcal{O}, H].
\tag{4}
$$

To describe optical processes in semiconductors we use the standard many-body Hamiltonian[1]

$$
H = H_{s-p} + H_C + H_I.
\tag{5}
$$

In Eq. (5),

$$
H_{s-p} = \sum_{\mathbf{k},e} \varepsilon_{\mathbf{k}}^{e} a_{e,\mathbf{k}}^{\dagger} a_{e,\mathbf{k}} + \sum_{\mathbf{k},h} \varepsilon_{\mathbf{k}}^{h} b_{h,-\mathbf{k}}^{\dagger} b_{h,-\mathbf{k}}
\tag{6}
$$

is the single-particle Hamiltonian, which contains the bandstructure (single-particle energy) of the electrons ($\varepsilon_{\mathbf{k}}^{e}$) and holes ($\varepsilon_{\mathbf{k}}^{h}$). H_C is the Coulomb interaction between the carriers and given by

$$
\begin{aligned}
H_C = \ &\frac{1}{2} \sum_{\mathbf{k},\mathbf{k}',\mathbf{q}\neq 0, e, e'} V_q a_{e,\mathbf{k}+\mathbf{q}}^{\dagger} a_{e',\mathbf{k}'-\mathbf{q}}^{\dagger} a_{e',\mathbf{k}'} a_{e,\mathbf{k}} \\
&+ \frac{1}{2} \sum_{\mathbf{k},\mathbf{k}',\mathbf{q}\neq 0, h, h'} V_q b_{h,\mathbf{k}+\mathbf{q}}^{\dagger} b_{h',\mathbf{k}'-\mathbf{q}}^{\dagger} b_{h',\mathbf{k}'} b_{h,\mathbf{k}} \\
&- \sum_{\mathbf{k},\mathbf{k}',\mathbf{q}\neq 0, e, h} V_q a_{e,\mathbf{k}+\mathbf{q}}^{\dagger} b_{h,\mathbf{k}'-\mathbf{q}}^{\dagger} b_{h,\mathbf{k}'} a_{e,\mathbf{k}}.
\end{aligned}
\tag{7}
$$

Here, the first two lines denote the repulsive interactions among electrons and among holes, respectively, and the last line describes the attractive interaction between electrons and holes. V_q denotes the Fourier transform of the Coulomb interaction potential. The interaction between the carriers and the classical electromagnetic field, H_I, is considered in dipole approximation via[1]

$$
H_I = -\mathbf{E}(t) \cdot \sum_{\mathbf{k},e,h} (\mathbf{d}^{eh} a_{e,\mathbf{k}}^{\dagger} b_{h,-\mathbf{k}}^{\dagger} + (\mathbf{d}^{eh})^{*} b_{h,-\mathbf{k}} a_{e,\mathbf{k}}),
\tag{8}
$$

with the electron-hole interband dipole matrix element $\mathbf{d}^{eh}$. The light field $\mathbf{E}(t)$ either creates or destroys *pairs* of electrons and holes.

For simplicity we neglect transitions from light-holes and energetically deeper valence bands throughout this chapter. Thus only the heavy-hole valence band and the lowest conduction band, both of which are two-fold spin-degenerate, are considered. The two heavy-hole bands ($h = 1, 2$) are characterized by the states $|-3/2, h\rangle$ and $|3/2, h\rangle$, and the conduction bands ($e = 1, 2$) by $|-1/2, e\rangle$ and $|1/2, e\rangle$, respectively.[25,26] For light propagating in the z-direction, i.e. perpendicular to the plane of the quantum well, we use the usual circularly polarized dipole matrix elements[25,26]

$$\mathbf{d}^{11} = d_0 \boldsymbol{\sigma}^+ = \frac{d_0}{\sqrt{2}} \begin{pmatrix} 1 \\ i \end{pmatrix},$$

$$\mathbf{d}^{12} = \mathbf{d}^{21} = 0, \tag{9}$$

$$\mathbf{d}^{22} = d_0 \boldsymbol{\sigma}^- = \frac{d_0}{\sqrt{2}} \begin{pmatrix} 1 \\ -i \end{pmatrix},$$

where d_0 is the modulus of $\mathbf{d}^{11}$ and $\mathbf{d}^{22}$. Due to these *diagonal* selection rules, i.e. $\mathbf{d}^{eh} \propto \delta_{eh}$, we have two separate subspaces of optical excitations, that are optically isolated. They are, however, coupled by the many-body Coulomb-interaction, since it is independent of the band indices (spin), see Eq. (7).

Evaluating the commutators in Eq. (4) one obtains the equation of motion for the microscopic polarization

$$\begin{aligned}
i\hbar \frac{\partial}{\partial t} P_{\mathbf{k}}^{eh} = &- \left(\varepsilon_{\mathbf{k}}^e + \varepsilon_{\mathbf{k}}^h \right) P_{\mathbf{k}}^{eh} + \sum_{\mathbf{q} \neq 0} V_q P_{\mathbf{k}-\mathbf{q}}^{eh} \\
&+ \left(\mathbf{d}^{eh} - \sum_{e'} f_{\mathbf{k}}^{ee'} \mathbf{d}^{e'h} - \sum_{h'} \mathbf{d}^{eh'} f_{\mathbf{k}}^{h'h} \right) \cdot \mathbf{E} \\
&- \sum_{\mathbf{q} \neq 0, \mathbf{k}', e'} V_q [\langle a_{e,\mathbf{k}}^\dagger a_{e',\mathbf{k}'}^\dagger b_{h,\mathbf{k}+\mathbf{q}}^\dagger a_{e',\mathbf{k}'-\mathbf{q}} \rangle \\
&\qquad\qquad - \langle a_{e,\mathbf{k}+\mathbf{q}}^\dagger a_{e',\mathbf{k}'}^\dagger b_{h,\mathbf{k}}^\dagger a_{e',\mathbf{k}'+\mathbf{q}} \rangle] \\
&+ \sum_{\mathbf{q} \neq 0, \mathbf{k}', h'} V_q [\langle a_{e,\mathbf{k}+\mathbf{q}}^\dagger b_{h',\mathbf{k}'+\mathbf{q}}^\dagger b_{h,\mathbf{k}}^\dagger b_{h',\mathbf{k}'} \rangle \\
&\qquad\qquad - \langle a_{e,\mathbf{k}}^\dagger b_{h',\mathbf{k}'+\mathbf{q}}^\dagger b_{h,\mathbf{k}-\mathbf{q}}^\dagger b_{h',\mathbf{k}'} \rangle].
\end{aligned} \tag{10}$$

In the two-band case, i.e. if only a single conduction and a single valence band is considered, we have $e = e' = 1$ and $h = h' = 1$ in Eq. (10). In the more general multiband configuration, summations over all the respective bands have to be considered. Since we restrict the analysis presented in this chapter to transitions from heavy-holes to the lowest conduction band, $P_{\mathbf{k}}^{eh}$ is nonvanishing only for $e = h = 1$ and $e = h = 2$, i.e. concerning the subband indices it is proportional to δ_{eh}, since the terms with $e \neq h$ have no sources. For similar reasons, also $f_{\mathbf{k}}^{ee'}$ and $f_{\mathbf{k}}^{h'h}$ are diagonal, i.e. proportional to $\delta_{ee'}$ and $\delta_{hh'}$, respectively. Therefore one can slightly reduce the notation by defining $f_{\mathbf{k}}^{e} \equiv f_{\mathbf{k}}^{ee}$ and $f_{\mathbf{k}}^{h} \equiv f_{\mathbf{k}}^{hh}$. For clarity we keep, however, $f_{\mathbf{k}}^{ee}$ and $f_{\mathbf{k}}^{hh}$ in what follows now, and use the notation $f_{\mathbf{k}}^{e}$ and $f_{\mathbf{k}}^{h}$ at a later stage, when the SBA is treated.

Equation (10) and the corresponding equations for the carrier occupation probabilities, $f_{\mathbf{k}}^{ee}$ and $f_{\mathbf{k}}^{hh}$, contain a coupling of the components of the single-particle density matrix among themselves (via $H_{s-p} + H_I$) and contributions that couple the two-operator terms to four-operator terms (via H_C).[1] This is the beginning of the well-known many-body hierarchy. If one computes equations of motion for $2n$ operator expectation values, due to the many-body Coulomb interaction one obtains a coupling to $2n + 2$ operator terms, and so on. In order to close the set of equations, approximations have to be used, i.e. truncate this hierarchy at some stage in a self-consistent fashion.

At the level of a dynamical Hartree–Fock like approximation one uses a decoupling scheme like, e.g.

$$\left\langle a_{e,\mathbf{k}}^{\dagger} a_{e,\mathbf{k}'-\mathbf{q}}^{\dagger} a_{e,\mathbf{k}-\mathbf{q}} a_{e,\mathbf{k}'} \right\rangle \simeq \left\langle a_{e,\mathbf{k}}^{\dagger} a_{e,\mathbf{k}} \right\rangle \left\langle a_{e,\mathbf{k}-\mathbf{q}}^{\dagger} a_{e,\mathbf{k}-\mathbf{q}} \right\rangle \delta_{\mathbf{k},\mathbf{k}'}, \tag{11}$$

where no other contribution appears since $\mathbf{q} \neq 0$.[1] Applying this factorization to all terms that appear in the equation of motion for the microscopic polarization, Eq. (10), and the corresponding equations for the carrier occupations, one obtains the well-known Hartree–Fock semiconductor Bloch equations (HF–SBE). For a system with dipole matrix elements that are diagonal in the band indices e and h, i.e. as discussed above $\mathbf{d}^{eh} \propto \delta_{eh}$, these equations can be written as[1,4]

$$\left[i\hbar \frac{\partial}{\partial t} - \epsilon_{\mathbf{k}}^{e}(t) - \epsilon_{\mathbf{k}}^{h}(t) \right] P_{\mathbf{k}}^{eh}(t) = \left[1 - f_{\mathbf{k}}^{ee}(t) - f_{\mathbf{k}}^{hh}(t) \right] \Omega_{\mathbf{k}}(t),$$

$$\frac{\partial}{\partial t} f_{\mathbf{k}}^{aa}(t) = -\frac{2}{\hbar} \mathrm{Im} \left[\Omega_{\mathbf{k}}(t) (P_{\mathbf{k}}^{eh}(t))^{*} \right] \tag{12}$$

where

$$\Omega_{\mathbf{k}}(t) = \mathbf{d}^{eh} \cdot \mathbf{E}(t) + \sum_{\mathbf{k}' \neq \mathbf{k}} V_{|\mathbf{k}-\mathbf{k}'|} P_{\mathbf{k}'}^{eh}(t) \tag{13}$$

is the renormalized field (Rabi frequency) and

$$\epsilon_{\mathbf{k}}^{a}(t) = \varepsilon_{\mathbf{k}}^{a} - \sum_{\mathbf{k}' \neq \mathbf{k}} V_{|\mathbf{k}-\mathbf{k}'|} f_{\mathbf{k}'}^{aa}(t) \qquad (14)$$

is the renormalized single-particle energy. If the Coulomb interaction ($V \equiv 0$) is ignored, Eqs. (12)–(14) become diagonal in $\mathbf{k}$. Thus in this case the optical response is described by a superposition of uncoupled two-level systems in $\mathbf{k}$ space and the electronic dispersion simply provides an inhomogeneous broadening. For finite Coulomb interaction, the Hartree–Fock, i.e. exchange renormalizations, couple all $\mathbf{k}$-states of the semiconductor material and introduce nonlinearities in the HF–SBE. Numerically, the renormalization contributions are not at all small corrections. In particular, the leading-order Coulomb effect gives rise to excitonic resonances, as shown in Sec. 2.3.1.

The nonlinearities in the HF–SBE, Eq. (12), are due to phase-space filling, which is represented by the terms proportional to the occupation probabilities times the field Ef, as well as energy and field renormalization. The phase-space filling arises as a consequence of the Fermionic nature of the electrons and holes (Pauli blocking), whereas the renormalizations are caused by anharmonic terms in the Hamiltonian (terms containing more than two operators, i.e. H_C). It should be noted, that the HF–SBE contain neither dephasing of the polarization nor screening of the interaction potential or relaxation of the carrier distributions. To include these effects, one has to go beyond the Hartree–Fock approximation and treat many-body correlations, as outlined below.

2.3.1. *Linear Exciton Absorption in Different Dimensions*

Starting from an unexcited semiconductor, i.e. no polarization or electron-hole population before the system is excited, one can linearize the polarization equation in the interaction with the external field. For parabolic bands (effective mass approximation) one obtains the equation of motion for the linear polarization as

$$\left[i\hbar \frac{\partial}{\partial t} - \epsilon_g - \frac{\hbar^2 k^2}{2m_r} \right] P_{\mathbf{k}}^{eh}(t) = \mathbf{d}^{eh} \cdot \mathbf{E}(t) + \sum_{\mathbf{k}'} V_{|\mathbf{k}-\mathbf{k}'|} P_{\mathbf{k}'}^{eh}(t), \qquad (15)$$

where ϵ_g is the bandgap energy and $m_r = (1/m_e + 1/m_h)^{-1}$ is the reduced mass which is given by the inverse of the sum over the inverse effective masses of the valence and conductions bands, m_h and m_e, respectively.

Via a Fourier transform to real space, one obtains

$$\left[i\hbar\frac{\partial}{\partial t} - \epsilon_g + \frac{\hbar^2\nabla_{\mathbf{r}}^2}{2m_r} - V(r)\right]P^{eh}(\mathbf{r},t) = -\mathbf{d}^{eh}\cdot\mathbf{E}(t)\delta(\mathbf{r}). \tag{16}$$

The homogeneous part of Eq. (16) is the *Wannier equation*. It is mathematically identical to the Schrödinger equation for the relative motion of the hydrogen atom. Hence, the solutions of the Wannier equation describe a discrete Rydberg series of bound eigenstates (energies smaller than ϵ_g) and also unbound solutions (energies bigger than ϵ_g). Whereas the bound states are the Wannier excitons, the unbound states represent the Coulomb interacting continuum excitations. From the solution of the inhomogeneous Eq. (16) one obtains the linear electron-hole pair susceptibility $\chi^{(1)}(\omega) = P(\omega)/E(\omega)$. Its imaginary part $Im[\chi^{(1)}(\omega)]$ is proportional to the linear absorption $\alpha^{(1)}(\omega)$ as described by the Elliott formula.[2]

The linear optical absorption spectra in the spectral vicinity of the band gap for three-, two-, and one-dimensional direct-gap semiconductors, corresponding to bulk, quantum well, and quantum wire structures, respectively, are shown in Fig. 1. The origin of the dashed lines is relatively simple to understand. We noted above, that without Coulomb interaction the microscopic polarizations $P_{\mathbf{k}}^{eh}$ for different $\mathbf{k}$ do not interact, see Eqs. (12) and (13). Thus, in this case, for each fixed $\mathbf{k}$ the interband transitions are described by noninteracting two-level system and the dispersion of the valence and conduction bands simply introduces an intrinsic inhomogeneous broadening. Therefore, without Coulomb interaction the absorption vanishes for frequencies that are smaller than the bandgap ϵ_g. Above the gap, the absorption is directly proportional to the joint density of states for the valence to conduction band transitions. Thus the absorption is proportional to $\Theta(\hbar\omega - \epsilon_g)(\hbar\omega - \epsilon_g)^{(d-2)/2}$ where d is the dimensionality.[1]

With Coulomb interaction the spectra are significantly modified due to excitonic effects. In any dimension, the strongly absorbing exciton states appear as discrete lines *below* the bandgap, see Fig. 1. Due to the optical selection rules, i.e. the coupling to the light field is proportional to $\delta(\mathbf{r})$ in Eq. (16), one typically observes in direct semiconductors only s-states of the hydrogenic series. Furthermore, the Coulomb interaction also modifies the absorption at and above ϵ_g. Instead of the joint density of states seen in the noninteracting case, the absorption due to transitions to interacting, but unbound electron-hole pairs is more or less constant as function of energy. More details and analytical evaluations for bulk, quantum-well, -wire, and also -dot systems can be found, e.g. in Ref. 1.

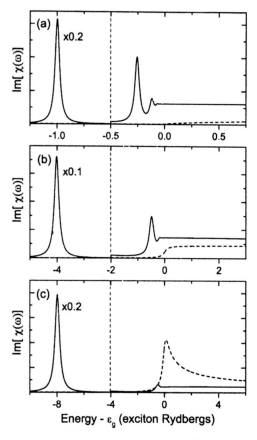

Fig. 1.   Imaginary part of the linear susceptibility, $Im[\chi^{(1)}(\omega)]$, which is proportional to the linear optical absorption spectra, $\alpha^{(1)}(\omega)$, in the vicinity of the bandgap for a (a) three-, (b) two-, and (c) one-dimensional direct semiconductor, corresponding to bulk, quantum well, and quantum wire structures, respectively. The dashed lines are calculated without excitonic effects (vanishing Coulomb interaction), whereas the solid lines result from a full calculation. For each case a homogeneous broadening corresponding to 1/20th of the exciton binding energy has been used. For better visibility of the absorption close to the band edge, the peak due to the 1 s exciton (left of dashed lines) has been reduced as indicated. Taken from Ref. 27.

2.4. Many-Body Correlation Effects and Their Influence on Experiments

In this section two theoretical approaches for the many-body correlation effects are reviewed. We first focus on the so-called second Born approximation (SBA) and then describe the dynamics controlled truncation (DCT) scheme in the coherent third-order ($\chi^{(3)}$) and fifth-order ($\chi^{(5)}$) limits. For each case, besides the derivation of the equations, examples for some experimentally observable consequences of the respective many-body correlations are presented and discussed.

2.4.1. *Second Born Approximation*

One can treat relaxation and dephasing on the simplest level by phenomenologically adding the terms

$$
\left. \frac{\partial}{\partial t} P_{\mathbf{k}}^{eh}(t) \right|_{corr} = -\frac{P_{\mathbf{k}}^{eh}(t)}{T_2},
$$

$$
\left. \frac{\partial}{\partial t} f_{\mathbf{k}}^{aa}(t) \right|_{corr} = -\frac{f_{\mathbf{k}}^{aa}(t) - F_{\mathbf{k}}^{aa}}{T_{rel}},
\tag{17}
$$

with $aa = ee$ and $aa = hh$ to the HF–SBE, Eq. (12). These terms introduce the exponential decay of the polarization with the decay time (dephasing time) T_2 and the thermalization of the actual carrier distribution towards the quasi-equilibrium distribution $F_{\mathbf{k}}^{aa}$ with the relaxation time T_{rel}. Even though such simple decay and relaxation processes are sometimes compatible with experimental information, the systematic derivation and computation of relaxation and dephasing times are not straightforward. This is to be expected, since, as is shown below, in the microscopic approach the decay and relaxation terms are **k**-dependent and furthermore depend on the excitations that are present in the systems, i.e. they are complicated functionals of $P_{\mathbf{k}}^{eh}$ and $f_{\mathbf{k}}^{aa}$.

A microscopically consistent way to obtain approximations beyond the Hartree–Fock level is to derive equations for the four-particle correlation functions and to use systematic factorization procedures at a higher level. This can be done, e.g. by using Green's function techniques, projection operator or equation of motion methods.[4–7,28] The SBA accounts for Coulomb effects beyond the mean-field Hartree–Fock level, by considering all correlation terms up to quadratic order in the screened Coulomb interaction. This scheme allows one to describe interaction-induced dephasing due to carrier–carrier scattering and higher-order polarization interaction as well as the corresponding dynamic energy renormalizations beyond the Hartree–Fock level. When combined with a Markov approximation, the correlation terms reduce in the incoherent limit ($P_{\mathbf{k}}^{eh} \equiv 0$) to Boltzmann scattering integrals as well as generalized Coulomb enhancement and bandgap renormalization terms.[15]

In the coherent regime, the SBA allows one to study the dynamics of the coherent excitonic polarization and the coherently driven carrier population under the influence of carrier–carrier scattering and nonlinear polarization interaction. Note that unlike the DCT, see below, the derived terms are not restricted to certain powers of the exciting field. On the other hand, the truncation in terms of powers of the screened Coulomb interaction excludes the consideration of bound biexciton states in the SBA.

Since the derivation of all correlation contributions that arise in second Born–Markov approximation is straightforward but tedious and can be found in Refs. 4–7 and 28, we restrict the discussion here to the most important steps using one correlation function as an example. Furthermore, for clarity we restrict the following analysis to the two-band case and include the additional terms arising in a multiband situation later. Using the 2×2 band situation with diagonal selection rules introduced above, the system reduces to the two-band situation when the incident fields have all the same circular polarization such that only one spin subsystem of the spin-degenerate electron and heavy-hole bands is excited.

Let us consider, e.g. $\langle a_{e,\mathbf{k}}^\dagger a_{e,\mathbf{k'}-\mathbf{q}}^\dagger a_{e,\mathbf{k}-\mathbf{q}} a_{e,\mathbf{k'}} \rangle$ which is one term that appears when the electron occupation $f_{\mathbf{k}}^{ee}$ is commuted with the Coulomb interaction H_C.[1] One starts by defining a reduced four-operator correlation function by subtracting the uncorrelated Hartree–Fock part via

$$\Delta\langle a_{e,\mathbf{k}}^\dagger a_{e,\mathbf{k'}-\mathbf{q}}^\dagger a_{e,\mathbf{k}-\mathbf{q}} a_{e,\mathbf{k'}} \rangle = \langle a_{e,\mathbf{k}}^\dagger a_{e,\mathbf{k'}-\mathbf{q}}^\dagger a_{e,\mathbf{k}-\mathbf{q}} a_{e,\mathbf{k'}} big\rangle$$
$$- \langle a_{e,\mathbf{k}}^\dagger a_{e,\mathbf{k}} \rangle \langle a_{e,\mathbf{k}-\mathbf{q}}^\dagger a_{e,\mathbf{k}-\mathbf{q}} \rangle \delta_{\mathbf{k},\mathbf{k'}}. \tag{18}$$

The equation of motion for this correlated part of the four-operator correlation function is given by

$$\frac{\partial}{\partial t}\Delta\langle a_{e,\mathbf{k}}^\dagger a_{e,\mathbf{k'}-\mathbf{q}}^\dagger a_{e,\mathbf{k}-\mathbf{q}} a_{e,\mathbf{k'}} \rangle = \frac{\partial}{\partial t}\langle a_{e,\mathbf{k}}^\dagger a_{e,\mathbf{k'}-\mathbf{q}}^\dagger a_{e,\mathbf{k}-\mathbf{q}} a_{e,\mathbf{k'}} \rangle$$
$$- \left(\frac{\partial}{\partial t}\langle a_{e,\mathbf{k}}^\dagger a_{e,\mathbf{k}} \rangle \right) \langle a_{e,\mathbf{k}-\mathbf{q}}^\dagger a_{e,\mathbf{k}-\mathbf{q}} \rangle \delta_{\mathbf{k},\mathbf{k'}}$$
$$- \langle a_{e,\mathbf{k}}^\dagger a_{e,\mathbf{k}} \rangle \left(\frac{\partial}{\partial t}\langle a_{e,\mathbf{k}-\mathbf{q}}^\dagger a_{e,\mathbf{k}-\mathbf{q}} \rangle \right) \delta_{\mathbf{k},\mathbf{k'}}. \tag{19}$$

The time derivates on the right-hand side of Eq. (19) can be obtained using the Heisenberg equation, Eq. (4). Evaluating the commutators, one finds that the resulting expression can be written as

$$\frac{\partial}{\partial t}\Delta\langle a^{\dagger}_{e,\mathbf{k}}a^{\dagger}_{e,\mathbf{k'-q}}a_{e,\mathbf{k-q}}a_{e,\mathbf{k'}}\rangle = \frac{\partial}{\partial t}\Delta\langle a^{\dagger}_{e,\mathbf{k}}a^{\dagger}_{e,\mathbf{k'-q}}a_{e,\mathbf{k-q}}a_{e,\mathbf{k'}}\rangle\Big|_{\text{hom}}$$

$$+ \frac{\partial}{\partial t}\Delta\langle a^{\dagger}_{e,\mathbf{k}}a^{\dagger}_{e,\mathbf{k'-q}}a_{e,\mathbf{k-q}}a_{e,\mathbf{k'}}\rangle\Big|_{\text{inhom}}, \tag{20}$$

with

$$i\hbar\frac{\partial}{\partial t}\Delta\langle a^{\dagger}_{e,\mathbf{k}}a^{\dagger}_{e,\mathbf{k'-q}}a_{e,\mathbf{k-q}}a_{e,\mathbf{k'}}\rangle\Big|_{\text{hom}}$$

$$= -\left(-\varepsilon^{e}_{\mathbf{k}} - \varepsilon^{e}_{\mathbf{k'-q}} + \varepsilon^{e}_{\mathbf{k-q}} + \varepsilon^{e}_{\mathbf{k'}}\right)\Delta\langle a^{\dagger}_{e,\mathbf{k}}a^{\dagger}_{e,\mathbf{k'-q}}a_{e,\mathbf{k-q}}a_{e,\mathbf{k'}}\rangle. \tag{21}$$

The homogeneous part of Eq. (20) contains simply differences of the kinetic energies. Its inhomogeneous part is purely given by the commutator with the Coulomb interaction, which leads to lengthy expressions containing coupling to six-operator terms.

In the second Born–Markov treatment one solves Eq. (20) in the adiabatic limit, i.e.,

$$\Delta\langle a^{\dagger}_{e,\mathbf{k}}a^{\dagger}_{e,\mathbf{k'-q}}a_{e,\mathbf{k-q}}a_{e,\mathbf{k'}}\rangle \approx \frac{\frac{\partial}{\partial t}\Delta\langle a^{\dagger}_{e,\mathbf{k}}a^{\dagger}_{e,\mathbf{k'-q}}a_{e,\mathbf{k-q}}a_{e,\mathbf{k'}}\rangle\Big|_{\text{inhom}}}{\frac{i}{\hbar}\left(-\varepsilon^{e}_{\mathbf{k}} - \varepsilon^{e}_{\mathbf{k'-q}} + \varepsilon^{e}_{\mathbf{k-q}} + \varepsilon^{e}_{\mathbf{k'}}\right) - \gamma}, \tag{22}$$

where γ is decay constant. In the expressions given below we use the limit $\gamma \to 0$. The SBA corresponds to factorizing all four- and six-operator terms that appear in the numerator on the right-hand side of Eq. (22) into products of expectation values of two operators, i.e. interband polarizations and carrier occupations.

At the level of the second Born–Markov approximation, memory effects are ignored and the analysis is restricted up to two-particle collisions, i.e. one keeps all terms up to quadratic order in the screened Coulomb potential. For a two-band situation one obtains the equation of motion for the microscopic polarization $P^{eh}_{\mathbf{k}}(t)$ as[5,7]

$$\left[i\hbar\frac{\partial}{\partial t} - \varepsilon^{e}_{\mathbf{k}}(t) - \varepsilon^{h}_{\mathbf{k}}(t)\right]P^{eh}_{\mathbf{k}}(t) = \left[1 - f^{e}_{\mathbf{k}} - f^{h}_{\mathbf{k}}\right]\Omega_{\mathbf{k}}(t)$$

$$- i\Gamma_{\mathbf{k}}P^{eh}_{\mathbf{k}}(t) + i\sum_{\mathbf{k'}}\Gamma_{\mathbf{k},\mathbf{k'}}P^{eh}_{\mathbf{k+k'}}(t), \tag{23}$$

where $\Omega_{\mathbf{k}}(t)$ and $\varepsilon^{a}_{\mathbf{k}}(t)$ with $(a = e, h)$ are the renormalized field and single-particle energies defined in Eqs. (13) and (14). Here, we use the short notation for the diagonal populations, i.e. $f^{e}_{\mathbf{k}} \equiv f^{ee}_{\mathbf{k}}$ and $f^{h}_{\mathbf{k}} \equiv f^{hh}_{\mathbf{k}}$.

The correlation contributions of Eq. (23) are either diagonal in the carrier momentum

$$\Gamma_{\mathbf{k}} = \sum_{\mathbf{k}',\mathbf{k}''} \sum_{a,b=e,h} g\left(\varepsilon_{\mathbf{k}}^a + \varepsilon_{\mathbf{k}'+\mathbf{k}''}^b - \varepsilon_{\mathbf{k}''}^b - \varepsilon_{\mathbf{k}'+\mathbf{k}}^a\right)$$
$$\times \left[W_{\mathbf{k}'}^2 - \delta_{ab} W_{\mathbf{k}'} W_{\mathbf{k}-\mathbf{k}''}\right]$$
$$\times \left[(1 - f_{\mathbf{k}'+\mathbf{k}''}^b) f_{\mathbf{k}''}^b f_{\mathbf{k}'+\mathbf{k}}^a + f_{\mathbf{k}'+\mathbf{k}''}^b (1 - f_{\mathbf{k}''}^b)(1 - f_{\mathbf{k}'+\mathbf{k}}^a)\right.$$
$$\left. - P_{\mathbf{k}'+\mathbf{k}''}^{eh}(P_{\mathbf{k}''}^{eh})^*\right], \tag{24}$$

or couple different states $\mathbf{k}$ and $\mathbf{k}'$

$$\Gamma_{\mathbf{k},\mathbf{k}'} = \sum_{\mathbf{k}''} \sum_{a,b=e,h} g\left(-\varepsilon_{\mathbf{k}}^a - \varepsilon_{\mathbf{k}'+\mathbf{k}''}^b + \varepsilon_{\mathbf{k}''}^b + \varepsilon_{\mathbf{k}'+\mathbf{k}}^a\right)$$
$$\times \left[W_{\mathbf{k}'}^2 - \delta_{ab} W_{\mathbf{k}'} W_{\mathbf{k}-\mathbf{k}''}\right]$$
$$\times \left[(1 - f_{\mathbf{k}}^a)(1 - f_{\mathbf{k}'+\mathbf{k}''}^b) f_{\mathbf{k}''}^b + f_{\mathbf{k}}^a f_{\mathbf{k}'+\mathbf{k}''}^b (1 - f_{\mathbf{k}''}^b)\right.$$
$$\left. - (P_{\mathbf{k}'+\mathbf{k}''}^{eh})^* P_{\mathbf{k}''}^{eh}\right], \tag{25}$$

where $W_{\mathbf{k}}$ is the screened Coulomb interaction which is evaluated in quasistatic RPA approximation (Lindhard formula), $g(\varepsilon) = \pi\delta(\varepsilon) + \mathcal{P}\frac{i}{\varepsilon}$, and $\mathcal{P}$ denotes that the principal value of the corresponding integral has to be taken.

The occupations for electrons and holes $f_{\mathbf{k}}^e(t)$ and $f_{\mathbf{k}}^h(t)$ obey the following equation

$$i\hbar \frac{\partial}{\partial t} f_{\mathbf{k}}^a(t) = -2\text{Im}\left[\Omega_{\mathbf{k}}(t)(P_{\mathbf{k}}^{eh}(t))^*\right]$$
$$+ i\{\Sigma_{\mathbf{k}}^{\text{in},a}(t)\left[1 - f_{\mathbf{k}}^a(t)\right] - \Sigma_{\mathbf{k}}^{\text{out},a}(t) f_{\mathbf{k}}^a(t) + \Sigma_{\mathbf{k}}^{\text{pol},a}(t)\}. \tag{26}$$

The correlation terms in Eq. (26) describe the changes of the carrier occupation probabilities $f_{\mathbf{k}}^a$ due to scattering of carriers into the state with momentum $\mathbf{k}$

$$\Sigma_{\mathbf{k}}^{\text{in},a} = 2\pi \sum_{\mathbf{k}',\mathbf{k}''} \sum_{b=e,h} \left[W_{\mathbf{k}'}^2 - \delta_{ab} W_{\mathbf{k}'} W_{\mathbf{k}-\mathbf{k}''}\right]$$
$$\times \delta\left(\varepsilon_{\mathbf{k}}^a + \varepsilon_{\mathbf{k}'+\mathbf{k}''}^b - \varepsilon_{\mathbf{k}''}^b - \varepsilon_{\mathbf{k}'+\mathbf{k}}^a\right)$$
$$\times (1 - f_{\mathbf{k}'+\mathbf{k}''}^b) f_{\mathbf{k}''}^b f_{\mathbf{k}'+\mathbf{k}}^a, \tag{27}$$

due to scattering of carriers out of the $\mathbf{k}$-state,

$$\Sigma_{\mathbf{k}}^{\text{out},a} = 2\pi \sum_{\mathbf{k}',\mathbf{k}''} \sum_{b=e,h} \left[W_{\mathbf{k}'}^2 - \delta_{ab} W_{\mathbf{k}'} W_{\mathbf{k}-\mathbf{k}''}\right]$$
$$\times \delta\left(\varepsilon_{\mathbf{k}}^a + \varepsilon_{\mathbf{k}'+\mathbf{k}''}^b - \varepsilon_{\mathbf{k}''}^b - \varepsilon_{\mathbf{k}'+\mathbf{k}}^a\right)$$
$$\times f_{\mathbf{k}'+\mathbf{k}''}^b (1 - f_{\mathbf{k}''}^b)(1 - f_{\mathbf{k}'+\mathbf{k}}^a), \tag{28}$$

as well as due to nonlinear polarization interaction,

$$
\begin{aligned}
\Sigma_{\mathbf{k}}^{\text{pol},a} = \sum_{\mathbf{k}',\mathbf{k}''} \sum_{b=e,h} & \left[W_{\mathbf{k}'}^2 - \delta_{ab} W_{\mathbf{k}'} W_{\mathbf{k}-\mathbf{k}''} \right] \\
\times & \left\{ g \left(\varepsilon_{\mathbf{k}}^a + \varepsilon_{\mathbf{k}'+\mathbf{k}''}^b - \varepsilon_{\mathbf{k}''}^b - \varepsilon_{\mathbf{k}'+\mathbf{k}}^a \right) \right. \\
\times & \left[\left(f_{\mathbf{k}}^a - f_{\mathbf{k}'+\mathbf{k}}^a \right) P_{\mathbf{k}'+\mathbf{k}''}^{eh} (P_{\mathbf{k}''*}^{eh}) + \text{c.c.} \right] \\
+ & g \left(\varepsilon_{\mathbf{k}}^{\bar{a}} + \varepsilon_{\mathbf{k}'+\mathbf{k}''}^b - \varepsilon_{\mathbf{k}''}^b - \varepsilon_{\mathbf{k}'+\mathbf{k}}^{\bar{a}} \right) \\
\times & \left. \left[\left(f_{\mathbf{k}''}^b - f_{\mathbf{k}'+\mathbf{k}''}^b \right) P_{\mathbf{k}}^{eh} (P_{\mathbf{k}'+\mathbf{k}}^{eh}) + \text{c.c.} \right] \right\}.
\end{aligned}
\tag{29}
$$

For details, see Refs. 7 and 15.

To gain information on the dynamics of photoexcited semiconductors, one often performs experiments where the system is excited by a sequence of short pulses. Four-wave mixing and pump probe are widely used examples for two-pulse experiments. In the following we focus on the theoretical description of the latter, since we compare below measurements and theory for this experiment.

To describe a pump-probe situation with two optical field components propagating in directions $\mathbf{k}_{\text{pump}}$ and $\mathbf{k}_{\text{probe}}$, i.e.

$$
\mathbf{E}^{\text{total}}(\mathbf{r}, t) = \mathbf{E}(\mathbf{r}, t) e^{i\mathbf{k}_{\text{pump}}\cdot\mathbf{r}} + \tilde{\mathbf{E}}(\mathbf{r}, t) e^{i\mathbf{k}_{\text{probe}}\cdot\mathbf{r}},
\tag{30}
$$

a spatial Fourier decomposition of the photoexcited material quantities can be used.[29] Assuming that the probe field is weak such that only terms linear in the probe field have to be considered, probe field induced changes of the carrier occupation probabilities $f_{\mathbf{k}}^{e,h}$ can be neglected. In this limit Eqs. (23)–(29) can be used to describe the nonlinear dynamics due to the pump pulse where $P_{\mathbf{k}}^{eh}(t)$ and $f_{\mathbf{k}}^a(t)$ are the pump-field induced microscopic polarizations and carrier occupations. Thus, $P_{\mathbf{k}}^{eh}(t)$ is the spatial Fourier component of the microscopic polarization which propagates in the direction of the pump pulse, i.e. $\mathbf{k}_{\text{pump}}$.[29] The combined action of pump and probe fields (and their induced polarization components) leads to Fourier components of the carrier population $\delta f_{\mathbf{k}}^{e,h} \propto e^{i(\mathbf{k}_{\text{probe}}-\mathbf{k}_{\text{pump}})\cdot\mathbf{r}}$ which are related to population pulsations.[30] These Fourier components enable scattering of the pump field into probe direction as well as coupling the pump-induced polarization to the probe field. They are also responsible for a diffracted four-wave mixing signal in the direction $2\mathbf{k}_{\text{pump}} - \mathbf{k}_{\text{probe}}$ which will not be considered here.

The equation of motion for the microscopic polarization which is linear in the probe $\tilde{\mathbf{E}}(t)$ is given by

$$\left[i\hbar\frac{\partial}{\partial t} - \varepsilon_{\mathbf{k}}^{e}(t) - \varepsilon_{\mathbf{k}}^{h}(t)\right]\tilde{P}_{\mathbf{k}}^{eh}(t) - \left[1 - f_{\mathbf{k}}^{e} - f_{\mathbf{k}}^{h}\right]\tilde{\Omega}_{\mathbf{k}}(t)$$

$$- \left[\tilde{\varepsilon}_{\mathbf{k}}^{e}(t) + \tilde{\varepsilon}_{\mathbf{k}}^{h}(t)\right]P_{\mathbf{k}}^{eh}(t) - \left[\delta f_{\mathbf{k}}^{e} + \delta f_{\mathbf{k}}^{h}\right]\Omega_{\mathbf{k}}(t)$$

$$= -i\Gamma_{\mathbf{k}}\tilde{P}_{\mathbf{k}}^{eh}(t) + i\sum_{\mathbf{k}'}\Gamma_{\mathbf{k},\mathbf{k}'}\tilde{P}_{\mathbf{k}+\mathbf{k}'}^{eh}(t)$$

$$- i\tilde{\Gamma}_{\mathbf{k}}P_{\mathbf{k}}^{eh}(t) + i\sum_{\mathbf{k}'}\tilde{\Gamma}_{\mathbf{k},\mathbf{k}'}P_{\mathbf{k}+\mathbf{k}'}^{eh}(t), \qquad (31)$$

where the renormalized probe field is

$$\tilde{\Omega}_{\mathbf{k}}(t) = \mathbf{d}_{cv}\cdot\tilde{\mathbf{E}}(t) + \sum_{\mathbf{k}'}V_{|\mathbf{k}-\mathbf{k}'|}\tilde{P}_{\mathbf{k}'}^{eh}(t). \qquad (32)$$

Via the last term on the left-hand side of Eq. (31), that contains the $\delta f_{\mathbf{k}}^{e,h}$ Fourier components of the carrier populations, the renormalized field of the *pump* directly influences the *probe* polarization. While the carrier occupation probabilities $f_{\mathbf{k}}^{e,h}$ lead to Hartree–Fock energy renormalizations according to Eq. (14), the additional Fourier components $\delta f_{\mathbf{k}}^{e,h}$ also introduce new renormalization terms

$$\tilde{\varepsilon}_{\mathbf{k}}^{a}(t) = -\sum_{\mathbf{k}'}V_{|\mathbf{k}-\mathbf{k}'|}\delta f_{\mathbf{k}'}^{a}(t), \qquad (33)$$

which mediate the direct entering of the pump-induced microscopic polarization $P_{\mathbf{k}}^{eh}(t)$ on the left-hand side of Eq. (31). The equation of motion for the $\delta f_{\mathbf{k}}^{e,h}$ Fourier components of the carrier populations is given by

$$i\hbar\frac{\partial}{\partial t}\delta f_{\mathbf{k}}^{a}(t) + \tilde{\Omega}_{\mathbf{k}}(t)(P_{\mathbf{k}}^{eh}(t))^{*} - \Omega_{\mathbf{k}}^{*}(t)\tilde{P}_{\mathbf{k}}^{eh}(t)$$

$$= i\Big\{ -\left[\Sigma_{\mathbf{k}}^{\mathrm{in},a}(t) + \Sigma_{\mathbf{k}}^{\mathrm{out},a}(t)\right]\delta f_{\mathbf{k}}^{a}(t)$$

$$+ \tilde{\Sigma}_{\mathbf{k}}^{\mathrm{in},a}(t)\left[1 - f_{\mathbf{k}}^{a}(t)\right] - \tilde{\Sigma}_{\mathbf{k}}^{\mathrm{out},a}(t)f_{\mathbf{k}}^{a}(t) + \tilde{\Sigma}_{\mathbf{k}}^{\mathrm{pol},a}(t)\Big\}, \qquad (34)$$

where the second and third terms on the left-hand side clearly reveal the combination of the *probe*-induced field renormalization with the pump-induced polarization and vice versa as the driving source of $\delta f_{\mathbf{k}}^{e,h}$.

To analyze the scattering terms in the 2×2 band situation with diagonal selection rules introduced above, in principle, additional sub-band (spin) indices and summations have to be added to the equations given above. To keep the notation as simple as possible, we discuss in the following only the results of such a consideration.

In Eqs. (24)–(29) we have assumed that the pump field is circularly polarized such that only one spin subsystem of the spin-degenerate electron and heavy-hole bands is excited. Then $f_\mathbf{k}^a$ and $P_\mathbf{k}^{eh}$ are the occupation probabilities and microscopic of this spin-subsystem.

If the pump field equally excites both spin-subsystems, a factor of 2 has to be added to $W_{\mathbf{k}'}^2$ in Eqs. (24)–(29) since the direct Coulomb interaction, described by these terms, allows also for scattering of carriers from one subsystem with carriers from the other subsystem, which effectively doubles the scattering probability. However, the exchange interaction, described by $\delta_{ab} W_{\mathbf{k}'} W_{\mathbf{k}-\mathbf{k}''}$, is limited to carriers of the same spin-subsystem.

When the probe field has the same circular optical polarization as the pump field, the first set of correlation contributions to the probe polarization, $\Gamma_\mathbf{k}$ and $\Gamma_{\mathbf{k},\mathbf{k}'}$, are given by Eqs. (24) and (25), respectively. Note that the last term of these equations describes a correlation-induced coupling of the pump-polarization $P_\mathbf{k}$ to the probe signal via direct and exchange Coulomb interactions, which appears in addition to the coupling via Hartree–Fock terms discussed above.

The $\delta f_\mathbf{k}^{e,h}$ Fourier components of the carrier populations lead to further correlations terms, $\tilde{\Gamma}_\mathbf{k}$ and $\tilde{\Gamma}_{\mathbf{k},\mathbf{k}'}$, which also allow for coupling of the pump-polarization to the probe signal in Eq. (31). In case of co-circular pump and probe polarization, one finds

$$
\begin{aligned}
\tilde{\Gamma}_\mathbf{k} = \sum_{\mathbf{k}',\mathbf{k}''} \sum_{a,b=e,h} & g\left(\varepsilon_\mathbf{k}^a + \varepsilon_{\mathbf{k}'+\mathbf{k}''}^b - \varepsilon_{\mathbf{k}''}^b - \varepsilon_{\mathbf{k}'+\mathbf{k}}^a\right) \\
& \times \left[W_{\mathbf{k}'}^2 - \delta_{ab} W_{\mathbf{k}'} W_{\mathbf{k}-\mathbf{k}''}\right] \\
& \times \left[-\delta f_{\mathbf{k}'+\mathbf{k}''}^b f_{\mathbf{k}''}^b f_{\mathbf{k}'+\mathbf{k}}^a + \delta f_{\mathbf{k}'+\mathbf{k}''}^b (1 - f_{\mathbf{k}''}^b)(1 - f_{\mathbf{k}'+\mathbf{k}}^a)\right. \\
& + (1 - f_{\mathbf{k}'+\mathbf{k}''}^b)\delta f_{\mathbf{k}''}^b f_{\mathbf{k}'+\mathbf{k}}^a - f_{\mathbf{k}'+\mathbf{k}''}^b \delta f_{\mathbf{k}''}^b (1 - f_{\mathbf{k}'+\mathbf{k}}^a) \\
& + (1 - f_{\mathbf{k}'+\mathbf{k}''}^b)f_{\mathbf{k}''}^b \delta f_{\mathbf{k}'+\mathbf{k}}^a - f_{\mathbf{k}'+\mathbf{k}''}^b (1 - f_{\mathbf{k}''}^b)\delta f_{\mathbf{k}'+\mathbf{k}}^a \\
& \left. - \tilde{P}_{\mathbf{k}'+\mathbf{k}''}^{eh} (P_{\mathbf{k}''}^{eh})^*\right],
\end{aligned}
\tag{35}
$$

$$
\begin{aligned}
\tilde{\Gamma}_{\mathbf{k},\mathbf{k}'} = \sum_{\mathbf{k}''} \sum_{a,b=e,h} & g\left(-\varepsilon_\mathbf{k}^a - \varepsilon_{\mathbf{k}'+\mathbf{k}''}^b + \varepsilon_{\mathbf{k}''}^b + \varepsilon_{\mathbf{k}'+\mathbf{k}}^a\right) \\
& \times \left[W_{\mathbf{k}'}^2 - \delta_{ab} W_{\mathbf{k}'} W_{\mathbf{k}-\mathbf{k}''}\right] \\
& \times \left[-\delta f_\mathbf{k}^a (1 - f_{\mathbf{k}'+\mathbf{k}''}^b)f_{\mathbf{k}''}^b + \delta f_\mathbf{k}^a f_{\mathbf{k}'+\mathbf{k}''}^b (1 - f_{\mathbf{k}''}^b)\right. \\
& - (1 - f_\mathbf{k}^a)\delta f_{\mathbf{k}'+\mathbf{k}''}^b f_{\mathbf{k}''}^b + f_\mathbf{k}^a \delta f_{\mathbf{k}'+\mathbf{k}''}^b (1 - f_{\mathbf{k}''}^b) \\
& + (1 - f_\mathbf{k}^a)(1 - f_{\mathbf{k}'+\mathbf{k}''}^b)\delta f_{\mathbf{k}''}^b - f_\mathbf{k}^a f_{\mathbf{k}'+\mathbf{k}''}^b \delta f_{\mathbf{k}''}^b \\
& \left. - (P_{\mathbf{k}'+\mathbf{k}''}^{eh})^* \tilde{P}_{\mathbf{k}''}^{eh}\right].
\end{aligned}
\tag{36}
$$

Note that in comparison to Eqs. (24) and (25) the Fourier component $\propto e^{i(\mathbf{k}_{\text{probe}}-\mathbf{k}_{\text{pump}})\cdot\mathbf{r}}$ is considered by means of $\delta f_{\mathbf{k}}^a$ as well as the combination of pump- and probe-polarizations. Similarly, the additional correlation terms in Eq. (34) are given by

$$
\begin{aligned}
\tilde{\Sigma}_{\mathbf{k}}^{in,a} = 2\pi \sum_{\mathbf{k}',\mathbf{k}''} \sum_{b=e,h} & \delta\big(\varepsilon_{\mathbf{k}}^a + \varepsilon_{\mathbf{k}'+\mathbf{k}''}^b - \varepsilon_{\mathbf{k}''}^b - \varepsilon_{\mathbf{k}'+\mathbf{k}}^a\big) \\
& \times \big[W_{\mathbf{k}'}^2 - \delta_{ab}W_{\mathbf{k}'}W_{\mathbf{k}-\mathbf{k}''}\big] \\
& \times \big[-\delta f_{\mathbf{k}'+\mathbf{k}''}^b f_{\mathbf{k}''}^b f_{\mathbf{k}'+\mathbf{k}}^a \\
& + (1 - f_{\mathbf{k}'+\mathbf{k}''}^b)\delta f_{\mathbf{k}''}^b f_{\mathbf{k}'+\mathbf{k}}^a + (1 - f_{\mathbf{k}'+\mathbf{k}''}^b)f_{\mathbf{k}''}^b \delta f_{\mathbf{k}'+\mathbf{k}}^a\big], \quad (37)
\end{aligned}
$$

$$
\begin{aligned}
\tilde{\Sigma}_{\mathbf{k}}^{out,a} = 2\pi \sum_{\mathbf{k}',\mathbf{k}''} \sum_{b=e,h} & \delta\big(\varepsilon_{\mathbf{k}}^a + \varepsilon_{\mathbf{k}'+\mathbf{k}''}^b - \varepsilon_{\mathbf{k}''}^b - \varepsilon_{\mathbf{k}'+\mathbf{k}}^a\big) \\
& \times \big[W_{\mathbf{k}'}^2 - \delta_{ab}W_{\mathbf{k}'}W_{\mathbf{k}-\mathbf{k}''}\big] \\
& \times \big[\delta f_{\mathbf{k}'+\mathbf{k}''}^b(1 - f_{\mathbf{k}''}^b)(1 - f_{\mathbf{k}'+\mathbf{k}}^a) \\
& - f_{\mathbf{k}'+\mathbf{k}''}^b \delta f_{\mathbf{k}''}^b(1 - f_{\mathbf{k}'+\mathbf{k}}^a) - f_{\mathbf{k}'+\mathbf{k}''}^b(1 - f_{\mathbf{k}''}^b)\delta f_{\mathbf{k}'+\mathbf{k}}^a\big], \quad (38)
\end{aligned}
$$

$$
\begin{aligned}
\tilde{\Sigma}_{\mathbf{k}}^{pol,a} = \sum_{\mathbf{k}',\mathbf{k}''} \sum_{b=e,h} & \big[W_{\mathbf{k}'}^2 - \delta_{ab}W_{\mathbf{k}'}W_{\mathbf{k}-\mathbf{k}''}\big] \\
& \times \Big\{ g\big(\varepsilon_{\mathbf{k}}^a + \varepsilon_{\mathbf{k}'+\mathbf{k}''}^b - \varepsilon_{\mathbf{k}''}^b - \varepsilon_{\mathbf{k}'+\mathbf{k}}^a\big) \\
& \times \big[(f_{\mathbf{k}}^a - f_{\mathbf{k}'+\mathbf{k}}^a)\tilde{P}_{\mathbf{k}'+\mathbf{k}''}^{eh}(P_{\mathbf{k}''}^{eh})^* \\
& + (\delta f_{\mathbf{k}}^a - \delta f_{\mathbf{k}'+\mathbf{k}}^a)P_{\mathbf{k}'+\mathbf{k}''}^{eh}(P_{\mathbf{k}''}^{eh})^* + \text{c.c.}\big] \\
& + g\big(\varepsilon_{\mathbf{k}}^{\bar{a}} + \varepsilon_{\mathbf{k}'+\mathbf{k}''}^b - \varepsilon_{\mathbf{k}''}^b - \varepsilon_{\mathbf{k}'+\mathbf{k}}^{\bar{a}}\big) \\
& \times \big[(f_{\mathbf{k}''}^b - f_{\mathbf{k}'+\mathbf{k}''}^b)\tilde{P}_{\mathbf{k}}^{eh}(P_{\mathbf{k}'+\mathbf{k}}^{eh})^* \\
& + (\delta f_{\mathbf{k}''}^b - \delta f_{\mathbf{k}'+\mathbf{k}''}^b)P_{\mathbf{k}}^{eh}(P_{\mathbf{k}'+\mathbf{k}}^{eh})^* + \text{c.c.}\big]\Big\}. \quad (39)
\end{aligned}
$$

For opposite circular optical polarization of pump and probe pulses various changes in the discussed equations are necessary:

(I) The $\delta f_{\mathbf{k}}^{e,h}$ Fourier components of the carrier populations do not occur since the electronic polarization driven in one spin-subsystem cannot couple to the optical field which excites the other spin-subsystem. Correspondingly, the diffraction of the pump field into probe direction and the nonlinear interaction of the pump-induced electronic polarization with the probe polarization are absent.

(II) From the spin-dependence of the involved carrier operators it follows that in Eq. (31), and in the Hartree–Fock and correlation terms of this equation, $f_{\mathbf{k}}^a$ and $f_{\mathbf{k}'+\mathbf{k}}^a$ are belonging to the spin-subsystem the probe pulse is interacting with. Since for cross-circular pump-probe excitation the pump field does not populate this spin-subsystem and the weak probe field does not lead to population effects, $f_{\mathbf{k}}^a$ and $f_{\mathbf{k}'+\mathbf{k}}^a$ are zero in Eq. (31) and in the Hartee–Fock and correlation terms of this equation. Hence there are no phase-space filling and energy renormalization contributions of the pump field to the probe signal and the interaction-induced dephasing will be reduced.

(III) The exchange contributions to the correlation terms $\propto \delta_{ab} W_{\mathbf{k}'} W_{\mathbf{k}-\mathbf{k}''}$ do not contribute in Eq. (31) since the exchange interaction does not couple different spin sub-systems.

Note that all Hartree–Fock and correlation terms entering Eqs. (23) and (26) are independent of the optical polarization of the probe field.

2.4.2. Density-Dependent Saturation and Broadening Effects

In the equation of motion for the microscopic polarization in SBA, one recognizes the structure of a Boltzmann scattering integral, however, the polarization itself, as well as the rates Γ are complex quantities. The term with $\Gamma_{\mathbf{k}}$ on the right-hand side of Eq. (23) can be considered as a microscopic diagonal dephasing rate, whereas the second term with $\Gamma_{\mathbf{k},\mathbf{k}'}$ represents nondiagonal dephasing. Diagonal and nondiagonal refer here to the matrix structure of the equation when the different $P_{\mathbf{k}}^{eh}$ are written as a vector in $\mathbf{k}$-space and the right-hand side of Eq. (23) as the product of the dephasing matrix times that vector. Solving this equation, it turns out, that strong cancellation effects between the diagonal and nondiagonal contributions occur such that one obtains very incorrect estimates of the dephasing time if one, e.g. ignores the off-diagonal parts.

This strong compensation among the diagonal and nondiagonal terms is visualized in Fig. 2 by showing numerical results for the excitonic resonance in the presence of an incoherent electron-hole plasma. The left part of Fig. 2 demonstrates that the full calculation yields a well-defined excitonic resonance. If, however, the off-diagonal dephasing is omitted one finds an almost complete saturation already at the rather low plasma density of $10^{10}\,\mathrm{cm}^{-2}$ at $T = 77\,\mathrm{K}$. The right part of the figure compares the full result to that obtained assuming a constant dephasing time. The constant dephasing approximation yields artificial, density-dependent shifts of the saturating exciton resonance. Altogether these results demonstrate clearly the danger of describing electronic polarization dephasing at the simple level of T_2 times. At best, such times can be regarded as effective quantities

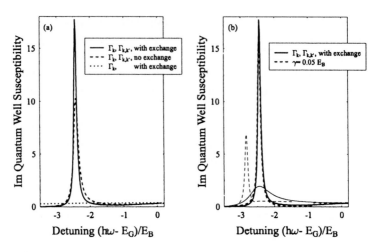

Fig. 2. (a) Computed excitonic absorption spectra (imaginary part of the optical susceptibility) using the parameters for an 8 nm GaAs-type quantum well and plasma den-
sity $10^{10}\,\text{cm}^{-2}$ at $T = 77\,\text{K}$. The solid line has been obtained with the full dephasing
terms, the dashed line without exchange contribution to the Coulomb matrix element,
the dotted line without the off-diagonal dephasing term in Eq. (23). (b) Comparison of
the full microscopic dephasing calculation (solid lines) with computations assuming con-
stant damping (dashed lines) for carrier density $10^{10}\,\text{cm}^{-2}$ (thick lines) and $10^{11}\,\text{cm}^{-2}$
(thin lines). Taken from Ref. 28.

that are empirically valid under some restrictive conditions. For a microscopic description of Coulombic dephasing, one, however, has to evaluate
the full polarization decay dynamics, according to Eq. (23).

The dependence of the effective dephasing on the level of optical excitation is investigated in Fig. 3. The top part of this figure shows the dynamical
evolution of the optical polarization after resonant excitation with a short fs
optical pulse. Besides the more rapid decay for higher excitation intensities,
the oscillations are the quantum beats between the excitonic resonances in
the system. The bottom part depicts the concomitant gradual saturation
of the exciton resonance in the absorption spectra as a consequence of
the *excitation-induced dephasing* which is the result of the dependence of
$\Gamma_{\mathbf{k}}$ and $\Gamma_{\mathbf{k},\mathbf{k}'}$ on the time-dependent population and polarization. One can
show directly that the $\mathbf{k}$ sum of the scattering rates Γ vanishes,[7] indicating that Coulomb dephasing of the optical polarization is a consequence of
destructive interference.[8]

Experimentally, excitation-induced dephasing has first been identified in
four-wave mixing studies performed with pre-pulses.[31] Excitonic saturation
and broadening has also been observed in pump-probe configuration, e.g.

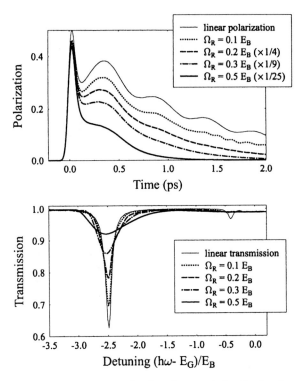

Fig. 3. Time-dependent polarization (top) and transmission (bottom) computed for an 8 nm GaAs single quantum-well structure. The curves are obtained assuming a 100 fs external laser pulse with increasing Rabi energy Ω_R (given in units of the exciton binding energy E_B). Taken from Ref. 28.

using resonant short pulse excitation. An example of the intensity dependent measured spectra for a multiple quantum-well structure is shown in Fig. 4.[15,32] The main qualitative features of the experimental results are in nice agreement with the microscopic theory, Fig. 3 (bottom part), i.e. the exciton resonance broadens with increasing intensity without the loss of oscillator strength and its spectral position is basically unchanged.

From a slightly different point of view the microscopic analysis of dephasing is nothing but a detailed lineshape theory. The SBA has also successfully been used to predict optical spectra in systems with elevated electron-hole-plasma densities where part of the absorption becomes negative, i.e. optical gain is realized. Such highly excited gain media are the basis for semiconductor lasers. The calculation of the proper gain lineshape of such structures has been a long standing problem since the use of a constant

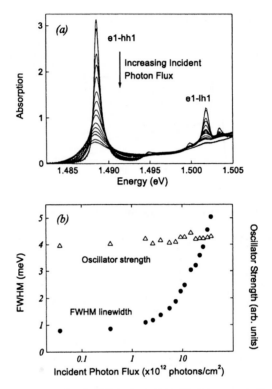

Fig. 4.   Exciton absorption (top), linewidth and oscillator strength (bottom) measured
for resonant excitation of a 20-quantum well sample using 100 fs pulses of increasing
intensity. Taken from Ref. 32.

dephasing approximation leads to the prediction of unphysical absorption
energetically below the gain region.[34] As it turns out, the numerical solutions of the semiconductor Bloch equations, where dephasing is treated
according to Eq. (23), provides a solution of the laser lineshape problem.
Such evaluations yield, furthermore, very good agreement with experimental results, as shown in Fig. 5. Much more background and details of the
microscopic semiconductor laser and gain theory, as well as many numerical
results, can be found in Ref. 34 and references cited therein.

2.4.3. Dynamics-Controlled Truncation Scheme: Coherent $\chi^{(3)}$-Approximation

Instead of using many-body methods that are nonperturbative in the fields
to close the dynamic equations for the reduced single-particle density

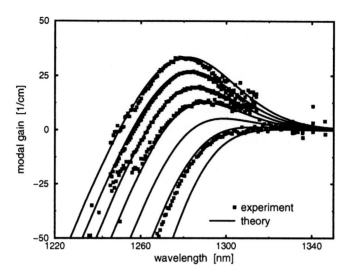

Fig. 5. Comparison of measured and calculated gain spectra of a (GaIn)(NAs)/GaAs ridge waveguide structure for injection currents of 7, 12, 14, 16 and 18 mA and sheet carrier densities of 0.625–1.625 × 10^{12} cm^2, respectively. Taken from Ref. 33.

matrix, Eq. (3), one can alternatively follow an approach where the nonlinear optical response is classified according to an expansion in powers of the applied field. Stahl and coworkers[12] were the first who recognized that this traditional nonlinear optics expansion[14] establishes a systematic truncation scheme of the Coulombic many-body correlations for purely coherent optical excitation configurations. In the following, we outline the basic steps that are involved in this procedure. More details can be found in Refs. 11, 12, 25, 35 and 36.

Studying the structure of the coupled equations for the correlation functions one finds that the four-operator terms which appear in the equation of motion for the microscopic polarization can be decomposed according to[11,35]

$$\left\langle a_{e,\mathbf{k}}^{\dagger} a_{e',\mathbf{k}'}^{\dagger} b_{h'',\mathbf{k}''} a_{e''',\mathbf{k}'''} \right\rangle = \sum_{\hat{\mathbf{k}},\hat{h}} \left\langle a_{e,\mathbf{k}}^{\dagger} a_{e',\mathbf{k}'}^{\dagger} b_{h'',\mathbf{k}''}^{\dagger} b_{\hat{h},\hat{\mathbf{k}}}^{\dagger} \right\rangle$$
$$\times \left\langle b_{\hat{h},\hat{\mathbf{k}}} a_{e''',\mathbf{k}'''} \right\rangle + O(E^5), \qquad (40)$$

where $\left\langle a_{e,\mathbf{k}}^{\dagger} a_{e',\mathbf{k}'}^{\dagger} b_{h'',\mathbf{k}''}^{\dagger} b_{\hat{h},\hat{\mathbf{k}}}^{\dagger} \right\rangle$ can be considered as an unfactorized product of polarization operators that is of second-order in the field, $\propto O(E^2)$ since two electron-hole pairs are created, whereas $\left\langle b_{\hat{h},\hat{\mathbf{k}}} a_{e''',\mathbf{k}'''} \right\rangle$ is linear in the field, $\propto O(E)$ since a single electron-hole pair is destroyed. Hence, the right-hand side of Eq. (40) is at least of third-order in the field. As explained below, the *mere correctness* of this decoupling scheme can be *verified* quite easily, whereas one needs more general considerations outlined below to see how to *find*, e.g. the right-hand side of Eq. (40).

In the most straightforward way one first notes that Eq. (40) is valid when the semiconductor is in its ground state, since in this case both sides vanish. Then one can take the time derivative of the lowest (third) order contributions of Eq. (40) which are given by

$$
\left\langle \frac{\partial}{\partial t}(a_{e,\mathbf{k}}^{\dagger} a_{e',\mathbf{k}'}^{\dagger} b_{h'',\mathbf{k}''}^{\dagger} a_{e''',\mathbf{k}'''}) \right\rangle
$$

$$
= \sum_{\hat{\mathbf{k}},\hat{h}} \left\langle \frac{\partial}{\partial t}(a_{e,\mathbf{k}}^{\dagger} a_{e',\mathbf{k}'}^{\dagger} b_{h'',\mathbf{k}''}^{\dagger} b_{\hat{h},\hat{\mathbf{k}}}^{\dagger}) \right\rangle \langle b_{\hat{h},\hat{\mathbf{k}}} a_{e''',\mathbf{k}'''} \rangle
$$

$$
+ \sum_{\hat{\mathbf{k}},\hat{h}} \langle a_{e,\mathbf{k}}^{\dagger} a_{e',\mathbf{k}'}^{\dagger} b_{h'',\mathbf{k}''}^{\dagger} b_{\hat{h},\hat{\mathbf{k}}}^{\dagger} \rangle \left\langle \frac{\partial}{\partial t}(b_{\hat{h},\hat{\mathbf{k}}} a_{e''',\mathbf{k}'''}) \right\rangle. \qquad (41)
$$

Computing the time derivatives by evaluating the commutators with the Hamiltonian according to the Heisenberg equation, inserting the obtained equations of motion up to the required order in the field, and performing the summations, Eq. (41) and thus Eq. (40) are readily verified.

Another, even simpler example for such a decoupling in a fully coherent situation is the expression for the occupation probabilities in terms of the microscopic polarizations up to second-order in the field

$$
\begin{aligned}
f_{\mathbf{k}}^{ee'} &= \sum_{h'} (P_{\mathbf{k}}^{eh'})^{*} P_{\mathbf{k}}^{e'h'} + O(E^4), \\
f_{\mathbf{k}}^{hh'} &= \sum_{e'} (P_{\mathbf{k}}^{e'h})^{*} P_{\mathbf{k}}^{e'h'} + O(E^4).
\end{aligned} \qquad (42)
$$

Analogous to what has been said above, also this conservation law, which is well known from a two-level system, can be verified by computing the time derivative and inserting the expressions up to the required order.

Besides just verifying the decoupling schemes one can also address the coherent dynamics of a many-body system on the level of a general analysis. As an example, we discuss in the following the two conservation laws given above as special cases and show how to generalize such expressions. Let us start by defining a normally ordered operator product as[11]

$$
\{N, M\} \equiv c^{\dagger}(\alpha_N) c^{\dagger}(\alpha_{N-1}) \cdots c^{\dagger}(\alpha_1) c(\beta_1) \cdots c(\beta_{M-1}) c(\beta_M), \qquad (43)
$$

where depending on α_i and β_j the operators $c^{\dagger}$ and c are electron or hole creation or annihilation operators for certain $\mathbf{k}_i$ and $\mathbf{k}_j$, respectively. The quantities $\{N, M\}$ contain the full information about the dynamics of the photoexcited system. For example, the microscopic polarization $P_{\mathbf{k}}^{eh}$ corresponds to $\{0, 2\}$, its complex conjugate $(P_{\mathbf{k}}^{eh})^*$ corresponds to $\{2, 0\}$, and the occupation probabilities $f_{\mathbf{k}}^{aa}$ to $\{1, 1\}$.

Next we consider the time derivative of the normally ordered operator products, which is given by

$$
\frac{\partial}{\partial t}\{N + 1, M + 1\} = \left(\frac{\partial c^\dagger(\alpha_{N+1})}{\partial t} \right) \{N, M\} c\,(\beta_{M+1})
$$

$$
+ c^\dagger(\alpha_{N+1}) \left(\frac{\partial \{N, M\}}{\partial t} \right) c(\beta_{M+1})
$$

$$
+ c^\dagger(\alpha_{N+1})\{N, M\} \left(\frac{\partial c(\beta_{M+1})}{\partial t} \right). \tag{44}
$$

To analyze to which correlations functions $\{N, M\}$ are coupled one can evaluate the equations of motion for the field operators, i.e. $\frac{\partial c^\dagger(\alpha_i)}{\partial t}$ and $\frac{\partial c(\beta_j)}{\partial t}$. By commuting with the Hamiltonian one finds that $\{1, 0\}$ ($\{0, 1\}$) is coupled to itself by the single-particle (bandstructure) part, coupled to $\{0, 1\}$ ($\{1, 0\}$) by the light-matter interaction, and coupled to $\{2, 1\}$ ($\{1, 2\}$) by the many-body Coulomb interaction.[11]

Inserting these results into Eq. (44) and restoring normal ordering, one can see that the light-matter interaction couples $\{N, M\}$ to $\{N-1, M+1\}$, $\{N+1, M-1\}$, $\{N-2, M\}$, and $\{N, M-2\}$, since it is given by pairs of creation and destruction operators. The many-body Coulomb interaction, however, couples $\{N, M\}$ to $\{N, M\}$ and $\{N+1, M+1\}$.[11] Therefore the four-operator part of the Hamiltonian generates the coupling to products that contain more operators, i.e. the many-body hierarchy. Clearly, for both parts of the Hamiltonian operators with $N-M$ being odd are only coupled to other operators where $N - M$ is odd, and operators with $N - M$ being even are only coupled to other operators where $N - M$ is even. Thus if we start from the ground state as the initial condition, i.e. $\langle\{0, 0\}\rangle = 1$ and all other expectation values are zero, the operators $\{N, M\}$ with $N - M$ being odd vanish at all times, since they contain no sources. These considerations are visualized in Fig. 6. This, furthermore, clearly shows that in any finite order in the external field, only a finite number of expectation values contribute to the optical response.[11,12] To find the lowest order in the light field of any term $\{N, M\}$, one only needs to evaluate the minimum number of dotted lines in Fig. 6, which are needed to connect it to $\{0, 0\}$.

As can be seen from Fig. 6 and can be proven rigorously,[11] the minimum order in the field in which $\langle\{N, M\}\rangle$ is finite is $(N + M)/2$ if N and M are

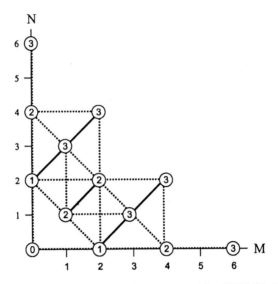

Fig. 6. Schematical drawing of relevant dynamic variables $\langle\{N, M\}\rangle$. The numbers in the symbols correspond to the minimum order in the external field in which $\langle\{N, M\}\rangle$ is finite. The dotted lines denote the coupling via the light-matter interaction, whereas the solid lines denote the coupling via the Coulomb interaction. After Ref. 11.

both even and $(N + M)/2 + 1$ if N and M are both odd, i.e.

$$\langle c^\dagger(\alpha_{2N}) \cdots c^\dagger(\alpha_1) c(\beta_1) \cdots c(\beta_{2M})\rangle = O(E^{N+M}), \tag{45}$$

and

$$\langle c^\dagger(\alpha_{2N+1}) \cdots c^\dagger(\alpha_1) c(\beta_1) \cdots c(\beta_{2M+1})\rangle = O(E^{N+M+2}). \tag{46}$$

If one starts initially from the ground state of the semiconductor and the dynamics is fully coherent, one can factorize expectation values containing a mix of creation and destruction operators to lowest order into products of expectation values which contain either only creation or only destruction operators. For our case we can write

$$\langle c^\dagger(\alpha_{2N}) \cdots c^\dagger(\alpha_1) c(\beta_1) \cdots c(\beta_{2M})\rangle$$
$$= \langle c^\dagger(\alpha_{2N}) \cdots c^\dagger(\alpha_1)\rangle\langle c(\beta_1) \cdots c(\beta_{2M})\rangle + O(E^{N+M+2}), \tag{47}$$

and

$$\langle c^\dagger(\alpha_{2N+1}) \cdots c^\dagger(\alpha_1) c(\beta_1) \cdots c(\beta_{2M+1})\rangle$$
$$= \langle c^\dagger(\alpha_{2N+1}) \cdots c^\dagger(\alpha_1)\rangle\langle c(\beta_1) \cdots c(\beta_{2M+1})\rangle + O(E^{N+M+4}). \tag{48}$$

To prove Eqs. (47) and (48) it is important to note that in the Heisenberg picture, where the operators are time dependent, all expectation values are taken with respect to the initial state, which is the ground state ($|0\rangle$) of the semiconductor, i.e.[11]

$$\langle c^\dagger(\alpha_N) c^\dagger(\alpha_{N-1}) \cdots c^\dagger(\alpha_1) c(\beta_1) \cdots c(\beta_M) \rangle$$
$$= \langle 0 | c_H^\dagger(\alpha_N, t) \cdots c_H^\dagger(\alpha_1, t) c_H(\beta_1, t) \cdots c_H(\beta_M, t) | 0 \rangle. \qquad (49)$$

Now one can define an interaction picture by decomposing the Hamiltonian $H = H_{s-p} + H_C + H_I$, see Eq. (5), into $H = H_0 + H_I$, where we introduced $H_0 = H_{s-p} + H_C$. In the interaction picture an operator $\mathcal{O}$ becomes time dependent according to

$$\mathcal{O}_I(t) = e^{iH_0 t} \mathcal{O} e^{-iH_0 t}. \qquad (50)$$

Inserting the interaction representation of the identity operator in between the creation and destruction operators of a ground state expectation value of Heisenberg operators, i.e. in between $c_H^\dagger(\alpha_1, t)$ and $c_H(\beta_1, t)$ of Eq. (49), one finds by iterating the exponentials $e^{\pm iH_0 t}$ of Eq. (50)[11]

$$\langle 0 | c_H^\dagger(\alpha_N, t) \cdots c_H^\dagger(\alpha_1, t) c_H(\beta_1, t) \cdots c_H(\beta_M, t) | 0 \rangle$$
$$= \langle 0 | c_H^\dagger(\alpha_N, t) \cdots c_H^\dagger(\alpha_1, t) | 0 \rangle \langle 0 | c_H(\beta_1, t) \cdots c_H(\beta_M, t) | 0 \rangle$$
$$+ \sum_{\delta_1} \langle 0 | c_H^\dagger(\alpha_N, t) \cdots c_H^\dagger(\alpha_1, t) c_I^\dagger(\delta_1, t) | 0 \rangle$$
$$\times \langle 0 | c_I(\delta_1, t) c_H(\beta_1, t) \cdots c_H(\beta_M, t) | 0 \rangle$$
$$+ \frac{1}{2} \sum_{\delta_1, \delta_2} \langle 0 | c_H^\dagger(\alpha_N, t) \cdots c_H^\dagger(\alpha_1, t) c_I^\dagger(\delta_2, t) c_I^\dagger(\delta_1, t) | 0 \rangle$$
$$\times \langle 0 | c_I(\delta_1, t) c_I(\delta_2, t) c_H(\beta_1, t) \cdots c_H(\beta_M, t) | 0 \rangle + \cdots. \qquad (51)$$

This general result can be used to verify the two special cases which are relevant for the coherent $\chi^{(3)}$-limit, i.e. Eqs. (40) and (42). The above given expressions allow one furthermore to extend the DCT scheme to higher orders. As an example, we present in Sec. 2.4.5 the equations of motion for the coherent $\chi^{(5)}$-limit schematically and discuss some results obtained at this level in Sec. 2.4.6.

We now apply the DCT truncation scheme to obtain the equations of motion which describe the optical semiconductor response in the coherent $\chi^{(3)}$-limit. In order to be able to distinguish between the uncorrelated Hartree-Fock part and the correlation contributions it is convenient to define a pure four-particle correlation function via

$$
\begin{aligned}
\bar{B}^{eh'e'h}_{\mathbf{k},\mathbf{k}',\mathbf{k}'',\mathbf{k}'''} &= \langle a^{\dagger}_{e,\mathbf{k}} b^{\dagger}_{h',\mathbf{k}'} a^{\dagger}_{e',\mathbf{k}''} b^{\dagger}_{h,\mathbf{k}'''} \rangle \\
&\quad - \langle a^{\dagger}_{e,\mathbf{k}} b^{\dagger}_{h',\mathbf{k}'} \rangle \langle a^{\dagger}_{e',\mathbf{k}''} b^{\dagger}_{e,\mathbf{k}'''} \rangle - \langle a^{\dagger}_{e,\mathbf{k}} b^{\dagger}_{h,\mathbf{k}'''} \rangle \langle a^{\dagger}_{e',\mathbf{k}''} b^{\dagger}_{h',\mathbf{k}'} \rangle \\
&= B^{eh'e'h}_{\mathbf{k},\mathbf{k}',\mathbf{k}'',\mathbf{k}'''} \\
&\quad - \langle a^{\dagger}_{e,\mathbf{k}} b^{\dagger}_{h',\mathbf{k}'} \rangle \langle a^{\dagger}_{e',\mathbf{k}''} b^{\dagger}_{e,\mathbf{k}'''} \rangle - \langle a^{\dagger}_{e,\mathbf{k}} b^{\dagger}_{h,\mathbf{k}'''} \rangle \langle a^{\dagger}_{e',\mathbf{k}''} b^{\dagger}_{h',\mathbf{k}'} \rangle.
\end{aligned}
\tag{52}
$$

Using the above introduced expansions up to third-order in the field, the polarization equation can be written as[25]

$$
\frac{\partial}{\partial t} P^{eh}_{\mathbf{k}} = \frac{\partial}{\partial t} P^{eh}_{\mathbf{k}}\Big|_{\text{hom}} + \sum_{n=1}^{3} \frac{\partial}{\partial t} P^{eh}_{\mathbf{k}}\Big|_{\text{inhom},n},
\tag{53}
$$

with

$$
i\hbar \frac{\partial}{\partial t} P^{eh}_{\mathbf{k}}\Big|_{\text{hom}} = -\left(\varepsilon^{e}_{\mathbf{k}} + \varepsilon^{h}_{\mathbf{k}}\right) P^{eh}_{\mathbf{k}} + \sum_{q} V_q P^{eh}_{\mathbf{k}-\mathbf{q}},
\tag{54}
$$

$$
\begin{aligned}
i\hbar \frac{\partial}{\partial t} P^{eh}_{\mathbf{k}}\Big|_{\text{inhom},1} &= \left(\mathbf{d}^{eh} - \sum_{e',h'} (P^{eh'}_{\mathbf{k}})^* P^{e'h'}_{\mathbf{k}} \mathbf{d}^{e'h} \right. \\
&\quad \left. - \sum_{h',e'} \mathbf{d}^{eh'} (P^{e'h'}_{\mathbf{k}})^* P^{e'h}_{\mathbf{k}} \right) \cdot \mathbf{E},
\end{aligned}
\tag{55}
$$

$$
\begin{aligned}
i\hbar \frac{\partial}{\partial t} P^{eh}_{\mathbf{k}}\Big|_{\text{inhom},2} &= -\sum_{q,e',h'} V_q \left[P^{eh'}_{\mathbf{k}} \left(P^{e'h'}_{\mathbf{k}}\right)^* P^{e'h}_{\mathbf{k}-\mathbf{q}} - P^{eh'}_{\mathbf{k}+\mathbf{q}} \left(P^{e'h'}_{\mathbf{k}+\mathbf{q}}\right)^* P^{e'h}_{\mathbf{k}} \right. \\
&\quad \left. + P^{eh'}_{\mathbf{k}+\mathbf{q}} \left(P^{e'h'}_{\mathbf{k}}\right)^* P^{e'h}_{\mathbf{k}} - P^{eh'}_{\mathbf{k}} \left(P^{e'h'}_{\mathbf{k}-\mathbf{q}}\right)^* P^{e'h}_{\mathbf{k}-\mathbf{q}} \right],
\end{aligned}
\tag{56}
$$

$$
\begin{aligned}
i\hbar \frac{\partial}{\partial t} P^{eh}_{\mathbf{k}}\Big|_{\text{inhom},3} &= \sum_{q,\mathbf{k}',e',h'} V_q \left(P^{e'h'}_{\mathbf{k}'}\right)^* \left[\bar{B}^{eh'e'h}_{\mathbf{k},\mathbf{k}',\mathbf{k}'-\mathbf{q},\mathbf{k}-\mathbf{q}} - \bar{B}^{eh'e'h}_{\mathbf{k}+\mathbf{q},\mathbf{k}',\mathbf{k}'-\mathbf{q},\mathbf{k}} \right. \\
&\quad \left. + \bar{B}^{eh'e'h}_{\mathbf{k}+\mathbf{q},\mathbf{k}'+\mathbf{q},\mathbf{k}',\mathbf{k}} - \bar{B}^{eh'e'h}_{\mathbf{k},\mathbf{k}'+\mathbf{q},\mathbf{k}',\mathbf{k}-\mathbf{q}} \right].
\end{aligned}
\tag{57}
$$

The homogeneous part of Eq. (53) contains the kinetic energies of electrons and holes plus their Coulomb attraction, see Eq. (54). This term is diagonal in the basis of exciton eigenstates. The contributions denoted with the subscript *inhom* in Eq. (53) are the different inhomogeneous *driving* terms, i.e. they are the sources for the microscopic polarizations. The direct coupling of the carrier system to the electromagnetic field is represented by the terms proportional to $\mathbf{d} \cdot \mathbf{E}$, see Eq. (55). These terms include the linear optical coupling $(\mathbf{d} \cdot \mathbf{E})$ and the phase-space filling contributions $(\mathbf{d} \cdot \mathbf{E}\, P^*P)$.

As a consequence of the many-body Coulomb interaction, Eq. (53) contains further optical nonlinearities. The contributions that are of first-order in the Coulomb interaction, which are proportional to VPP^*P, are given by Eq. (56). Those, together with the phase-space filling terms, correspond to the Hartree–Fock limit of Eq. (53). The *correlation contributions* to the polarization equation, see Eq. (57), consist of four terms of the structure $VP^*\bar{B}$, where $\bar{B}$ is the genuine four-particle correlation function defined in Eq. (52). It is clearly demonstrated, that due to the many-body Coulomb interaction the two-particle electron-hole amplitude P is coupled to higher-order correlation functions $\bar{B}$, which is just the beginning of the general many-body hierarchy mentioned earlier.

The Hartree–Fock part of Eq. (53) can be obtained as a third-order expansion of Eqs. (12)–(14) using the conservation law, Eq. (42), which is valid for a coherent system. This procedure allows one to identify the first-order Coulomb terms as the lowest-order parts of the renormalized field and transition energy.

If the calculation of the coherent nonlinear optical response is limited to a finite order in the applied field, the many-particle hierarchy is truncated automatically and one is left with a finite number of correlation functions.[11,12,35,36] This truncation occurs since one considers here purely optical excitation where only the optical field exists as linear source which in first-order induces a linear polarization, in second-order leads to carrier occupations, and so on. In the more general situation where incoherent populations may be present, as, e.g. by pumping via carrier injection or in the presence of relaxed occupations, a classification of the nonlinear optical response in terms of powers of a field is no longer meaningful. For such cases one can use a cluster expansion.[22] This scheme generalizes both the second Born and the dynamics-controlled truncation approaches but includes exciton occupations as well. For the purpose of this chapter, we do not consider exciton population effects and restrict the analysis to the level provided by the second Born and the dynamics-controlled truncation schemes.

Equation (53) has been derived within the coherent $\chi^{(3)}$-limit. If this condition is fulfilled, the nonlinear optical response is fully determined by the dynamics of single and two electron-hole-pair excitations, P and $\bar{B}$, respectively.[11,12,36] The equation for $\bar{B}$'s can be written as[25]

$$\frac{\partial}{\partial t}\bar{B}^{eh'e'h}_{\mathbf{k},\mathbf{k}',\mathbf{k}'',\mathbf{k}'''} = \frac{\partial}{\partial t}\bar{B}^{eh'e'h}_{\mathbf{k},\mathbf{k}',\mathbf{k}'',\mathbf{k}'''}\bigg|_{\mathrm{hom}} + \frac{\partial}{\partial t}\bar{B}^{eh'e'h}_{\mathbf{k},\mathbf{k}',\mathbf{k}'',\mathbf{k}'''}\bigg|_{\mathrm{inhom}}, \quad (58)$$

with

$$i\hbar\frac{\partial}{\partial t}\bar{B}^{eh'e'h}_{\mathbf{k},\mathbf{k}',\mathbf{k}'',\mathbf{k}'''}\bigg|_{\text{hom}} = -\left(\varepsilon^e_{\mathbf{k}} + \varepsilon^{h'}_{\mathbf{k}'} + \varepsilon^{e'}_{\mathbf{k}''} + \varepsilon^h_{\mathbf{k}'''}\right)\bar{B}^{eh'e'h}_{\mathbf{k},\mathbf{k}',\mathbf{k}'',\mathbf{k}'''}$$

$$+ \sum_{\mathbf{q}'} V_{\mathbf{q}'}\left[\bar{B}^{eh'e'h}_{\mathbf{k}+\mathbf{q}',\mathbf{k}'+\mathbf{q}',\mathbf{k}'',\mathbf{k}'''} - \bar{B}^{eh'e'h}_{\mathbf{k}+\mathbf{q}',\mathbf{k}',\mathbf{k}'',\mathbf{k}'''-\mathbf{q}'}\right.$$

$$+ \bar{B}^{eh'e'h}_{\mathbf{k}+\mathbf{q}',\mathbf{k}',\mathbf{k}'',\mathbf{k}'''+\mathbf{q}'} + \bar{B}^{eh'e'h}_{\mathbf{k},\mathbf{k}'+\mathbf{q}',\mathbf{k}''+\mathbf{q}',\mathbf{k}'''}$$

$$\left. - \bar{B}^{eh'e'h}_{\mathbf{k},\mathbf{k}'+\mathbf{q}',\mathbf{k}'',\mathbf{k}'''-\mathbf{q}'} + \bar{B}^{eh'e'h}_{\mathbf{k},\mathbf{k}',\mathbf{k}''+\mathbf{q}',\mathbf{k}'''+\mathbf{q}'}\right], \quad (59)$$

$$i\hbar\frac{\partial}{\partial t}\bar{B}^{eh'e'h}_{\mathbf{k},\mathbf{k}',\mathbf{k}'',\mathbf{k}'''}\bigg|_{\text{inhom}} = -V_{|\mathbf{k}-\mathbf{k}'''|}\left(P^{eh}_{\mathbf{k}'''} - P^{eh}_{\mathbf{k}}\right)\left(P^{e'h'}_{\mathbf{k}'} - P^{e'h'}_{\mathbf{k}''}\right)$$

$$+ V_{|\mathbf{k}-\mathbf{k}'|}\left(P^{eh'}_{\mathbf{k}'} - P^{eh'}_{\mathbf{k}}\right)\left(P^{e'h}_{\mathbf{k}'''} - P^{e'h}_{\mathbf{k}''}\right). \quad (60)$$

The homogeneous part of the equation for $\bar{B}$, given by Eq. (59), contains the kinetic energies as well as the attractive and repulsive interactions between two electrons and two holes, i.e. the *biexciton problem*. The sources of Eq. (58) consist of the Coulomb interaction potential times nonlinear polarization terms, VPP, see Eq. (60).

The third-order limit of the SBA (see previous section) can be obtained from Eq. (58) if one neglects the Coulombic interactions in Eq. (59), i.e. if one ignores bound biexcitons. In this limit, many-body correlations are described up to second-order in the Coulomb interaction.

So-called Coulombic memory effects are automatically included if one solves the coupled Eqs. (53) and (58). The appearance of the memory can be seen most easily by integrating Eq. (58). The formal solution for $\bar{B}(t)$ is schematically obtained as

$$\bar{B}(t) = \frac{i}{\hbar}V\int_{-\infty}^{t} dt' e^{i\bar{\varepsilon}(t-t')/\hbar}P(t')P(t'). \quad (61)$$

Inserting this expression for $\bar{B}(t)$ into Eq. (53) it is clear that $P(t)$ depends on contributions $P(t')$ with $t' < t$, i.e. on terms with earlier time arguments. Such memory effects would be neglected if one solves Eq. (58) adiabatically within the Markov approximation.

When comparing the dynamics-controlled truncation approach with the SBA for the Coulomb correlation contributions it should be noted that the SBA is generally not restricted to the coherent regime or to a finite order in the interaction with the field, and, furthermore, also includes the effect of screening of the Coulomb interaction. However, since the SBA includes neither exciton populations nor bound biexcitons, it is particularly well suited to describe excitation conditions where electron-hole plasma effects dominate the nonlinear response. On the other hand, the DCT, when used up to third-order, describes the dynamics of bound and unbound biexciton resonances explicitly by following the evolution of the four-particle correlation function $\bar{B}$. It is therefore well suited to analyze the influence of such biexciton correlations on the nonlinear optical semiconductor response in coherent situations.

Since the equations of motion in the coherent $\chi^{(3)}$-limit are quite lengthy, it is for the clarity of discussions often useful to reduce the notation and to consider a schematical set of equations of motion. These can be obtained from Eqs. (53) and (58) by neglecting all indices and superscripts and the vector character, and have the following structure[37]

$$i\hbar \frac{\partial}{\partial t} P = -\varepsilon_P P + (d - bP^*P)E - V_{\mathrm{HF}} P^* PP + V_{\mathrm{corr}} P^* \bar{B} \qquad (62)$$

and

$$i\hbar \frac{\partial}{\partial t} \bar{B} = -\varepsilon_B \bar{B} - \bar{V}_{\mathrm{corr}} PP. \qquad (63)$$

Here, ε_P and ε_B are the energies of single- and biexciton states, and b denotes the Pauli blocking, V_{HF} the first-order (Hartree–Fock) Coulomb terms, and V_{corr} as well as $\bar{V}_{\mathrm{corr}}$ Coulomb correlation contributions. If one uses instead of $\bar{B}$, the full biexciton correlation function B given by $\bar{B}$ plus the Hartree–Fock contribution, see Eq. (52), one obtains[37]

$$i\hbar \frac{\partial}{\partial t} P = -\varepsilon_P P + (d - bP^*P)E + V_{\mathrm{corr}} P^* B \qquad (64)$$

and

$$i\hbar \frac{\partial}{\partial t} B = -\varepsilon_B B - \tilde{d} PE. \qquad (65)$$

Here, $\tilde{d}$ is the optical matrix element for the exciton to biexciton transition.

At the end of this section we show how the coherent $\chi^{(3)}$-equations are used to calculate pump-probe signals, i.e. the differential absorption (four-wave mixing can be described very similarly, see, e.g. Ref. 39). For clarity we use the schematical equations, i.e. Eqs. (62) and (63). The optical excitation consists of the sum of two pulses, which within the rotating wave approximation are given by

$$\mathbf{E}(t) = E_1(t)\mathbf{e}_1 e^{i(\mathbf{k}_1 \cdot \mathbf{r} - \omega_1 t)} + E_2(t)\mathbf{e}_2 e^{i(\mathbf{k}_2 \cdot \mathbf{r} - \omega_2 t)}. \qquad (66)$$

Here, $E_1(t) \propto e^{-((t+\tau)/\bar{t}_1)^2}$ $(E_2(t) \propto e^{-(t/\bar{t}_2)^2})$ denotes the temporal envelope of the Gaussian pump (probe) pulse which is centered at $t = -\tau$ $(t = 0)$, $\mathbf{e}_1$ ($\mathbf{e}_2$) its polarization direction, $\mathbf{k}_1$ ($\mathbf{k}_2$) its propagation direction, and ω_1 (ω_2) the central frequency. In this notation, a positive time delay τ corresponds to the pump pulse arriving before the probe pulse.

Considering the kinematic directions of the pulses, i.e. $e^{i\mathbf{k}_1\cdot\mathbf{r}}$ and $e^{i\mathbf{k}_2\cdot\mathbf{r}}$, the exciton and biexciton transitions, P and B, respectively, will be proportional to products of the type $e^{i(n\mathbf{k}_1+m\mathbf{k}_2)\cdot\mathbf{r}}$. We thus expand them into a Fourier series[37-39]

$$P = \sum_{n,m} P^{(n,m)} e^{i(n\mathbf{k}_1+m\mathbf{k}_2)\cdot\mathbf{r}}, \tag{67}$$

$$B = \sum_{n,m} B^{(n,m)} e^{i(n\mathbf{k}_1+m\mathbf{k}_2)\cdot\mathbf{r}}. \tag{68}$$

We have to apply this scheme up to third-order to obtain $\delta\mathbf{P}(t,\tau)$, which is the time-domain polarization in differential absorption geometry. This can be done by considering all contributions which (i) propagate in the direction of the probe pulse ($\mathbf{E}_2$), (ii) include two interactions with the pump pulse ($\mathbf{E}_1$), and (iii) are linear in the probe pulse. Explicitly we have in first-order

$$i\hbar\frac{\partial}{\partial t}P^{(1,0)} = -\varepsilon_P P^{(1,0)} + dE_1, \tag{69}$$

$$i\hbar\frac{\partial}{\partial t}P^{(0,1)} = -\varepsilon_P P^{(0,1)} + dE_2, \tag{70}$$

which are the exciton transitions that are linear in each of the two pulses.

In second-order, the biexciton transition that is excited by the pump and the probe has to be considered, i.e.

$$i\hbar\frac{\partial}{\partial t}\bar{B}^{(1,1)} = -\varepsilon_B \bar{B}^{(1,1)} - \tilde{V}_{\text{corr}}P^{(1,0)}P^{(0,1)} \tag{71}$$

which are the exciton transitions that are linear in each of the two pulses. In third-order the differential polarization in probe direction is induced by the following exciton transition

$$i\hbar\frac{\partial}{\partial t}\delta P^{(0,1)} = -\varepsilon_P \delta P^{(0,1)}$$
$$- b((P^{(1,0)})^* P^{(1,0)} E_2 + (P^{(1,0)})^* P^{(0,1)} E_1)$$
$$- V_{HF}(P^{(1,0)})^* P^{(1,0)} P^{(0,1)} + V_{\text{corr}}(P^{(1,0)})^* \bar{B}^{(1,1)}. \tag{72}$$

The differential polarization is given by

$$\delta\mathbf{P}(t,\tau) = \mathbf{d}^* \delta P^{(0,1)}(t,\tau). \tag{73}$$

If the probe pulse is very short, i.e. spectrally flat in the region of interest, the differential absorption spectrum $\delta\alpha(\omega,\tau)$ can be obtained via

$$\delta\alpha(\omega,\tau) \propto \text{Im}\left[\int dt\,(\mathbf{e}_2)^* \cdot \delta\mathbf{P}(t,\tau)e^{i\omega t}\right], \tag{74}$$

where $\mathbf{e}_2$ denotes the polarization-direction of the probe pulse.

Due to the additivity of the three types of nonlinear sources in Eq. (62), the total differential absorption can be written as the sum of three contributions[25,37,39]

$$\delta\alpha(\omega,\tau) = \delta\alpha_{pb}(\omega,\tau) + \delta\alpha_{CI,1st}(\omega,\tau) + \delta\alpha_{CI,corr}(\omega,\tau). \qquad (75)$$

Here, pb denotes the optical nonlinearity induced by Pauli-blocking. The terms denoted with CI are due to Coulomb-interaction-induced nonlinearities. $CI, 1st$ is the first-order (Hartree–Fock) term, and $CI, corr$ the higher-order correlation contribution.

2.4.4. *Signatures of Four-Particle Correlations*

On the basis of the coherent $\chi^{(3)}$-equations one can analyze a variety of optical excitation configurations such as low intensity pump-probe or four-wave mixing experiments, see, e.g. Refs. 9, 16, 25, 37, 38, 40–43 and references therein. To get a feeling for the importance of four-particle correlations, we focus in the following on low intensity pump-probe configurations considering pumping either resonantly at or spectrally below the exciton. For simplicity, we assume that the energetic splitting between the heavy-hole and light-hole excitons is much larger than the spectral width of the pump pulse and its possible detuning which allows us to neglect the light-hole transitions.[37,42] For results obtained including the light-hole transitions, see, e.g. Refs. 9, 43–45.

To keep the numerical complexity within reasonable limits, we have calculated the differential absorption for a one-dimensional model system.[37,42] Here, the single-particle dispersions are described in the tight-binding approximation, i.e. by a cosine dispersion in k space. Homogeneous broadening is introduced phenomenologically through decay rates $\gamma_P = 1/3$ ps and $\gamma_B = 1/1.5$ ps, respectively. As shown in Ref. 42, the differential absorption spectra obtained using the one-dimensional model system are in qualitative agreement with two-dimensional calculations and experiments performed on high-quality quantum wells.

The differential absorption spectrum for resonant excitation at the exciton resonance with co-circularly polarized pump and probe pulses is shown in Fig. 7(a). $\delta\alpha(\omega)$ is negative in the vicinity of the exciton resonance corresponding to a pump-pulse-induced bleaching of the exciton. For energies larger than the exciton energy we see in Fig. 7(a) positive contributions to $\delta\alpha(\omega)$, which are explained below. Besides the total signal in the upper part

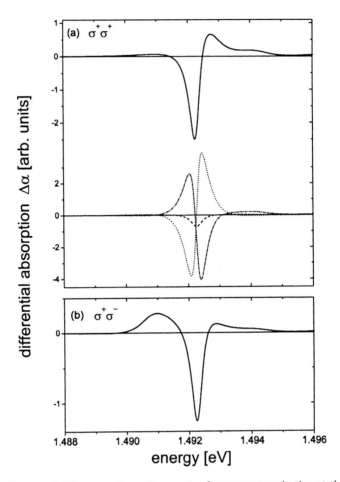

Fig. 7.   Calculated differential absorption spectra for resonant excitation at the exciton and $\tau = 0$ ps. (a) Co-circularly polarized pump- and probe-pulses ($\sigma^+\sigma^+$), the lower panel shows the three contributions to the signal (solid: Pauli blocking, dashed: first-order Coulomb, and dotted: Coulomb correlation). (b) Opposite-circularly polarized pump- and probe-pulses ($\sigma^+\sigma^-$). Taken from Ref. 42.

of Fig. 7(a), also the three individual contributions to $\delta\alpha(\omega)$, see Eq. (75), arising from Pauli blocking, and first- and higher-order Coulomb contributions are displayed separately in the lower part. It is shown that $\delta\alpha_{pb}$ is weak and corresponds to a pure pump-induced bleaching of the exciton resonance. This is what is to be expected, since the Pauli blocking contribution corresponds to approximating the exciton as a simple two-level system.[37]

As shown by the dashed line in the lower part of Fig. 7(a) $\delta\alpha_{CI,1st}$ is strong and antisymmetric around the exciton resonance. It has a dispersive shape corresponding to a blue shift of the exciton, i.e. increased (decreased) absorption above (below) the resonance. As for the Pauli blocking this lineshape of the first-order Coulomb term can be understood on the basis of a simple calculation.[37]

$\delta\alpha_{CI,\text{corr}}$ is shown as the dotted line in the lower part of Fig. 7(a). Its shape is again mainly dispersive around the exciton resonance, but with opposite sign compared to $\delta\alpha_{CI,1st}$, i.e. this term corresponds to a red-shift. Besides contributions at the exciton energy, $\delta\alpha_{CI,\text{corr}}$ also includes terms with resonances at the energies corresponding to transitions from excitons to unbound two-exciton states. These transitions induce some positive differential absorption, so-called excited-state absorption, at energies above the exciton resonance.

When adding up the three contributions to the differential absorption, strong compensations occur between $\delta\alpha_{CI,1st}$ and $\delta\alpha_{CI,\text{corr}}$. The resulting differential absorption exhibits a predominantly absorptive spectral shape around the exciton resonance. Figure 7(a) shows furthermore, that the resulting bleaching at the exciton resonance is dominated by Coulomb-interaction-induced nonlinearities and only weakly enhanced by $\delta\alpha_{pb}$.[37,42]

We now consider excitation with opposite circularly polarized pulses. For this polarization geometry both, $\delta\alpha_{pb}$ and $\delta\alpha_{CI,1st}$ vanish as long as the system is spatially homogeneous.[37] This is due to the fact that none of these contributions introduces any coupling between the subspaces of different spin states. The origin of this is that the Hartree–Fock approximation introduces no interaction among electrons and heavy-holes from the different spin-degenerate sub-bands. Therefore, for this polarization geometry, in the coherent $\chi^{(3)}$-limit the total signal is given purely by the correlation contribution, i.e. $\delta\alpha = \delta\alpha_{CI,\text{corr}}$. As for co-circular excitation, also for opposite circular excitation we find bleaching at the exciton resonance and excited-state absorption due to transitions to unbound two-exciton states appearing energetically above the exciton resonance, see Fig. 7(b). Whereas for co-circularly polarized excitation only contributions from unbound two-exciton states are present, now there is a clear signature of a bound biexciton in the differential absorption spectrum, appearing about 1.4 meV below the excitonic resonance.[37,42] As shown in Ref. 42, the differential absorption spectra of Fig. 7 are in good qualitative agreement with experiments performed on high-quality InGaAs quantum wells.

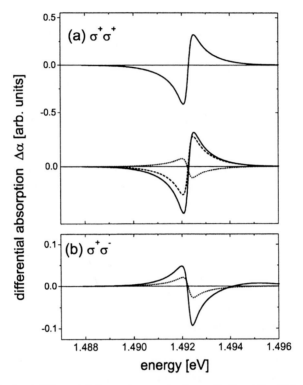

Fig. 8.   Calculated differential absorption spectra for excitation 4.5 meV below the exciton and zero time delay. (a) Co-circularly polarized pump- and probe-pulses ($\sigma^+\sigma^+$), the lower panel shows the three contributions to the signal (solid: Pauli blocking, dashed: first-order Coulomb, and dotted: Coulomb correlation). (b) Opposite-circularly polarized pump- and probe-pulses ($\sigma^+\sigma^-$). The dotted line represents the result of a calculation within the second Born approximation including memory effects. After Ref. 42.

Calculated results for the excitonic optical Stark effect, i.e. the induced spectral changes around the exciton for detuned optical pumping below the resonance, are displayed in Fig. 8. To investigate the fundamental light-polarization dependence of the transient exciton shifts in semiconductors, we concentrate on the case of circularly polarized pump and probe pulses. In Fig. 8(a) computed differential absorption spectrum for co-circularly polarized pulses together with the separate contributions from the three different types of optical nonlinearities are shown. One can see that the phase-space filling nonlinearity and the first-order Coulomb terms, i.e. the Hartree–Fock contributions induce a blue-shift, whereas the correlations alone would yield a red-shift.[42] However, for the $\sigma^+\sigma^+$ configuration the magnitude of the correlation-induced contribution is relatively small and in turn, the total signal is dominated by the blue-shift of the Hartree–Fock terms.

The situation is, however, very different if opposite circularly polarized $(\sigma^+\sigma^-)$ pump and probe pulses are considered. As discussed above, for this configuration the Hartree–Fock contributions vanish as long as only heavy-hole and no light-hole transitions are relevant. Therefore, for moderate pump detunings, i.e. excitation below the heavy-hole exciton with an energetic separation much less than the heavy-hole light-hole exciton splitting,[43,45] the influence of light-hole transitions is negligible and the response is dominated by correlations involving heavy-hole excitons only.

As for $\sigma^+\sigma^+$ also for $\sigma^+\sigma^-$ excitation, Fig. 8(b), the correlation part of the differential absorption corresponds to a red shift. However, since the Hartree–Fock contributions vanish for this configuration, the red-shift is not compensated by other terms and thus survives. The occurrence of this red-shift is not directly related to the existence of a bound biexciton that can be excited with $\sigma^+\sigma^-$ polarized pulses. This can be seen already from the analysis of the $\sigma^+\sigma^+$ excitation configuration, where no bound biexciton can be excited but still the correlation term alone amounts to a red-shift of the exciton resonance.

Further insight is gained by additional calculations where the contributions from bound biexciton transitions are artificially eliminated, i.e. in the third-order limit of the SBA including memory effects.[42] As shown in Fig. 8(b) these calculations also yield a red-shift where only the magnitude is somewhat reduced. If, however, the SBA is used together with a Markov approximation the red-shift disappears for reasonable values of homogeneous broadening.[9] Therefore, one can conclude that the red-shift is a genuine consequence of the dynamics (memory character) of the many-body correlations.

Experimental and theoretical results for the light-polarization dependent excitonic optical Stark effect are displayed in Fig. 9. For co-circularly polarized and for linearly polarized pump and probe beams one always obtains a spectral blue-shift of the exciton reflected in the dispersive-type absorption changes seen in the six top frames of Fig. 9. This blue-shift is the *classical* result well known from a simple two-level system that is pumped spectrally below its resonance frequency. It can be qualitatively understood in terms of level repulsion. For cross-circularly polarized pump and probe pulses, however, the sign of the shift is reversed, i.e. a spectral shift of the exciton *towards* the pump (red-shift) is found both in experiment (bottom section of left column) and theory (bottom section of right column).

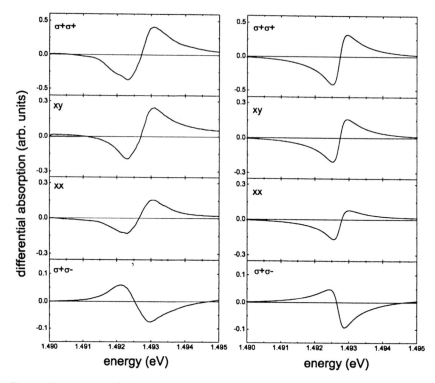

Fig. 9. Experimental (left column) and theoretical (right column) differential absorption spectra for excitation 4.5 meV below the exciton for various polarization directions of the pump and probe pulses and zero time delay. After Ref. 42.

2.4.5. *Dynamics-Controlled Truncation Scheme: Coherent* $\chi^{(5)}$*-Approximation*

Here, we show schematically how one can extend the DCT scheme to the next nontrivial order, which is the coherent $\chi^{(5)}$-limit. This can be done by following the general scheme introduced in Sec. 2.4.3. In the following, we briefly sketch how one has to extend the decoupling schemes to include the fifth order and give the resulting equations of motion schematically. More details can be found in Refs. 35 and 46–48.

Up to fifth-order, the four-operator terms in the equation of motion for P are given by

$$\langle a^\dagger a^\dagger b^\dagger a \rangle = P^* B + (B^* - P^* P^*)(PB - W) + \mathcal{O}(E^7). \qquad (76)$$

At this point, the three-exciton transition $W \equiv \langle a^\dagger b^\dagger a^\dagger b^\dagger a^\dagger b^\dagger \rangle$ is introduced. Furthermore, additional terms in the conservation laws for the electron and hole occupations, f^{ee} and f^{hh} have to be considered. Up to fourth-order one obtains

$$\langle a^\dagger a \rangle = P^*P + (B^* - P^*P^*)(B - PP) + \mathcal{O}(E^6), \tag{77}$$

and an identical schematical equation for $\langle b^\dagger b \rangle$.

In the coherent $\chi^{(5)}$-limit we thus have to follow the dynamics of single-, double-, and triple-electron-hole pair transitions, P, B and W, respectively. The schematic equations of motion for these quantities (without $\bar{}$, i.e. without removing the lower-order correlations) read

$$i\hbar\frac{\partial}{\partial t}P = -\varepsilon_P P + (d - b(P^*P + (B^* - P^*P^*)(B - PP)))E$$
$$+ V_{\text{corr}}(P^*B - (B^* - P^*P^*)(PB - W)), \tag{78}$$

$$i\hbar\frac{\partial}{\partial t}B = -\varepsilon_B B + (\tilde{d}P - \tilde{b}P^*B)E, +\tilde{V}_{\text{corr}}P^*W \tag{79}$$

$$i\hbar\frac{\partial}{\partial t}W = -\varepsilon_W W + \tilde{\tilde{d}}BE. \tag{80}$$

The equation of motion for the three-exciton amplitude W represents a full quantum-mechanical six-body problem (five-body in homogeneous systems, where the center-of-mass momentum can be ignored). Solving these equations is thus numerically extremely demanding.

2.4.6. *Signatures of Four-Particle Correlations at Higher Intensities*

Whereas in the coherent $\chi^{(3)}$-limit the optical response of semiconductors is fairly well understood, regarding the influence of many-exciton correlations, only little knowledge is available beyond this low-intensity regime. This is partly due to the fact, that already in the coherent $\chi^{(5)}$-limit on top of the dynamics of one- and two-excitons, in principle, as is shown in Sec. 2.4.5, three-exciton states W also need to be considered. Microscopic calculations of the nonlinear optical response including three-exciton resonances represent a full quantum-mechanical six-body problem. Numerical evaluations of the resulting equations are currently only possible for very small model systems, e.g. small and narrow semiconductor nanorings,[49] but not feasible for extended semiconductors. For systems like quantum wires and well, to the best of our knowledge, calculations that fully include three-exciton resonances have not been performed yet. It has, however, been possible to describe some experimental observations in four-wave mixing and pump probe that arise with increasing intensity by neglecting the dynamics of three-exciton states or by treating them approximately.[47,48,50–52]

Here, we analyze the importance of four-particle correlations beyond the third-order limit, by investigating the intensity dependence of the differential absorption considering pumping either resonantly at or spectrally below the exciton. For that purpose, we neglect in our solutions of the microscopic versions of the coherent $\chi^{(5)}$-equations the three-exciton resonances W. As shown below, the signals calculated using this approximation are in good qualitative agreement with measurements.[47]

Numerical results for resonant excitation and 2 *ps* time delay are shown in Fig. 10. For these conditions the lineshape of the pure $\chi^{(5)}$ terms correspond basically to the negative of the $\chi^{(3)}$ result. Consequently, as in the experiment, see Ref. 47, the bleaching and the induced absorption decrease with increasing intensity in the normalized spectra. The calculations also result in a small shift of the bleaching maximum towards higher energies with increasing intensity, see Fig. 10, corresponding to a change in the lineshape of the absorption peak. This shift of the bleaching maximum can be understood to be the result of the induced absorption due to unbound two-exciton resonances. Figure 10 furthermore demonstrates that the biexciton-induced polarization dependence of the optical response remains present also for increased pump intensity.

Differential absorption spectra for off-resonant excitation (4.5 meV below the exciton) and various pump intensities are displayed in Fig. 11. For low intensity the exciton blue shift for the co-circular and the red-shift for opposite-circular polarized pump and probe pulses, as discussed above at the $\chi^{(3)}$ level, are reproduced.[42] For co-circular excitation with a slightly increased pump intensity, the negative part of the dispersive signal gains whereas the positive part loses in amplitude. By subtracting the two lowest intensity differential absorption spectra from each other, one obtains an experimental estimate of the pure $\chi^{(5)}$ contribution. As shown in Fig. 11(a), the fifth-order differential absorption has a lineshape corresponding to exciton broadening in contrast to the dispersive shape, for low intensity, i.e. to $\chi^{(3)}$. Results obtained for opposite-circularly polarized pulses are displayed in Fig. 11(b). For this geometry we find that by increasing the pump intensity the amplitude of the shift is gradually reduced. Therefore the pure $\chi^{(5)}$ contribution corresponds to a blue-shift which at elevated pump intensity partly compensates the $\chi^{(3)}$ red-shift in the normalized spectra.

Numerical results on the polarization and pump-intensity dependent absorption changes for off-resonant excitation are shown in Fig. 12. In agreement with the experiment, see Fig. 11(a), for co-circular excitation we find exciton broadening in $\chi^{(5)}$. This broadening leads to distinct

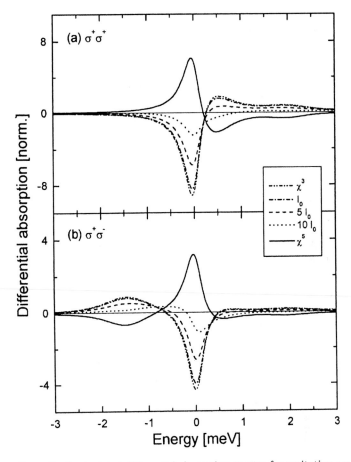

Fig. 10. Normalized calculated differential absorption spectra for excitation resonant at the exciton resonance and $\tau = 2\,\text{ps}$ for various pump pulse intensities. (a) Co-circularly and (b) opposite-circularly polarized pump and probe pulses. Also the lineshape of the calculated pure $\chi^{(5)}$ contribution is shown; its amplitude has been scaled. The zero of the energy scale coincides with the position of the exciton in the linear absorption. Taken from Ref. 47.

signatures in the lineshape of the differential absorption spectra with increasing pump intensity. For this excitation configuration both the $\chi^{(3)}$ and $\chi^{(5)}$ results can be qualitatively understood already in the Hartree–Fock limit. Describing the 1s-exciton as a simple two-level system, which corresponds to neglecting all four- and six-particle correlations proportional to B and W, respectively, in the coherent $\chi^{(5)}$-equations, it is sufficient to reproduce both the shift in $\chi^{(3)}$ as well as the broadening in $\chi^{(5)}$. Thus our

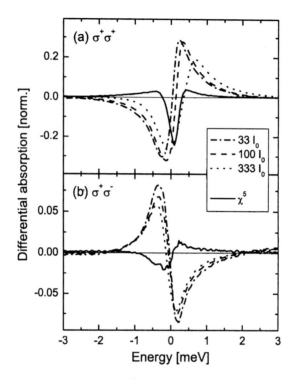

Fig. 11. Normalized experimental differential absorption spectra at $\tau = 0$ ps for off-resonant excitation 4.5 meV below the $1s$ hh-exciton resonance for various pump pulse intensities and the deduced $\chi^{(5)}$ contribution. (a) Co-circularly and (b) opposite-circularly polarized pump and probe pulses. Taken from Ref. 47.

results confirm those of Ref. 29, where it was shown that due to phase-space filling effects, i.e. Pauli blocking, an off-resonant pump may induce exciton broadening type signatures in absorption if pump and probe pulses overlap.

For off-resonant and opposite-circularly polarized excitation the calculated normalized spectra show a decrease of the correlation-induced red-shift with increasing pump intensity, see Fig. 12(b). As in the experiment, the pure $\chi^{(5)}$ contribution corresponds to a blue-shift. Altogether, the experimentally measured polarization and intensity dependence of the optical Stark effect plus the difference in amplitude between co- and opposite-circular excitation is very well reproduced by our model calculations. Even the ratio of the pump intensities necessary to observe $\chi^{(5)}$ effects in the resonant and nonresonant cases agrees quite well with the experiment.[47]

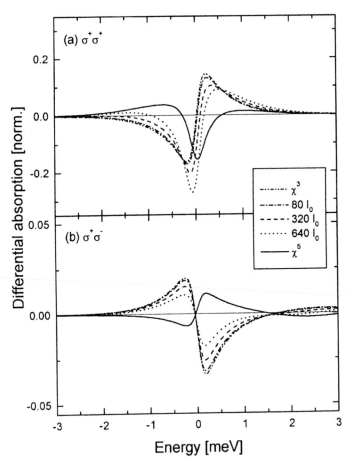

Fig. 12. Normalized calculated differential absorption spectra for excitation 4.5 meV below the exciton and $\tau = 0$ ps for various pump pulse intensities. (a) Co-circularly and (b) opposite-circularly polarized pump and probe pulses. Also the lineshape of the calculated pure $\chi^{(5)}$ contribution is shown; its amplitude has been scaled. The zero of the energy scale coincides with the position of the exciton in the linear absorption. Taken from Ref. 47.

3. Optical Properties of Semiconductor Quantum Wells in Microcavities

3.1. *Introduction*

When a quantum-well exciton transition is resonant with a single mode of a high Q microcavity, the linear optical response of the coupled system may be described in terms of normal modes or cavity polaritons.[53] In fact, the linear response has been shown to be quantitatively accounted for by linear dispersion theory, where the (disorder-averaged) excitonic lineshape determines the normal-mode response of the cavity.[54] The nonlinear optical response of the normal-mode-coupling (NMC) or cavity-polariton system has also been a subject of considerable interest.[15,55] In the earliest experiments showing a nonlinear response, the quantum-well exciton was strongly inhomogeneously broadened, so the nonlinear response for either resonant pumping or due to free carriers from an off-resonant pump could be understood using a phenomenological saturation picture.[56,57] Later experiments using higher-quality samples with large normal-mode splitting-to-linewidth ratios showed that broadening rather than saturation was the dominant nonlinear effect induced by a free-carrier plasma; the experiments were quantitatively accounted for by a microscopic electron-hole theory via the SBA.[28]

More recently, the resonant nonlinear response has been studied in considerable detail. A number of these experiments has been interpreted in terms of various models in which the nonlinearities are treated as exciton–exciton interactions, so in the coupled quantum well-cavity system one has a picture of polariton–polariton interactions. A model based on parametric scattering of cavity polaritons was introduced to explain the polarization dependence in frequency-domain degenerate four-wave mixing measurements.[58] Angle-resonant stimulated polariton amplification was demonstrated,[59] and a theory based on an effective interaction between polaritons was developed to account for the results.[60] Parametric scattering in microcavities has been investigated with regards to nonlinear emission.[61–64] The nonlinear emission of microcavity polaritons has been studied in a nonresonant pumping scheme, and the nonlinear behavior of polariton emission over a large range of carrier density was simulated by rate equations for a polariton population.[65] Signatures of the biexciton transition in a microcavity have also been observed in pump-probe measurements with cross-circular polarization,[66–68] and a transfer of oscillator strength from the exciton to the biexciton transition has been seen.[69]

In this section we are concerned with the microscopic physical description of the resonant nonlinear optical response of the NMC system. As outlined in Sec. 2, tremendous progress has been made in the past decade in the development of microscopic many-body theories which can account for the details of the nonlinear response of semiconductors, within the frameworks of the SBA and the DCT schemes.

With the second-Born approach and the coherent $\chi^{(3)}$ and $\chi^{(5)}$ expansions, it is possible to account microscopically for the main features of nonlinear-optical experiments on quantum wells, including the dependence of differential transmission or four-wave mixing time-resolved spectra on polarization, detuning and density. In this section, these theoretical methods are applied to normal-mode microcavities and compared with an extensive series of experiments in which these parameters are systematically varied. Although simpler models are often used to provide intuitive insight and avoid the need for extensive numerical calculations, it is important to extend the microscopic quantum-well theories to the normal-mode coupling case to obtain a more complete picture of the physics underlying the nonlinear response. Few-level models suffer from the limitation that they cannot account self-consistently for polarization dependent nonlinear spectra at nonzero detuning (which is particularly important for the microcavity case, since the effect of the cavity is to filter the excitation so that the underlying quantum-well exciton transition is excited off-resonance).[70] Models incorporating only exciton–exciton interactions do not include fundamental effects such as the interaction with incoherent carriers. Finally, the microscopic theories of the microcavity response have recently been extended to include the interaction of quantum-well carriers with quantized light fields, leading to the proper description of nonlinear luminescence in quantum wells and microcavities,[17,18,71] and the observation of novel carrier-field quantum correlations in the nonlinear response of microcavities.[20,72] The further development of fully quantum effects in semiconductor microcavities will benefit from a systematic comparison of semiclassical many-body theories with experiments.

In the following, we discuss the nonlinear optical response of a quantum-well microcavity in the NMC regime. The microscopic treatment of Coulomb many-body correlations has already been introduced and discussed in detail in Sec. 2. Here, we therefore only briefly describe how one treats the electromagnetic field in the microcavity in Sec. 3.2 and focus afterwards on the description and interpretation of the calculated results, and their comparison with experiments. Pump-probe measurements investigating the nonlinear optical properties of semiconductor microcavities are described in Sec. 3.3. The spectral response and the temporal dynamics following excitation with circularly polarized pulses are discussed in Secs. 3.3.1 and 3.3.2, respectively. In Sec. 3.4 we present a number of measurable signatures in density- and intensity-dependent measurements which can be described using the SBA. These are, on the one hand, changes to normal-mode spectra induced by an incoherent electron-hole plasma, see Sec. 3.4.1. Furthermore, changes in the reflection spectrum of a weak probe pulse are analyzed for various pumping conditions in Sec. 3.4.2. Here, also the dependence on the pump intensity is studied for both co- and cross-circular polarization of the pump and probe pulses, and furthermore, the dependence of the nonlinear signal on the pump-probe delay is investigated. It is shown that many experimental findings of Sec. 3.3 are in good agreement with the results of the microscopic theory. Section 3.5 presents applications of the DCT scheme to semiconductor microcavities. Biexcitonic signatures in polarization dependent pump-probe measurements using low intensities are discussed in Sec. 3.5.1 and the contributions of four-particle correlations to intensity- and polarization-dependent pump-probe experiments are analyzed in Sec. 3.5.2. The comparison between experiment and numerical results obtained within microscopic theories based on the microscopic electron-hole Hamiltonian in the second-Born and coherent-$\chi^{(5)}$ approximation, demonstrates that the prominent features of the experimental data can be well described by these theoretical approaches.

3.2. Description of Light Propagation and Normal-Mode Splitting

Whereas in Sec. 2, it was basically sufficient to consider the excitation of the polarization **P** and the further relevant material quantities by *external* pulses **E**, for the case of a quantum well placed inside a microcavity, the radiative coupling of **P** and **E** has to be described explicitly. Due to the multiple scattering of the light, a self-consistent treatment of both quantities is required. In the following we briefly outline how one can obtain the transverse electric field at the quantum-well position considering a one-dimensional light-propagation in z-direction, which is the growth direction of the heterostructure. More details can be found in Ref. 15.

As already outlined in Sec. 1, in the semiclassical picture the interaction between the optical polarization $\mathbf{P} = \sum_{\mathbf{k},e,h}(\mathbf{d}^{eh})^* P_{\mathbf{k}}^{eh} + \text{c.c.}$ and the transverse electric field **E** is governed by Maxwell's equations. Restricting ourselves here to the case of perpendicular incidence, the problem reduces to describing a one-dimensional light-propagation. The corresponding one-dimensional wave equation reads

$$\left[\frac{\partial^2}{\partial z^2} - \frac{n^2(z)}{c_0^2}\frac{\partial^2}{\partial t^2}\right] E(z,t) = -d_o^{eh}\frac{\partial^2}{\partial t^2}P(z,t). \tag{81}$$

For a quantum well of finite thickness, the material polarization is given by

$$P(z,t) = P_{QW}(t)|\xi(z)|^2, \tag{82}$$

where we considered only the resonant interaction with the lowest sub-band and $\xi(z)$ is the confinement wavefunction. Further, it is a sensible approximation to neglect the extension of the quantum well in z-direction, since the typical well width is very small compared to the optical wavelength, i.e. to use

$$P(z,t) = P_{QW}(t)\delta(z), \tag{83}$$

for a quantum well at $z = 0$.

For an incident field propagating in forward direction E_0^+, i.e. towards positive z, Maxwell's equations together with Eq. (83) are solved by[7,73]

$$E^+\left(t - \frac{z}{c}\right) = E_0^+\left(t - \frac{z}{c}\right) - \frac{d_o^{eh}c}{2}\frac{\partial}{\partial t}P_{QW}\left(t - \frac{z}{c}\right), \tag{84}$$

$$E^-\left(t + \frac{z}{c}\right) = -\frac{d_o^{eh}c}{2}\frac{\partial}{\partial t}P_{QW}\left(t + \frac{z}{c}\right). \tag{85}$$

As a result, the transverse field at the position of the quantum well is obtained as

$$E_{QW}(t) = E_0(t) - \frac{d_o^{eh}c}{2}\frac{\partial}{\partial t}P_{QW}(t). \tag{86}$$

In the following treatment, we apply the slowly varying envelope approximation (SVEA),[73] in order to use a transfer matrix technique for solving Eqs. (84) and (85). The microcavity structure consists of two dielectric mirrors on a substrate, i.e. alternating $\frac{\lambda}{4}$-layers of materials with a large difference in the refractive index n. In between the two mirrors a cavity is present which usually has a spatial extension of a multiple of $\frac{\lambda}{2}$. At each interface i between layers of refractive index n_{i-1} and n_i, the fields propagating in forward (E^+) and backward (E^-) directions are connected by a transfer matrix, via

$$\begin{pmatrix} E_i^+(t) \\ E_{i-1}^-(t) \end{pmatrix} = M_i \begin{pmatrix} E_{i-1}^+(t - \Delta t) \\ E_i^-(t - \Delta t) \end{pmatrix}, \tag{87}$$

where

$$M_i = \frac{1}{n_i + n_{i-1}}\begin{pmatrix} 2n_{i-1} & n_i - n_{i-1} \\ n_{i-1} - n_i & 2n_i \end{pmatrix}. \tag{88}$$

Clearly, for the case of $n_i = n_{i-1}$, M_i is a unit matrix and the free propagation without any scattering is recovered.

The linear reflection spectrum of a bare Bragg mirror is characterized by a *stop band*, i.e. a spectral region with a reflectivity close to unity ($R \approx 1$) around the center frequency ω_0, which is defined by the structural parameter λ. Further away from ω_0, oscillatory structures are present in the reflectivity spectrum. Introducing a perturbation into this mirror structure, leads to a discrete long-living resonance inside the system, which can clearly be distinguished from the stop band. In Fig. 13(a), we consider

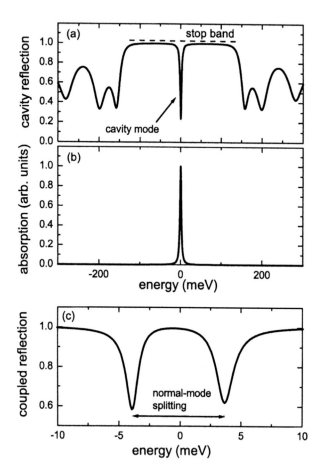

Fig. 13.   (a) Calculated linear reflection spectrum of a microcavity structure with a $\frac{3}{2}\lambda$ cavity. The zero of the energy scale is chosen to coincide with the energy of the cavity mode. The dashed line indicates the width of the stop band. (b) Calculated absorption spectrum of a homogeneously broadened two-level system which is resonant with the cavity mode. (c) Calculated linear reflection spectrum of the coupled system; note the change of the energy scale.

a cavity of length $\frac{3}{2}\lambda$, which is sandwiched in between the top and bottom Bragg reflectors. This breaking of the translation invariance of the system results in a cavity mode which has two antinodes of the field within the $\frac{3}{2}\lambda$ layer.

If we place an optical transition, e.g. a quantum-well exciton, at the position of one of the antinodes of the cavity mode, both resonances couple via Maxwell's equations. For the resonant case, where the optical transition and the cavity mode have the same frequency, the coupling among the resonances may result in a splitting of the originally degenerate resonances into two peaks, provided that one is in the nonperturbative coupling regime, where the coupling among the resonances is bigger than their widths, see Fig. 13(c). These new eigenresonances are referred to as the normal modes of the coupled matter–light system. The existence of these coupled modes is often referred to in the literature as *strong coupling*, in analogy with atom-cavity quantum-electrodynamical systems. It should be noted, however, that the normal-mode coupling arises in the quantum-well-microcavity system due to the large number of effective oscillators (exciton states) interacting with the cavity mode, whereas true strong coupling arises when coupled modes are the result of the cavity interacting with a single oscillator, e.g. a single atom or quantum dot. Hence we are concerned here with quantum-well-microcavity systems which are *nonperturbatively* rather than *strongly* coupled.[15]

An experimental example for the nonperturbative coupling regime realized in a microcavity structure which contains a quantum well with the excitonic resonance at the same energy as the cavity mode is shown in Fig. 14. Here, the normal-mode splitting is about 6.4 meV and thus much bigger than the linewidths of about 0.4 meV. If one plots the energetic positions of the two normal modes as function of the detuning, one obtains in the nonperturbative coupling regime data which show the anticrossing of the two coupled resonances.[53] For a detuning exceeding the coupling between the exciton and the cavity mode, the mixing of the exciton and cavity resonances in the normal modes is basically absent.

In what follows we concentrate on the analysis and description of the nonlinear optical properties of quantum-well-microcavity systems. A more detailed discussion of the linear optical properties of these systems can be found in Ref. 15 and on a more basic and introductory level in Ref. 74.

In the linear optical regime, numerical calculations in the time domain are not necessary, since one can solve the problem analytically in the frequency domain.[15] For the nonlinear case investigated below, such a

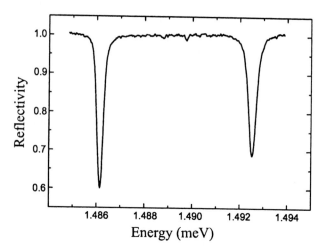

Fig. 14. Normal-mode coupling at zero detuning with ten quantum wells placed in a $11\lambda/2$ microcavity (sample NMC66) with mirrors designed for 99.94% reflectivity. Taken from Ref. 15.

treatment is, however, not possible. In the following sections we focus on the modeling of pump-probe experiments in reflection geometry. For this purpose we ignore the transmission on the substrate side of the cavity. Furthermore, both pulses are incident at small angles, meaning that they are treated as orthogonal to the layers for the propagation problem, but their slightly different directions are still considered in the Fourier decomposition of the material excitations, as introduced in Secs. 2.4.1 and 2.4.3. Within the coherent $\chi^{(3)}$ and $\chi^{(5)}$-schemes, the various orders of the field are propagated separately.[42] By a weighted summation of the linear, third- (linear in the pump intensity), and fifth-order (quadratic in the pump intensity) terms, the intensity dependence of the optical response can be simulated.

3.3. *Nonlinear Optical Properties of Semiconductor Microcavities Viewed by Pump-Probe Experiments*

Here, we present experimental results of the nonlinear optical response of a quantum-well microcavity in the nonperturbative coupling regime. The changes in the reflection spectrum of a weak probe pulse are analyzed for various pumping conditions, where the selective excitation of only the lower or upper branch is compared with the case of both-mode excitation. The dependence on the pump intensity is studied for both co- and cross-circular polarization of the pump and probe pulses. Furthermore, the dependence of the nonlinear signal on the pump-probe delay is investigated. In the following sections, the experimental findings presented here are compared to numerical results obtained within microscopic theories based on the microscopic electron-hole Hamiltonian in the second-Born and coherent-$\chi^{(5)}$ approximations. As a result of this comparison we find that the prominent features of the experimental data are well-described by these theoretical approaches.

The sample used in the experiments has two $In_{0.04}Ga_{0.96}As$ quantum wells placed at the antinodes of a $3\lambda/2$ cavity formed by two 99.6%-reflectivity distributed Bragg reflectors with 14 and 16.5 periods of GaAs/AlAs for top and bottom mirrors, respectively. The quantum-well exciton transition at 1.492 eV (linewidth 0.7 meV) is resonant with the cavity mode at 10 K, where all experiments were performed. The normal-mode splitting on resonance is 4.5 meV. The output of 85 MHz mode-locked Ti:sapphire laser producing 75 fs pulses centered at 831 nm has been used as the light source of the pump-probe set up. The pump pulses were generated by spectrally filtering the laser output. The pump pulse was selectively resonant with only the upper or lower mode, with pulse durations of 190 fs, or excited both modes simultaneously. The spectral filter provided a very high contrast ratio, so that the intensity of the pump at the upper (lower) mode was less than 10^{-3} of its intensity at the lower (upper) mode. The pump and probe beams were overlapped on the sample at near normal incidence; the pump (probe) diameter was 35(20) μm, and the probe fluence was always less than 10^{-2} that of the pump. Probe reflectance spectra were recorded on an optical multichannel analyzer.

3.3.1. *Measured Spectra Using Circularly Polarized Pulses*

We investigate the optical nonlinear response of the normal modes in the coherent regime. The pump spectrum was filtered to excite either the lower or the upper normal mode, or both modes; as we shall see, substantially different behavior is observed under these different pump conditions. Pumping the lower mode only is particularly important for coherent control, since such experiments have shown a dephasing time which is longer than the ones obtained for other schemes.[75,76]

Figure 15 shows the probe reflectance spectra for co-circular polarization when only the lower mode is excited. The probe time delay was set around 0 ps. The data were measured for various cavity detuning ($\delta \equiv E_{cavity} - E_{exciton}$) and pump fluence conditions. Some features of

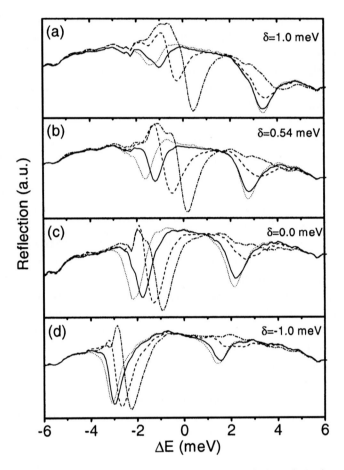

Fig. 15. Measured probe reflectance spectra for co-circularly polarized pump and probe when the lower mode is excited. The probe time delay is 0 ps. Thin dotted-lines represent probe spectra without pump. (a) Detuning $\delta = 1.0$ meV, pump intensity $I_p = 0.4$ (solid), 2.0 (dashed), and 6.8 (dashed-dot) $\mu J/cm^2$ per pulse. (b) $\delta = 0.54$ meV, $I_p = 0.8$ (solid), 2.7 (dashed), and 8.2 (dashed-dot) $\mu J/cm^2$ per pulse. (c) $\delta = 0.0$ meV, $I_p = 0.8$ (solid), 3.4 (dashed), and 7.2 (dashed-dot) $\mu J/cm^2$ per pulse. (d) $\delta = -1.0$ meV, $I_p = 0.4$ (solid), 2.0 (dashed), and 6.4 (dashed-dot) $\mu J/cm^2$ per pulse.

the nonlinear response are noteworthy. First, the lower mode undergoes a substantial lineshift to the high energy side. This blue-shift and the mode amplitude increase with pump fluence. The lineshift is more prominent in positive detuning than in negative detuning. Second, the upper mode also shows a blue-shift, although it is much smaller. Unlike the lower mode, the upper mode saturates and the linewidth broadens at high pump fluence.

We observe that a significant portion of the observed spectral broadening is produced by the coherent nonlinearities because the broadening is reversible in coherent control experiments.[75,76] Overall, the normal-mode splitting reduces at high carrier density, which occurs due to phase space filling and screening. Third, a gain peak appears at the red side of the lower mode. The peak amplitude increases with the pump fluences. Its size relative to the lower mode is larger and the linewidth is wider for positive detuning than for negative detuning.

The normal-mode spectra for cross-circular polarization, presented in Fig. 16, reveal distinctive differences. The probe spectra for cross-circular configuration are less sensitive to the pump fluence than for co-circular polarization. Furthermore, the nonlinear response is weaker in negative detuning than in positive detuning. The gain peak for the co-circular polarization is absent for cross-circular excitation and a slight red-shift occurs. The prominent feature of the cross-circular configuration is the appearance of the new resonance at the red side of the upper mode. The new mode can be accounted for by a biexcitonic transition. Due to the excitation-induced spectral broadening, the biexcitonic peak is not well resolved from the upper mode. The biexcitonic mode is not noticeable up to a certain pump fluence. The critical pump fluence increases with the cavity detuning, because the biexcitonic shift increases with the cavity detuning. It is clearly shown in the data that the energy difference between the upper normal mode and the biexciton-induced signature is bigger for positive detuning than for negative detuning. With increasing carrier density, the biexcitonic mode grows as the upper mode fades away, because of the depletion of the ground states. When the positive detuning is sufficiently large ($> 0.9\,\text{meV}$), the biexcitonic mode does not appear even at high excitation level. Instead, a new spectral feature emerges at the red side of the lower mode.

3.3.2. *Measured Temporal Dynamics for Circularly Polarized Pulses*

The normal-mode spectrum itself reflects the frequency components contributing to the time-resolved probe signal. The dynamics of the normal modes can be accessed by varying the delay between the pump and probe pulses. Figure 17 shows the time-resolved differential probe reflection (DR) spectra for various spectral shapes of the pump pulse. The pump spectra are shown in Fig. 17(a). When both modes are pumped, Fig. 17(b), fast temporal oscillations of the DR around zero pump-probe delay can

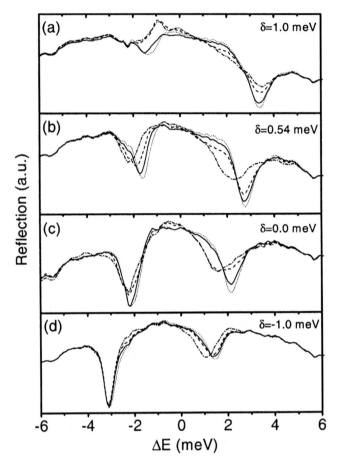

Fig. 16.   Measured probe reflectance spectra for cross-circularly polarized pump and probe when the lower mode is excited. The probe time delay is 0 ps. Thin dotted-lines represent probe spectra without pump. (a) Detuning $\delta = 1.0$ meV, pump intensity $I_p = 0.8$ (solid), 4.1 (dashed), and 8.2 (dashed-dot) μJ/cm^2 per pulse. (b) $\delta = 0.54$ meV, $I_p = 0.8$ (solid), 2.0 (dashed), and 8.2 (dashed-dot) μJ/cm^2 per pulse. (c) $\delta = 0.0$ meV, $I_p = 0.8$ (solid), 4.1 (dashed), and 8.2 (dashed-dot) μJ/cm^2 per pulse. (d) $\delta = -1.0$ meV, $I_p = 0.4$ (solid), 0.8 (dashed), and 6.4 (dashed-dot) μJ/cm^2 per pulse.

be observed. The oscillation period of 0.9 ps corresponds to the normal-mode splitting. The oscillation amplitude is much stronger in the lower mode than in the upper mode since in connection with an asymmetric exciton spectrum the upper mode exhibits a stronger broadening.[15] For lower-mode excitation, Fig. 17(c), the DR signal decays within a few picoseconds and the upper mode signal is much weaker than the lower-mode signal.

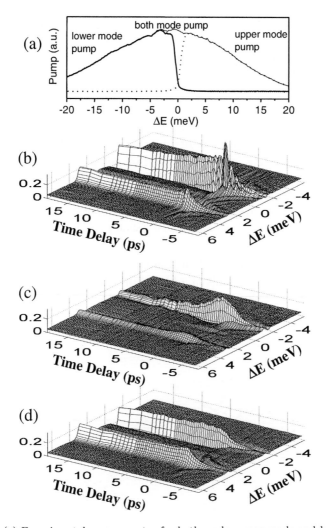

Fig. 17.   (a) Experimental pump spectra for both-mode, upper-mode and lower-mode pumping. Time-resolved differential reflectivity spectra of polariton modes in the quantum-well microcavity when (b) both modes, (c) the lower mode, and (d) the upper mode is pumped. Pump fluence is $4.8\,\mu\mathrm{J/cm^2}$ per pulse before the spectral filtering. The pump and probe pulses are co-circularly polarized. The pulse durations of both- and single-mode pump pulses are 190 fs and 75 fs, respectively.

Since in this case the pump spectrum has strongly reduced overlap with the quantum-well exciton absorption, the frequency components in the pump-induced polarization corresponding to absorptive states are much weaker and carrier scattering is strongly suppressed. Only a small portion of incoherent carriers can be excited and the major part of the coherent pump-induced polarization decays radiatively within 5 ps. At longer time delay, the nonlinear response for both and upper-mode pumping, Figs. 17(b) and 17(d), is much stronger than for lower mode pumping because of the higher incoherent population.

The fast oscillatory dynamics of the gain peak at the lower mode for both-mode pumping is shown in Fig. 18 for various pump-probe delay times between −4.7 and 1.3 ps in steps of 0.1 ps. Since both pump and probe fields and the corresponding induced polarization components undergo normal-mode oscillations in the microcavity, their constructive or destructive interference (depending on the pump-probe delay) can enhance or suppress the probe nonlinearities. Around zero delay, the temporal evolution of the gain peak at the red side of the lower mode follows this normal-mode interference. For larger positive delay times, the polariton modes are shifted to the blue side, because the exchange interaction terms contributing to the nonlinear signal always result in a blue-shift of the exciton resonance.

Figure 19 shows the dynamics of the biexcitonic mode for cross-circular polarization when the lower mode is pumped. The biexcitonic dip at the

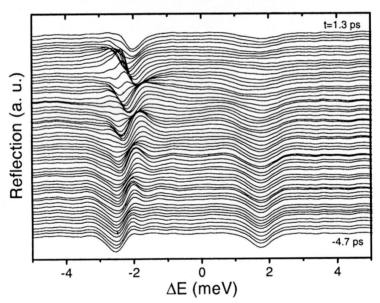

Fig. 18. Co-circular reflection spectra of the polariton modes for various time delays between the pump and probe pulses from −4.7 to 1.3 ps when both modes are pumped. Each spectrum is vertically displaced, and the time delay between the spectra is 0.1 ps. Pump fluence is $1.4\,\mu\mathrm{J/cm^2}$ per pulse.

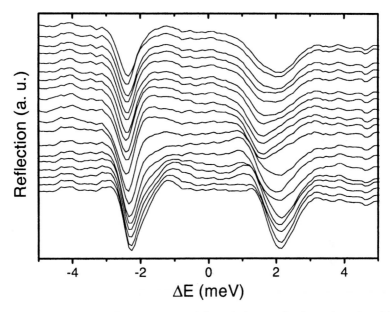

Fig. 19.   Cross-circular reflection spectra of the polariton modes for various time delays between the pump and probe pulses from −4 to 2 ps when the lower mode is pumped. Each spectrum is vertically displaced, and the time delay between the spectra is 0.33 ps. Pump fluence is $4.8\,\mu\mathrm{J/cm}^2$ per pulse.

red side of the upper mode rises and fades away within a few picoseconds. This indicates that the biexcitonic transition is a coherent process; we do not observe a biexcitonic transition at long time delay which shows the lack of incoherent step-wise transitions relying on exciton population.

In Fig. 20, we present the time evolution of the DR at the upper and lower normal-mode resonance for lower-mode pumping when the cavity detuning is −1.5 or +1.6 meV. In the case of negative (positive) detuning, the contribution of the cavity photon (exciton) mode to the lower-mode polariton is larger than that of the exciton (cavity photon) mode. Thus, the radiative decay time of the lower-mode polariton for negative detuning is faster than for positive detuning.

3.4. *Second Born Approximation in Semiconductor Microcavities*

Here, we present and discuss numerical results obtained for the nonlinear optical response of semiconductor microcavities using the SBA. In Sec. 3.4.1 we review some results on the influence of an electron-hole plasma on the

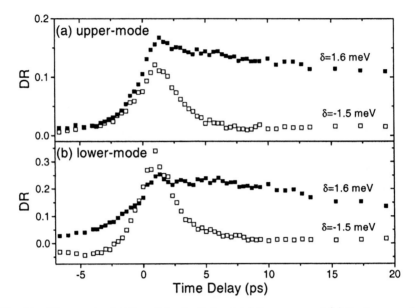

Fig. 20.   Time-resolved differential reflectivity at (a) the upper and (b) the lower mode resonance wavelength for lower-mode pumping. The cavity detuning (δ) is -1.5 meV (open square) and 1.6 meV (solid square). Pump fluence is $1.2\,\mu J/cm^2$ per pulse for $\delta = -1.5$ meV and $1.5\,\mu J/cm^2$ per pulse for $\delta = 1.6$ meV. The resonance wavelength of the lower mode is 832.72 nm ($\delta = -1.5$ meV) and 831.52 nm ($\delta = 1.6$ meV), and the resonance wavelength of the upper mode is 829.93 nm ($\delta = -1.5$ meV) and 828.81 nm ($\delta = 1.6$ meV).

normal-mode spectra. In Sec. 3.4.2 we present results for polarization-dependent pump-probe experiments, which are compared with the measurements presented in Sec. 3.3.

3.4.1. Plasma-Induced Changes of the Normal-Mode Splitting

We start by demonstrating that the theory is able to reproduce the Density-dependent absorption spectra of bare quantum wells. Figure 21(a) shows the saturation of the 1s-exciton resonance with increasing plasma density computed within the SBA. The numerical results are in good agreement with Fig. 22(a), which depicts the measured absorption of a 20 quantum-well sample.[28]

The NMC spectrum for a microcavity containing quantum wells is obtained by using the quantum-well susceptibility within a transfer-matrix calculation for the microcavity design.[15] Here, we consider a cavity with a $\frac{3}{2}\lambda$ GaAs spacer between GaAs/AlAs mirrors. For the left mirror (exposed

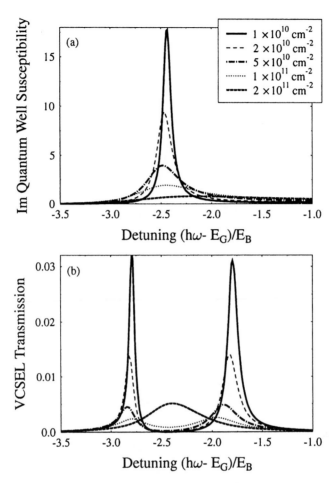

Fig. 21. (a) Imaginary part of the optical susceptibility for an 8 nm quantum well and plasma excitation with various densities at carrier temperature 77 K. (b) Calculated transmission of the quantum-well microcavity for increasing plasma density and bleaching of the exciton according to Fig. 21(a). The cavity resonance has been tuned from −2.05 (full line) to −2.14 (short dashed line) to compensate for the small numerical exciton shift. Taken from Ref. 28.

to air) and right mirror (on a substrate) a reflectivity of 99.6% is obtained with 14 and 16.5 quarterwave pairs. A 8 nm $In_{0.04}Ga_{0.96}As$ quantum well is placed in each of the two cavity antinodes. The cavity wavelength is chosen to coincide with the 1 s-exciton transition of the quantum wells. The calculated microcavity transmission for various plasma densities is shown in Fig. 21(b). Due to the increasing broadening of the 1 s-exciton resonance

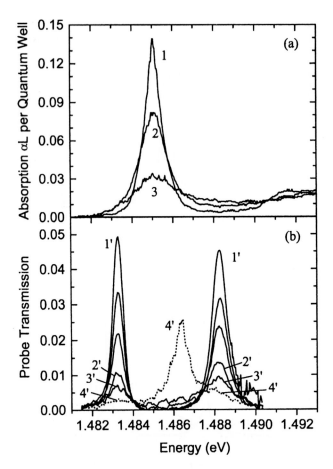

Fig. 22. Experimental probe transmission spectra with increasing pumping at 787 nm for (a) exciton absorption of 20 quantum wells like the two in the microcavity and (b) microcavity normal-mode coupling. Since absolute densities were not measured, curves in (a) and (b) cannot be compared directly; however, Kramers–Kronig transfer-matrix microcavity calculations using the nonlinear data in (a) show that 2′ corresponds closely to 2 and 3′ to 3. Noise from photoluminescence prevented determining the probe transmission when the exciton is completely saturated (4′). Stronger pumping in (b) results in lasing at a wavelength close to the 4′ peak. Taken from Ref. 28.

with increasing carrier density, we find a strong reduction of the NMC peak-height, but only small reduction of the NMC splitting. The reduced trans-mission and the increased widths of the individual NMC peaks are the result of the strong broadening of the exciton resonance, i.e. excitation-induced dephasing, whereas the small reduction of the splitting clearly reveals the minor reduction of the exciton oscillator strength within a large plasma density range. Since the renormalized band edge approaches the energetically stable 1 s-exciton resonance with increasing plasma density, there is a rather abrupt replacement of the NMC doublet by a single transmission peak, which occurs when the cavity resonance becomes degenerate with the band edge. This corresponds to the transition from the nonperturbative regime to the perturbative regime, where the coupling between the exciton and the cavity mode is smaller than the linewidth.

The dramatic effect of excitonic broadening on NMC has been measured using 8 nm $In_{0.04}Ga_{0.96}As$ quantum wells within the above discussed cavity design.[28,77] In these samples the In concentration is sufficiently large for the heavy-hole exciton peak to be around 834 nm at 4 K, such that the GaAs substrate does not have to be removed for transmission studies. Concomitantly the strain shifts the light-hole exciton peak to 826 nm, 13 meV above the heavy-hole exciton, where it does not interfere with NMC studies with the heavy-hole exciton. The small exciton HWHM linewidth ($\gamma_{ex} \approx 0.5$ meV $= 0.3$ nm at 4 K) results in splitting-to-linewidth (NMC HWHM) ratios of 41 (13.6; 4.6) for Bragg mirrors consisting of 19/21.5(14/16.5;5/12.5) periods for the left and right mirrors, respectively, with 99.94% (99.6 %; 97.7%) calculated reflectivity, see Ref. 15.

Experimental results of continuous wave pump-probe excitation of quantum wells and the corresponding exciton saturation are shown in Fig. 22(a). In agreement with Fig. 21(a), with increasing excitation, there is little change in oscillator strength (integrated absorption) but considerable broadening.[31,78] The broadening induced by many-body Coulomb correlations increases the absorption at the energies of the two peaks thereby decreasing their transmission as shown in Fig. 22(b), in agreement with Fig. 21(b). When the exciton is completely saturated, the transmission of the almost-empty cavity opens up close to the middle of the two normal modes. Figure 22(b) was measured with the pump wavelength at the first transmission minimum above the stop band. Reduction in transmission without reduction in splitting is also obtained when the pump wavelength coincides with either one of the original peaks or is chosen in between them. However, the power dependence is different for each of the wavelengths.

The loss of oscillator strength dominated the first nonlinear NMC experiments because the structural-disorder-induced inhomogeneous broadening masked the carrier-density-dependent excitonic broadening.[56,57] In Fig. 22 loss of oscillator strength certainly occurs but only in the transition from strong to weak coupling at densities above the regime where transmission goes down with little change in splitting.

3.4.2. *Intensity- and Polarization-Dependent Pump-Probe Studies*

Results of calculations with Coulomb correlations in second-order Born approximation for intensity- and polarization-dependent pump-probe experiments are shown in Fig. 23. The Rabi energy of the pump pulse — in units of the Rydberg energy — refers to the external pulse incident on the cavity before the blue side of the spectrum (about 45%) is blocked. In order to compare the theoretical results with the experimental ones of Fig. 15, the Rabi energy (Ω_{Pump}) can be converted to the pump fluence (I_p) as $I_p(\mu\text{J}/\text{cm}^2) = 4(\Omega_{\text{Pump}}(E_B))^2$. The calculated results reproduce the gain at the red side of the lower mode for co-circular polarization. As in the experiment, this gain is absent for cross-circular polarization. It has its origin in the coupling of the optical pump field and the pump-induced coherent

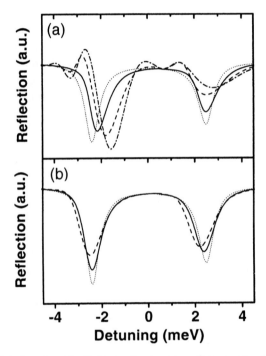

Fig. 23.   Calculated probe reflectivity spectra based on the second-order Born approximation for lower-mode excitation, zero detuning, and increasing Rabi-energy of the pump pulse Ω_{Pump}. (a) Co-circular; Ω_{Pump} is 0 E_B (0 $\mu\text{J}/\text{cm}^2$, dotted), 0.4 E_B (0.64 $\mu\text{J}/\text{cm}^2$, solid), 0.8 E_B (2.6 $\mu\text{J}/\text{cm}^2$ dashed), and 1.0 E_B (4.0 $\mu\text{J}/\text{cm}^2$, dashed-dotted). (b) Cross-circular; Ω_{Pump} is 0 E_B (0 $\mu\text{J}/\text{cm}^2$, dotted), 0.4 E_B (0.64 $\mu\text{J}/\text{cm}^2$, solid), and 0.8 E_B (2.6 $\mu\text{J}/\text{cm}^2$, dashed). E_B is the 3D excitonic Rydberg energy.

polarization to the probe polarization (which determines the reflected and transmitted probe fields) via coherent wave mixing. As can be seen from Eq. (31) for the probe polarization, the Rabi energy of the pump field $\Omega_{\mathbf{k}}$ enters by means of $\delta f_{\mathbf{k}}^{e,h}$ Fourier components of the carrier population which describe population pulsations.[30] This coupling gives rise to a diffraction of the pump contributions into probe directions. A similar effect has been reported in Ref. 79.

The oscillations in the probe reflection spectra for co-circular polarization and strong pumping are due to the interference of propagating modes involved in the above-discussed coherent wave-mixing process. This effect is not present for cross-circular excitation due to the absence of population pulsations and self-diffraction for this geometry.

In the case of co-circular polarization, the saturation of the upper mode is stronger than for cross-circular polarization (in agreement with the corresponding experimental findings), because, for cross-circular polarization, part of the dephasing terms do not contribute as a result of their spin dependence (see Sec. 2.4.1). For co-circular excitation, the carrier operators, $f_{\mathbf{k}}^a$ and correlation terms, $f_{\mathbf{k'}+\mathbf{k}}^a$, in Eq. (31) give rise to population effects such as phase-space filling, energy renormalization, and the interaction-induced dephasing while the terms are zero for cross-circular excitation; thus the population effects are significantly reduced. For strong pumping and cross-circular polarization, two reflection dips remain clearly present. In this case, broadening with nearly constant splitting corresponds to dominant interaction-induced dephasing. Co-circular excitation leads to a pronounced blue-shift of the normal-mode peaks whereas for cross-circular excitation only a weak red-shift occurs (in agreement with the experiment and previous bare quantum-well results[42]). The blue-shift for co-circular polarization originates from the exchange contributions to the correlation terms proportional to $\delta_{ab} W_{\mathbf{k'}} W_{\mathbf{k}-\mathbf{k''}}$ which are absent in cross-circular excitation since the exchange interaction does not couple different spin subsystems. On the other hand, this model cannot describe the emergence of a sidepeak of the upper mode for cross-circular polarization. This confirms the biexcitonic nature of this additional resonance since the biexciton bound state is not included within the second-order Born approximation. The additional resonance is clearly reproduced in $\chi^{(5)}$ calculations presented below where bound biexcitons are included.

In the following, we demonstrate for some cases that the rich variety of experimental features in the temporal dynamics of the pump-probe spectra, see Sec. 3.3.2 can be reproduced by microscopic theory. The calculations in

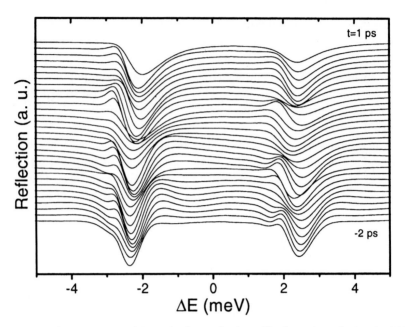

Fig. 24. Reflection spectra for co-circular excitation of both modes calculated within the second-order Born approximation. The Rabi energy of the pump pulse is 0.4 E_B and pump-probe delays vary from -2 ps (bottom) to $+1$ ps (top) in steps of 0.1 ps.

second-order Born approximation corresponding to the situation of Fig. 18 are shown in Fig. 24. The oscillatory behavior of the gain at the red side of the lower normal mode is well reproduced. For positive delay, a small blue-shift of the lower mode is obtained while the upper mode shows practically no shift in agreement with the experiment. For negative delay, coherent oscillations of the excitonic polarization lead to a slight temporal oscillation of the normal-mode peak positions. We suppose that due to inhomogeneous broadening these oscillations of the upper mode are less prominent in the experiment.

3.5. *Dynamics-Controlled Truncation Scheme in Semiconductor Microcavities*

Here, we present and discuss numerical results obtained for the nonlinear optical response of semiconductor microcavities using the DCT scheme. In Sec. 3.5.1 we review some results on the appearance of biexciton-induced resonances in the normal-mode spectra. In Sec. 3.5.2 we present results for polarization-dependent pump-probe experiments, which are compared to the measurements presented in Sec. 3.3.

3.5.1. *Biexciton Signatures for Low Excitation Intensities*

Here, we treat the many-body correlations on the coherent $\chi^{(3)}$-level, see Sec. 2.4.3. The microcavity system considered in the following calculations consists of two distributed Bragg mirrors with alternating GaAs/AlAs $\frac{\lambda}{4}$-layers (index of refraction $n = 3.61$ and $n = 2.95$, respectively) on a GaAs-substrate. The $\frac{3}{2}\lambda$ spacing between the mirrors leads to two anti-nodes of the field inside the cavity. The semiconductor material, which interacts resonantly with the cavity field, is placed at one of these anti-nodes. The frequency of the cavity mode is taken to coincide with the $1s$-exciton resonance of the system. For small broadening, the coupling between the cavity mode and the exciton leads to the normal-mode splitting.

The propagation of the pulses along the cavity is again modeled using the transfer matrix technique, see Sec. 3.2. The pump-probe signal is calculated for resonant excitation where the center of the pump frequency is equal to the free-cavity and free-exciton frequencies. Furthermore, we assume zero delay $\tau = 0$ between the pulses. We separately solve for the propagation of the fields in $\mathbf{k}_1$ (pump) direction, as well as first- and third-order components of the field in $\mathbf{k}_2$ (probe) direction. In the frequency-domain spectrum of the reflected field $\mathbf{E}_R(\omega)$ corresponding to the probe-direction $\mathbf{k}_2$, we distinguish between linear and third-order non-linear contributions[37]

$$\mathbf{E}_R(\omega) = \mathbf{E}_R^{(1)}(\omega) + a\mathbf{E}_R^{(3)}(\omega), \tag{89}$$

where a depends on the intensity of the pump pulse. It is chosen such that the third-order field $a\mathbf{E}_R^{(3)}(\omega)$ causes only a small change of the linear field $\mathbf{E}_R^{(1)}(\omega)$, i.e. such that the perturbation expansion in orders of the field is valid.

In order to determine the strength of the third-order term, one can compare the total intensity $|\mathbf{E}_R(\omega)|^2$ at the normal-mode peaks with the first-order contribution $|\mathbf{E}_R^{(1)}(\omega)|^2$. The reflection $R(\omega) = |\mathbf{E}_R(\omega)|^2/|\mathbf{E}_0(\omega)|^2$ is normalized with respect to the spectrum of the applied probe field and the differential reflection $\delta R(\omega)$ is given by the difference of the total reflection and the linear reflection

$$\delta R(\omega) = \frac{|\mathbf{E}_R(\omega)|^2 - \left|\mathbf{E}_R^{(1)}(\omega)\right|^2}{|\mathbf{E}_0(\omega)|^2}. \tag{90}$$

If the third-order field amplitude is much smaller than the first-order one and if the probe pulse is very short, i.e. spectrally broad, one obtains[37]

$$\delta R(\omega) \propto Re\left[\mathbf{E}_R^{(1)}(\omega)^* \mathbf{E}_R^{(3)}(\omega)\right]. \tag{91}$$

For our numerical analysis, we use 14 and 16.5 mirror layer pairs for the top and bottom mirrors, respectively. The system parameters of the one-dimensional model system are chosen to yield exciton and biexciton binding energies of 8.0 meV and 1.4 meV, respectively.[37] These values are similar to those of the experimental system studied in Ref. 66.

The left part of Fig. 25 shows the microcavity reflectivity spectrum for co-circular $(\sigma^+\sigma^+)$ excitation with a 1 ps long pump pulse. In Fig. 25(a), the reflections with and without pump are compared, ensuring that the effect of the third-order contributions is small so that the $\chi^{(3)}$ treatment is valid. Note that the spectra have been vertically displaced relative to each other in order to display the dashed lines. Figure 25(b) shows the resulting

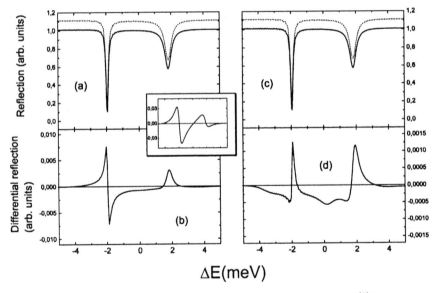

Fig. 25. (a) Probe-reflection $R(\omega)$ (solid) and linear probe-reflection $R^{(1)}(\omega)$ (dotted) for excitation with co-circularly polarized pump and probe pulse and $\tau = 0$ ps. Note that $R^{(1)}(\omega)$ is shifted by 0.1. (b) Resulting differential reflection $\delta R(\omega)$ (Inset: calculation of the differential reflection without correlations, i.e. in Hartree–Fock approximation, using the same horizontal scale). (c) and (d) show the same as (a) and (b) for cross-circularly polarized pump and probe pulses and $\tau = 0$ ps. Taken from Ref. 37.

differential reflection spectra $\delta R(\omega)$. While the energetically higher normal-mode peak exhibits bleaching, the lower peak shifts to higher energies, which gives rise to an overall decrease of the normal-mode splitting. The inset shows that this shift remains when no correlations are included in the calculation, i.e. in the Hartree–Fock limit. However, unlike the results with correlations, in Hartree–Fock the bleaching of the upper peak is accompanied by a blue-shift. This demonstrates that inside a microcavity the biexciton correlation effects lead to characteristic signatures.

Figures 25(c) and 25(d) display results for cross-circular $(\sigma^+\sigma^-)$ excitation. Note that also within the cavity, for this choice of polarization directions the signal is solely due to the correlation contributions. For this configuration, we again find bleaching of the upper and a dispersive shape at the lower resonance. In this case, however, the lower resonance shows a red-shift corresponding to an increase of the normal-mode splitting. Qualitatively, these results[37] agree well with the experimental findings of X. *Fan, et al.*[66] For the cross-circular polarization, we find additional structure in the differential reflection spectra, see Fig. 25(d), which is analyzed in the following.

In order to more clearly investigate the peak structure appearing for cross-circular polarization the system parameters are modified. We reduce the reflectivity of the mirrors by changing the number of top and bottom layer pairs to 10 and 12.5, respectively. Furthermore, the dephasing times are doubled ($\gamma_p^{-1} = 6\,\text{ps}$ and $\gamma_B^{-1} = 3\,\text{ps}$) corresponding to a reduced homogeneous broadening and excitation with a short pump pulse ($\bar{t}_1 = 100\,\text{fs}$) is considered.

In Fig. 26, full reflection spectra (solid) for $\sigma^+\sigma^-$-excitation are compared with the linear spectra (dashed) (note that in Fig. 26 the same constant a is used as in Fig. 25). In Fig. 26(a), the same normal-mode splitting was taken as in Fig. 25, while in (b) this has been enhanced by enlarging the modulus of the dipole matrix element $|\mathbf{d}^{eh}|$. The peak in between the normal-mode resonances, which can also be seen in Fig. 25(d), is somewhat stronger for the present parameters. Additionally, in Figs. 26(a) and 26(b) a similar shoulder can be seen below the lower normal-mode resonance. Further calculations of the optical response of our model system without the microcavity confirm that the energetic distance of these peaks to the normal-mode resonances coincides with the biexciton binding energy. This appearance of biexcitons in the reflection spectra is to be expected. Since the cavity mode is resonant with the exciton, the coupling leads to two normal modes. The biexciton states, however, are not directly coupled to

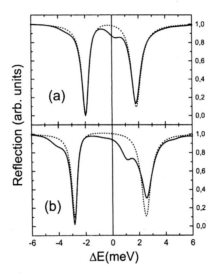

Fig. 26.   Total probe-reflection $R(\omega)$ (solid) and linear probe-reflection $R^{(1)}(\omega)$ (dotted) using cross-circularly polarized pump and probe pulses and $\tau = 0$ ps. Two different normal-mode splittings of about 4 meV in (a) and 6 meV (b) are considered. Taken from Ref. 37.

the cavity field and thus remain unchanged inside the cavity (no biexciton normal-mode coupling exists). Therefore, the transition from the normal modes to the biexciton has to show up one biexciton binding energy below both normal-mode-coupled resonances. Another feature that is apparent in Fig. 26(b) is connected to the intensity of the peaks. Whereas the dip in the reflection at the lower normal-mode resonance is stronger than at the upper normal-mode resonance, the intensity of the biexciton-induced features appearing below these resonances behaves oppositely. This can be explained by the following considerations: for our system parameters the normal-mode splitting is about four times larger than the biexciton binding energy. Therefore the frequency difference between the upper normal-mode resonance and the biexciton is close to the frequency of the lower resonance, and vice versa. Thus, when the pump excites the upper normal-mode resonance, a probe frequency close to the lower resonance is needed in order to spectrally excite the biexciton, which explains the obvious reversal of the intensities of the biexciton-induced peaks seen in Fig. 26(b). The results in the biexciton-induced features in the spectra of semiconductor microcavities[37] are in good qualitative agreement with experimental findings.[66,68]

3.5.2. *Four-Particle Correlations in Intensity- and Polarization-Dependent Pump-Probe Studies*

The second-order Born theory, making use of a complete factorization into two-particle functions, is not able to describe bound states consisting of more than two particles. For a bare quantum well it has been shown that both biexciton and unbound two-exciton states significantly influence the excitonic optical Stark effect, see Sec. 2.4.4, which can even change its sign depending on the polarization directions of the incident pulses. Signatures of two-exciton states and especially the biexciton for NMC systems have been discussed in Refs. 37, 66 and 68, see also the preceding Sec. 3.5.1.

To be able to describe the intensity and polarization dependence of the NMC response including biexcitons, we use the coherent $\chi^{(5)}$-expansion introduced in Sec. 2.4.5. It is restricted to the coherent (transition-like) contributions, and to low orders in the optical field. Therefore, this procedure limits us to short timescales and a certain range of not too strong fields.

In order to solve the microscopic coherent $\chi^{(5)}$-equations within reasonable numerical limits, we use a one-dimensional tight-binding model, which has been described and evaluated in Refs. 37, 39, 42 and 48, and also Secs. 2.4.4, 2.4.6 and 3.5.1. It has been shown that this model is capable of reproducing experimental results obtained for quantum wells, e.g. two-dimensional structures, at least in a good qualitative fashion.

Two pulses are propagated through the cavity structure to model the experimental pump-probe situation, see Sec. 3.3. The equations are solved separately for different orders and (pump and probe) directions of the optical field. We assume the probe pulse to be weak, and therefore neglect the influence of the probe on the pump field, and also all terms higher than linear in the probe field.[37] The signal is computed by adding the contributions up to fifth order. Since the reflected field is calculated linearly in the probe field, the pump field contributes up to fourth order.

Figure 27 shows results obtained within the $\chi^{(5)}$ formalism for the case of lower-mode pumping and zero delay. The gain peak at the red side of the lower mode is well reproduced for co-circular polarization. As in the second Born theory, this gain peak originates from the diffraction of pump-induced polarization into probe direction. An increase in pumping yields a blue-shift of both modes, which is more pronounced in the lower than in the upper mode. However, the magnitude of the shifts, and also of the upper-mode saturation is underestimated by the $\chi^{(5)}$ results. For the cross-circular configuration, the biexcitonic mode appears at the red side of the upper

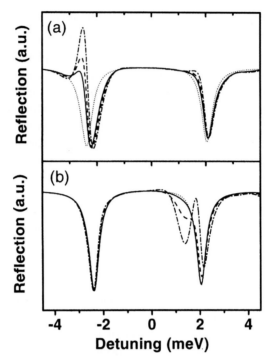

Fig. 27. Probe reflectivity spectra calculated within the $\chi^{(5)}$ formalism for lower-mode excitation, zero detuning, and increasing energy of the pump pulse. (a) Co-circular; pump intensity 0 (dotted), 0.42 E_B (solid), 0.47 E_B (dashed), 0.52 E_B (dashed-dotted). (b) Cross-circular; pump intensity 0 (dotted), 0.23 E_B (solid), 0.52 E_B (dashed), 0.74 E_B (dashed-dotted). E_B is the 3D excitonic Rydberg energy.

mode, while the lower mode shifts slightly to the red. The biexcitonic mode is more clearly visible in Fig. 27(b) than in the experimental data because the scattering terms are significantly reduced in the one-dimensional $\chi^{(5)}$ formalism, where the excitation-induced spectral broadening is very small. That the biexcitonic mode appears on the red side of the upper mode may be understood simply as as a consequence of energy conservation: one requires one pump photon (resonant with the lower mode) plus one probe photon (at the upper mode minus the binding energy) in order to create a biexciton in the quantum well. It is interesting to note that the main spectral features (direction of shifts, gain peak, biexcitonic mode) are already present in a third-order calculation (not shown in figure) see also Ref. 37. There are, however, quantitative differences between the results of third- and fifth-order calculations for higher intensities.

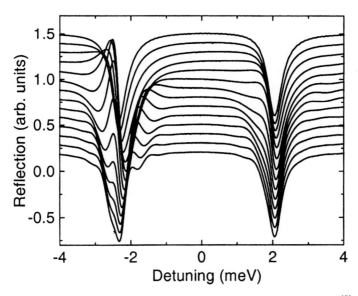

Fig. 28.   Reflection spectra for co-circular excitation calculated within the $\chi^{(5)}$ formalism. Pump intensity 0.52 E_B, and pump delays of -4 ps (bottom) to $+2.5$ ps (top). Delayed spectra have been displaced vertically.

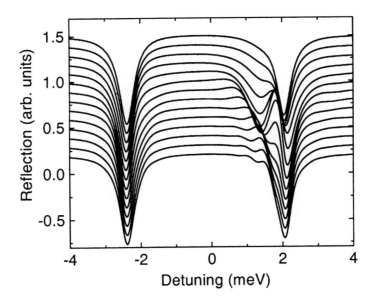

Fig. 29.   Same as Fig. 28 for cross-circular excitation and pump intensity 0.74 E_B.

Also some of the experimental features on the temporal dynamics of the pump-probe spectra, see Sec. 3.3.2 can be reproduced by the $\chi^{(5)}$ formalism. The calculated time evolution of the probe reflection spectrum is displayed in Fig. 28 for the case of lower-mode pumping and co-circular excitation, corresponding to the experimental results in Fig. 17(c). At the lower mode, for negative delays spectral oscillations with a period that is inversely proportional to the modulus of the delay are present.[37] Out of these oscillations, the gain peak evolves and decays again rapidly after maximum pulse overlap. Because only one mode is excited, no additional temporal oscillations occur.

In Fig. 29, the same is shown for cross-circular excitation. The rise and subsequent decay of the biexcitonic mode are clearly evident, while the lower mode remains essentially unchanged. The dynamics of the biexcitonic mode agrees well with the experimental result illustrated in Fig. 19. At zero delay, the energetic separation between upper normal mode and the transition to the biexciton is most prominent.

4. Summary and Conclusions

What are the elementary excitations of a semiconductor microcavity that dominate the optical properties? The answer seems to be nontrivial if one follows the current discussion on coherent nonlinear optical properties of semiconductor microcavities, as summarized in the introduction to Sec. 3. While the optical field generates electrons and holes, the many-body Coulomb interaction between these carriers is responsible for various correlation effects. The attraction between an electron and a hole leads to excitonic states and the interaction between two excitons generated with opposite circular polarization can result in biexcitonic excitations. In this sense, the Coulomb interaction results in various elementary excitations.

On the other hand, the Coulomb interaction also provides the dominant interaction processes between the elementary excitations in addition to the interaction with the optical field. (For simplicity the interaction with phonons and structural disorder is not considered here.) Furthermore, the many-body interaction between electrons and/or holes causes dephasing, energy renormalization, and screening, and also electron–exciton and exciton–exciton interactions additionally contribute to these effects. In a microcavity, the large oscillator strength of the excitonic states allows modifications of their radiative lifetime and polariton effects.

At this point it should be clear that a consistent microscopic description of all the discussed effects is rather challenging. The *full* problem can be formulated in terms of a Hamiltonian for electrons and holes containing the Coulomb interaction between the carriers and the dipole interaction with optical fields. While the application of this Hamiltonian results in an infinite hierarchy of correlation functions, different approximation schemes have been developed to derive closed and tractable theories. Here, two well-established approaches, the second-Born approximation and the dynamics-controlled truncation scheme, have been introduced and analyzed.

The comparison with experimental results shows that the second-Born approximation is able to describe excitation-dependent optical nonlinearities in connection with the exciton saturation, the line broadening and shifts at elevated pump intensities very well. At the same time, coherent nonlinear effects like the observed probe-signal gain are reproduced within this method, which is nonperturbative in the field. Despite these advantages, this method does not include biexcitonic effects since they require the explicit consideration of higher-order correlation functions.

The treatment of Coulomb correlations using the dynamics-controlled truncation scheme, up to the coherent $\chi^{(3)}$ and $\chi^{(5)}$ contributions, is able to reproduce the observed biexciton signatures in the normal-mode spectra for cross-circular polarization and zero pump-probe delay. Furthermore, the results on the excitation-dependent shifts are in agreement with the low-intensity contributions of the second-Born approximation. On the other hand, the theory is restricted to weak fields such that, $\chi^{(7)}$ and higher-order contributions can be neglected. Hence the high-intensity saturation behavior is expected to be underestimated as seen in our theory-experiment comparison. Furthermore, this approach is restricted to the coherent regime, which limits the applicable timescale, since incoherent occupations cannot be classified this way.

Although polariton–polariton interactions are sometimes invoked to account for nonlinear optical experiments on semiconductor microcavities, it is not necessary and often problematic to do so. Some effects, such as carrier–carrier scattering and Coulomb correlations are not easily included in a polaritonic picture. Furthermore, it has been shown here, that microscopic theories based on the electron-hole-Hamiltonian are able to provide a good agreement with the presented experimental data. Thus the fundamental carrier interactions responsible for the nonlinear optical response of semiconductor quantum wells are also adequate to account for the polarization- and density-dependent nonlinear response of semiconductor microcavities.

Acknowledgments

The Marburg group acknowledges support from the Deutsche Forschungs-gemeinschaft (DFG), a DFG Heisenberg fellowship (TM), the Center for Optodynamics of the Philipps-University Marburg, and the Max-Planck prize of the Humboldt Foundation and the Max-Planck Society. F. Jahnke and T. Meier acknowledge the John von Neumann Institut für Computing (NIC), Forschungszentrum Jülich, Germany, for grants of CPU time on their supercomputer systems. The Michigan portion of the work was supported by AFOSR and NSF through the Center for Ultrafast Optical Science. The Arizona group was supported by AFOSR, NSF AMOP, NSF ECS EPDT, JSOP, COEDIP, and the Humboldt Research program (HMG).

References

1. H. Haug and S.W. Koch, *Quantum Theory of the Optical and Electronic Properties of Semiconductors*, 4th ed., World Scientific Publ., Singapore, 2004.
2. R.J. Elliott, in *Polarons and Excitons*, eds. C.G. Kuper and G.D. Whitefield, Oliver and Boyd, 269 pp., 1963.
3. H. Haug and S. Schmitt-Rink, *Prog. Quant. Electron.* **9**, 3, 1984.
4. M. Lindberg and S.W. Koch, *Phys. Rev.* **B38**, 3342, 1988.
5. W. Schäfer, *J. Opt. Soc. Am.* **B13**, 1291, 1996.
6. R. Binder and S.W. Koch, *Prog. Quant. Electron.* **19**, 307, 1995.
7. F. Jahnke, M. Kira and S.W. Koch, *Z. Physik* **B104**, 559, 1997.
8. S.W. Koch, T. Meier, F. Jahnke, and P. Thomas, *Appl. Phys.* **A71**, 511, 2000.
9. T. Meier and S.W. Koch, in *Ultrafast Physical Processes in Semiconductors*, ed. K.T. Tsen, Semiconductors and Semimetals, Vol. 67, Academic Press, 231 pp., 2001.
10. E.P. Wigner, *Phys. Rev.* **46**, 1002, 1934.
11. M. Lindberg, Y.Z. Hu, R. Binder and S.W. Koch, *Phys. Rev.* **B50**, 18060, 1994.
12. V.M. Axt and A. Stahl, *Z. Phys.* **B93**, 195, 1994; *ibid.*, 205, 1994.
13. D.S. Chemla and J. Shah, *Nature* **411**, 549, 2001.
14. N. Bloembergen, *Nonlinear Optics*, Benjamin Inc., New York, 1965.
15. G. Khitrova, H.M. Gibbs, F. Jahnke, M. Kira and S.W. Koch, *Rev. Mod. Phys.* **71**, 1591, 1999.
16. S.W. Koch, M. Kira and T. Meier, invited review, *J. Opt.* **B3**, R29, 2001.
17. M. Kira, W. Hoyer, F. Jahnke and S.W. Koch, *Prog. Quant. Electron.* **23**, 189, 1999.
18. M. Kira, F. Jahnke and S.W. Koch, *Phys. Rev. Lett.* **81**, 3263, 1998.
19. M. Kira, F. Jahnke and S.W. Koch, *Phys. Rev. Lett.* **82**, 3544, 1999.
20. Y.S. Lee, T.B. Norris, M. Kira, F. Jahnke, S.W. Koch, G. Khitrova and H.M. Gibbs, *Phys. Rev. Lett.* **83**, 5338, 1999.

21. M. Kira, W. Hoyer, T. Stroucken and S.W. Koch, *Phys. Rev. Lett.* **87**, 176401, 2001.
22. W. Hoyer, M. Kira and S.W. Koch, *Phys. Rev.* **B67**, 155113, 2003.
23. W. Schäfer, R. Lövenich, N.A. Fromer and D.S. Chemla, *Phys. Rev. Lett.* **86**, 344, 2001.
24. R. Lövenich, C.W. Lai, D. Hgele, D.S. Chemla and W. Schäfer, *Phys. Rev.* **B66**, 045306, 2002.
25. W. Schäfer, D.S. Kim, J. Shah, T.C. Damen, J.E. Cunningham, K.W. Goosen, L.N. Pfeiffer and K. Köhler, *Phys. Rev.* **B53**, 16429 1996.
26. K. Bott, O. Heller, D. Bennhardt, S.T. Cundiff, P. Thomas, E.J. Mayer, G.O. Smith, R. Eccleston, J. Kuhl and K. Ploog, *Phys. Rev.* **B48**, 17418, 1993.
27. S.W. Koch, T. Meier, W. Hoyer and M. Kira, invited review, *Physica* **E14**, 45, 2002.
28. F. Jahnke, M. Kira, S.W. Koch, G. Khitrova, E.K. Lindmark, T.R. Nelson, D.V. Wick, J.D. Berger, O. Lyngnes, H.M. Gibbs and K. Tai, *Phys. Rev. Lett.* **77**, 5257, 1996
29. R. Binder, S.W. Koch, M. Lindberg, W. Schäfer and F. Jahnke, *Phys. Rev.* **B43**, 6520, 1991.
30. P. Meystre and M. Sargent III, *Elements of Quantum Optics*, Springer, New York, 1991.
31. H. Wang, K.B. Ferrio, D.G. Steel, Y.Z. Hu, R. Binder and S.W. Koch, *Phys. Rev. Lett.* **71**, 1261, 1993.
32. O. Lyngnes, J.D. Berger, J.P. Prineas, S. Park, G. Khitrova, H.M. Gibbs, F. Jahnke, M. Kira and S.W. Koch, *Solid State Commun.* **104**, 297, 1997.
33. M. Hofmann, A. Wagner, C. Ellmers, C. Schlichenmeier, S. Schäfer, F. Höhnsdorf, J. Koch, W. Stolz, S.W. Koch, W.W. Rühle, J. Hader, J.V. Moloney, E.P. O'Reilly, B. Borchert, A.Yu. Egorov and H. Riechert, *Appl. Phys. Lett.* **78**, 3009, 2001.
34. W.W. Chow and S.W. Koch, *Semiconductor-Laser Fundamentals*, Springer, Berlin, 1999.
35. W. Schäfer and M. Wegener, *Semiconductor Optics and Transport Phenomena*, Springer, Berlin, 2002.
36. T. Östreich, K. Schönhammer and L.J. Sham, *Phys. Rev. Lett.* **74**, 4698, 1995.
37. C. Sieh, T. Meier, A. Knorr, F. Jahnke, P. Thomas and S.W. Koch, *Europ. Phys. J.* **B11**, 407, 1999.
38. M. Lindberg, R. Binder and S.W. Koch, *Phys. Rev.* **A45**, 1865, 1992.
39. S. Weiser, T. Meier, J. Möbius, A. Euteneuer, E.J. Mayer, W. Stolz, M. Hofmann, W.W. Rühle, P. Thomas and S.W. Koch, *Phys. Rev.* **B61**, 13088, 2000.
40. P. Kner, S. Bar-Ad, M.V. Marquezini, D.S. Chemla and W. Schäfer, *Phys. Rev. Lett.* **78**, 1319, 1997.
41. P. Kner, W. Schäfer, R. Lövenich and D.S. Chemla, *Phys. Rev. Lett.* **81** 5386, 1998.

42. C. Sieh, T. Meier, F. Jahnke, A. Knorr, S.W. Koch, P. Brick, M. Hübner, C. Ell, J. Prineas, G. Khitrova and H.M. Gibbs, *Phys. Rev. Lett.* **82**, 3112, 1999.

43. S.W. Koch, C. Sieh, T. Meier, F. Jahnke, A. Knorr, P. Brick, M. Hübner, C. Ell, J. Prineas, G. Khitrova and H.M. Gibbs, *J. Lumin.* **83/84**, 1, 1999.

44. T. Meier, S.W. Koch, M. Phillips and H. Wang, *Phys. Rev.* **B62**, 12605, 2000.

45. P. Brick, C. Ell, G. Khitrova, H.M. Gibbs, T. Meier, C. Sieh and S.W. Koch, *Phys. Rev.* **B64**, 075323, 2001.

46. K. Victor, V.M. Axt and A. Stahl, *Phys. Rev.* **B51**, 14164, 1995.

47. T. Meier, S.W. Koch, P. Brick, C. Ell, G. Khitrova and H.M. Gibbs, *Phys. Rev.* **B62**, 4218, 2000.

48. W. Langbein, T. Meier, S.W. Koch and J.M. Hvam, *J. Opt. Soc. Am.* **B18**, 1318, 2001.

49. T. Meier, C. Sieh, E. Finger, W. Stolz, W.W. Rühle, P. Thomas and S.W. Koch, *Phys. Stat. Sol.* (b) **238**, 537, 2003.

50. G. Bartels, V.M. Axt, K. Victor, A. Stahl, P. Leisching and K. Köhler, *Phys. Rev.* **B51**, 11217, 1995.

51. T.F. Albrecht, K. Bott, T. Meier, A. Schulze, M. Koch, S.T. Cundiff, J. Feldmann, W. Stolz, P. Thomas, S.W. Koch and E.O. Göbel, *Phys. Rev.* **B54**, 4436, 1996.

52. H.P. Wagner, H.-P. Tranitz, M. Reichelt, T. Meier and S.W. Koch, *Phys. Rev.* **B64**, 233303, 2001.

53. C. Weisbuch, M. Nishioka, A. Ishikawa and Y. Arakawa, *Phys. Rev. Lett.* **69**, 3314, 1992.

54. C. Ell, J. Prineas, T.R. Nelson, Jr., S. Park, H.M. Gibbs, G. Khitrova, S.W. Koch and R. Houdre, *Phys. Rev. Lett.* **80**, 4795, 1998.

55. M. Koch, J. Shah and T. Meier, *Phys. Rev.* **B57**, 2049, 1998.

56. J.-K. Rhee, D.S. Citrin, T.B. Norris, Y. Arakawa and M. Nishioka, *Sol. State Commun.* **97**, 941, 1996.

57. R. Houdre, J.L. Gibernon, P. Pellandini, R.P. Stanley, U. Oesterle, C. Weisbuch, J. O'Gorman, B. Roycroft and M. Ilegems, *Phys. Rev.* **B52**, 7810, 1995.

58. M. Kuwata-Gonokami, S. Inouye, H. Suzuura, M. Shirane, R. Shimano, T. Someya and H. Sakaki, *Phys. Rev. Lett.* **79**, 1341, 1997.

59. P.G. Savvidis, J.J. Baumberg, R.M. Stevenson, M.S. Skolnick, D.M. Whittaker and J.S. Roberts, *Phys. Rev. Lett.* **84**, 1547, 2000.

60. C. Ciuti, P. Schwendimann, B. Deveaud and A. Quattropani, *Phys. Rev.* **B62**, R4825, 2000.

61. R.M. Stevenson, V.N. Astratov, M.S. Skolnick, D.M. Whittaker, M. Emam-Ismail, A.I. Tartakovskii, P.G. Savvidis, J.J. Baumberg and J.S. Roberts, *Phys. Rev. Lett.* **85**, 3680, 2000.

62. A.I. Tartakovskii, D.N. Krizhanovskii and V.D. Kulakovskii, *Phys. Rev.* **B62**, R13298, 2000.

63. J.J. Baumberg, P.G. Savvidis, R.M. Stevenson, A.I. Tartakovskii, M.S. Skolnick, D.M. Whittaker and J.S. Roberts, *Phys. Rev.* **B62**, R16247, 2000.

64. G. Messin, J.Ph. Karr, A. Baas, G. Khitrova, R. Houdre, R.P. Stanley, U. Oesterle and E. Giacobino, *Phys. Rev. Lett.* **87**, 127403, 2001.
65. P. Senellart and J. Bloch, *Phys. Rev. Lett.* **82**, 1233, 1999.
66. X. Fan, H. Wang, H.Q. Hou and B.E. Hammons, *Phys. Rev.* **B57**, R9451, 1998.
67. P. Borri, W. Langbein, U. Woggon, J.R. Jensen and J.M. Hvam, *Phys. Rev.* **B62**, R7763, 2000.
68. U. Neukirch, S.R. Bolton, N.A. Fromer, L.J. Sham and D.S. Chemla, *Phys. Rev. Lett.* **84**, 2215, 2000.
69. M. Saba, F. Quochi, C. Ciuti, U. Oesterle, J.L. Staehli, B. Deveaud, G. Bongiovanni and A. Mura, *Phys. Rev. Lett.* **85**, 385, 2000.
70. M. Reichelt, C. Sieh, T. Meier and S.W. Koch, *Phys. Stat. Sol.* (b) **221**, 249, 2000.
71. M. Kira, F. Jahnke, S.W. Koch, J.D. Berger, D.V. Wick, T.R. Nelson, Jr., G. Khitrova and H.M. Gibbs, *Phys. Rev. Lett.* **79**, 5170, 1997.
72. C. Ell, P. Brick, M. Hübner, E.S. Lee, O. Lyngnes, J.P. Prineas, G. Khitrova, H.M. Gibbs, M. Kira, F. Jahnke, S.W. Koch, D.G. Deppe and D.L. Huffaker, *Phys. Rev. Lett.* **85**, 5392, 2000.
73. T. Stroucken, A. Knorr, P. Thomas and S.W. Koch, *Phys. Rev.* **B53**, 2026, 1996.
74. T.B. Norris, in *Semiconductor Quantum Optoelectronics: From Quantum Physics to Smart Devices*, eds. A. Miller and D. Finlayson, Institute of Physics, 1999, pp. 121.
75. Y.-S. Lee, T.B. Norris, A. Maslov, D.S. Citrin, J. Prineas, G. Khitrova and H.M. Gibbs, *Appl. Phys. Lett.* **78**, 3941, 2001.
76. Y.-S. Lee, T.B. Norris, J. Prineas, G. Khitrova and H.M. Gibbs, *Phys. Stat. Sol.* (b) **221**, 121, 2000.
77. H.M. Gibbs, D.V. Wick, G. Khitrova, J.D. Berger, O. Lyngnes, T.R. Nelson, Jr., E.K. Lindmark, S. Park, J. Prineas, M. Kira, F. Jahnke, S.W. Koch, W.W. Rühle, S. Hallstein and K. Tai, Festkörperprobleme/*Advances in Solid State Phys.* **37**, 227, 1997.
78. G.W. Fehrenbach, W. Schäfer, J. Treusch and R.G. Ulbrich, *Phys. Rev. Lett.* **49**, 1281, 1982.
79. F. Quochi, C. Ciuti, G. Bongiovanni, A. Mura, M. Saba, U. Oesterle, M.A. Dupertuis, J.L. Staehli and B. Deveaud, *Phys. Rev.* **B59**, R15594, 1999.

CHAPTER 7

POLYMER MICRORING RESONATORS

Payem Rabiei and William H. Steier

Department of Electrical Engineering
University of Southern California
Los Angeles, CA 90089-0483, USA

The advantages of polymers in photonics, such as their low optical loss, potential low cost, and an electro-optic effect make them good candidates for materials for microresonators. In this chapter expressions are derived for the parameters of microresonators including the free-spectral range (FSR) and the quality factor (Q). Computer simulations are presented for the mode profiles, bending loss, effective index, and the scattering and material loss for conditions typical in polymers. Simulations are also included for calculating the vertical coupling and the fabrication accuracy required between polymer waveguides and microrings. The results for passive polymer resonators for a range of ring diameters and FSR demonstrate what can be achieved with polymers. Electro-optic polymer ring resonators are reviewed along with a discussion of the modulation bandwidth limits and data on modulators fabricated with current electro-optic polymers. The chapter concludes with results on coupled microresonators and their use as widely tunable mirrors and a review of some possible applications for polymer resonators.

1. Introduction

Polymers have been used for many applications in photonics and guided wave optics where their unique properties can be used to advantage. The applications include very low-loss optical waveguides,[1] electro-optical devices,[2] thermo-optic switches,[3] optical light emitting diodes,[4] grating wavelength filters,[5] and arrayed-waveguide grating filters.[6] The promise of polymer components is ruggedness, low-cost fabrication particularly of complex high-density circuits, the possibility of integration with semiconductor based components,[7] the combining of different polymers with selected properties within the same optical circuit[8] and, in the case of electro-optics (EO), high-speed, low voltage modulators and switches.[9] Some of this promise has been realized with particular success in low loss halogenated waveguides of 0.036 dB/cm @ 840 nm[1] and 0.1 dB/cm @ 1300 nm[10] and in thermo-optic waveguide switches with 0.6 dB insertion loss.[3] Polymer EO modulators have been demonstrated with a V_π of 1.8 V @ 1550 nm and $\sim$20 GHz bandwidth (3 dBe).[11] Modulators with $V_\pi < 1\,V$[12] and devices that operate at over 100 GHz[13] have also been reported. Polymer modulators hold significant promise but have not found commercial applications as yet because of the difficulty in displacing the LiNbO$_3$ technology and remaining issues on their long term stability.

Polymers offer a wide range of indices of refraction, and a number of processes for fabricating optical waveguides have been developed. By mixing different but compatible polymers, one can tune the index of refraction to a desired value. Typical indices of refraction range from 1.3 to 1.7. The lowest loss materials contain halogens and therefore typically have the lower indices. In Table 1 are listed some of the polymers we have used along with their properties. There are numerous polymers available commercially other than the ones listed and chemists in industry and university laboratories have the ability to synthesize an almost limitless number of new materials with engineered properties.

Optical waveguides, both single mode and multimode, have been fabricated by reactive ion etching,[14] photolithography,[15] optical bleaching,[16] laser ablation,[17] molding,[18] embossing,[19] and, in EO polymers, by electric field poling.[20] Many of the processing methods can produce long-term

Table 1. The refractive index and the loss of optical polymers.

Material	Index 1300 nm	Loss (dB/cm) 1300 nm	Index 1550 nm	Loss (dB/cm) 1550 nm
SU-8[a]	1.567	0.5	1.565	4.0
NOA 61[b]	1.545	0.3	1.541	1.1
UFC170[c]	1.490	0.5	1.488	3.0
UV15[d]	1.510	0.9	1.504	4.2
Teflon AF 1601[e]	1.300	—	1.297	—
CLD1/APC[f] (Active)	1.614 (TE)	1.0	1.612 (TE)	1.7
ZPU12-R1[g]	—	0.06	1.45–1.47	0.35
ZPU13-R1[g]		0.06	1.43–1.45	0.35

[a]MicroChem Co., [b]Norland Products, Inc., [c]URAY Co, Korea, [d]MasterBond Co., [e]Dupont, Inc., [f]See Ref. 9, [g]Zen Photonics, Korea.

stable, single mode, low loss waveguides. The molding or embossing approaches have the potential for being low cost and are therefore of interest in a manufacturing process.

The advantages of polymers, such as their low optical loss, potential low cost, and the EO effect, make them good candidates for applications in microresonators. The first demonstration of polymer microresonators was by Hida *et al.*[21] High Q devices are possible because of the low loss, and tunable or switchable devices are possible with EO materials. The spin technology has the promise of good control of the layer thickness and therefore good control of the vertical coupling from waveguides into the resonators. One of the limitations of polymer technology is the limited index difference (∼0.4) that is possible between the ring core and the cladding and this limits how small the rings can be and therefore their free spectral range (FSR).

In this chapter we will present an analysis of microresonators and calculations of their properties for materials parameters that are characteristic of polymers. Our results with passive polymers include experimental results on rings of various sizes and materials, temperature tuning effects, and possible fabrication accuracies. We then extend this to rings made with EO polymers and provide experimental results of voltage tuning along with a discussion of the limits of tuning speed and tuning voltage. The EO rings are wavelength selective modulators and we will discuss some interesting WDM applications. Finally we will review and present some experimental results of the wide band tuning that is possible using the vernier effect in double ring resonators that have slightly different diameters. In the conclusion we will present some applications and our view of the possible future and the limitations of the technology.

2. Analysis of Microring Resonators

In this section we will present an analysis of microring resonators as an introduction to the discussion of polymer microrings. Similar analyses have been published and this work is not new, but we will use our notation and results as a framework to discuss the properties of the polymers and how they affect microring performance.

2.1. *Microresonator Transfer Function*

A general picture of the microring resonator with a coupling waveguide is shown in Fig. 1. Using the coupled mode theory[22] we can relate the electric

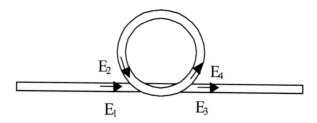

Fig. 1.   Geometry of single waveguide coupled to resonator.

fields in the coupling region as:

$$E_3 = rE_1 + itE_2$$
$$E_4 = itE_1 + rE_2 \qquad (1)$$

where t is the electric field transmission coefficient and r is the electric field reflection coefficient of the coupling. Assuming the coupling is lossless, t and r are related by:

$$|t|^2 + |r|^2 = 1 \qquad (2)$$

One round trip in the ring can be written as

$$E_2 = \exp(-\alpha\pi R)\exp(i2\pi\beta R)E_4 = a\exp(i2\pi\beta R)E_4 \qquad (3)$$

where β is the mode propagation constant in the ring given by:

$$\beta = \frac{2\pi n_{\text{eff}}}{\lambda} \qquad (4)$$

a is the E field transmission for one round trip in the ring

$$a = \exp(-\alpha\pi R) \qquad (5)$$

and α is the intensity loss coefficient given by:

$$\alpha = \alpha_m + \alpha_b + \alpha_w \qquad (6)$$

In the following expressions:

n_{eff} = effective index of refraction for the ring mode
α_m = absorption loss
α_b = radiation loss due to bending
α_w = scattering loss due to wall roughness
R = radius of the ring

By replacing E_2 from (3) to (2) and solving the equation for E_4 we obtain:

$$\frac{E_4}{E_1} = \frac{ita\exp(i\phi)}{1 - ra\exp(i\phi)} \tag{7}$$

In the typical case, the coupling waveguide is phase-matched to the ring waveguide and therefore r and t are real numbers. In this case:

$$\frac{I_4}{I_1} = \frac{(1 - r^2)a^2}{1 - 2ra\cos(\phi) + r^2a^2} \tag{8}$$

where

$$\phi = 2\pi\beta R.$$

We can also obtain:

$$\frac{E_3}{E_1} = \exp(i(\pi + \phi))\frac{a - r\exp(-i\phi)}{1 - ra\exp(i\phi)} \tag{9}$$

and again if r and t are real:

$$\frac{I_3}{I_1} = \frac{a^2 - 2ra\cos(\phi) + r^2}{1 - 2ra\cos(\phi) + r^2a^2} \tag{10}$$

At resonance, $\phi = 2\,m\pi$ (m an integer) the through-put becomes

$$\frac{I_3}{I_1} = \frac{(a - r)^2}{(1 - ra)^2} \tag{11}$$

Critical coupling ($I_3 = 0$) occurs when $a = r$. If phase-matching is not achieved in the coupling and r and t are not real, this will shift the resonance to some other value of ϕ but will not otherwise affect the microresonator response.

If the microring has a second output waveguide as shown in Fig. 2, the round trip expression [Eq. (3)] is modified by substituting a by r_2a. The output intensity transmission (I_5/I_1) can then be calculated as;

$$\frac{I_5}{I_1} = \frac{t_1^2 t_2^2}{1 - 2r_1r_2a\cos(\phi) + r_1^2r_2^2a^2} \tag{12}$$

where r_1, t_1 are the E field reflection and transmission coefficients for the input coupling and r_2, t_2 are the E field reflection and transmission coefficients for the output coupling. The through-put intensity transmission

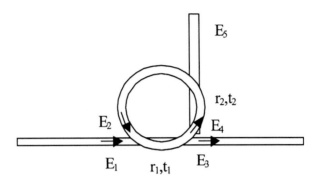

Fig. 2. Geometry of two waveguides coupled to a microresonator.

(I_3/I_1) and output intensity transmission (I_5/I_1) at resonance $(\phi = 2m\pi)$ can be calculated as:

$$\frac{I_3}{I_1} = \frac{(ar_2 - r_1)^2}{(1 - r_1 r_2 a)^2} \tag{13}$$

and

$$\frac{I_5}{I_1} = \frac{t_1^2 t_2^2}{(1 - r_1 r_2 a)^2} \tag{14}$$

Critical coupling $(I_3 = 0)$ and the maximum of (I_5/I_1) occur when $r_1 = r_2 a$ and in this case

$$\left(\frac{I_5}{I_1}\right)_{\text{max}} = \frac{1 - r_2^2}{1 - (r_2 a)^2} \tag{15}$$

If there is loss in the ring $(a < 1)$ then $\left(\frac{I_5}{I_1}\right)_{\text{max}} < 1$ is always the case.

2.2. *Microresonator Parameters*

There are a number of parameters that define the properties of microresonators. The first parameter is Q, which is defined as

$$Q = \frac{\lambda_0}{\Delta\lambda_{1/2}} \tag{16}$$

where $\Delta\lambda_{1/2}$ is the full-width half-maximum of the dropped intensity or transmitted intensity and λ_0 is the resonant wavelength. To calculate $\Delta\lambda_{1/2}$

we can rewrite (12) as:

$$\frac{I_5}{I_1} = \frac{t_1^2 t_2^2}{(1 - r_1 r_2 a)^2 + 4 r_1 r_2 a \sin^2\left(\frac{\phi}{2}\right)} \tag{17}$$

The values of ϕ at half-maximum, $\phi_{\pm 1/2}$, is given as

$$\sin\left(\frac{\phi_{\pm 1/2}}{2}\right) = \pm \frac{(1 - r_1 r_2 a)}{\sqrt{4 r_1 r_2 a}} \tag{18}$$

Since resonance is at $\phi = 2m\pi$ (m an integer) we can approximate

$$\frac{\phi_{\pm 1/2}}{2} = \pm \frac{(1 - r_1 r_2 a)}{\sqrt{4 r_1 r_2 a}} \tag{19}$$

$\Delta\phi$ at FWHM is

$$\Delta\phi = 2 \frac{(1 - r_1 r_2 a)}{\sqrt{4 r_1 r_2 a}} \tag{20}$$

Using the definitions of ϕ and β we can obtain

$$\Delta\lambda_{1/2} = \frac{(1 - r_1 r_2 a)\lambda_0^2}{2R\pi^2 n_{\text{eff}} \sqrt{r_1 r_2 a}} \tag{21}$$

where we have neglected the dispersion of n_{eff} over $\Delta\lambda$. From (16), the loaded Q_L (coupling losses included):

$$Q_L = \frac{2\pi^2 R n_{\text{eff}} \sqrt{r_1 r_2 a}}{(1 - r_1 r_2 a)\lambda_0} \tag{22}$$

Q_L for a single coupling waveguide can be obtained by setting $r_2 = 1$ and the unloaded Q_U can be obtained by setting $r_1 = r_2 = 1$. Since the round trip loss is typically small and $a \approx 1$

$$1 - a = 1 - \exp(-\alpha\pi R) \approx \alpha\pi R \tag{23}$$

and the unloaded Q can be approximated as

$$Q_U \approx \frac{2\pi n_{\text{eff}}}{\alpha\lambda_0} \tag{24}$$

For practical purposes where α is specified in dB/cm and λ_0 is specified in microns we obtain:

$$Q_U \approx \frac{2.73 \times 10^5 n_{\text{eff}}}{\alpha\lambda_0} \tag{25}$$

In this expression, Q_U is not dependent on R but in practice α can depend on R since the radiation loss, α_b, will depend on R. It is interesting to note that even with loss of 1dB/cm, a Q_U of more than 10^5 is possible.

The free spectral range (FSR) is the distance between two resonances of the resonator where m changes by one.

$$m = \frac{n_{\text{eff}}(\lambda_m)2\pi R}{\lambda_m}, \qquad m+1 = \frac{n_{\text{eff}}(\lambda_{m+1})2\pi R}{\lambda_{m+1}} \qquad (26)$$

If we neglect the dispersion of n_{eff}, the free spectral range is:

$$FSR = \lambda_{m+1} - \lambda_m = \frac{\lambda^2}{n_{\text{eff}}(\lambda)2\pi R} \qquad (27)$$

The finesse, F, is given by:

$$F = \frac{FSR}{\Delta\lambda_{1/2}} = \frac{\pi\sqrt{r_1 r_2 a}}{1 - r_1 r_2 a} \qquad (28)$$

The finesse of an unloaded resonator is:

$$F_U = \frac{\pi\sqrt{a}}{1 - a} \qquad (29)$$

and if α is in dB/cm and R is in microns the unloaded finesse is given by:

$$F_U = \frac{8.68 \times 10^4}{\alpha R} \qquad (30)$$

By measuring the finesse, $\frac{I_3}{I_1}$, and $\frac{I_5}{I_1}$ at resonance one can use Eqs. (13), (14) and (28) to calculate a, r_1 and r_2.

3. Simulations of Polymer Microresonators and Calculation of Parameters

In this section we will analyze polymer microring resonators by computer simulations and calculate some of the important parameters. First the mode profile and the effective index of the microresonator mode are calculated. In the next part we will discuss the loss mechanisms including bending loss, scattering loss due to surface roughness, and material loss in the available polymers. Based on these results we can make some conclusions regarding polymer microresonators.

3.1. Mode Profiles

Marcatilli[23] was the first to analyze the properties of curved dielectric waveguides and this work has been followed by a number of papers on techniques to compute, α_b, the radiation or bending loss. The conformal transformation approach that transforms the curved slab waveguide into the equivalent straight slab waveguide provides some insight into the mode

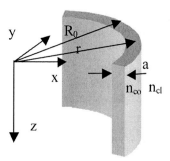

Fig. 3.   Geometry of curved cylindrical waveguide.

properties. However the analysis of the rectangular cross-section curved waveguides and the calculation of the bending loss and mode profiles is best done by using a commercial semi-vectoral finite difference software package.

For the confocal transformation approach, the geometry of the curved slab waveguide is shown in Fig. 3. The scalar wave equation for $E_z(x, y)$ in this geometry can be written as (assuming no z variation):

$$\Delta_{xy}^2 E_z(x, y) + k(x, y)^2 E_z(x, y) = 0 \tag{31}$$

where

$$k(x, y) = \frac{2\pi n(x, y)}{\lambda} \tag{32}$$

By using the conformal transformation[24]:

$$W = R_0 \ln \left(\frac{Z}{R_0} \right) \tag{33}$$

where

$$Z = x + iy$$
$$W = u + iv \tag{34}$$

the wave equation becomes:

$$\nabla_{uv}^2 E_z(u, v) + k(x, y)^2 e^{2u/R_0} E_z(u, v) = 0 \tag{35}$$

From Eq. (33) it can be shown that

$$u = R_0 \ln \left(\frac{r}{R_0} \right) \tag{36}$$

$$v = R_0 \theta \tag{37}$$

where r, θ are the cylindrical coordinates.

Therefore the inner edge of the ring at $r = R_0$ is $u = 0$ line in the W plane and the outer edge at $r = R_0 - a$ is at $u = R_0 \ln(R_0 - a)/R_0$. The curved structure is transformed to a straight structure in the W plane. However the index varies across the transformed straight waveguide as shown in Fig. 4. From this analysis, one can see that the equivalent index is higher on the outer surface and therefore the E field of the mode is skewed with higher magnitude on the outer side of the ring. It can also be seen that the mode always radiates from the outer edge since at some $r > R_0$ the equivalent cladding index is higher than the highest equivalent index in the core. We have used TempSelene,[25] which is a commercially available package to calculate mode profiles, effective indices, and the bending loss in the 3D structure. This software uses semi-vectorial finite difference method in cylindrical coordinates with absorbing boundary conditions. Figure 5 shows

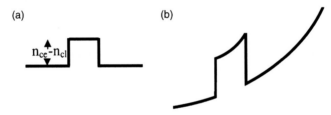

Fig. 4.   (a) Original refractive index profile for curved waveguide, (b) equivalent straight waveguide refractive index profile.

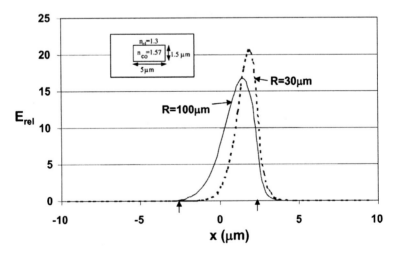

Fig. 5.   Horizontal profiles for the lowest order TM mode at two different radii of curvature. The ring waveguide dimensions and indices are shown in the inset. The arrows on the horizontal axis show the edges of the waveguide.

the horizontal mode profiles for the lowest order TM (polarized perpendicular to the plane of the ring) mode. The index and dimensions of the ring waveguide are shown in the figure. The mode profile is shown for two ring radii; 100 μm and 30 μm. As expected, the E field is largest at the outer edge of the ring and becomes larger as the ring radius decreases. The modes are hybrid but the other field components are at least an order of magnitude smaller. This curved waveguide is multimode but at these small radii of curvature, the higher modes have significantly higher radiation loss from the outer edge.

3.2. *Bending Loss*

Using the TempSelene software, the bending or radiation loss can also be calculated for the lowest mode and Fig. 6 shows the minimum radius to limit the bending loss to 1 dB/cm as a function of the index contrast between the core and cladding for a typical polymer waveguide @ 1550 nm in the TM polarization. The material loss and the scattering loss are typically on the order of or larger than 1 dB/cm so this limit on ring radius essentially keeps the bending loss from becoming the dominant factor. The index assumed for the core corresponds to SU8, a passive polymer, and is reasonably close to the index of an EO polymer, CLD/APC ($n = 1.61$). The dimensions of the waveguide are typical of the ring resonators. This data therefore provides a guide to the limits on the size of polymer microresonators and their FSR. The lowest index cladding polymer available to us is Teflon

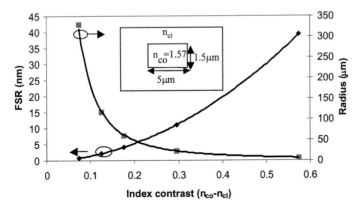

Fig. 6. Minimum radius required to achieve 1 dB/cm loss and the corresponding FSR calculated for 1.55 μm as a function of difference of index of refraction of core to cladding for the structure shown in Fig. 5.

with an index of 1.3 so the maximum index contrast is ~0.3. Therefore in the polymer technology, the smallest practical microresonator is ~50 μm in diameter and the maximum FSR rang is ~10nm.

3.3. *Effective Index*

The coupling between a microresonator and the coupling waveguides is optimized when the interaction is phase-matched or when there is a velocity match between the curved and straight waveguides. To analyze the coupling, an important parameter is therefore the effective index of the ring mode. In a curved waveguide, the wave velocity and the effective index will vary across the waveguide due to the curvature. Since the coupling between the waveguides is maximum where the E field is the largest we define the effective index of the ring mode as the following:

$$n_{\text{eff}} = \frac{\lambda}{2\pi} \frac{m}{R_{\text{max}}} \tag{38}$$

where R_{max} is the radius at which the E field amplitude is largest. Figure 7 shows the calculated effective index as a function of the index contrast for the same typical polymer waveguide as in Fig. 5. At each index contrast the radius of the ring is set to the minimum value given in Fig. 6. Our coupling calculations confirm that one should design the coupling waveguide to have an effective index close to the values defined by Eq. (38).

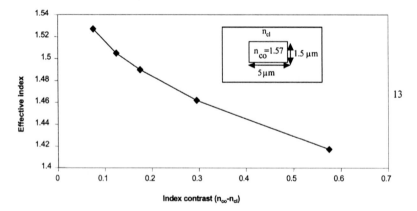

13

Fig. 7.   Calculated effective index of the curved waveguide at 1.55 μm for the fundamental mode. At each value of the refractive index difference between core and cladding the radius is set to give a bending loss of 1dB/cm.

3.4. *Scattering Loss*

Scattering loss from the rough surface of straight buried waveguides has been analyzed by Lacey and Payne.[26] If the radius of curvature of the ring waveguide is much larger than the features of the surface roughness, the scattering from curved waveguides should not be significantly different from a straight waveguide. In the straight waveguide the mode profile is symmetric with respect to the cross-section and the scattering from the inner and outer walls are the same. For the curved waveguide, the E field is larger on the outer wall and smaller on the inner wall so scattering from the outer wall should be larger and that from the inner wall should be smaller. However, the total scattering per unit length from the curved waveguide should not be significantly different from the straight waveguide.

From scattering theory, it is known that the scattering loss coefficient, α_w, is proportional to $\frac{(n_{cl}-n_{co})^2 \delta\rho^2 L_c}{\lambda^3}$. The scattering coefficient for the small rings with higher values of $(n_{cl}-n_{co})$ should be much larger than that from the larger rings. The parameters of the scattering surface, the standard deviation $(\delta\rho)$ and the correlation length (L_c), are determined by the photo mask quality and in our work, $\delta\rho \approx 30\,\text{nm}$ and $L_c \approx 100\,\text{nm}$. Based on the calculate mode patterns at 1550 nm, we expect $\alpha_w \approx 20\,\text{dB/cm}$ for the small rings (25 μm rad.) where the index difference is $\sim$0.3 and $\alpha_w \approx 2\,\text{dB/cm}$ for the larger rings where the index difference is $\sim$0.1. The scattering loss at 1300 nm should be larger than at 1550 nm.

3.5. *Material Loss*

The measured optical loss of several polymers at 1300 nm and 1550 nm are given in Table 1. In our work we used SU8 for the passive rings and this loss is low (0.5 dB/cm) at 1300 nm but higher (4.0 dB/cm) at 1550 nm. For the active rings we used the electro-optic polymer CLD/APC which has a loss ranging from 1 dB/cm @ 1300 nm to 1.4 dB/cm @ 1550 nm.

3.6. *Expected Parameters of Polymer Microresonators*

Based on these analyses and material measurements, one can make some conclusions about polymer microresonators and their limits. The limits on ring size and therefore the FSR are set by the index of refraction difference possible in polymers. Because $\Delta n = n_{core} - n_{cladding} < 0.3$, the maximum FSR is $\sim$10 nm and the smallest ring size is $\sim$50 μm dia. For small rings with a relatively high Δn, Q is likely to be limited by the side wall scattering.

For larger rings (300 μm diameter or greater) with small Δn, the scattering loss can be made comparable to the material loss and Q will be determined by a combination of the two losses. The greatest finesse will be with the smaller rings and using our current fabrication methods and photo-masks the maximum finesse will be ∼200. The highest unloaded Q will be with the larger rings and using polymers this should be ∼10^5 @ 1300 nm. The Q of the smaller rings should be 5–10 times smaller. If smoother side walls can be fabricated and lower material loss polymers are used, both the finesse and Q could be significantly increased.

4. Waveguide Coupling to Microresonators

Waveguide coupling to microresonators has been demonstrated for the case of the waveguide beside the ring (side coupling)[27] and for the case of the waveguide above or below the ring (vertical coupling).[28] Good coupling between the ring and the waveguide requires the gap between the two to be small. In the case of small rings, where the index contrast is large, the gap is typically <1 μm and to achieve critical coupling the gap must be controlled to ∼0.1 μm. The fabrication of side coupling is done by photolithography and it is very difficult to achieve the 0.1 μm accuracy by lithography in polymers. On the other hand, vertical coupling requires close control of the layer thickness and the index of the polymer. For the size and design of the microresonators we have investigated, vertical coupling is the most practical approach for polymers because the thickness of films can be reasonably well controlled by the spin conditions and the index can be controlled by the polymer used and the curing conditions. In addition, vertical coupling is used for the electro-optic resonators coupled to passive waveguides since it is very difficult to fabricate resonators and waveguides of different materials in the same plane. In this section, we will calculate some examples of vertical coupling in polymers to show the gap dimensions required and the accuracy of the dimensions and the index.

(a)   (b)

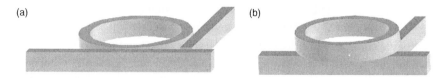

Fig. 8.   (a) Side coupling to microresonator, (b) vertical coupling to the microresonator.

4.1. *Beam Propagation Coupling Calculations*

We have used Olympios[29] 3D beam propagation software for the calculation of the coupling from a straight waveguide into a curved waveguide for the geometry shown in Fig. 9. The simulation monitors the power through-put of the straight waveguide from which the power coupled up into the curved waveguide can be determined. In the simulation, the coupled power to the curved waveguide is completely lost because of the absorbing boundaries defined in the beam propagation method.

Figure 10 shows a typical result for the mode profiles in the coupling section. The mode in the straight waveguide is symmetric and the mode in the curved waveguide mode is skewed to the outer edge as expected. The waveguide dimensions and coupling conditions are the same as in Fig. 12.

To put the following calculated examples of coupling in perspective, it is of interest to see how well the coupling must be controlled to achieve critical coupling and to see how coupling errors affect the microring performance. Figure 11 is a plot of the through-put intensity transmission I_3/I_1 for a single coupled resonator as a function of a and r/a. For the polymer microrings, the parameter a, the E field round-trip transmission in the

Fig. 9.   The structure modeled for calculating the coupling to the resonator.

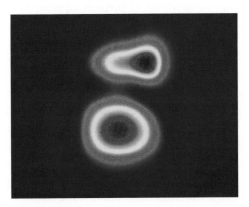

Fig. 10.   Field profile for coupling between the waveguide and the microresonator. The upper profile is the ring resonator mode and the lower is the coupling waveguide mode.

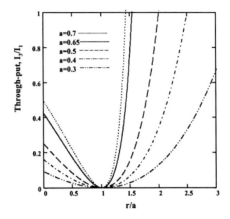

Fig. 11. The through-put, I_3/I_1 for the microring coupled to a single waveguide. The electric field reflection coefficient of the coupling is r and the electric field round-trip transmission of the ring is a. Critical coupling is at $r = a$. If $r < a$, the resonator is over-coupled and if $r > a$, it is under-coupled.

ring, is close to one and we can write $a = 1 - \Delta$ where $\Delta \ll 1$. Critical coupling requires $r = a$ or the power coupling from the waveguide to the ring, $t^2 = 1 - a^2 \approx 2\Delta$. At this coupling the power through-put, (I_3/I_1) for the single coupled ring is zero. A rough rule of thumb is that if there is an error in the power coupled by $\pm 1.2\Delta$, I_3/I_1 increases to ~ 0.1. The increase is slightly more if the coupling is less than critical coupling and slightly less if the coupling is more than critical coupling. For example, in a $25\,\mu$m radius ring with a loss of $15\,$dB/cm, $Q_U = 1.8 \times 10^4$ and $\Delta \approx 0.03$. Critical power coupling is ~ 0.06. If there is an error in the power coupled by ± 0.036 ($t^2 = 0.096$ or 0.024), I_3/I_1 will increase from 0 to ~ 0.1. From Fig. 11, one can see that the through-put is greater than 0.1 if the coupling is less than critical ($r > a$) and less than 0.1 if the coupling is more than critical ($r < a$).

To design for a given coupling and to get some insight into the fabrication accuracies required, the coupling as a function of various parameters has been calculated. Figure 12 shows the power coupling coefficient as a function of the refractive index of the channel waveguide. As expected, the coupling is maximum when the effective index of the coupling waveguide matches the effective index of the curved waveguide. We can observe that one needs to control the refractive index to ± 0.001. It is interesting to note that the coupling is not as sensitive to the effective index mismatch as coupling between two straight waveguides.[30] This is due to the circular

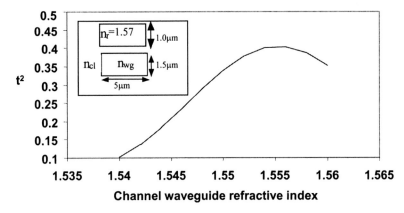

Fig. 12. Calculated power coupling coefficient to the microresonator as a function of refractive index of waveguide for the structure shown. The maximum occurs at the point where the waveguides are phase-matched. In this calculation, $n_{cl} = 1.49$, $R = 300\,\mu$m, spacing between the waveguides is $1\,\mu$m, and $\lambda = 1550$ nm.

geometry of the resonator, which provides a range of effective indexes since the phase front of the waves travel with different speeds depending on their distance from the center of the resonator. It is possible by adjusting the refractive index of the channel waveguide to excite higher order modes in the microresonator. This will appear as another maxima at lower values of the refractive index in Fig. 12.

Next, we consider the effect of misalignment of the waveguide and the resonator. In this case we assume that the effective indices are matched. Figure 13 shows the power coupling coefficient as a function of the alignment mismatch x defined in Fig. 14.

The maximum coupling occurs when $x = -1\,\mu$m. This is probably due to the fact that for negative x there is a larger interaction length between the two waveguides. With a waveguide width of $5\,\mu$m, a misalignment of $\pm 0.5\,\mu$m changes the coupling by $\pm 10\%$.

Next we calculate the amount of dropped power as a function of the gap, d, between the coupling waveguide and the microresonator. This example, shown in Fig. 15 is calculated for a small ring with a high index contrast and we have assumed optimum alignment and phase-matching. The index of the cladding is 1.3 and the core index is 1.57. The radius of the device simulated in this case is $40\,\mu$m. For this high index difference, the vertical mode confinement is larger and significant coupling requires $d < 1\,\mu$m. Reasonable accuracy of the coupling requires the accuracy of d to be $\pm 0.1\,\mu$m.

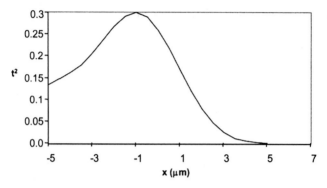

Fig. 13. Calculated power coupling coefficient as a function of misalignment of the resonator and waveguide. The conditions are the same as Fig. 12 except the coupling gap has been increased to $1.5\,\mu$m.

Fig. 14. Geometry for misalignment of the resonator and waveguide.

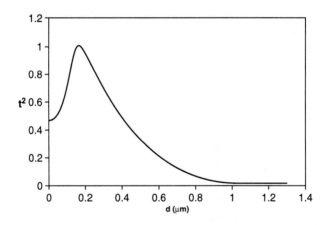

Fig. 15. Calculated power coupling coefficient as a function of d, the spacing between the coupling waveguide and the ring waveguide. In this example, $R = 40\,\mu$m, n_{cl}=1.3, and $n_{co} = 1.57$. The coupling waveguide is phase-matched to the ring waveguide and the alignment offset, x, is optimized. The waveguide dimensions are the same as in Fig. 12.

Also it is interesting to note that almost complete power transfer occurs at a distance of 0.2 μm.

4.2. *Required Fabrication Accuracy*

By looking at the various examples calculated we can arrive at some general conclusions regarding the accuracy required to control the coupling. For reasonable accuracy in the coupling to polymer microresonators (i) the index of refraction should be controlled to ±0.001, (ii) the alignment between the coupling and the ring waveguide should be ±0.5 μm, and (iii) the accuracy of the gap should be ±0.1 μm.

5. Passive Polymer Microresonators

In this section, we will describe the fabrication and measurement of passive polymer microresonators. In the first case polymers with a relatively small index difference are used and the rings are therefore relatively large. In the second case polymers with a larger index difference are used and smaller rings with larger FSR are demonstrated.

5.1. *Larger Devices*

The cross-section of the device and a photo of the passive microresonator are shown in Fig. 16. The microring is coupled vertically to the input and output channel waveguides. SU-8, a relatively low loss material which can

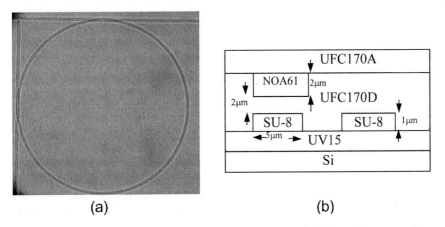

(a)   (b)

Fig. 16.   Large passive device. (a) Fabricated device picture, (b) schematic cross-section.

be patterned using photolithography, is used for the core of the microring. The optical properties of the materials used in this device are summarized in Table 1. The index difference between the core (SU-8) and cladding (UFC170A) is 0.07. To keep the bending loss low, the minimum ring radius is 220 μm @ 1300 nm and 330 μm @ 1550 nm (see Fig. 6). The material loss for SU-8 is 4 dB/cm @ 1550 nm and 0.5 dB/cm @ 1300 nm. If we assume 2 dB/cm as scattering loss and 1 dB/cm as bending loss, the total ring waveguide loss is approximately 7 dB/cm for 1550 nm and 3.5 dB/cm for 1300 nm. Based on this total loss, the expected Q_u is 4×10^4 @ 1550 nm and 9×10^4 @ 1300 nm.

Critical coupling into a 330 μm radius resonator without a drop port at 1550 nm requires 28% intensity coupling. For the smaller 220 μm radius ring at 1300 nm, critical coupling is 11% intensity coupling. The coupling waveguide was designed to have the same effective index, 1.54, as the lowest-order ring waveguide mode. The coupling gap was designed to be close to critical coupling.

The fabrication steps are shown in Fig. 17. The 2.6 μm thick UV15 lower cladding is spin coated and cured using UV light for 30 sec and the sample is baked for 1 hr at 160°C to fully cure the UV15. Next a 1 μm layer of the negative photoresist, SU-8, is spin coated and patterned using mask aligner to form the microring resonator. No etching is required to form the ring structure. The sample is baked for 1 hr at 160°C. The ring structure is covered with a 5 μm UFC-170D layer that serves as the coupling and cladding layer and also planarizes the sample. The sample is cured by UV light for 10 sec and baked for 1 hr at 160°C. The UFC-170D layer is patterned and RIE etched to form the trenches for the waveguides. Next, a thick layer of NOA61 is spin coated to over fill the waveguide trenches. The excess NOA61 is removed by RIE and a 3 μm upper cladding of UFC-170A is spin coated on the NOA61. Finally the device was cut using dicing saw for fiber coupling.

A New Focus tunable laser with a linewidth of less than 300 kHz was fiber coupled into the device and the drop port power was collected by a fiber and detected while the through port power was collected by a lens and detected. Figure 18 shows the TM response of the two ports of the device for 1300 nm. The bandwidth of the microresonator is 2 GHz and 3 GHz for TM and TE polarization respectively. This bandwidth corresponds to 0.01 nm and 0.015 nm and a Q of 1.3×10^5 and 0.9×10^5 respectively. The free spectral range of the device is 0.8 nm for 220 μm radius device and the finesse is 80 (TM). The response shown in Fig. 17 indicates the device is

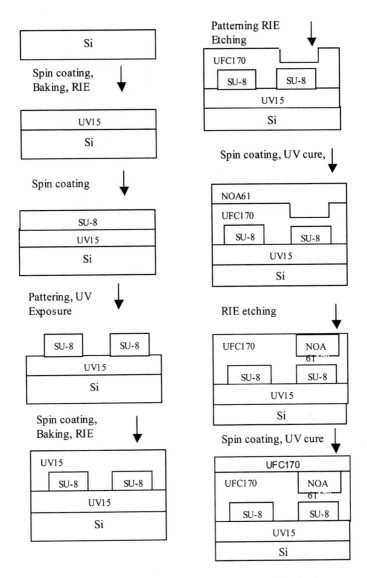

Fig. 17.   The fabrication procedure for large passive devices.

under-coupled and assuming that Q observed is close to unloaded Q, the waveguide loss in the ring for TM is ~2.5 dB/cm. The lower Q for the TE polarization is typically attributed to the higher coupling for TE.

The $\Delta\nu_{FWHM}$ of the 330 μm radius device for 1550 nm is 5 GHz(TM) and 10 GHz(TE) corresponding to 0.0375 nm and 0.075 nm linewidths

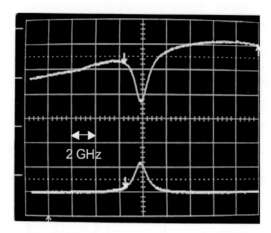

Fig. 18.   Measured response of 220 μm radius device as a function of the input frequency for both the drop and through port at 1300 nm.

and Q's of 3.4×10^4 (TM) and 1.7×10^4 (TE). The FSR is 0.74 nm and hence the TM finesse for this wavelength is 19. Assuming the measured Q is close to the unloaded Q, the loss of the ring waveguide is 8 dB/cm for 1550 nm. This is in reasonable agreement with our calculation.

5.2. *Higher FSR Microresonators*

As pointed out earlier, higher FSR devices must be smaller and this requires a larger index contrast between the core and cladding of the microring. In this section we demonstrate devices that are near the limit in FSR for polymer devices. For the smaller devices we used a spinnable Teflon[1] material for the cladding. Teflon has a refractive index of 1.3 and is the lowest index material available. This makes an index difference of ∼0.3 and a maximum FSR @ 1550 nm of ∼10 nm possible.

For these devices, the core material was SU-8 and the refractive index difference is 0.275 (1.575 − 1.3) and this limits the ring diameter to ∼50 μm for 1550 nm (Fig. 6) and ∼40 μm for 1300 nm. For 1300 nm, assuming 0.5 dB/cm material loss, 1 dB/cm bending loss, and 14 dB/cm scattering loss, we expect a total waveguide loss of 15.5 dB/cm. The predicted Q_u is ∼2×10^4 and critical power coupling for a single coupling waveguide is ∼0.06. At 1550 nm the material loss increases while the scattering loss is expected to decrease. The estimated total waveguide loss is therefore approximately the same as at 1300 nm and the expected Q_u is the same.

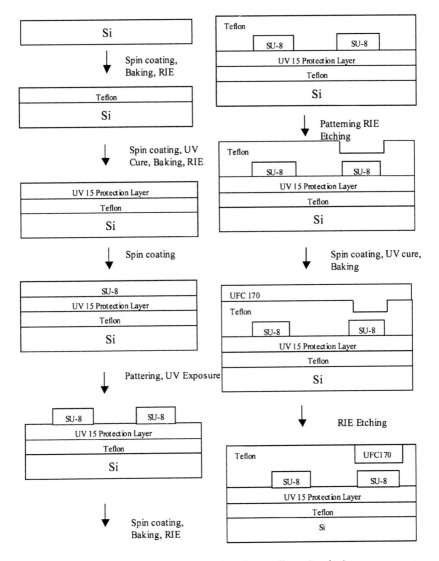

Fig. 19.   Fabrication procedure for small passive device.

The fabrication procedure is shown in Fig. 19. The fabrication of the device starts with spin coating Teflon on a silicon substrate. Using 11% solution of Teflon AF 1600 in 3 M FC-40 solvent the 2.8 μm layer is spun. The surface is etched by RIE for 2 min in oxygen to improve the adhesion of the Teflon. Next a 0.5 μm thick UV 15 protection layer is spin coated on the Teflon and etched down to 0.1 μm so that it does not affect the optical properties of the microring. The protection layer is used to protect the Teflon from the solvents used in later layers. The 1.5 μm thick SU-8 layer is spin coated on the substrate and is patterned to form the microring. Rings of 55 μm and 65 μm diameter, were fabricated. A 4.5 μm Teflon cladding layer is spin coated and again etched to improve the adhesion. The Teflon layer is patterned using photoresist and etched to form trenches for the channel waveguides. The UFC 170 A layer is spin coated to fill out the trenches, UV cured, baked, and the excess UFC 170 layer is removed by RIE. Finally the device is cut using dicing saw. Figure 20 shows the cross-section of the device and the top picture of the fabricated device.

Figure 21 shows the measured power @ 1550 nm for a 55 μm diameter device in the drop port. The FSR is 8.1 nm and the linewidth is 0.1 nm or 12 GHz. The same device @ 1300 nm has FSR of 5.4 nm and a linewidth of

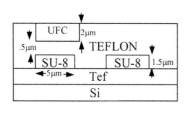

Fig. 20.   Small passive devices cross-section schematic and the fabricated device picture.

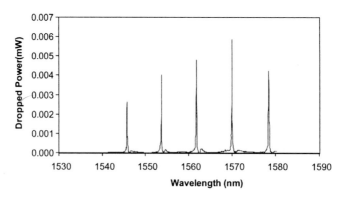

Fig. 21.   Measured power at drop port as a function of laser wavelength for 55 μm diameter device.

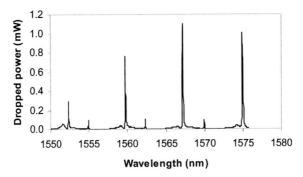

Fig. 22. Measured power at drop port as a function of laser wavelength for 64 μm diameter device.

0.06 nm. The finesse is 84 for 1550 nm and 92 for 1300 nm. Judging from the power in the through port, the device is under-coupled and these parameters are close to those for the unloaded resonator.

Figure 22 shows the response @ 1550 nm for a 64 μm diameter device. In this case the Q is higher and the FSR is smaller. As mentioned earlier, the ring waveguide is multimode but the higher order modes have higher bending and scattering loss and a different effective index and are therefore not often observed. A higher-order mode can be seen in Fig. 21.

5.3. Summary of Passive Devices

Table 2 summarizes the performances of the passive devices. As expected, the loss in the larger devices is largely a combination of bending and material loss while the loss in the smaller rings with a higher index contrast is dominated by scattering. It can also be seen that the longer wavelength has smaller scattering loss. A maximum finesse of 141 is achieved for a 64 μm diameter device at 1.3 μm.

Table 2. The measured FSR and Finesse for different devices.

Index Difference	Device Diameter (μm)	BW (GHz) 1300 nm	Finesse 1300 nm	Loss* (dB/cm)	BW (GHz) 1550 nm	Finesse 1550 nm	Loss* (dB/cm)
0.3	64	6.2	141	8.0	8	117	6.9
0.3	55	10.3	92	13.55	12	84	12.3
0.1	440	1.7	80	1.8			
0.1	660				5	20	2.82

*Excluding material loss.

5.4. *Resonant Wavelength Control by Temperature Tuning*

The index of refraction of polymers has a large thermal dependence and this effect is used in polymer waveguide thermal switches. This effect can be used to tune the resonant wavelength of polymer microresonators. Figure 23 shows the thermal tuning of the 55 μm diameter microresonator @ 1550 nm. The device was placed on a thermoelectric plate for these measurements. The resonance wavelength tunes at a rate of 14 GHz/°C. The change in the resonance wavelength is due to the change in the refractive index of the material. The measured tuning corresponds to a thermal change in the index of refraction of the SU-8 of $-1.1 \times 10^{-4}/°C$, which is consistent with published data on polymers.[31] This effect could be used to tune the resonance if the temperature can be well controlled. Alternately, the thermal tuning can be reduced by using an a thermal design.[32]

5.5. *Resonance Wavelength Control by Fabrication*

A practical WDM system will require several ring resonators at slightly different resonant wavelengths and radii and one would need to control the resonant wavelength in fabrication as closely as possible. It is important therefore to get some idea of the accuracy limits possible in the fabrication of polymer microresonators. For example, if 25 GHz channel spacing is required, the radius difference between adjacent microresonators must be

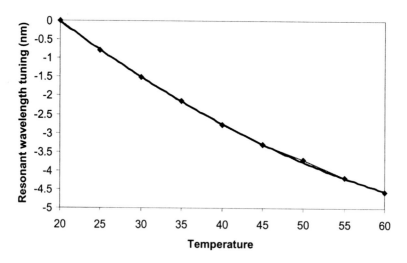

Fig. 23.   Temperature tuning of the microresonator filter.

about 4 nm for a device with a diameter of 40 μm at 1550 nm. It is not obvious that the lithographically defined structures can achieve this kind of resolution. We fabricated an array of microresonators with a nominal radius of 70 μm in which the radius of adjacent rings increases by 10 nm. Figure 24 shows the picture of the array, which used SU-8 as the core material and Teflon as the cladding. Figure 25 shows the measured resonance wavelength for different microresonators in the array.

From this graph one can conclude that by changing the radius on the mask, the resonance wavelength of adjacent rings can be controlled to <1 nm. However the change in wavelength is not consistent with our

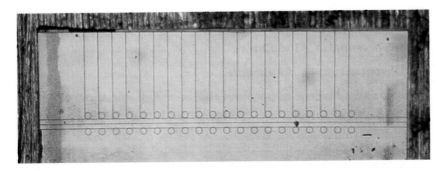

Fig. 24.   Array of microresonators chip.

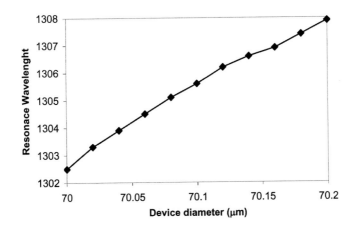

Fig. 25.   Measured resonance wavelength as a function of device diameter.

calculations. The resonance wavelength shift is given by:

$$\frac{\delta\lambda}{\lambda} = \frac{n_{\text{eff}}}{n_{\text{g}}}\left(\frac{n_{\text{eff}}}{n_{\text{eff}}} + \frac{\delta R}{R}\right)$$

where δn_{eff} is the change in the effective index, δR is the change in the radius of the device and n_{g} is the group index defined as

$$n_g(\lambda) = n_{\text{eff}}(\lambda) - \lambda\frac{dn_{\text{eff}}(\lambda)}{d\lambda}$$

Based on the measurement, the slope of tuning is 0.00042. Since the $\delta R/R$ is 0.00028 we conclude that $\delta n_{\text{eff}}/n_e$ is equal to 0.00013 for two adjacent devices. Based on the simulations, the change in effective index should be smaller than this result. Perhaps this is a limitation of the software or the assumptions made.

6. Microresonator Modulators

6.1. *Sensitivity and Bandwidth*

The electro-optic effect in polymers has been well documented and has been used for high bandwidth traveling wave optical modulators.[9] By using EO polymers in the microresonators, a relatively low voltage modulator is possible although with some sacrifice in bandwidth. The electro-optic tensor of polymers poled in the z direction can be written as[33]:

$$[r_{ij}] = \begin{pmatrix} 0 & 0 & r_{13} \\ 0 & 0 & r_{13} \\ 0 & 0 & r_{33} \\ 0 & r_{13} & 0 \\ r_{13} & 0 & 0 \\ 0 & 0 & 0 \end{pmatrix} \tag{39}$$

where $r_{33} \approx 3r_{13}$.

If a modulating field, E_z, is applied, the change in index for the TM polarization is

$$\Delta n_z = \frac{1}{2}n^3 r_{33} E_z \tag{40}$$

and for the TE polarization is

$$\Delta n_x = \Delta n_y = \frac{1}{2}n^3 r_{13} E_z. \tag{41}$$

Electro-optic modulation in microresonators fabricated in the cubic III–V semiconductors such as GaAs and GaP is not possible because of the crystal

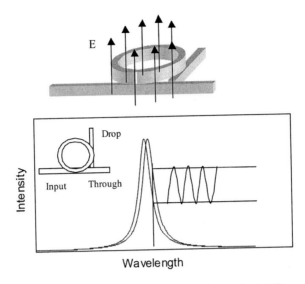

E

Fig. 26. Light modulation concept using a single MR.

symmetry.[34] In these materials, for an applied E_z, the average index around the ring does not change.

For the resonator coupled to a single waveguide, the through-put port transmission was derived earlier:

$$\frac{I_3}{I_1} = \frac{a^2 - 2ra\cos(\phi) + r^2}{1 - 2ra\cos(\phi) + r^2 a^2} \tag{42}$$

If we assume critical coupling ($a = r$) and operation close to a resonant frequency, ω_0, the transmission can be approximated as

$$\frac{I_3}{I_1} = \frac{\frac{r^2}{(1-r^2)^2}\frac{(2\pi R)^2 n_{\text{eff}}^2}{c^2}(w - w_0)^2}{1 + \frac{r^2}{(1-r^2)^2}\frac{(2\pi R)^2 n_{\text{eff}}^2}{c^2}(w - w_0)^2} \tag{43}$$

or

$$\frac{I_3}{I_1} = \frac{(\omega - \omega_0)^2}{(\Delta\omega_{\text{FWHM}}/2)^2 + (\omega - \omega_0)^2} \tag{44}$$

$$\frac{\Delta\omega_{\text{FWHM}}}{2} = \frac{c(1 - r^2)}{2\pi r R n_{\text{eff}}} \tag{45}$$

To find the maximum sensitivity to the index change we have

$$\frac{\partial^2\left(\frac{I_3}{I_1}\right)}{\partial\omega^2} = 0 \rightarrow (\omega - \omega_0) = \frac{\sqrt{3}}{3}\frac{\Delta\omega_{\text{FWHM}}}{2} \tag{46}$$

Notice that this point is the maximum sensitivity point and also the linear point at the same time. In other words if the modulator is biased at this point the variation of the output intensity will be a linear function of the index change and consequently of the voltage change. At this point we have

$$\frac{I_3}{I_1}\bigg|_{(\omega-\omega_0)=\frac{\sqrt{3}}{3}\frac{\Delta\omega}{2}} = \frac{1}{4} \tag{47}$$

and

$$\frac{\partial\left(\frac{I_3}{I_1}\right)}{\partial n}\bigg|_{\max} = \frac{3\sqrt{3}}{4n_{\mathrm{eff}}}\frac{\omega}{\Delta\omega_{\mathrm{FWHM}}} = \frac{3\sqrt{3}Q}{4n_{\mathrm{eff}}} \tag{48}$$

The sensitivity, S, of the modulator can be written as:

$$S = \frac{\partial\left(\frac{I_3}{I_1}\right)}{\partial V_z} = \frac{3\sqrt{3}QKn_{\mathrm{eff}}^2 r_{33}}{8d} \tag{49}$$

where $E_z = \frac{V_z}{d}$ and K is the confinement factor which expresses the overlap of the optical field, the EO polymer, and the rf field. The higher Q device will have higher sensitivity as expected.

In traveling-wave electro-optic polymer modulators, the bandwidth is usually determined by the rf loss of the microstrip line since the velocity mismatch between the optical and the microwave signals is not an issue up to frequencies over $100\,\mathrm{GHz}$.[11] Since the microresonator modulator is usually small compared to the modulation wavelength, the rf electrode loss does not determine the bandwidth but rather the bandwidth is set by the optical response of the microresonator. For a given transfer function $H(i\omega)$ the group delay is given by:

$$\tau_{\mathrm{g}} = \frac{\partial}{\partial\omega}(\arg(H(\omega))) \tag{50}$$

In this equation $\arg(H(\omega))$ is the phase response of the resonator and the group delay is basically the time it takes for a photon to pass through the resonator. This can be considered as the time the photons spend in the resonator. For a single pole resonator

$$H(\omega) = \frac{i\frac{\omega-\omega_0}{\Delta\omega_{\mathrm{FWHM}}/2}}{1+i\frac{\omega-\omega_0}{\Delta\omega_{\mathrm{FWHM}}/2}} \tag{51}$$

and

$$\tau_{\mathrm{g}} = \frac{\frac{1}{\Delta\omega_{\mathrm{FWHM}}/2}}{1+\frac{(\omega-\omega_0)^2}{\Delta\omega_{\mathrm{FWHM}}^2/4}} \tag{52}$$

At the maximum sensitivity point this group delay is given by:

$$\tau_g = \frac{3}{2\Delta\omega_{\text{FWHM}}} \tag{53}$$

Since the electrode is small, we can consider it as a lumped circuit and therefore the photon will experience the integrated average of the electric field over the time, t_g, it spends in the resonator. The average value for a sinusoidal electrical signal can be written as:

$$\overline{V} = \frac{1}{\tau_g} \int_{t}^{t+\tau_g} V\sin(\omega_{\text{rf}}t)dt \approx \frac{\sin(\omega_{rf}\tau_g)}{\omega_{rf}\tau_g} V\sin(\omega_{\text{rf}}t) \tag{54}$$

where ω_{rf} is the modulation frequency. The 3dB electrical bandwidth, f_{3dBe}, of an EO modulator is defined as the frequency at which the received rf power drops to 0.5 of the low frequency value or when the effective voltage on the modulator drops to 0.707 of the low frequency value. By setting the sine function in Eq. (51) to 0.707 and using Eq. (50) we find:

$$f_{\text{3dBe}} \approx \frac{\Delta\omega_{\text{FWHM}}}{2\pi} = \Delta\nu_{\text{FWHM}} = \frac{c}{\lambda Q} \tag{55}$$

The bandwidth of the modulator is inversely proportional to Q. However the product of the sensitivity and the bandwidth is not related to Q and is given by:

$$f_{\text{rf(3dB)}}S = 0.65\frac{n^2 r_{33}K}{d}\frac{c}{l} \tag{56}$$

One can improve the sensitivity but at the cost of reduced bandwidth. The only way to improve the sensitivity (keeping bandwidth constant) is to increase the electro-optic coefficient. This fundamental trade-off was recently pointed out by Williamson.[35]

6.2. *Larger Diameter Modulators*

The schematic of larger diameter EO devices is shown in Fig. 27 and is similar to the passive resonators except that the EO polymer CLD1/APC[9] has been used in the ring and a gold ground plane and upper electrode are used. The refractive indexes of the materials used in the device are summarized in Table 1.

UV15 is used as the lower cladding with a sufficient thickness ($5\,\mu$m) to assure negligible optical loss due to the bottom Au electrode. A $1\,\mu$m CLD1/APC layer is coated and etched using RIE in oxygen to form a channel waveguide in the form of a microring. The width of the ring waveguide is

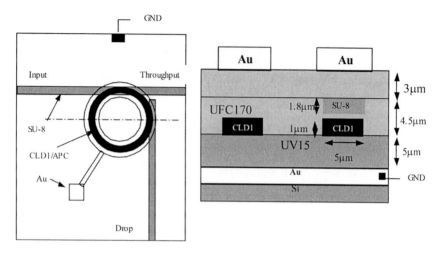

Fig. 27.   Large diameter electro-optic microresonators.

5 μm. Different radius rings are fabricated on the same substrate. Based on the calculation and measurement, the bending loss for a cladding of UFC-170 is negligible for devices larger than 300 μm in diameter for 1550 nm and 200 μm for 1300 nm. The UFC-170 is spun, patterned, and the trenches for the coupling waveguides are etched by RIE. The film thickness and the etch depths are set to provide 10% coupling between the ring and the waveguides. Next the SU-8 layer fills the waveguide trenches and the excess is removed by RIE. The effective refractive index of the SU-8 waveguide matches the effective refractive index of the mode of the CLD1/APC microring. Finally a 3 μm UFC-170 upper layer is spin coated on the device. The CLD1/APC polymer is corona poled at 145°C and an applied voltage is 10 kV for 30 min. The upper gold electrode is deposited and patterned and the devices are separated with a dicing saw for fiber coupling.

Cleaved fiber was butt coupled to the input and output waveguides and a tunable New Focus laser at 1300 nm was used as the source. Figure 28(a) shows the drop port power as a function of wavelength for a 1500 μm diameter device. As can be seen from this figure, $\Delta\nu_{FWHM}$ @ 1300 nm is ~3 GHz for TM polarization (~4 GHz for TE). Hence the Q is 7.6×10^4 and 6.2×10^4 and for TM and TE respectively. The Q @ 1500 nm is approximately half the values at 1300 nm. Figure 28(b) shows the modulated TM light at the drop port of the device when a sawtooth voltage is applied to the electrode. One figure of merit for the EO microresonator is the voltage, V_{FWHM}, required to shift the resonant frequency by $\Delta\nu_{FWHM}$. This is measured to be 4.86 volt

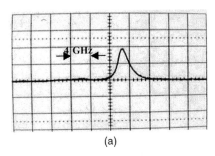 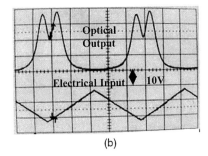

(a) (b)

Fig. 28. (a) Measured dropped power at zero applied voltage as a function of wavelength. (b) Modulated dropped power and the applied voltage on the electrode at 1300 nm for a 1500 μm diameter resonator at a fixed laser wavelength.

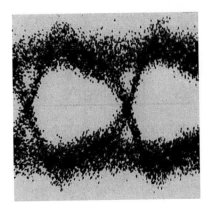

Fig. 29. Eye diagram for data transmission at 1 Gb/sec using electro-optic MR.

@ 1300 nm and this corresponds to an r_{33} of 33 pm/V. The typical r_{33} of CLD1/APC in Mach–Zehnder modulators is 50 pm/V @ 1300 nm and is probably less in the microresonators because of less efficient poling. V_{FWHM} for 1550 nm for this device is 16 V. For a smaller 600 μm diameter device Q is slightly lower (4.7×10^4 and 5.8×10^4 for TE and TM, respectively) and the V_{FWHM} was 9 V.

To demonstrate digital modulation, a 1 Gb/sec, 1 V data stream was applied to the electrode of the 1500 mm diameter ring. The relatively clear eye diagram is shown in Fig. 29.

6.3. *High FSR Modulators*

One potential application for microresonator modulators is in WDM communication systems where each wavelength channel is modulated by a microresonant modulator tuned to that wavelength. The number of WDM channels will be partly determined by the FSR of the microresonators. To increase the FSR we have fabricated a series of smaller devices by using Teflon as the cladding material. Figure 30 is the schematic cross-section and a photograph of the device. The various layer dimensions are different from the larger devices in order to index match the microrings and the coupling waveguides. The fabrication is similar to the larger devices except that the CLD1/APC material was corona poled before the Teflon

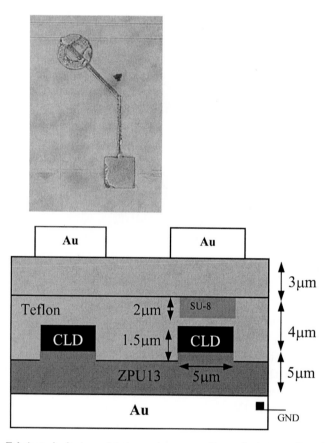

Fig. 30. Fabricated device picture and cross-section of the smaller, high FSR modulators.

layer was applied. It is difficult to pole an EO polymer when a relatively high resistivity material such as Teflon is used as a cladding.

Figure 31 shows the power in the drop port of a 50 μm diameter device as a function of wavelength at 1300 nm. The FSR is 6.5 nm, the $\Delta\nu_{\text{FWHM}}$ is 16 GHz, and the finesse is 67. Figure 32 shows the modulated TM light

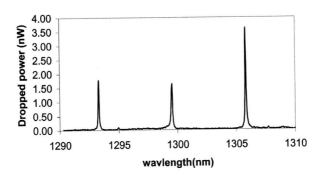

Fig. 31. Response of the electro-optic device as a function of laser wavelength.

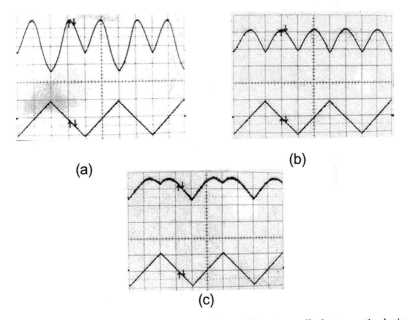

(a)

(b)

(c)

Fig. 32. The modulation performance for three different small electro-optic devices. (a) 150 μm, (b) 70 μm, (c) 50 μm (diameter). In each case the upper trace is the power in the drop port and the lower is the applied 20 Vpp sawtooth voltage.

Table 3. The performance of three different devices at 1300 nm.

Device Diameter (µm)	FSR (GHz)	$\Delta\nu_{FWHM}$ (GHz)	V_{FWHM} (V)	Tuning (GHz/V)
150	300	12	10	1.2
70	770	16	20	0.8
50	1100	18	36	0.5

at the drop port of three different radii microresonators when a sawtooth voltage is applied to the electrode. Table 3 summarizes these results. V_{FWHM} is consistently larger for the smaller devices due to the decrease of Q from the higher scattering loss and also due to the reduction of the confinement of the light in the core of the modulator. For a 150 µm diameter ring the tuning rate is 1.2 GHz/V and for the 50 µm diameter ring it drops to 0.5 GHz/V.

7. Coupled Double Microresonators for Wide Band Tuning

Tunable filters and lasers with wide tuning bands have many applications in optical measurements and in optical networks that use multiple wavelengths such as wavelength division multiplexing systems. The most successful demonstrations of compact semiconductor lasers with wide tunability have used sampled gratings and super structure gratings. Both concepts are based on achieving wavelength periodicity in the reflectivity and using the vernier effect of two mirrors with different periodicities.[36,37] In tunable semiconductor lasers the gratings are tuned by injecting current into the Bragg section of the device and changing the index and the absorption of the material.

Bin Liu *et al.*[38,39] introduced the idea of using waveguide ring microresonators (MR) instead of Bragg gratings to fabricate widely tunable lasers. The advantages of the MR devices are that the frequency response is periodic and the peak reflectivity is the same over a wide wavelength band in contrast to sampled gratings in which the peak reflection decreases far from the Bragg condition. This means that if two MR's with slightly different diameters are used in series as shown in Fig. 33, a larger vernier effect or a larger tuning enhancement factor, M, is possible. The two coupled MR's pass only the wavelengths at which both rings are resonate. The tuning enhancement factor for a double microring structure, DMR, is given by:

$$M = \frac{1}{1 + \frac{r_1}{r_2}} \tag{57}$$

where r_1 and r_2 are the radii of the two rings. M is the ratio of the spacing of reflectivity peaks in the combination of two rings to the spacing of reflectivity peaks for one of the rings. If the tuning is done by changing the index of refraction in one of the rings, M is the ratio of the change in wavelength of peak reflectivity of combination of two rings to the change in wavelength of peak reflectivity of one of the rings for a given index change. The tuning is discontinuous and jumps at the spacing of the reflectivity peaks of one ring. M is typically limited to ∼10 in sampled gratings[39] but can be 40 or greater in high Q DMR's.

A second advantage of the DMR is that the tuning can potentially be done by the EO effect that is fast and does not change the material absorption. On the other hand, the speed of carrier injection tuning is limited by the carrier recombination time and there is some change in the material absorption. The electro-optic effect can achieve very rapid tuning but the index changes are very small and therefore a wide tuning range will require a high M. We have fabricated tunable coupled DMR's in polymers with $M = 40$. The tuning is done by the thermo-optic effect or by the electro-optic effect.

7.1. *Thermo-Optic Device*

The fabrication of the device is very similar to previous structures. The cross-section and top view of the device is shown in Fig. 33. The fabrication of thermo-optic devices starts with spin coating a lower cladding of UV15 epoxy. The lower cladding is 15 μm thick to provide thermal isolation between the thin film heater and the bottom substrate. A 1.8 μm layer of SU-8 is spin coated and patterned to form the 240 μm and 246 μm radius MR's. The middle cladding of UFC170A is spin coated and the waveguide trenches are etched. The distance between the ring waveguide and the coupling waveguide is 1 μm to give 10% power coupling. The waveguides' core layer (NOA72) is spin coated, the excess is removed by RIE, and the upper cladding is spin coated on the device. Finally 100 nm thick, 20 μm wide gold electrodes (heaters) are deposited. The radius of the MR's (240 μm and 246 μm) gives a tuning enhancement factor, M of 40 and the FSR of the single MR is approximately 120 GHz. The tuning will be discrete with 120 GHz spacing.

To characterize a single resonator of the device, a TM polarized tunable laser was coupled into port 1 and the response measured at port 2. Based on the measured transmission as a function of wavelength, the coupling to

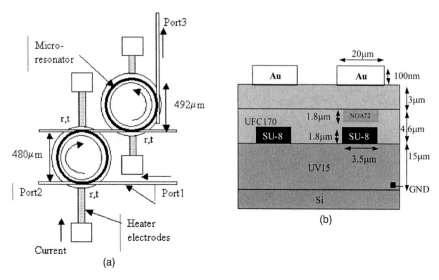

Fig. 33.   (a) Schematic of the thermally tuned DMR. (b) Cross-section of the device.
SU-8 is the MR waveguide and NOA72 is the coupling waveguide.

the ring and the round-trip loss in the ring can be measured as discussed
earlier. From these measurements at 1550 nm and assuming the input and
output coupling are the same, the power coupling between the waveguide
and the ring is 6%, the round-trip loss in the ring is 30%, and the unloaded
Q is $\sim 3 \times 10^4$. This corresponds to a loss in the SU-8 ring waveguide of
10 dB/cm. The material loss in SU-8 is 4 dB/cm and the additional loss
is due to a combination of scattering and radiation from the ring. The
resonance transmission of a single ring from input to output waveguide is
-10 dB. The measured transmission through both rings (port 1 to port 3)
at the wavelength where both are at resonance is -20 dB indicating that
both rings are very close to the same power coupling and Q.

To measure the tuning performance, a Newport EDFA (FPA-35) was
used as a broadband spontaneous emission source and as a broadband gain
medium. The EDFA has a small signal gain of 40 dB. A polarizer and polarization controller was used to assure the TM polarization. Low-loss input
and output coupling to the DMR was achieved by butt coupling from small
core fiber to the waveguides. In the first measurement, the EDFA was used
as a broadband source and coupled into port 1. The output from port 3
was measured using an optical spectrum analyzer. The results are shown in
Fig. 34 along with the measured output spectra of the EDFA. The vernier

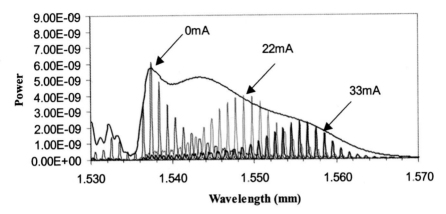

Fig. 34. The filtered spontaneous emission of the device for different values of the current to the electrode heater and the EDFA spontaneous emission spectrum.

effect of the coupled resonators can be seen; the FSR of a single ring is 120 GHz (0.96 nm), the finesse, F, is 19, and the spacing between the peak responses (where both rings are resonant) is M times FSR or ~39 nm. The thermal tuning was demonstrated by passing current through the electrode on top of the smaller ring. The resistance of the electrode, from one side to the other, is 20 Ω. The thermo-optic coefficient of polymers is relatively large ($\sim -10^{-4}/°C$) and this, combined with high M, leads to tuning rates of 4 nm/°C or 0.6 nm/mW. Notice that as the current increases the index decreases and the highest transmission mode moves to longer wavelengths; this is consistent with tuning of the smaller ring.

Next the EDFA and the device were used as a ring laser by connecting port 3 to the input of the amplifier as shown in Fig. 35. Polarizer and polarization controllers were included in the ring and the isolator in the EDFA assured oscillation in only one direction. The laser output was measured from port 1. Figure 36 shows the output spectra measured by the OSA. The suppression of the adjacent ring resonator side mode is 30 dB and the output power is 1 mW. Using our spectrum analyzer with 0.08 nm resolution only a single mode is observed. However the total optical pathlength around the laser is long because of the length of fiber within the amplifier and the longitudinal mode spacing is therefore very small and it is likely there are several longitudinal modes lasing. Figure 37 shows the thermal tuning of the laser that is consistent with the thermal tuning of the double ring structure. The tuning curve is not smooth but consists of jumps at the ring mode spacing of 0.96 nm. There are regions without lasing due to gain

P. Rabiei & W. H. Steier

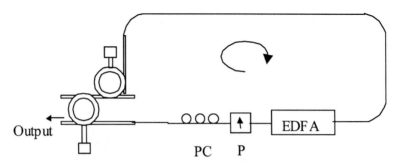

Fig. 35.   The optical circuit for the laser. P: Polarizer, PC: Polarization controller,
EDFA: Erbium doped fiberamplifier.

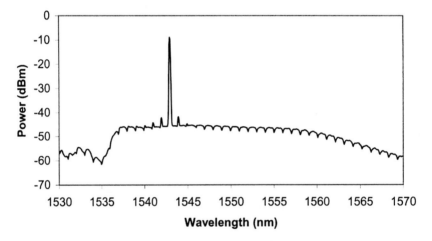

Fig. 36.   The lasing spectrum of the DMR thermo-optic device.

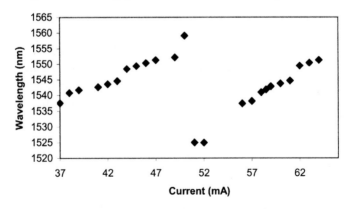

Fig. 37.   The tuning characteristic of the DMR device as a function of the current in
the electrodes.

features of EDFA. We also observed the time constant of the thermal tuning by applying a step current function. The time constant for increasing the temperature is 1 msec and for decreasing the temperature is 2 m sec.

7.2. *Electro-Optic Tuning*

Electro-optic tuning is much faster than thermal tuning but the achievable index of refraction change is less and the tuning range is smaller. We fabricated coupled ring electro-optic resonators of the same diameters and tuning enhancement factor as the thermal devices. Details of the device fabrication were described earlier. The dimensions and cross-section of the device, shown in Fig. 38 are similar to the thermo-optic device except for a lower gold electrode between the Si substrate and the UV15 lower cladding, the thickness of the lower cladding is reduced to $5\,\mu m$, and the nonlinear polymer CLD/APC replaces the SU-8 as the ring waveguide material. The measured coupling is 6%, the waveguide loss is $11\,dB/cm$, and the finesse is 18. The tuning enhancement factor is 40.

Figure 39 shows the filtered spontaneous emission and the tuning by changing the voltage across the smaller ring. The tuning rate of the peak resonance is $8\,GHz/V$ while the tuning rate of a single ring is $1/M$ of this value ($0.25\,GHz/V$). This tuning rate corresponds to an electro-optic coefficient, r_{33}, of the CLD/APC of $12\,pm/V$. This value is lower than that demonstrated in traveling wave modulators because the poling conditions

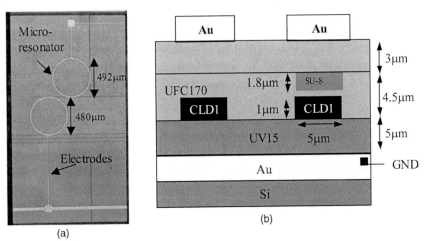

(a)

(b)

Fig. 38. (a) The electro-optic device picture. The waveguides and the electrodes can be seen. The MR is under the electrode. (b) The cross-section of the E-O DMR device.

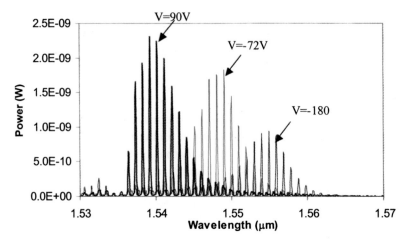

Fig. 39. The filtered spontaneous emission of the device for different values of voltage applied to the electrodes of the device.

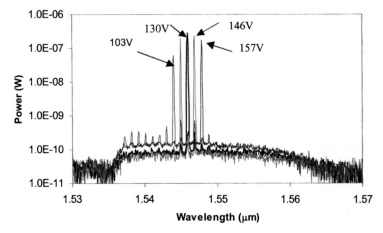

Fig. 40. The lasing spectrum for four different voltages applied to the electrodes of the device.

for the polymer were not optimized in this device. In a similar electro-optic polymer we have achieved r coefficients about three times this value.[5] Since the measured loss in the ring waveguide is considerable larger than that measured in straight waveguides we expect higher Q resonators and hence a higher M are possible with improved processing. The higher r coefficient and higher M could lead to a 10-fold increase in the tuning per volt.

Next the laser cavity was closed by feeding the output of port 3 into the input of the EDFA. The tunable single resonance lasing is shown in Fig. 40.

The side mode suppression is 30 dB and the tuning rate is consistent with the measures tuning rate of the coupled rings. The measured output power for the device is 1 mW. We measured the tuning speed of the electro-optic DMR/EDFA laser by applying a step voltage to the device. The output at the new wavelength shows a damped oscillation with a period of 32 μsec and a settling time of 200 μsec. The response time of the electro-optic DMR is limited by the photon lifetime in the resonators and is a few nanoseconds. The observed tuning speed is due to the response time of this relatively long laser cavity and the EDFA gain. By using shorter cavities and a semiconductor optical amplifier, much faster tuning speeds can be expected with the electro-optic DMR.

7.3. *Summary of Double Microresonators*

This is the first demonstration of the tuning of a laser by using two coupled microrings and a tuning enhancement factor of 40. The lasing wavelength can be thermally or electro-optically tuned with 30 dB side mode suppression over the 35 nm bandwidth of the EDFA. By selecting the ring diameters, the discrete tuning can be set to match the wavelength spacing of a wavelength division multiplexing system (ITU standard). If the polymer technology can be integrated with a semiconductor gain section, fast tunable lasers with cavity lengths small enough to achieve single-mode lasing should be possible. The tuning speed for the thermo-optic version was limited by the thermo-optic tuning of the ring resonators and is in the millisecond range. The tuning speed for the electro-optic version, in our experiment, was limited by the round-trip time in the cavity. Since the electro-optic effect is very fast, shorter cavities could lead to sub-nanosecond tuning and this might open the door for new modulation schemes based on rapid shifts in the laser wavelength. The tuning of the EO version could be enhanced by using voltages of the opposite polarity for the poling (push–pull poling) on the two rings, by higher Q through improved processing, and by increasing the EO coefficient by better poling.

8. Applications

Perhaps the most interesting application of microresonator modulators is a multiwavelength transmitter as shown in Fig. 41. In a typical WDM system a different laser source and modulator are required for each wavelength. Each modulated wavelength is then multiplexed onto a single fiber for transmission. In the proposed scheme, a multimode laser is the source

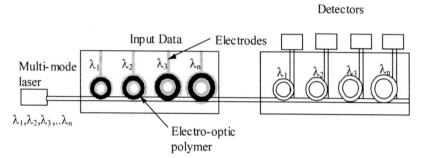

Fig. 41. Multiple wavelength channel transmit and receive modules for WDM systems.

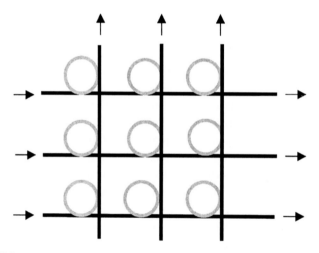

Fig. 42. Voltage controlled optical waveguide cross-connect in which the voltage controlled microresonators at each corner act to route the signal. The horizontal street waveguides are below the rings and the vertical avenue waveguides are above the rings.

and a resonant microresonator modulator is used to modulate each wavelength. Each modulator will modulate one wavelength and will leave any other wavelength unchanged. In the receiver, passive microresonators are used to separate the wavelength channels.

A second potential application for the EO microresonators is for a reconfigurable cross-connect as shown in Fig. 42. In this application the voltage controlled resonators are the corner switches in a street-avenue waveguide interconnect. The street waveguides are fabricated above the rings and the avenue waveguides are fabricated below. If an applied voltage tunes the microresonator to resonance, the street waveguide is connected to the avenue waveguide at that corner.

Cohen *et al.*[40] and Ilchenko *et al.*[41] have demonstrated that a resonant modulator can be used as a high frequency but narrow band modulator by using a modulation frequency that matches the FSR or some multiple of the FSR. Thus a $300\,\mu$m radius ring can modulate at $100\,$GHz but with a bandwidth of $\Delta\nu_{\text{FWHM}}$ which is typically 2–5 GHz. At this modulation frequency the ring electrode cannot be treated as a lumped circuit since the circumference of the ring will be close to the modulation wavelength. The electrode must become a traveling wave electrode and phase matching between optical and modulation waves must be considered.

Post fabrication trimming by laser ablation and bleaching of polymer Mach Zehnder modulators and polymer directional couplers has been demonstrated in EO polymers.[42,43] It may be possible to trim the resonant wavelength and the coupling of polymer microresonators by similar techniques. We have trimmed the resonant wavelength of EO microresonators using an $850\,$nm laser by bleaching. However it is difficult to change the coupling to the ring by laser bleaching without simultaneously changing the resonant wavelength.

In many system applications it will be critical to make the resonant wavelength as temperature stable as possible to avoid costly temperature controllers or feedback stabilization. If the temperature of a polymer microresonator increases, the index of refraction of the ring waveguide decreases and the circumference of the ring increases. These two effects will tend to cancel each other since the optical pathlength is the product of the two. In all of our experimental work the microresonators are fabricated on a Si substrate that is by far the main bulk of the structure. Thus the thermal expansion of the rings is determined by the thermal expansion of the Si but the index of refraction change is determined by $\frac{dn}{dT}$ of the polymer. In this case the change of the index of refraction is the dominant effect. It is possible to use another polymer as the substrate material whose coefficient of thermal expansion is selected so that the length increase cancels the index of refraction decrease and, over some temperature span, the resonant wavelength is not a function of temperature. This is the technique used in polymer AWG's to achieve stable athermal operation.[32]

Acknowledgments

The authors wish to acknowledge the support of the Air Force Office of Scientific Research, the National Science Foundation, and Intel Corporation that made it possible for us to perform the research.

References

1. L. Eldada and L. W. Shacklette, "Advances in polymer integrated optics," *IEEE J. Sel. Topics Quant. Electron.* **6**, 54–68, 2000.
2. C. C. Teng, "Traveling wave polymeric optical intensity modulator with more than 40 GHz of 3 dB electrical bandwidth," *Appl. Phys. Lett.* **60**, 1538–1540, 1992.
3. Y. Hida, H. Onose and S. Imamura, "Polymer waveguide thermooptic switch with low electric power consumption at 1.3 μm," *IEEE Photon. Technol. Lett.* **5**, 782–784, 1993. Also M. H. M. Klein Koerkamp, M. C. Donckers, B. H. M. Hams and W. H. G. Horsthuis, "Design and fabrication of pigtailed thermo-optic 1 × 2 switch," *Proc. Integr. Photon. Res. Conf.* **3**, 274–276, 1994.
4. G. E. Jabbour, S. E. Shaheen, M. M. Morrell, J. D. Anderson, P. Lee, S. Thayumanavan, S. Barlow, E. Bellmann, R. H. Grubbs, B. Kippelen, S. Marder, N. R. Armstrong and N. Peyghambarian, "High Tg hole transport polymers for the fabrication of bright and efficient organic light-emitting devices with an air-stable cathode," *IEEE J. Quant. Electr.* **36**, 12–17, 2000.
5. L. Eldada, S. Yin, C. Poga, C. Glass, R. Blomquist and R. A. Norwood, "Integrated multi-channel OADM's using polymer Bragg grating MZI's," *IEEE Photon. Tech. Lett.* **10**, 1416–1418, 1998.
6. Y. Hida, Y. Inoue and S. Imarura, "Polymeric arrayed-waveguide grating multiplexer operating around 1.3 μm," *Electron. Lett.* **30**, 959–960, 1994.
7. S. Kalluri, M. Ziari, A. Chen, V. Chuyanov, W. H. Steier, D. Chen, B. Jalali, H. Fetterman and L. R. Dalton, "Monolithic integration of waveguide polymer electrooptic modulators on VLSI circuitry," *Phot. Tech. Lett.* **8**, 644–646, 1996.
8. S.-W. Ahn, W. H. Steier, Y.-H. Kuo, M.-C. Oh, H.-J. Lee, C. Zhang and H. R. Fetterman, "Integration of electro-optic polymer modulators with low-loss fluorinated polymer waveguides," *Opt. Lett.* **27**, 2109, 2002.
9. M.-C. Oh, H. Zhang, A. Szep, W. H. Steier, C. Zhang, L. R. Dalton, H. Erlig, Y. Chang, Boris Szep and H. R. Fetterman, "Recent advances in electro-optic polymer modulators incorporating phenyltetraene bridged chromophore," *J. Quant Electr. Sel. Topics Organ. Photon.*, **7** 826–835, 2001.
10. S. W. Imamur, R. Yosimura and T. Izawa, "Polymer channel waveguides with low loss at 1.3 mm," *Electron. Lett.* **27**, 1342–1343, 1991.
11. Hua Zhang, M.-C., Oh, A. Szep, W. H. Steier, C. Zhang, L. R. Dalton, H. Erlig, Y. Chang, D. H. Chang and H. R. Fetterman, "Push pull electro-optic polymer modulators with low half-wave voltage and low loss at both 1310 nm and 1550 nm," *Appl. Phys. Lett.* **78**, 3116–3118, 2001.
12. Y. Shi, W. Lin, D. J. Olson, J. H. Bechtel. H. Zhang, W. H. Steier, C. Zhang and L. R. Dalton, "Electro-optic polymer modulators with 0.8 V half-wave voltage," *Appl. Phys. Lett.* **77**, 2000.
13. D. Chen, H. R. Fetterman, A. Chen, W. H. Steier, L. R. Dalton, W. Wang and Y. Shi, "Demonstration of 110 GHz electro-optic polymer modulators," *Appl. Phys. Lett.* **70**, 3335–3337, 1997.

14. R. Yoshimura, M. Hikita, S. Tomaru and S. Imamura, "Low-loss polymeric optical waveguides fabricated with deuterated polyfluoromethacrylate," *IEEE J. Lightwave Tech.* **16**, 1030–1037, 1998.

15. T. Li, S. Tang, R. Wickman, R. Wu, F. Li, M. Dubinovsky and R. T. Chen, "Polymer waveguide based high speed clock signal distribution system," *Proc. SPIE, Optoelectronic Interconnects and Packaging IV*, **3005**, 128–135, 1997.

16. K. B. Rochford, R. Zanoni, Q. Gong and G. I. Stegeman, "Fabrication of integrated optical structures in polydiacetylene films by irreversible photoinduced bleaching," *Appl. Phys. Lett.* **55**, 195–202, 1989.

17. L. Eldada, C. Xu, K. M. T. Stengel, L. W. Shacklatte and J. T. Yardly, "Laser-fabricated low-loss single-mode raised-rib waveguiding devices in polymers," *IEEE J. Lightwave Tech.* **14**, 1704–1713, 1996.

18. L. Wu, F. Li, S. Tang, B. Bihari and R. T. Cheng, "Compression-molded three-dimensional tapered polymeric waveguides for low loss opto-electronic packaging," *IEEE Photon. Tech. Lett.* **9**, 1601–1603, 1997.

19. C.-G. Choi, S.-P. Han, B. C. Kim, S.-H. Ahn and M.-Y. Jeong, "Fabrication of large-core 1X16 optical power splitters in polymers using hot-embossing process," *IEEE Phot. Tech. Lett.* **15**, 825–827, 2003.

20. S. Kim, K. Geary, D. Chang, H. Zhang, C. Wang, W. Steier and H. Fetterman, "TM-pass polymer modulators with poling-induced waveguides and self-aligned electrodes," *IEEE Phot. Tech. Lett.*, accepted

21. Y. Hida, S. Imamura and T. Izawa, "Ring resonator composed of low loss polymer waveguides at $1.3\,\mu$m," *Electron. Lett.* **28**, 1314–1316, 1992.

22. Reinhard Marz, *Optical Waveguide Theory*, Artech House, Boston, 1995.

23. E. A. J. Marcatili, "Bends in optical dielectric waveguides," *Bell Syst. Tech. J.* **48**, 2103–2132, 1969.

24. S. Ramo, J. R. Whinnery and T. VanDuzer, *Fields and Waves in Communication Electronics*, John Wiley, NY, 1984.

25. TempSelene is available from Kymata Software, Netherlands.

26. J. P. R. Lacey and F. P. Payne, "Radiation loss from planar waveguides with random wall imperfections," *IEE Proc.* **137**, 282–288, 1990.

27. B. E. Little, S. T. Chu, H. A. Haus, J. Foresi and J.-P. Laine, "Microring resonator channel dropping filters," *J. Lightwave Tech.* **15**, 998–1005, 1997.

28. D. V. Tishinin, P. D. Dapkus, A. E. Bond, I. Kim, C. K. Lin and J. O'Brien, "Vertical resonant couplers with precise coupling efficiency control fabricated by wafer bonding," *IEEE Photon. Tech. Lett.* **11**, 1003–1005, 1999.

29. Olympios is available from C2V Software, Netherlands.

30. R. Ulrich, "Theory of the prism-film coupler by plane-wave analysis," *J. Opt. Soc. America* **60**, 1337–1350, 1970.

31. M. B. J. Diemeer, J. J. Brons and E. S. Trommel, "Polymeric optical waveguide switch using the thermooptic effect," *J. Lightwave Tech.* **7**, 449–453, 1989.

32. N. Keil, H. H. Yao, C. Zawadzki, J. Bauer, M. Bauer, C. Dreyer and J. Schneider, "Athermal all-polymer arrayed-waveguide grating multiplexer," *Electron. Lett.* **37**, 579–580, 2001.

33. K. D. Singer, M. G. Kuzyk and J. E. Sohn, "Second-order nonlinear-optical processes in orientationally ordered materials: relationship between molecular and macroscopic properties," *J. Opt. Soc. America B (Optical Physics)* **4**, 968–976, 1987.

34. A. Yariv and P. Yeh, *Optical Waves in Crystals*, Wiley-Interscience, NY, 1984.

35. R. C. Williamson, "Sensitivity-bandwidth product for electro-optic modulators," *Opt. Lett.* **26**, 1362–1363, 2001.

36. V. Jayaraman, Z. M. Chuang and L. A. Coldren, "Theory, design, and performance of extended tuning range semiconductor lasers with sampled gratings," *IEEE J. Quant. Electron.* **29**, 1824–1834, 1993.

37. L. Zhang and J. C. Cartledge, "Fast wavelength switching of three-section DBR lasers," *IEEE J. Quant. Electron.* **31**, 75–81, 1995.

38. Bin Liu; Shakouri, A.; Bowers, J.E., "Wide tunable double ring resonator coupled lasers," *IEEE Photon. Tech. Lett.* **14**, 600–602, 2002.

39. Bin Liu, Ali Shakouri and John E. Bowers, "Passive microring-resonator-coupled lasers," *Appl. Phys. Lett.* **79**, 3561–3563, 2001.

40. D. A. Cohen, M. Hossein-Zadeh and A. F. J. Levi, "Microphotonic modulator for microwave receiver," *Electron. Lett.* **37**, 300–301, 2001.

41. V. S. Ilchenko, A. A. Savchenkov, A. B. Matsko and L. Maleki, "Whispering-gallery-mode electro-optic modulator and photonic microwave receiver," *J. Opt. Soc. America* **B20**, 333–342, 2003.

42. A. Chen, V. Chuyanov, F. I. Marti-Carrera, S. Garner, W. H. Steier, S. S. H. Mao, Y. Ra and L. R. Dalton, "Trimming of polymer waveguide Y-junctions by rapid photobleaching for tuning the power splitting ratio," *Photon. Tech. Lett.* **9**, 1499–1501, 1997.

43. Y.-H. Kuo, W. H. Steier, S. Dubovitsky and B. Jalali, "Demonstration of wavelength insensitive biasing using an electro-optic polymer modulator," *IEEE Photon. Tech. Lett.* **15**, 813–815, 2003.

CHAPTER 8

ATOMS IN MICROCAVITIES: QUANTUM ELECTRODYNAMICS, QUANTUM STATISTICAL MECHANICS, AND QUANTUM INFORMATION SCIENCE

Andrew C. Doherty and H. Mabuchi

Physics and Control & Dynamical Systems,
California Institute of Technology, USA

Modern cavity quantum electrodynamics (cavity QED), with cold atoms in microcavities, provides a rich setting for theoretical and experimental investigations in quantum optics. The growing importance of cavity QED as a model system stems largely from the fact that recent experiments have achieved the condition of *strong coupling*, and have demonstrated its utility in novel strategies for quantum measurement and control. In the strong coupling regime, coherent radiative dynamics of one atom and a single electromagnetic field mode can dominate over the decohering effects of energy loss and dephasing, which in turn allows the atom-cavity system to evolve into highly nonclassical states far from thermodynamic equilibrium. In this chapter we provide a tutorial overview of the theory of quantum dynamics and decoherence in cavity QED, and discuss some applications of strong coupling in the contexts of quantum statistical mechanics and quantum information science.

1. Introduction and Overview

Historically, research in cavity quantum electrodynamics (cavity QED) has focused on the modifications of atomic and molecular radiative processes that result from imposition of boundary conditions for the electromagnetic field.[1-3] Material structures ranging in complexity from simple dielectric interfaces to high-quality resonators can profoundly affect spatiotemporal properties of the quantized electromagnetic field; radiative processes involving systems of bound charges are consequently impacted through changes in the spectral and spatial distribution of vacuum fluctuations, and by related changes in the effective electric field per photon. In recent years the core subject of cavity QED has grown to encompass radiative processes in condensed matter systems, such as controlled photon generation using quantum dots in microcavities[4] and the formation of cavity-polariton modes in semiconductor heterostructures.[5] A number of important formal analogies have also been identified between cavity QED and fundamental phenomena in systems such as trapped ions[6] and low-temperature electrical circuits,[7] lending renewed importance to studies of the canonical model of atoms in cavities.

Researchers now distinguish between two broad parameter regimes for cavity QED, corresponding to strong and weak atom-field coupling. Roughly speaking, in the weak coupling regime the introduction of electromagnetic boundary conditions leads only to perturbative effects on radiative processes. For atoms or molecules near simple dielectric interfaces, for example, these take the form of energy-level shifts and modified spontaneous emission rates.[1,8,3] In the strong coupling regime, however, "exotic" boundary conditions induce major modifications of the electromagnetic mode structure and qualitatively new dynamics emerge. The classic example of this is the phenomenon of vacuum Rabi oscillation, in which an initially excited atom coupled to a high-finesse electromagnetic resonator undergoes damped oscillations between its excited and ground states, rather than the usual monotonic exponential decay (spontaneous emission). While the demonstration of strong coupling in cavity QED experiments has been a relatively recent achievement,[9] it already seems to be widely acknowledged that the novel dynamics accessible in this regime could be a crucial enabling factor for future progress in fields such as nonequilibrium statistical mechanics, precision measurement and quantum information science.

In this chapter, the scope of our review will be limited to selected topics in optical strong-coupling cavity QED with neutral gas-phase atoms. The related field of microwave cavity QED has seen tremendous progress in recent years, but for an overview we refer the reader to Refs. 10 and 11. We should similarly point out that there has been a wealth of activity in cavity QED in the perturbative (weak coupling) regime,[3,12] and that there is growing momentum towards cavity QED with trapped ions.[13,14] We note that a more condensed review of recent literature on strong coupling cavity QED can be found in Ref. 15, and that more complete discussions of the field's status and history can be found in references.[3,16]

Our aim here will be to provide the reader with a sense of the depth and detail of our modern understanding of cavity QED, and to illustrate how this points to scientific and technological "applications" for strong coupling that drive most current research in the field. We specifically hope to motivate a central role for cavity QED in the future development of nonequilibrium quantum statistical mechanics and quantum information science. The core of this chapter is a tutorial review of the theory of cavity QED as an open quantum system, mingled with some brief perspectives on scientific experiments and technology demonstrations that now seem feasible according to this theory. In the remainder of this introduction, however, we will try to provide a high-level overview of the basic phenomenology of strong coupling and to sketch its promising role in enabling new lines of research.

1.1. *Cavity QED and Strong Coupling*

In the strong coupling regime, cavity QED can be conceptualized in terms of a quasi-coherent interaction between an atom and an eigenmode of an optical resonator. If the frequency of the optical mode coincides with that of an allowed transition between atomic states, and if the optical polarization satisfies its selection rules, it is possible for the atom and cavity mode to undergo a cyclic exchange of energy via absorption and emission of photons. In a scenario in which the atom is initially excited and the cavity mode is in its ground (vacuum) state, this cyclic energy exchange corresponds to the aforementioned vacuum Rabi oscillation and its rate is given by 4π times a geometry-dependent[9] quantity called the vacuum Rabi frequency, g. The vacuum Rabi frequency provides a universal measure of the strength of the atom–cavity coupling. In optical cavity QED, it can be maximized by minimizing the effective volume of the optical eigenmode and thus maximizing the electric field per photon. (In microwave cavity QED, it is also possible to utilize atomic transitions with extremely large dipole moment.) However, the full value of g for a given optical mode can only be realized in practice if a way is found to confine atoms at spatial maxima of the mode eigenfunction. For atoms located at positions where the eigenfunction has less than its maximal value, g must be scaled accordingly.

The vacuum Rabi frequency represents one of the three basic parameters of cavity QED. The other two are the atomic spontaneous emission rate (to noncavity modes) $4\pi\gamma$ and the cavity (energy) decay rate $4\pi\kappa$. Whereas g characterizes the coherent atom–cavity interaction, γ and κ are dissipative rates associated with processes that inhibit the buildup and preservation of atom–cavity coherence. For example, either an atomic spontaneous emission event or the escape of a photon from the cavity mode will obviously terminate vacuum Rabi oscillation. While there does not seem to be a universally accepted definition, cavity QED systems with $g > (\gamma, \kappa)$ are

generally considered to be in the strong coupling regime. In essence, the
term "strong coupling" is meant to denote physical conditions such that
the coherent atom–cavity interaction dominates over dissipative processes
in determining the qualitative nature of the system dynamics.

A somewhat more refined characterization of strong coupling follows
from the definition of two dimensionless parameters[9]: the critical photon
number

$$m_0 \equiv \frac{\gamma^2}{2g^2}, \tag{1}$$

and the critical atom number

$$N_0 \equiv \frac{2\gamma\kappa}{g^2}. \tag{2}$$

The critical photon number is directly related to the electric field per photon and to the atomic transition moment. It provides a rough measure
of the number of photons required in the cavity mode before the atomic
response begins to saturate, and therefore becomes nonlinear. The critical
atom number bears an additional dependence on the cavity finesse (quality factor), and it provides a rough measure of the number of atoms that
must be resonantly coupled to the cavity mode before its spectral line shape
becomes drastically altered. Phenomenologically speaking, we are clearly in
a regime of strong coupling if $m_0 < 1$, as this would imply that it should be
possible to observe nonlinear optical effects with just one photon per mode.
Strong coupling would likewise be indicated by $N_0 < 1$, as this would imply
that continuous optical probing of cavity input–output properties should
provide a very sensitive means of single atom detection.

Over the past decade, advances in electromagnetic resonator technology and in the control and cooling of atomic motion have enabled experi-
menters to delve deep into the strong coupling regime. Direct observational
signatures of vacuum Rabi oscillation have been obtained in both optical[17]
and microwave[18] cavity QED experiments. Nonlinear-optical response of
an atom–cavity system with one photon per mode has likewise been
demonstrated in both frequency domains.[19–21] In optical experiments with
$N_0 \ll 1$, continuous monitoring of cavity transmission has been used to
detect the passage of individual atoms through the cavity[22] and to perform real-time estimation of their spatial trajectories.[23] To date, definitive
demonstrations of strong coupling have been limited to experiments involving gas-phase neutral atoms.

From a broad scientific perspective, the main import of achieving strong
coupling in cavity QED is that one thus realizes an experimental paradigm

in which a well-defined quantum *system* with rich internal dynamics (the atom plus an intracavity field mode) is coupled weakly (by spontaneous emission and cavity decay) to a well-defined *reservoir* (optical modes other than the relevant cavity eigenmode). The fact that the atom–cavity system has basic Markovian coupling to a standard bosonic reservoir makes cavity QED a clean setting for fundamental studies in nonequilibrium quantum statistical mechanics (exploring the physics of "open" quantum systems), while the fact that the system-reservoir coupling can be made so weak (relative to the "internal" atom–cavity coupling) makes cavity QED an important test bed — and potential technology base — for emerging ideas in quantum information science.

1.2. *Applications of Strong Coupling*

The recent explosion of interest in quantum information science has brought about renewed enthusiasm for the study of *decoherence* in open quantum systems. Hypothetically, truly isolated quantum systems would preserve coherence perfectly during time evolution because of the unitary nature of Schrödinger dynamics. But as real-world physical systems always have some coupling to an environmental reservoir (heat bath), any excess coherence that might be pumped into them by nonequilibrium processes will have a very strong tendency (unless protected by some special symmetry) to decay via dissipation and dephasing. While formally situated as a special topic in quantum statistical mechanics, the subject of decoherence has assumed quite a broad intellectual footprint in contemporary physics because of its presumed centrality to such lingering issues as the quantum-classical interface, quantum measurement, and the microscopic origins of irreversibility. And as mentioned above, scrupulous analyses of decoherence mechanisms and the formulation of strategies to curtail or evade them are of great strategic interest for the implementation of advanced quantum information technologies.

Cavity QED has emerged as one of a handful of model open quantum systems in which one can productively pursue exacting agreement between subtle theoretical predictions and definitive experimentation. In this regard, the relative simplicity of cavity QED (as opposed, e.g. to systems in condensed matter) is a prime asset. Implementations with single atoms are especially attractive since the intracavity dynamics can often be well described in a truncated Hilbert space of reasonable (of order 10–100) dimension, amenable to computational analysis.

The fundamental decoherence mechanisms in cavity QED, corresponding to atomic spontaneous emission and cavity decay, are likewise well described by simple Lindblad terms in the Master Equation.

Despite tremendous recent progress in tailoring laser cooling and trapping techniques for integration with cavity QED,[24] however, residual atomic motion relative to the field eigenmode remains a significant complicating factor in optical experiments. Atomic center-of-mass motion couples with atom–cavity dynamics through the position dependence of g; in experiments with ultra-cold atoms, this leads not only to variation of the coupling strength but also creates the potential for entanglement among the atomic center-of-mass position, the atomic internal state, and the cavity mode. If the atomic motion is uncontrolled this essentially represents an additional decoherence mechanism, but it has been shown theoretically that the position dependence of g could also be used for the controlled creation of nonclassical states of atomic motion.[25]

It has long been recognized that intriguing quantum and semi-classical phenomena can arise from the interplay of coherent (driven) and dissipative dynamics in open quantum systems. In addition to nonequilibrium steady state behavior, optical cavity QED provides access to novel quantum stochastic processes when the system's output channels (spontaneous emission and/or cavity decay) are subjected to continuous observation by highly efficient photodetectors. In theory, continuous observation of optical reservoir modes can (at least partially) convert decoherence into *conditional evolution*. Perfect monitoring of the reservoir state would convert the natural buildup of system-reservoir entanglement (responsible for decoherence) into a classical correlation between the measurement record and random unitary evolutions of the system state. Accordingly, partial observation of the reservoir (through measurements with imperfect efficiency, extrinsic noise, or incomplete bandwidth) should lead to a partial "unravelling" of the system evolution.

Cavity QED is proving to be a rich setting for investigations of this type of conditional evolution theory for open quantum systems, which has come to be known in the quantum optics literature as *quantum trajectory theory*. Quantum trajectory theory has been used to predict a wide range of novel phenomena in cavity QED, and experimental observations of such phenomena would provide important validations of the underlying theory. There appear to be new types of multistability and bifurcation in continuously observed cavity QED systems,[26,27] which substantially expand the purview of nonlinear dynamics in open quantum systems.[2] It should be possible to utilize conditional dynamics for the preparation of nonclassical states of the optical field or of atomic motion,[25,28,29] and also for dynamic generation of entangled spin states in an atomic ensemble.[30] Looking beyond schemes that utilize conditional evolution in a passive way, several groups have begun to investigate real-time feedback control of quantum dynamics in cavity QED.[31,32] Potential applications of quantum feedback control include quantum metrology, stabilization of systems such as atom lasers, and robust storage and processing of quantum information.

Strong-coupling cavity QED has pervasive ties to quantum information science. The struggle to promote coherent coupling over dissipative and dephasing processes is central to both fields, as is an emphasis on purposeful control of quantum dynamics to construct highly-nonclassical states and entanglement. In cavity QED the direct coupling is between atoms and photons, which can be viewed as carriers of quantum information via atomic internal states (electronic, hyperfine, or Zeeman) and photon number or polarization. Given the current technology in laser cooling and trapping, atomic internal states can provide long-term storage of quantum information.[33] Photonic states on the other hand provide a unique medium for long-distance transmission of quantum information through optical fiber interconnects. By creating strong coupling between atoms and photons, cavity QED enables coherent mapping of quantum information (quantum states) between formats appropriate for storage or transmission,[34] and would thus seem to be a critical base technology for quantum communication systems.[35] The advanced understanding of decoherence processes in cavity QED is also an important advantage, and indeed cavity QED has been the setting for some of the most sophisticated fault tolerance protocols yet created in quantum information science.[36,37]

Utilized in second-order, strong atom–photon coupling can also serve as a physical basis for quantum logic interactions between pairs of photons, or between pairs of atoms. For the purpose of all-optical quantum logic, the possibility of single-photon saturation of atomic response (low critical photon number) can be used to modulate a target optical mode based on either the occupation number or polarization (via atomic selection rules) of a control optical mode. A cross-Kerr modulation of this type was in fact demonstrated in a seminal experiment on quantum logic in cavity QED.[19] Similarly, the internal state of one intracavity atom could be mapped to that of the cavity mode, with a transformation of the state of a second atom then conditioned on the state of the cavity mode to achieve quantum logic operation between atoms. While this type of process has yet to be demonstrated in the laboratory, quantum logic schemes based on it have been analyzed theoretically[38] and seem to provide the greatest promise for scalable implementations.[39] There is also growing interest in the development of feedback control and stabilization schemes for robust implementations of intracavity quantum logic with atomic qubits.[40]

2. The Atom–Cavity System as an Open Quantum System

Dedicated experimental effort over the past two decades has led to current implementations of optical cavity QED in which strong coupling can be achieved together with excellent suppression of technical noise and minimal uncertainty in the critical physical parameters. At the same time, theoretical models for the atom–cavity dynamics and reservoir interactions have been refined to match extremely closely with realistic experimental conditions and measurement protocols. Cavity QED has thus developed into an exemplary setting for the investigation of strongly coupled open quantum systems, in which researchers should be able to have high confidence in all aspects of their theory's correspondence with reality. In this section we provide a theoretical overview of the atom–cavity system as an open quantum system and provide a brief survey of current and forthcoming experiments. We have incorporated a fairly detailed discussion of the underlying system-reservoir theory, which provides a deep connection between cavity QED and modern topics in quantum statistical mechanics.

For most purposes it is sufficient to describe the atom–cavity system in terms of a rather simple approximate version of nonrelativistic quantum electrodynamics.[41] In this section we will describe this standard model rather than derive it, focusing on the physical ideas that indicate its regime of validity. We will refer the reader to more careful treatments at the appropriate points. Most of these ideas are standard in the field of quantum optics[42–44] and the resulting simplified models are in fact investigated in the general context of open quantum systems[45] and continuous measurement theory.[46]

As discussed above we will describe the dynamics of a single atom inside an optical Fabry–Perot resonator. We will be interested in deriving equations for the internal states and center of mass of the atom coupled to a single mode (the TEM_{00} mode) of the optical cavity. Since this system is coupled to other modes of the electromagnetic field we will have to average over these modes and the resulting equations of motion will not propagate a pure state of the system under the Schrödinger equation but will take the form of a master equation propagating a density matrix for the system.

The entropy of the state will no longer be a constant of the motion due to this coupling to the outside world. In later sections we will be interested in the quantum state of the light transmitted through the cavity and specifically in the statistics of photodetection on this light beam. As a result we also wish to develop a description of this light field in the same approximations used to derive the master equation. The resulting theory is usually known in quantum optics as input–output theory[42,47] and is basically scattering theory for the free electromagnetic field outside the cavity (see Ref. 48 for an explicit comparison).

The motivation for the model to be used results from identifying differences of timescales in the problem. The fastest of the relevant scales is the optical frequency of the cavity and atomic resonance $\omega/2\pi$ (for the cesium D-line transition at 852 nm this is $\simeq 3 \times 10^{14}$ Hz) followed by the frequency spacing between the modes of the Fabry–Perot cavity ($\delta \simeq 7 \times 10^{11}$ Hz for transverse modes and $\simeq 2 \times 10^{13}$ Hz for longitudinal modes in Ref. 23). The dynamics that are in fact of interest are much slower, the single photon Rabi frequency describing the interaction of the cavity mode with the atom (as yet no more than $\simeq 10^8$ Hz[23]) and the cavity and spontaneous energy loss rates (both $\simeq 5 \times 10^6$ Hz in Ref. 23). Characteristic frequencies for the atomic center of mass motion are slower again, no more than $\simeq 5 \times 10^5$ Hz. Note that while the coupling is strong in the sense that the coherent interaction of the atom and cavity mode dominates the loss processes caused by the coupling of this system to other modes of the electromagnetic field, it is still possible to consider a perturbative treatment for g with respect to the optical timescales. Given this, it should be possible to find an effective description of these (relatively) slow dynamics having averaged over the other modes using perturbation theory in small parameters like g/ω. Similarly given that $g/\delta \ll 1$ it should be possible to restrict attention to a single mode description of the electromagnetic field inside the cavity.

2.1. *Model for the Optical Cavity*

To begin with we will adopt a simple model for a cavity of length l with an imperfect mirror at $z = 0$ and a perfect mirror at $z = -l$ and disregard the atom. If we solve Maxwell's equations to find the eigenmodes of this problem all of the modes will be infinitely extended. Writing quantum electrodynamics for the free electromagnetic field with imperfect dielectric mirrors will result in a model with a single, noninteracting, quantum harmonic oscillator for each classical mode. We are, however, always free to consider nonrelativistic quantum electrodynamics in terms of spatial modes that are not eigenmodes of the problem at hand, the resulting Hamiltonian will lead to coupling between these modes. These are often termed *quasimodes*, see Refs. 49 and 50, for example, for a discussion of such an approach. In our case it is useful to write the Hamiltonian in terms of modes that are solutions to Maxwell's equations (in the paraxial approximation) for a perfect Fabry–Perot cavity in $-l < z < 0$ and of the free electromagnetic field in a quantization box of length L for $0 < z < L$. Since the mirrors are not in fact perfect these modes will be coupled by the QED Hamiltonian. A heuristic argument for this interaction is that in a narrow spatial region around $z = 0$ the electromagnetic fields due to these modes will overlap and the QED Hamiltonian $H = \int d\mathbf{r}(\mathbf{E}(\mathbf{r})^2 + \mathbf{B}(\mathbf{r})^2)$ will couple the different modes. Thus we posit the model

$$H = H_{0c} + H_{int} \tag{3}$$
$$H_{0c} = \hbar \omega a^\dagger a + \hbar \sum \omega_n a_n^\dagger a_n$$
$$H_{int} = \hbar \sum g_n \left(a a_n^\dagger + a^\dagger a_n \right).$$

Here a is the lowering operator for the TEM$_{00}$ mode of a lossless Fabry–Perot cavity. Similarly the operators a_n add and remove photons from the modes of the electromagnetic field in $0 < z < L$. The coupling constants g_n will be left unspecified for the moment.

In order to describe the interaction of the atom with the field and also to calculate the photodetection statistics of the experiment, it is necessary to have an approximate description of the electric field operator for this model. Outside of the mirror region (assumed to be narrow) we can write the electric field operators[41]

$$\mathbf{E}(\mathbf{r}) \simeq i\sqrt{\frac{\hbar \omega}{2\epsilon_0 V}} \cos(kz) \exp[-(x^2 + y^2)/2w_0^2](a\mathbf{e}_+ - a^\dagger \mathbf{e}_+^*) \quad z < 0,$$

$$\mathbf{E}(\mathbf{r}) \simeq i\sum_n \sqrt{\frac{\hbar \omega_n}{2\epsilon_0 AL}} \exp[(x^2 + y^2)/2w_0^2] \tag{4}$$
$$\times (e^{ik_n z} a_n \mathbf{e}_+ - e^{-ik_n z} a_n^\dagger \mathbf{e}_+^*) \quad z > 0.$$

We have only written the contribution of a single circularly polarized mode of the cavity for $-l < z < 0$ on the expectation that this is the only mode that will interact significantly with the atom due to the very large transverse mode spacing δ and the physical assumption that the atom has been pumped into a state such that the dipole selection rules will only allow an interaction with a single circular polarization of the field. This is fairly straightforward to realize in practice. (Such a single mode description also disregards any birefringence of the cavity mirrors.) Similarly we have only written modes of the field outside the cavity with the same Gaussian profile as the cavity mode, thus effecting a reduction to an essentially one-dimensional scalar electrodynamics. In practice the mirror will not have just transmission but also scattering losses resulting from coupling to modes left out of H_{int} for the moment. The modes resulting in scattering losses and imperfections of the other mirror can be treated exactly analogously to the discussion below and will be returned at the end of the calculation. Similarly we have disregarded for the time being field modes not propagating along the cavity axis, since these do not directly couple to the cavity mode although they will result in spontaneous emission of the atom. Close to the mirror there will also be contributions from modes of the field essentially localized inside the dielectric that will result in van der Waals forces (see Ref. 51 for a recent exposition). These effects will typically be very short ranged and rather weak, we will assume throughout that the atom is not within a half wavelength or so of the cavity mirrors and neglect van der Waals forces.

In writing these expressions we have assumed that the paraxial approximation is well satisfied in solving Maxwell's equations for the cavity field. For small cavities this may be the weakest point of the treatment. There is, however, unlikely to be any significant effect on the master equation that results apart from some change in the parameters g, κ. On the other hand, if the eigenmodes of the Fabry–Perot cavity in the limit of perfect mirrors depart significantly from those predicted by the paraxial approximation then this may have an effect on the modeshapes of the light propagating out of the cavity and also on motional effects that depend on the spatial variation of the electromagnetic fields inside the cavity. We have implicitly followed the standard procedure of solving the classical electrodynamics equations in the paraxial approximation and then performing canonical quantization of the resulting modes,[52] more sophisticated approaches are possible that derive this description directly as an approximation to quantum electrodynamics.[53]

In writing the interaction term in Eq. (3) we have already made the rotating wave approximation for the interaction of the cavity mode with the propagating modes outside the cavity. The Hamiltonian $\mathbf{E(r)}^2 + \mathbf{B(r)}^2$ would lead to contributions to H_{int} like $-(a - a^\dagger)(a_n - a_n^\dagger)$ but in the interaction picture with respect to H_{0c} the terms $aa_n, (aa_n)^\dagger$ would pick up a time dependence like $\exp(i2\omega t)$ that is very fast compared to the time evolution at rates $\simeq \kappa$ that results from H_{int}. Thus these energy nonconserving terms approximately average to zero and may be disregarded.

2.2. Master Equation for Cavity Decay

We will now outline the derivation of the master equation describing cavity decay from the Hamiltonian (3). Note that the linear coupled oscillator model (3) has a long history in the field of open quantum systems and quantum statistical mechanics (see for example Ref. 54 and references therein), and may in fact be solved with or without the rotating wave approximation.[55,56] The interest of the methods used here is that they highlight the simple physics of the system in the parameter regime of relevance in cavity QED and generalize easily to the nonlinear, strong coupling situation that we wish to consider.

Our derivation is standard in the quantum optics literature[44,57,58] and is originially due to Senitzky.[59,60] We will follow Carmichael closely.[44] The arguments here work in general for systems like Eq. (3), so in the following we will often refer to the cavity mode as the system and the modes of the field outside as the bath. The bath states will be assumed to be in the thermal equilibrium state ρ_B at temperature T and we aim to find a differential equation for the system state $\rho = \mathrm{Tr}_B \chi$ where χ is the joint state of the system and bath and Tr_B indicates the partial trace or average over the bath modes. Since our bath is made up of oscillators this assumption of thermal equilibrium gives $\langle a_n \rangle = \langle a_n^2 \rangle = 0$, $\langle a_n^\dagger a_n \rangle = n(\omega_n, T)$ and $\langle a_n a_n^\dagger \rangle = n(\omega_n, T) + 1$, where the thermal occupation number is $n(\omega_n, T) = \exp(-\hbar\omega_n/kT)/(1 - \exp(-\hbar\omega_n/kT))$. The physical assumptions are that the coupling between the system and bath are weak in the sense that evolution due to the coupling is slow compared to the timescales of the system dynamics ($\kappa \ll \omega$) and that the bath is made up of very many modes and thus remains essentially in thermal equilibrium at all times.

The first step is to go into an interaction picture with respect to the uncoupled Hamiltonian H_{0c}. Since H_{0c} is so easy to solve we readily obtain the interaction picture operators $a(t) = e^{-i\omega(t-t_0)}a$ and $a_n(t) = e^{-i\omega_n(t-t_0)}a_n$ where t_0 is an initial time at which the interaction and Schrödinger pictures coincide. The resulting evolution of the interaction picture density matrix can be written exactly as an integro-differential equation

$$\dot{\chi}(t) = \frac{1}{i\hbar}\left[H_{\mathrm{int}}(t), \chi(0)\right] - \frac{1}{\hbar^2}\int_0^t dt'\, \left[H_{\mathrm{int}}(t), \left[H_{\mathrm{int}}(t'), \chi(t')\right]\right]. \quad (5)$$

It is assumed that at the initial time the system and bath are only weakly correlated and so we can write $\chi(0) = \rho(0) \otimes \rho_B$ and it easy to check from Eq. (3) that $\mathrm{Tr}_B H_{\mathrm{int}}(t)\,(\rho(0) \otimes \rho_B) = 0$. In general this can always be achieved by modifying the bath Hamiltonian H_0. Performing the partial trace gives an equation for $\rho(t)$

$$\dot{\rho}(t) = -\frac{1}{\hbar^2} \int_0^t dt' \text{Tr}_{\text{B}} \left[H_{\text{int}}(t), [H_{\text{int}}(t'), \chi(t')] \right]. \tag{6}$$

Since the system and bath are assumed to be only weakly correlated at all times with the bath state essentially unnaffected by the system, we have $\chi(t) = \rho(t) \otimes \rho_{\text{B}} + O(H_{\text{int}}(t))$ and then accepting second-order perturbation theory we get

$$\dot{\rho}(t) = -\frac{1}{\hbar^2} \int_0^t dt' \text{Tr}_{\text{B}} \left[H_{\text{int}}(t), [H_{\text{int}}(t'), \rho(t') \otimes \rho_{\text{B}}] \right]. \tag{7}$$

After this Born approximation, the second major simplification is to argue that $\rho(t')$ may be replaced by $\rho(t)$ on the right-hand side. This is termed the Markov approximation since it means that the evolution of the state at time t does not depend on the state at earlier times. A careful investigation (see Ref. 44 for example) shows that this approximation requires that bath correlation functions like $\langle \Gamma^\dagger(t) \Gamma(t') \rangle$, where $\Gamma(t) = \sum g_n a_n(t)$, decay to zero for values of $t - t'$ much smaller than $1/\kappa$. This in turn requires that $|g_n|^2$ and $n(\omega_n, T)$ do not vary too rapidly as functions of ω_n. In the cavity model we expect $|g_n|^2$ to become roughly constant near ω as the length of interaction region becomes short (essentially so long as the time for light to traverse the mirror is small compared to relevant timescales). So this is well satisfied in our case. See Ref. 44 for consideration of this approximation for atomic spontaneous emission where $|g_n|^2$ is proportional to ω^3. The so-called Ohmic bath where $|g_n|^2$ goes like ω^2 is often considered in the condensed matter literature. The frequency dependence of the thermal occupation number can become sharp for low temperature, the relevant timescale for decay of these correlations is $\tau_{\text{B}} = h/kT$. At room temperature $1/\tau_{\text{B}} \simeq 6 \times 10^{12}$ Hz is again very much greater than the rates associated with evolution due to H_{int}. In any case, at our optical frequency the thermal occupation $n(\omega, T) \simeq 2 \times 10^{-22}$ is miniscule and so only the $T = 0$ contribution to the evolution will prove to be relevant. The resulting Markovian master equation is

$$\dot{\rho}(t) = -\frac{1}{\hbar^2} \int_0^t dt' \text{Tr}_{\text{B}} \left[H_{\text{int}}(t), [H_{\text{int}}(t'), \rho(t) \otimes \rho_{\text{B}}] \right]. \tag{8}$$

Now by substituting $H_{\text{int}}(t)$ from Eq. (3) into the master equation Eq. (8) and retaining only the $T = 0$ terms we obtain

$$\dot{\rho}(t) = \beta \left(a\rho a^\dagger - a^\dagger a \rho \right) + \text{h.c.} \tag{9}$$

where

$$\beta = \int_0^t dt' \sum_n |g_n|^2 e^{-i(\omega_n - \omega)t'} = \int_0^t dt' \int_0^\infty d\omega' r(\omega') |g(\omega')|^2 e^{-i(\omega' - \omega)t'}$$

$$= \pi r(\omega) |g(\omega)|^2 + iP \int_0^\infty d\omega' \frac{r(\omega') |g(\omega')|^2}{\omega - \omega'} \equiv \kappa + i\Delta. \tag{10}$$

We have replaced the sum over bath modes by an integral, sending the quantization length L to infinity. The density of states $r(\omega')$ is such that the number of modes between ω' and $\omega' + d\omega'$ is $r(\omega')d\omega'$. In the third equality, since we wish to describe dynamics on timescales very much longer than the time over which the integrand decays to zero we replace $\int_0^t dt'$ by $\lim_{t \to \infty} \int_0^t dt'$. P indicates the Cauchy principal value of the integral. We have made the definition of κ anticipating that this will correspond to the loss rate of the cavity. Our master equation is then

$$\dot{\rho}(t) = -i\Delta[a^\dagger a, \rho] + \kappa(2a\rho a^\dagger - a^\dagger a\rho - \rho a^\dagger a). \tag{11}$$

The principle value term is simply a modification of ω, the cavity resonance frequency (recall that this master equation is in the interaction picture). In practice, the frequency and loss rate will be determined by physical measurements on the cavity so we simply absorb this term into the definition of ω and write the Hamiltonian from now on in terms of the physical resonance frequency of the cavity. The second term does not correspond to unitary evolution and describes light leaking out of the cavity into the transmitted field. These terms give the desired Markovian model of a lossy cavity.

The equation for ρ is a Markovian master equation in the Lindblad form[61] and as such it generates a completely positive, trace preserving map. Like the Schrödinger equation, Eq. (11) always maps quantum states to quantum states. It is straightforward to calculate differential equations for expectation values of system operators X from a master equation using $\langle \dot{X} \rangle = \text{Tr} X \dot{\rho}$. So, for example, we can determine the decay of the photon number in the cavity using the commutation relation $[a, a^\dagger] = 1$ and the cyclic property of the trace to find

$$\frac{d}{dt} \langle a^\dagger a \rangle = -2\kappa \langle a^\dagger a \rangle. \tag{12}$$

As expected, the model describes exponential decay of the light inside the cavity with time constant $1/2\kappa$. To further check that we have obtained a sensible model for a damped harmonic oscillator note that an initial coherent state remains a coherent state with a decaying amplitude. It follows from $a|\alpha\rangle = \alpha|\alpha\rangle$ that $\rho(t) = |\alpha e^{-\kappa t}\rangle\langle\alpha e^{-\kappa t}|$ solves Eq. (11) with $\Delta = 0$ if $\rho(0) = |\alpha\rangle\langle\alpha|$.

It is also possible to derive correlation functions of system observables from the master equation. It is not hard to show using the arguments given above and the cyclic property of the trace[44,42] that if we write the master equation $\dot{\rho} = \mathcal{L}\rho$ then

$$\langle X(\tau)Y(0)\rangle = \text{Tr}\left[X \exp(\mathcal{L}\tau)\left\{Y\rho(0)\right\}\right] \tag{13}$$

$$\langle X(0)Y(\tau)\rangle = \text{Tr}\left[Y \exp(\mathcal{L}\tau)\left\{\rho(0)X\right\}\right], \tag{14}$$

for arbitrary system observables X and Y. The notation $\exp(\mathcal{L}\tau)\left\{Y\rho(0)\right\}$ represents the solution of the master equation with initial condition $Y\rho(0)$.

In fact we wish to model a cavity with an imperfect mirror at $z = -l$ through which the cavity mode can be driven by a laser at frequency $\omega_L \simeq \omega$. This mirror will contribute to the loss rate κ, as will the scattering losses. In the following we will use κ to refer to the total cavity (field) decay rate. The output coupling losses which go into the transmitted light field will be designated κ_{oc}. Typically the loss rates due to the two mirrors are equal and scattering losses can be made a small fraction of κ, hence $\kappa_{oc} \simeq \kappa/2$. Going through the above arguments with the freely propagating field in a coherent state as a model of laser driving of the optical cavity it is possible to show that the resulting master equation is

$$\dot{\rho}(t) = -i(\omega - \omega_L)[a^\dagger a, \rho] - i[Ea^\dagger + E^*a, \rho] + \kappa\left(2a\rho a^\dagger - a^\dagger a\rho - \rho a^\dagger a\right). \tag{15}$$

The laser amplitude and phase are indicated by E and in order to have a time-independent Hamiltonian we have moved into an interaction picture with respect to $H_{0c} - (\omega - \omega_L)a^\dagger a$ and so the system operators rotate at angular frequency ω_L. The detuning is again a small quantity of order g, κ, γ.

2.3. *Master Equation for Spontaneous Emission*

Turning now to the atom, we will write the master equation appropriate for a two-level atom in free space. The two-level atom will be an adequate approximation for an optically pumped alkali metal atom such as cesium. The atom of mass m has center of mass position and momentum described by the operators $\mathbf{r}, \mathbf{p}$ and a dipole given by $2F_g + 1$ Zeeman sublevels of the ground state and $2F_e + 1$ sublevels for the excited state. F_g is the total angular momentum of the hyperfine groundstate involved in the transition. The relevant Hamiltonian in the dipole approximation has a similar structure to the equation for cavity decay[62]

$$H = H_{0a} + H_{int} \tag{16}$$

$$H_{0a} = \frac{\mathbf{p}^2}{2m} + \frac{1}{2}\hbar\omega_0 \left(\sum_e |e\rangle\langle e| - \sum_g |g\rangle\langle g| \right) + \hbar \sum_{\mathbf{k},e} \omega_n a^\dagger_{\mathbf{k},e} a_{\mathbf{k},e}$$

$$H_{int} = -\mathbf{D} \cdot \mathbf{E}(\mathbf{r}).$$

Hence, for an atom in free space, the quantized field is

$$\mathbf{E}(\mathbf{r}) = i \sum_{\mathbf{k},e} \sqrt{\frac{\hbar\omega_n}{2\epsilon_0 L^3}} e^{i\mathbf{k_n}\cdot\mathbf{r}} a_{\mathbf{k},e} \mathbf{e} + \text{h.c.} \tag{17}$$

The sum is over wavevectors $\mathbf{k_n}$ with angular frequency $\omega_n = c|\mathbf{k_n}|$ and over polarization vectors $\mathbf{e} \perp \mathbf{k_n}$. The atomic dipole operator $\mathbf{D}$ can be written in terms of the reduced dipole matrix element d as $\mathbf{D} = d(\mathbf{S} + \mathbf{S}^\dagger)$. The components of $\mathbf{S}$ lower states from the excited to the ground state manifold and thus in the interaction picture $\mathbf{S}(t) = e^{-i\omega_0 t}\mathbf{S}$. The atomic resonance frequency is $\omega_0 \simeq \omega$.

Proceeding precisely as before to make the rotating wave and Born–Markov approximations we may derive the following master equation[44,62] (in the interaction picture with respect to H_{0a})

$$\dot{\rho} = 2\gamma \int \frac{d^2\hat{\mathbf{k}}}{8\pi/3} \sum_{\mathbf{e}\perp\hat{\mathbf{k}}} (\mathbf{S}\cdot\mathbf{e}^*)\exp(-i\mathbf{k}\cdot\mathbf{r})\rho\exp(i\mathbf{k}\cdot\mathbf{r})(\mathbf{S}^\dagger\cdot\mathbf{e})$$
$$- \gamma\left(\mathbf{S}^\dagger\cdot\mathbf{S}\rho + \rho\mathbf{S}^\dagger\cdot\mathbf{S}\right). \tag{18}$$

The unit vector $\hat{k}$ is in direction of the emitted photon. The spontaneous emission rate is $2\gamma = d^2\omega^2/3\pi\epsilon_0\hbar c^3$. The neglected principal value term is the nonrelativistic expression for the Lamb shift. If the atom is optically pumped such that it occupies only the Zeeman sublevels $m_{F_g} = F_g$, $m_{F_e} = F_e$ we can label these states $|0\rangle$ and $|1\rangle$ respectively and introduce the standard lowering operator $\sigma = |0\rangle\langle 1|$ to obtain

$$\dot{\rho} = -\gamma\left(\sigma^\dagger\sigma\rho + \rho\sigma\sigma\right) + 2\gamma \int \frac{d^2\hat{\mathbf{k}}}{8\pi/3}\beta(\hat{\mathbf{k}}\cdot\hat{\mathbf{z}})\sigma\exp(-i\mathbf{k}\cdot\mathbf{r})\rho\exp(i\mathbf{k}\cdot\mathbf{r})\sigma^\dagger.$$
$$\tag{19}$$

The pattern of the dipole radiation is accounted for by the angular factor[63] $\beta(\hat{\mathbf{k}}\cdot\hat{\mathbf{z}}) = [1 + (\hat{\mathbf{k}}\cdot\hat{\mathbf{z}})^2]/2$.

2.4. *Master Equation for the Atom–Cavity System*

When the atom is in the cavity the dipole interaction $-\mathbf{D} \cdot \mathbf{E}(\mathbf{r})$ is as before except that for a sufficiently small cavity the TEM$_{00}$ mode of the cavity can become the largest contribution to the quantized field $\mathbf{E}(\mathbf{r})$ as in Eq. (4). We assume though that the most significant part of the spontaneous emission results from modes of the field with $\hat{\mathbf{k}}$ sufficiently different from $\hat{\mathbf{z}}$ that they do not see the cavity mirrors. A more careful calculation of the modes of the field inside the cavity would simply result in a reduction of γ by a factor of the solid angle subtended by the cavity mode, the so-called cavity-inhibited spontaneous emission.[64] The rate γ then describes radiation from the atom out of the side of the cavity and in practice it is not usually much reduced from the free space value. Substituting from Eq. (4) the internal Hamiltonian of the atom cavity system has, in the rotating wave approximation, the additional term

$$V = -ig\cos(kz)\exp[-(x^2 + y^2)/2w_0^2]\left(a\sigma^\dagger - a^\dagger\sigma\right) \qquad (20)$$

where $g = d\sqrt{\frac{\hbar\omega}{2\epsilon_0 V_m}}$ (the vacuum Rabi frequency) depends on the reduced dipole matrix element and the electric field per photon.[9]

Repeating the derivation of the master equation, the internal Hamiltonian of this new system now has transition frequencies ω_i (for zero detunings) of $\omega \pm (\sqrt{n} - \sqrt{n-1})g$, $\omega \pm (\sqrt{n} + \sqrt{n-1})g$ where n is the excitation of the cavity and we have ignored the small splittings due to the center of mass. The energy shifts are due to the Jaynes–Cummings coupling V. Transition energies result in δ-functions picking out the strength of the coupling to the bath at the frequency in the derivation of the master equation. Each of these transitions will decay at a rate given by $\pi r(\omega_i)|g(\omega_i)|^2$. The standard approximation is to assume $r(\omega_i)|g(\omega_i)|^2 \simeq r(\omega)|g(\omega)|^2$ and hence the coupling of the atom and cavity does not affect the Lindblad terms describing the relaxation of the system. This approximation amounts to perturbation theory in $\sqrt{n}g/\omega$ and so we expect it to be a very good approximation for any achievable excitation of the cavity. Master equations for strongly coupled open systems where this is not the case are discussed in Ref. 65 where solutions beyond this standard perturbative approach are given. The fact that the loss terms for a coupled system are not the same as for the uncoupled system was recently emphasized in the context of noise models for quantum computers.[66] One way to derive the simplest form of the master equation is to include V in H_{int} rather than H_0 which means that V is treated in second-order perturbation theory in $g\sqrt{n}/\omega$. With this final simplification, the resulting master equation is

$$\dot{\rho} = -i[H, \rho] + \kappa \left(2a\rho a^\dagger - a^\dagger a\rho - \rho a^\dagger a\right) - \gamma \left(\sigma^\dagger \sigma \rho + \rho \sigma \sigma\right)$$

$$+ 2\gamma \int \frac{d^2\hat{k}}{8\pi/3} \beta(\hat{k} \cdot \hat{z}) \sigma \exp(-i\mathbf{k} \cdot \mathbf{r}) \rho \exp(i\mathbf{k} \cdot \mathbf{r}) \sigma^\dagger \tag{21}$$

$$H = \frac{\mathbf{p}^2}{2m} - i\hbar g \cos(kz) \exp[-(x^2 + y^2)/2w_0^2] \left(a\sigma^\dagger - a^\dagger \sigma\right) + \hbar(Ea^\dagger + E^*a)$$

$$+ \hbar(\omega - \omega_L)a^\dagger a + \hbar(\omega_0 - \omega_L)\sigma^\dagger \sigma. \tag{22}$$

This, finally, is the standard model for the atom–cavity system, written in an interaction picture with respect to $H_0 = \hbar\omega_L a^\dagger a + \hbar\omega_L \sigma^\dagger \sigma$. All the rates appearing in Eq. (21) are very slow compared to the optical frequency ω_L as discussed above. The two loss terms describe cavity decay and spontaneous emission respectively. The Hamiltonian terms represent the kinetic energy of the center of mass of the atom, the coherent interaction of the atomic dipole and the cavity mode, the laser driving of the optical cavity and the detuning of this driving from the cavity and atomic resonances.

While the above master equation is believed to provide an accurate and detailed description of the atom-field dynamics, direct comparisons with experiments have been complicated somewhat by the effects of residual atomic motion. As the Jaynes–Cummings Hamiltonian depends on the atomic center-of-mass position through $g(\mathbf{r})$, experiments can be adversely affected by uncontrolled atomic motion on timescales comparable to integration times for data acquisition. In optical cavity QED most quantitative experiments to date have been spectroscopic in nature, probing the intracavity system via measurements of the steady-state transfer function from optical driving fields to the field transmitted through the cavity. Atomic motion basically leads to an averaging of the master equation's predictions over a distribution of values for g, which can wash out some details of interest.

In the early days of strong coupling, experiments were conducted using fast (thermal) atomic beams. Atom transit times through the cavity were in general only marginally longer than dynamical timescales ($g, \kappa, \gamma_\perp$), and much shorter than the integration times utilized in acquiring transmission spectra of the atom–cavity system. In such scenarios it made sense to formulate the theory in terms of an average effective atom number, and to compare experimental data to predictions of the master equation after averaging over appropriate fluctuations in g.[44] By this approach it was possible to establish excellent agreement between theory and experiment in measurements of the weak-field spectrum (vacuum Rabi splitting),[17] and of photon anti-bunching.[67] An important study of nonlinear-optical response in the strong coupling regime,[68] which culminated in one of the first experimental demonstrations of the conditional dynamics required for implementing quantum logic gates,[19] was also performed with a thermal atomic beam. However, atom number fluctuations proved to be a serious obstacle for further attempts to characterize strong-field behavior in cavity QED with strong coupling.[69]

More recent experiments have capitalized on progress in laser cooling and trapping to deliver low-velocity atoms into the cavity mode volume,[22] and even to trap atoms inside the cavity using optical dipole forces.[21,70,71] The use of cold atoms has provided much cleaner access to the strong-field regime of cavity QED, as meaningful data points could be acquired on timescales shorter than the transit time of a single atom through the cavity.[21,72] Even so, without true atomic localization at suboptical wavelength scales, residual motion continues to complicate the formulation and implementation of ambitious schemes for quantum control.[73] Very recently, Kimble and co-workers have demonstrated a noninvasive optical trapping technique that confines atoms inside the cavity without imposing AC Stark shifts,[24] and have utilized this new capability to construct a very clean single-atom laser.[74] While there appears to be some uncertainty about the severity of residual (translational or oscillatory) motion of atoms along the cavity axis while confined in this new type of trap, it seems clear that this new trapping technique could enable nearly-ideal realizations of the Jaynes–Cummings model in future experiments.

It should be noted that the inherent coupling of atomic center-of-mass motion to the dynamics of the atomic internal state and cavity field could in theory be exploited for the controlled engineering of entanglement involving all three degrees of freedom.[75] Prospects for such experiments are perhaps best in scenarios with state-conditioning induced by continuous observation (see below).[25] Information about the atomic motion can likewise be recovered from measurements of the transmitted optical field, as shown theoretically in Ref. 76. Several experiments have been conducted,[21–23,77] to explore this aspect of strong coupling. The atom–cavity system can thus be thought of as a type of "microscope" for tracking single-atom motion,[23] with intriguing prospects of achieving quantum-limited precision.[78]

2.5. *Input–Output Formalism: The Transmitted Field*

In practice measurements are made on the light transmitted through the cavity. If we hope to find correlation functions for these measurements or states conditioned on particular measurement outcomes it is necessary to avoid averaging over the states of the transmitted light beam. This is achieved by finding expressions for the dynamics of the atom–cavity system *and* the transmitted field in the same approximations as were used to derive the master equation. It turns out that the bath mode that interacts with the system at time t corresponds in the Markov approximation to the field mode incident on a detection plane at $z = z_0$ at time $t + z_0/c$. This is generally referred to as the input–output formalism. Our discussion essentially parallels the standard quantum optics approach.[42,79] In particular, we follow the treatment in Ref. 79 closely. We focus on just the terms involving the cavity and transmitted field. The other degrees of freedom are replaced at the end.

In this section we will be rather explicit about the picture in which the equations of motion are being written. Unless otherwise stated operators with an explicit time argument $X(t)$ are in the interaction picture and operators without time arguments are in the Schrödinger picture.

Firstly we wish to replace the sum over modes in Eq. (3) by an integral at this stage of the calculation. We define bath operators $a(\omega')$ satisfying the commutation relation

$$\left[a(\omega'), a^\dagger(\omega'')\right] = \delta(\omega' - \omega''). \tag{23}$$

It is straightforward to check that the commutation relations are preserved by the replacement

$$\sum_n g_n a_n = \int_0^\infty \sqrt{r(\omega')} g(\omega') a(\omega') \, d\omega'. \tag{24}$$

The interaction picture Hamiltonian is then

$$\begin{aligned}
H(t) &= \hbar \int_0^\infty d\omega' \sqrt{r(\omega')} g(\omega') \left(a^\dagger(\omega', t) a(t) + a(\omega', t) a(t)^\dagger\right) \\
&= \hbar \int_0^\infty d\omega' \sqrt{r(\omega')} g(\omega') \left(a^\dagger(\omega') a e^{i(\omega' - \omega)(t - t_0)} + \text{H.c.}\right) \\
&= \hbar \sqrt{2\kappa_{\text{oc}}} \left(b_{\text{in}}(t)^\dagger a + b_{\text{in}}(t) a_1^\dagger\right)
\end{aligned} \tag{25}$$

where κ_{oc} is the contribution to the cavity decay due to the output coupling as before. In the second line we have expressed the interaction picture Hamiltonian in terms of Schrödinger picture operators, which turns out to be convenient. In the final line we define the so-called input field $b_{\text{in}}(t)$

such that

$$b_{\text{in}}(t) \equiv \frac{1}{\sqrt{2\kappa_{\text{oc}}}} \int \sqrt{r(\omega')} g(\omega') a(\omega') e^{-i(\omega'-\omega)(t-t_0)} d\omega'. \qquad (26)$$

This standard notation is a little confusing, the input operator $b_{\text{in}}(t)$ should be considered to be a linear combination of *Schrödinger picture* operators, with the phase factors of the coefficients depending on t.

In the Markov approximation, which we discussed in the previous section, the coupling to the bath is assumed to roughly independent of frequency near the resonance ω, so $\sqrt{r(\omega')} g(\omega') \simeq \sqrt{r(\omega)} g(\omega)$. The commutator of the input field is then

$$[b_{\text{in}}(t), b_{\text{in}}(t')^\dagger] \simeq \frac{1}{2\pi} \int_0^\infty d\omega' e^{-i\omega'(t-t')} \simeq \delta(t-t'). \qquad (27)$$

We again disregard the principal value term. Similarly if the bath is in the vacuum state we get $\langle b_{\text{in}}(t) b_{\text{in}}^\dagger(t') \rangle = \delta(t-t')$. This δ-function correlation is a result of the Markovian approximation and suggests that the input fields are a quantum analog of the white noise terms in Langevin equations. For this reason the operator equations of motion arising from $H(t)$ are often called quantum Langevin equations. We will see that $b_{\text{in}}(t)$ can be thought of as a mode of the electromagnetic field interacting with the cavity at time t and these commutation relations imply that modes at distinct times t are independent.

As a result of the singular commutator Eq. (27) it is necessary to be careful in integrating the equations of motion determined by $H(t)$. This is usually taken into account in the language of quantum stochastic differential equations.[42,46] These can be defined rigorously and guarantee that the singular limit for the commutators is well defined. We will not adopt this level of rigor and explicitly consider the time evolution operator

$$U(t_1, t_0) = T \exp\left[\sqrt{2\kappa_{\text{oc}}} \left(a \int_{t_0}^{t_1} b_{\text{in}}^\dagger(t') dt' - a^\dagger \int_{t_0}^{t_1} b_{\text{in}}(t') dt' \right) \right], \qquad (28)$$

T represents time-ordering. The cavity operator a factors out the time integrals since it has no time dependence. In finding the small time limit of the evolution, $t_1 \to t_0 + dt$, note that the second-order term in perturbation theory is of order dt because

$$\left[\int_0^{\Delta t} b_{\text{in}}(t') dt', \int_0^{\Delta t} b_{\text{in}}^\dagger(t'') dt'' \right] = \Delta t. \qquad (29)$$

The usefulness of the input field is that it can be related to the electromagnetic field outside the cavity. From Eq. (4) we define

$$\mathbf{E}^{(+)}(\mathbf{r}) = i \exp[-(x^2 + y^2)/2w_0^2] \int_0^\infty \sqrt{\frac{\hbar\omega' r(\omega')}{2\epsilon_0 AL}} e^{ik'z} a(\omega')\mathbf{e}_+ \qquad (30)$$

with $\mathbf{E}^{(-)} = \mathbf{E}^{(+)\dagger}$ and $\mathbf{E} = \mathbf{E}^{(+)} + \mathbf{E}^{(-)}$ for $z > 0$. A photodetection device (whether a photon-counter or a homodyne or heterodyne detection setup) is placed at $z = z_0$. Then the photon-counting statistics or the statistics of homodyne or heterodyne detection may be written in terms of $\mathbf{E}^{(+)}$ and its correlations using the Kelley–Kleiner formula, see Refs. 80 and 81 for a full account.

Whatever kind of detection is available it will be sensitive to light at frequencies close to ω but the rapid time dependence of $\langle \mathbf{E}^{(+)}(\mathbf{t}) \rangle$ will not be directly detectable. For this reason it is useful to consider the Heisenberg picture field operator at $z = z_0$ and time t_1 having removed this fast time dependence at frequency ω

$$e^{i\omega(t_1 - t_0)}\mathbf{E}^{(+)H} = E_0 \mathbf{e}_+ + \int_0^\infty d\omega' \left[\sqrt{\frac{\omega' r(\omega')}{2\pi\omega r(\omega)}} e^{-i(\omega' - \omega)(\bar{t}_1 - t_0)} \right.$$

$$\left. \times U^\dagger(t_1, t_0) a(\omega') U(t_1, t_0) \right]$$

$$\simeq E_0 \mathbf{e}_+ U^\dagger(t_1, t_0) b_{\text{in}}(\bar{t}_1) U(t_1, t_0). \qquad (31)$$

The first line uses the definition of the Heisenberg picture field operator in terms of full evolution operator $\exp[iH_0(t_1 - t_0)/\hbar]U(t_1, t_0)$. The retarded time is $\bar{t}_1 = t_1 - z_0/c$. The approximate equality in the final line holds in the Markov approximation and may be checked by comparing with the definition of $b_{\text{in}}(t)$. The overall constant field is $E_0 = i\sqrt{\pi\hbar\omega r(\omega)/2\epsilon_0 AL} \exp[-(x^2 + y^2)/2w_0^2] \exp(i\omega z_0/c)$. So it is indeed the case that the field outside the cavity is simply related to the input field mode $b_{\text{in}}(t)$ that interacts with the system at time t.

This motivates the definition of an output field operator

$$b_{\text{out}}(t) \equiv U^\dagger(t_1, t_0) b_{\text{in}}(t) U(t_1, t_0). \qquad (32)$$

The definition proves to be independent of t_1 so long as $t_1 > t$. For $t_1 > t > t_0$ we can write $U(t_1, t_0) = U(t_1, t + dt)U(t + dt, t)U(t, t_0)$ and then from the commutation relations Eq. (27) for $b_{\text{in}}(t)$ and the definition of the evolution operator Eq. (28) it is clear that the time evolution takes place only around time t since $[b_{\text{in}}(t), U(t, t_0)] = 0$ and $[b_{\text{in}}(t), U(t_1, t + dt)] = 0$.

Physically this is a result of the Markov approximation: after interacting with the system at time t the subsequent dynamics of the mode $b_{in}(t)$ are unaffected by the interaction. In fact we can see from Eq. (31) that after this infinitesimal period of interaction the mode $b_{in}(t)$ propagates freely away from the cavity. Hence we can write

$$
\begin{aligned}
b_{out}(t) &= U^\dagger(t+dt, t_0)b_{in}(t)U(t+dt, t_0) \\
&= U^\dagger(t, t_0)(b_{in}(t) + \sqrt{2\kappa_{oc}}a)U(t, t_0) \\
&= b_{in}(t) + \sqrt{2\kappa_{oc}}U^\dagger(t, t_0)aU(t, t_0).
\end{aligned}
\tag{33}
$$

The second line expanded $U(t+dt, t)$ to first-order and employed the commutation relations Eq. (27). This equation is precisely what we expect from the boundary condition for the fields at $z = 0$ due to the imperfect mirror. The reflected field is related to the incident field and the transmitted field as we would expect in the limit of a very high reflectivity of the mirror. This provides some justification for our simple model of the lossy cavity. It is also possible to show that the output field, like the input field, is δ-correlated.

Recall that the Heisenberg picture field operator at the detector $e^{i\omega(t_1-t_0)}\mathbf{E}^{(+)H}(\mathbf{z_0}, t_1)$ is proportional to $b_{out}(t_1 - z_0/c)$, so b_{out} should be regarded as a Heisenberg picture operator and expectation values should be taken against the initial state $\rho(t_0)$. This is convenient since the modes of the bath are initially assumed to have thermal statistics and expectation values of the first term in (33) are easy to evaluate. On the other hand

$$
\text{Tr}[U^\dagger(t, t_0)aU(t, t_0)\rho(t_0)] = \text{Tr}[a\rho^I(t)] = \exp\left[i\omega(t - t_0)\right]\langle a(t)\rangle
\tag{34}
$$

where $\rho^I(t)$ emphasizes that this is the interaction picture density matrix of the system since it is evolved by $U(t, t_0)$. The demodulation picks out the component of the field oscillating at ω. It is clearly very desirable to divide off the fast rotation of the field expectation values as we have done here. This leads to the ubiquitous practice in quantum optics of evaluating expectation values by tracing Schrödinger picture observables against interaction picture states. In the following we will also adopt this convenient convention, so that $\langle X(t)\rangle = \text{Tr}X\rho^I(t)$.

The commutation relations Eq. (27) and the evolution operator U make it possible to evaluate correlations of the output field in terms of correlations of the system observables, see Refs. 42 and 81. These can in turn be calculated from the master equation Eq. (21) as discussed above. So for example, from (33) we find $\langle b_{out}(t)\rangle = \sqrt{2\kappa_{oc}}\langle a(t)\rangle$ and it is possible to show for $n(\omega, T) = 0$ that $\langle b_{out}^\dagger(t)b_{out}(t')\rangle = \sqrt{2\kappa_{oc}}\langle a^\dagger(t)a(t')\rangle$ and

$\langle b_{\text{out}}(t) b_{\text{out}}(t') \rangle = \sqrt{2\kappa_{\text{oc}}} \langle T[a(t)a(t')] \rangle$. This provides us with all the information we need about the statistics of measurements on the light transmitted through the cavity. We will see in the next section that it is also possible to write evolution equations for the quantum state of the atom–cavity system conditioned on specific sequences of measurement results by projecting onto eigenstates of appropriate observables of the fields $b_{\text{in}}(t)$.

3. Dissipation, Decoherence, and Continuous Observation

3.1. *Conditional Evolution and Quantum Trajectories*

In the previous section we discussed the statistics of optical measurements on the light transmitted through the cavity using the input–output formalism to derive correlation functions of the measured signal in terms of correlations of the cavity field. The input–output theory is essentially a scattering theory description of light reflecting off the cavity mirror. On the other hand, the master equation provides a dynamical description of the atom and the cavity mode, having averaged over the transmitted field. From the master equation it is possible to calculate the correlation functions of the cavity field that appear in expressions for the transmitted field correlation functions. However, more than simply describing the statistics of measurement results, or the state of the atom–cavity system averaged over the states of the transmitted field, quantum mechanics makes it possible to describe the state of the system given a specific measurement result, by means of the von Neumann projection postulate. In this section we introduce evolution equations for such conditioned states. In this case the measurement is continuous in time since the system interacts with the independent modes $b_{\text{in}}(t)$ at each time t and these modes are subsequently measured.

A theory of measurements continuous in time, initiated by Davies[82] was developed in the mathematical physics literature, particularly by Barchielli and collaborators[83–85] and Belavkin[86–88] who was the first to write down quantum trajectory equations of the kind we discuss here in the context of continuous measurements. These developments are discussed in the recent monograph.[46] There are related, much less general, discussions at this time by Caves and Milburn[89] and Diosi.[90] The theory of quantum trajectories[79,91,92] was developed in quantum optics to handle optical measurements such as those we consider in this review and large simulations of systems such as laser-cooled atomic gases, largely independent of results in the mathematical physics literature. Several earlier discussions of related issues in the quantum optics literature were motivated by consideration of

electron shelving experiments in ion traps, for example,[93–95] as reviewed
in Ref. 96. Here we will give a simple argument for the quantum trajectory equations essentially due to Wiseman[97] which highlights the fact that,
rather than being outside conventional quantum mechanics, quantum trajectory equations simply arise from the unitary interaction of the system
of interest with a series of independent bath modes followed by projective
measurement on these modes. This point of view is also emphasized in the
recent review by Brun.[98] Our discussion is rather informal and in particular
we avoid the language of quantum stochastic differential equations required
to rigorously interpret Eq. (25). For such a discussion we refer the reader
to Refs. 42 and 46.

3.2. *Quantum Trajectories*

We recall that the interaction picture Hamiltonian (considering for the
moment only the coupling of the cavity to the transmitted field) was given
in Eq. (25)

$$H(t) = \hbar\sqrt{2\kappa_{\text{oc}}}\big(b_{\text{in}}(t)^\dagger a + b_{\text{in}}(t)a_1^\dagger\big).$$

The singular field commutation relations (27) require that the input fields
at different times should be regarded as independent modes. Physically
the input field will have a nonzero commutator for times of the order of
the bath correlation time discussed in the previous section. If κ_{oc} is small
enough we should be able to do perturbation theory in $H(t)$ by expanding
the interaction picture time evolution operator (28). As discussed above, the
singular commutation relations require that we expand up to second-order.

In this perturbation theory $b_{\text{in}}(t)$ will be replaced by a time coarse
grained version and it is useful to define $c_t \equiv \int_t^{t+\Delta t} b_{\text{in}}(t)dt/\sqrt{\Delta t}$. Then we
get the commutation relation (for equal times, the others are also easy to
work out)

$$\begin{aligned}\left[c_t, c_t^\dagger\right] &= \frac{1}{\Delta t}\int_{t-\Delta t/2}^{t+\Delta t/2}\int_{t-\Delta t/2}^{t+\Delta t/2}\left[b_{\text{in}}(t'), b_{\text{in}}(t'')^\dagger\right]dt'dt'' \\ &= \frac{1}{\Delta t}\int_{t-\Delta t/2}^{t+\Delta t/2}dt' = 1.\end{aligned}$$

Essentially c_t is a conventional lowering operator for a harmonic oscillator
describing the field mode at time t having coarse-grained over Δt. The
perturbation expansion shows that in each timestep the system interacts

with an independent mode of the transmitted field described by c_t. At time t the state of the system and bath is something like

$$R(t) = \rho(t) \otimes \mu_t \otimes \mu_{t+\Delta t} \otimes \mu_{t+2\Delta t} \cdots$$

where the subscripts indicate the time at which each bath system will interact with the system.

The Dyson expansion of Eq. (28) gives the state at time $t + \Delta t$

$$
\begin{aligned}
R(t + \Delta t) &= R(t) - \frac{i}{\hbar}\left[H(t), R(t)\right]\Delta t - \frac{1}{\hbar^2}\left[H(t), \left[H(t), R(t)\right]\right]\Delta t^2 \\
&= R(t) - i\sqrt{2\kappa_{oc}}\left[c_t^\dagger a + a^\dagger c_t, R(t)\right]\sqrt{\Delta t} \\
&\quad - 2\kappa_{oc}\left[c_t^\dagger a + a^\dagger c_t, \left[c_t^\dagger a + a^\dagger c_t, R(t)\right]\right]\Delta t
\end{aligned}
$$

To make further progress it is necessary to specify the bath states μ. For convenience we will specialize to the case where each of the bath modes is in the ground state. Given our application in the optical regime this is no great restriction. The following treatment is sufficient to deal with any Gaussian state of the bath, thus it is possible to describe coherent driving of the cavity, thermal states of the bath and driving of the bath by broadband squeezed light.[42] The general theory of quantum stochastic differential equations[46] also allows the system to be coupled to the intensity of a coherent state field of the bath but we do not consider this possibility here. So we are assuming that

$$c_i \mu = 0 = \mu c_i^\dagger.$$

This simplifies our expression to

$$
\begin{aligned}
R(t + \Delta t) &= \rho(t) \otimes \mu_t - i\sqrt{2\kappa_{oc}}\left(a\rho(t) \otimes c_t^\dagger \mu_t - \rho(t)a^\dagger \otimes \mu_t c_t\right)\sqrt{\Delta t} \\
&\quad + \kappa_{oc}\left(2a\rho(t)a^\dagger \otimes c_t^\dagger \mu_t c_t - a^\dagger a\rho(t) \otimes c_t c_t^\dagger \mu_t\right. \\
&\quad \left. - \rho(t)a^\dagger a \otimes \mu_t c_t c_t^\dagger\right)\Delta t \\
&= \rho(t) \otimes |0\rangle\langle 0| - i\sqrt{2\kappa_{oc}}\left(a\rho(t) \otimes |1\rangle\langle 0| - \rho(t)a^\dagger \otimes |0\rangle\langle 1|\right)\sqrt{\Delta t} \\
&\quad + \kappa_{oc}\left(2a\rho(t)a^\dagger \otimes |1\rangle\langle 1| - a^\dagger a\rho(t) \otimes |0\rangle\langle 0|\right. \\
&\quad \left. - \rho(t)a^\dagger a \otimes |0\rangle\langle 0|\right)\Delta t,
\end{aligned}
\tag{35}
$$

where we have omitted explicit mention of the bath modes for $t' \neq t$. It is easy to see that tracing over the bath modes and taking $\Delta t \to 0$ limit leads to the cavity master equation derived in the previous section.

The reason to consider this rederivation is that it is now straightforward to describe the effect of measurements on the bath by projecting onto bath states corresponding to measurement outcomes.

3.3. *Counting Photons*

As we discussed in the previous section, the bath oscillators propagate away at the speed of light after interacting with the system and are eventually incident on a detector. We saw that $b_{\text{out}}(t) = U^\dagger b_{\text{in}}(t) U$ represents the field mode at the detector at time $t + z_0/c$. Recall that all the time evolution for $b_{\text{in}}(t)$ occurs in a time interval arbitrarily close to t. In our interaction picture description photon counting measurements of $\mathbf{E}^{(-)H} \cdot \mathbf{E}^{(+)H} = E_0^2 b_{\text{out}}(t)^\dagger b_{\text{out}}(t)$ by the detector at $z = z_0$ and time $t + z_0/c$ will correspond to projection onto eigenstates of $c_t^\dagger c_t$. We imagine that a measurable burst of photoelectrons results every time a photon hits the detector no matter how short a time has elapsed since the last photodetection event. Such a detector is essentially a projective measurement of the field reflected off the cavity in the number state basis for each mode. Detectors with quantum efficiency less than one are straightforward to incorporate and more realistic properties of detectors such as dark counts, dead times, and finite bandwidth, are discussed in Refs. 99–101. The only two outcomes of this measurement at time $t + z_0/c$ are zero photons or one photon. The two projection operators corresponding to these measurement outcomes are

$$P_0 = I_S \otimes |0\rangle\langle 0|, \qquad P_1 = I_S \otimes |1\rangle\langle 1|,$$

where I_S is the identity operator on the system. The probability that no photon hits the detector in this timestep is

$$p_0 = \text{Tr}\left[P_0 R(t + \Delta t)\right]$$
$$= 1 - 2\kappa_{\text{oc}}\langle a^\dagger a\rangle \Delta t,$$

on the other hand, the probability that a photon does strike the detector is

$$p_1 = \text{Tr}\left[P_1 R(t + \Delta t)\right]$$
$$= 2\kappa_{\text{oc}}\langle a^\dagger a\rangle \Delta t.$$

Using the projection postulate the state if no photon is detected should be projected onto $|0\rangle\langle 0|$ for the bath oscillator

$$R_0\left(t + \Delta t\right) = P_0 R\left(t + \Delta t\right) P_0$$
$$= \left[\rho(t) - \kappa_{\text{oc}}\left(a^\dagger a\rho(t) - \rho(t)a^\dagger a\right)\Delta t\right] \otimes |0\rangle\langle 0|.$$

So we can define a state for the system alone, conditioned on not detecting a photon in the output field

$$\bar{\rho}_0\left(t + \Delta t\right) = \rho(t) - \kappa_{oc}\left(a^\dagger a\rho(t) - \rho(t)a^\dagger a\right)\Delta t.$$

The overbar is to indicate that in fact this state is not normalized since

$$\mathrm{Tr}\bar{\rho}_0\left(t + \Delta t\right) = 1 - 2\kappa_{oc}\langle a^\dagger a\rangle\Delta t = p_0.$$

The trace of the unnormalized conditioned state is equal to the probability of the measurement outcome on which it is conditioned.

On the other hand, if a photon is detected it should be clear that the projected state is

$$\begin{aligned}
R_1\left(t + \Delta t\right) &= P_1 R\left(t + \Delta t\right) P_1 \\
&= 2\kappa_{oc}\Delta t a\rho(t)a^\dagger \otimes |1\rangle\langle 1|.
\end{aligned}$$

Once again we may define a system state conditioned on detecting a photon

$$\bar{\rho}_1\left(t + \Delta t\right) = 2\kappa_{oc}\Delta t a\rho(t)a^\dagger$$

and again its trace is equal to the probability of the appropriate outcome

$$\mathrm{Tr}\bar{\rho}_1\left(t + \Delta t\right) = p_1.$$

Note that when normalized this state is precisely the state of the system at time t with exactly one photon removed. This corresponds to the intuitive idea of what the photon loss process should do to the state. The measurement process just determines the times at which photons leave the cavity. Note that averaging over the measurement results gives the master equation evolution back again

$$\bar{\rho}_0\left(t + \Delta t\right) + \bar{\rho}_1\left(t + \Delta t\right) = \rho(t) + \kappa_{oc}\left(2a\rho(t)a^\dagger - a^\dagger a\rho(t) - \rho(t)a^\dagger a\right)\Delta t.$$

This has to be the case since measuring the bath after the interaction with the system can have no effect on the system itself.

There is now a fairly well defined $\Delta t \to 0$ limit. Since the probability of not detecting a photon goes to one in each infinitesimal time increment there is a differential equation for the state of the system conditioned on not detecting a photon up to time τ

$$\frac{d}{d\tau}\bar{\rho}_0\left(t + \tau\right) = -\kappa_{oc}\left(a^\dagger a\bar{\rho}_0\left(t + \tau\right) - \bar{\rho}_0\left(t + \tau\right)a^\dagger a\right). \tag{36}$$

By making an inductive argument it can be shown that the probability of no photodetection up to time $t + n\Delta t$ obeys the identity

$$p_0\left(t + n\Delta t\right) = \text{Tr}\bar{\rho}_0\left(t + n\Delta t\right)$$

and hence in the continuous limit the probability of no photon detection occurring between t and τ is just $p_0\left(\tau\right) = \text{Tr}\bar{\rho}_0\left(t + \tau\right)$ where $\bar{\rho}_0\left(t + \tau\right)$ obeys the above differential equation. If the state is initially a pure state $\rho(t) = |\psi(t)\rangle\langle\psi(t)|$ then the unnormalized pure state satisfying

$$\frac{d}{d\tau}|\bar{\psi}(t + \tau)\rangle = -\kappa_{oc}a^\dagger a|\bar{\psi}(t + \tau)\rangle \tag{37}$$

gives a solution of Eq. (36) $\bar{\rho}_0\left(t + \tau\right) = |\bar{\psi}(t + \tau)\rangle\langle\bar{\psi}(t + \tau)|$. When a photodetection occurs at $t + \tau + z_0/c$ we can update the state by removing a photon and normalizing

$$\rho_1(t + \tau) = a\bar{\rho}_0(t + \tau)a^\dagger / \text{Tr}[a^\dagger a\bar{\rho}_0(t + \tau)]. \tag{38}$$

If $\bar{\rho}_0(t + \tau)$ is a pure state we likewise have

$$|\bar{\psi}_1(t + \tau)\rangle = a|\bar{\psi}_1(t + \tau)\rangle / \sqrt{\langle\bar{\psi}(t + \tau)|a^\dagger a|\bar{\psi}(t + \tau)\rangle}. \tag{39}$$

So initial pure states remain pure under this evolution and we may propagate the state of the system by considering the state vector only. This is a result of continually projecting the bath onto eigenstates and therefore destroying the entanglement between the system and the bath that leads to mixed states of the system when the bath is averaged over. However, when we add the terms corresponding to spontaneous emission and other imperfections of the atom–cavity system the conditioned state will no longer be pure even if the initial state is pure which is the motivation for considering density matrices throughout this discussion. With this procedure we may continually update the state of the system based on the measured record of photodetector clicks. It is important to note that given the initial state of the system, and the record of detector clicks, an experimenter is able to propagate the conditioned state for all later times.

Although the evolution of different measurement trajectories is interspersed by discontinuous jumps, the probability of each of these jumps is infinitesimal in any infinitesimal timestep. This equation is a form of stochastic differential equation driven by a jump process. The jump and no-jump possibilities may be written in one equation by defining the point process stochastic increment dN which has the properties $\text{E}[dN(t)] = 2\kappa_{oc}\langle a^\dagger a(t)\rangle dt$ and $dN^2 = dN$. That is the increment is almost always equal to zero, but events with $dN = 1$ occur with a finite rate given by $2\kappa_{\text{oc}}\langle a^\dagger a\rangle$. We may now write a stochastic master equation

$$d\rho_{\text{c}} = dN(t)\mathcal{G}[a]\rho_{\text{c}}(t) - \kappa_{\text{oc}}dt\mathcal{H}[a^\dagger a]\rho_{\text{c}}(t). \tag{40}$$

This equation is simply a compact notation for the $\Delta t \to 0$ limit for the stochastic recursion relations (36) and (38) described in the proceeding paragraph. We have defined the following superoperators

$$\mathcal{G}[s]\rho = \frac{s\rho s^\dagger}{\text{Tr}[s\rho s^\dagger]} - \rho,$$

$$\mathcal{H}[s]\rho = s\rho + \rho s^\dagger - \text{Tr}[s\rho + \rho s^\dagger]\rho,$$

where s is an arbitrary system operator. The times where $dN = 1$ are just the times at which photodetections take place. Important properties of the stochastic master equation (40) are that it preserves trace and positivity of the density matrix, gives the master equation when averaged over the stochastic process $dN(t)$ and that it is nonlinear but only as a result of the need to normalize the state after the application of the projectors P_0 or P_1 in each timestep.

Once again the special properties of coherent states make it possible to solve the evolution. From $a|\alpha\rangle = \alpha|\alpha\rangle$ we find that that $\rho(t) = |\alpha e^{-\kappa_{\text{oc}}t}\rangle\langle\alpha e^{-\kappa_{\text{oc}}t}|$ solves Eq. (40) if $\rho(0) = |\alpha\rangle\langle\alpha|$ and the probability of detecting a photon between t and $t+dt$ is $p_1(t)dt = \exp(-2\kappa_{\text{oc}}t)|\alpha|^2 dt$. So again we have the natural picture of photons leaking out of the cavity mirror at rate $2\kappa_{\text{oc}}$ being detected at random intervals with the appropriately decaying intensity. The very special property of the coherent states that they are eigenstates of a means that the conditioned state is independent of measurement results.

On the other hand these trajectory equations can be used as a means of simulating the dynamics of the master equation or the results of measurements on the system by deciding at random when photodetections should occur. The basic idea is to choose the times for the jumps at random from the correct distribution of measurement results. Then the property that the master equation is recovered after averaging over the measurement results leads to a kind of pure state Monte Carlo simulation of the master equation. To do this we should pick a random number q between zero and one with a uniform distribution and integrate the differential equation (37) for no detection until the time τ at which $\text{Tr}\bar{\rho}_0\,(t+\tau)$ reaches this threshold value. At this point we specify that a photodetection occurs by applying the annihilation operator to remove a photon and then normalizing as described by Eq. (39). Another random number is drawn and the simulation is carried out for the desired time. In order to recover master equation averages one simply averages over many runs of the simulation. This is a so-called "quantum trajectory simulation." The particular form described here estimates the time for a photodetection to all orders in Δt and is the form of the photon-counting trajectory that is implemented in Ref. 102. The implementation of this kind of simulation in quantum optics was driven by the desire to simulate large quantum mechanical systems, particularly the behavior of laser cooled atomic samples at or near the recoil limit where the quantum mechanical nature of the atomic center of mass motion (rather than just the quantum mechanical nature of its electronic states) becomes important, see for example Ref. 92. Quantum trajectories facilitated simulations that, for example, achieved quantitative agreement with experiment in regimes where effects of quantized atomic can be distinguished experimentally,[103] and also simulations that quantized all three directions of atomic motion for laser cooled atoms in optical molasses and obtained similarly good agreement with experiment.[104]

Several cavity QED experiments have been performed that illustrate conditional evolution associated with photon counting. Although not couched in the language of quantum trajectories, the intensity autocorrelation function measurement by Rempe *et al.*[67] implicitly demonstrates the type of conditional state-collapse of multiple intracavity atoms discussed theoretically by Carmichael and co-workers.[105] Conditional preparation of an intracavity state by photodetection has been investigated explicitly in experiments by Orozco and co-workers.[106]

The quantum trajectory theory of photon counting has been utilized to formulate schemes for quantum system identification,[107] quantum jump inversion,[108] and quantum feedback control of atomic dressed-state cascades.[31] It likewise plays a central role in many recent proposals for quantum state preparation and quantum information processing via cavity QED, as in Refs. 35, 38 and 109.

3.4. *Measuring Field Quadratures*

The quantum trajectories that can result from a given master equation depend on the measurements that are made on the bath modes. Carmichael refers to these as different unravellings of the master equation.[91] In the case of optical cavity QED there is one class of measurements, other than direct photodetection, that may be implemented in experiments and these are the so-called homodyne and heterodyne detection measurements. In these experiments the light transmitted through the cavity is interfered at a beam splitter with a strong laser field that is at or near the optical frequency of the cavity. The desired signal is then the difference in intensity of the two light beams propagating away from the beam splitter. The strong laser beam is often called a local oscillator by analogy with the radio frequency measurements with which the technique is analogous. In practice the local oscillator beam is drawn from the same laser source as the light driving the cavity so that the experimental set-up is essentially an unbalanced Mach–Zehnder interferometer with the cavity in the arm with lower light intensity.

Making an arbitrary choice of phases for transmission and reflection from the beam splitter, the positive frequency components of the electric field operators of the two output ports of the beam splitter are $\mathbf{E}_1^{(+)}(z_0, t) \equiv [i\mathbf{E}^{(+)}(z_0, t) + e^{i\phi}\mathbf{E}_{\mathrm{LO}}^{(+)}(0, t)]/\sqrt{2}$ and $\mathbf{E}_2^{(+)}(z_0, t) \equiv [\mathbf{E}^{(+)}(z_0, t) + ie^{i\phi}\mathbf{E}_{\mathrm{LO}}^{(+)}(0, t)]/\sqrt{2}$. We will use z' to label position along the axis of the local oscillator beam and assume $z' = 0$ at the beam splitter. The detected intensity at either port of the beam splitter depends on the choice of local oscillator phase ϕ, this gives the measurement access to the transmitted field quadrature amplitude and not just the transmitted intensity. We assume that the local oscillator is in a coherent state and it is convenient to separate the field into coherent and vacuum parts $\mathbf{E}_{\mathrm{LO}}^{(+)}(z', t) = E_{\mathrm{LO}}\mathbf{e}_+ e^{-i(\omega(t-t_0)-kz')} + \delta\mathbf{E}_{\mathrm{LO}}^{(+)}(z', t)$. Physically the position dependence of the local oscillator amplitude corresponds to the need to lock the pathlength of the two arms of the interferometer to be equal, which is the physical content of the fact that $z' = 0$ can be set at the beam splitter. Similarly the choice of overall phase depending on t_0 requires an appropriate constant phase relationship between the local oscillator and transmitted beams, leading to the Mach–Zehnder design of actual experiments. We can choose E_0 to be real since any phase may be absorbed into ϕ. Then in the strong local oscillator limit the observed difference intensity is, by substituting from Eq. (31) into the definition of $\mathbf{E}_1^{(+)}$ and $\mathbf{E}_2^{(+)}$,

$$\mathbf{E}_1^{(-)H} \cdot \mathbf{E}_1^{(+)H} - \mathbf{E}_2^{(-)H} \cdot \mathbf{E}_2^{(+)H} \simeq E_0 E_{\mathrm{LO}}\big(ie^{-i\phi}b_{\mathrm{out}} - ie^{i\phi}b_{\mathrm{out}}^\dagger\big), \quad (41)$$

where terms independent of E_{LO} have been dropped on account of the assumed strength of the local oscillator. The intensity difference operator is written in the Heisenberg picture and the suppressed time arguments on the right-hand side are suitably retarded by the propagation time between the cavity and the detectors both before and after the beam splitter. Notice that the time dependence of the local oscillator field has led to the appearance of the demodulated output fields we defined in the previous section. In heterodyne detection we simply allow the local oscillator phase ϕ to vary linearly with time, that is the local oscillator is detuned from the cavity resonance by a small amount. There are several more careful discussions of the statistics of homodyne and heterodyne detection in the quantum optics literature.[80,81,110,111] However, this treatment is sufficient to see that by measuring the difference current in a homodyne detection experiment the detected quantum mechanical bath observables are of the form $ie^{-i\phi}b_{\text{out}} - ie^{i\phi}b_{\text{out}}^\dagger$. By using Eq. (32) to return to the interaction picture we see that rather than projecting onto number states of the bath modes after the interaction with the system we wish to project onto the observables $ie^{-i\phi}c_t - ie^{i\phi}c_t^\dagger$. Since we are interested in the overall signal to noise and the quantum state conditioned on detected values of the homodyne current we will freely rescale the measured current by an overall factor in the following, starting by dropping the prefactor $E_0 E_{\text{LO}}$.

The forgoing discussion is supposed to show that the idea of the derivation of the homodyne detection master equation is precisely the same as for the case of photon counting. In particular we emphasize that the trajectories correspond to the measurement of suitable observables of the bath and the application of the von Neumann projection postulate. However, there are some technical difficulties associated with the continuous spectrum of possible measurement outcomes in each timestep and the $\Delta t \to 0$ limit. Once again our treatment will have no concern for mathematical rigor, this discussion is based on the argument in the appendix of Ref. 97. Rather than considering projections on the bath observables $ie^{-i\phi}c_t - ie^{i\phi}c_t^\dagger$ directly, it is also possible to derive the homodyne detection master equation by performing the previous derivation for photon counting trajectories corresponding to measurements of $\mathbf{E}_1^{(-)} \cdot \mathbf{E}_1^{(+)}$ and $\mathbf{E}_2^{(-)} \cdot \mathbf{E}_2^{(+)}$ and then to take a strong local oscillator limit in which many photodetection events take place in any timestep and thus the Poissonian statistics of the photocounts approach an appropriate Gaussian distribution. There are versions of this argument in both the mathematical physics[85] and the quantum optics literature.[91,111]

Once again we wish to project the system-bath state $R(t + \Delta t)$ after the interaction at t, given by Eq. (35), onto eigenstates of the bath. Since we want to project onto eigenstates of the position it will be useful to use the fact that μ is the vacuum state

$$c_t\mu = 0 = \mu c_t^\dagger$$

to write $\Delta R = R(t + \Delta t) - R(t)$ in terms of $X = ie^{-i\phi}c_t - ie^{i\phi}c_t^\dagger$,

$$\Delta R = \sqrt{2\kappa_{oc}}\left(e^{-i\phi}a\rho(t) \otimes X|0\rangle\langle 0| + \rho(t)e^{i\phi}a^\dagger \otimes |0\rangle\langle 0|X\right)\sqrt{\Delta t}$$
$$+ \kappa_{oc}\left(2a\rho(t)a^\dagger \otimes X|0\rangle\langle 0|X - a^\dagger a\rho(t) \otimes |0\rangle\langle 0| - \rho(t)a^\dagger a \otimes |0\rangle\langle 0|\right)\Delta t.$$

It is now possible to investigate the statistics of the homodyne measurement. We will label the eigenvalues of $X = ie^{-i\phi}c_t - ie^{i\phi}c_t^\dagger$ by x. The eigenstates will have the property that $X|x\rangle = x|x\rangle$. Recall that the measured value x results from a difference of the light intensity from the two output channels of the beamsplitter in the homodyne detection set-up. So x relates directly to the difference in the charge released in the two photodetectors by the incident light field during the time Δt. For this reason we define the scaled charge increment $\Delta Q = \sqrt{\kappa_{oc}\Delta t}\,x$. By taking a trace it is easy to evaluate the mean value of this signal

$$\mathrm{E}[\Delta Q] = \sqrt{2\kappa_{oc}\Delta t}\,\mathrm{Tr}\left[X R(t + \Delta t)\right] = 2\kappa_{oc}\Delta t\langle e^{-i\phi}a + e^{i\phi}a^\dagger\rangle.$$

We will use the symbol $\mathrm{E}[]$ to indicate a classical average over the measurement results and reserve angular brackets to indicate quantum mechanical expectation values. This mean value shows clearly that the measurement gives information about the quadrature observables of the cavity field. However for small Δt this signal is swamped in noise which is to do with the uncertainty of the ground state wavefunction of bath oscillator

$$\mathrm{E}\left[(\Delta Q)^2\right] = \kappa_{oc}\Delta t\,\mathrm{Tr}\left[X^2 R(t + \Delta t)\right] = 2\kappa_{oc}\Delta t.$$

This is a reflection of the fact that optical photocurrents have an approximately white noise background which may be associated with the vacuum fluctuations of the electromagnetic field.

Using the projector $P_x = I_S \otimes |x\rangle\langle x|$ we can find the state corresponding to the measurement result ΔQ,

$$P_x R(t + \Delta t)P_x = |\langle 0|x\rangle|^2 \left[\rho + \Delta Q\left(e^{-i\phi}a\rho + \rho e^{i\phi}a^\dagger\right) + \Delta Q^2 a\rho a^\dagger\right.$$
$$\left. -\kappa_{oc}\left(a^\dagger a\rho + \rho a^\dagger a\right)\Delta t\right] \otimes |x\rangle\langle x|.$$

Given that the distribution $|\langle 0|x\rangle|^2$ is a Gaussian of mean zero and variance one, it is straightforward to show that by averaging over the measurement results we recover the master equation as was the case for photon counting trajectories.

The probability distribution for x is

$$p(x) = \mathrm{Tr} P_x R(t + \Delta t) P_x$$
$$= |\langle 0|x\rangle|^2 \left(1 + \sqrt{2\kappa_{oc}\Delta t}x\langle e^{-i\phi}a + e^{i\phi}a^\dagger\rangle + 2\kappa_{oc}\left(x^2 - \Delta t\right)\langle a^\dagger a\rangle\right)$$

and as $\Delta t \to 0$ this approaches a Gaussian distribution with mean $\sqrt{2\kappa_{oc}\Delta t}\langle e^{-i\phi}a + e^{i\phi}a^\dagger\rangle$ and variance 1. So for sufficiently small Δt we can regard ΔQ as a Gaussian of mean $2\kappa_{oc}\Delta t\langle e^{-i\phi}a + e^{i\phi}a^\dagger\rangle$ and variance $2\kappa_{oc}\Delta t$. So we can write

$$\Delta Q = 2\kappa_{oc}\Delta t\langle e^{-i\phi}a + e^{i\phi}a^\dagger\rangle + \sqrt{2\kappa_{oc}}\Delta W$$

were ΔW is of mean zero and variance Δt. It is important to bear in mind that ΔQ is the measurement result directly available to the experimenter and ΔW can be computed from ΔQ and the current system state $\rho(t)$. In the language of control theory ΔW is the innovation or residual, that describes the degree to which the measurement result is inconsistent with previous information about the system. Also note that due to its root mean square amplitude we should scale ΔW like $\sqrt{\Delta t}$ and retain terms up to second-order in ΔW and first-order in Δt.

We can define an unnormalized state conditioned on the measurement result ΔQ

$$\bar{\rho}_{\Delta Q}(t + \Delta t) \otimes |x\rangle\langle x| \equiv P_x R(t + \Delta t) P_x / |\langle 0|x\rangle|^2$$
$$\bar{\rho}_{\Delta Q}(t + \Delta t) = \rho + \Delta Q \left(e^{-i\phi}a\rho - \rho e^{i\phi}a^\dagger\right) + \Delta Q^2 a\rho a^\dagger$$
$$- \kappa_{oc}\left(a^\dagger a\rho + \rho a^\dagger a\right)\Delta t$$
$$= \rho + \sqrt{2\kappa_{oc}}\Delta W \left(e^{-i\phi}a\rho + \rho e^{i\phi}a^\dagger\right)$$
$$+ 2\kappa_{oc}\Delta t\langle e^{-i\phi}a + e^{i\phi}a^\dagger\rangle\left(e^{-i\phi}a\rho + \rho e^{i\phi}a^\dagger\right)$$
$$+ \kappa_{oc}\left(2a\rho a^\dagger - a^\dagger a\rho - \rho a^\dagger a\right)\Delta t$$
$$+ 2\kappa_{oc}\left(\Delta W^2 - \Delta t\right)a\rho a^\dagger.$$

In order to normalize the state, note that

$$\mathrm{Tr}\bar{\rho}_{\Delta Q}(t + \Delta t) = 1 + \sqrt{2\kappa_{oc}}\Delta W\langle e^{-i\phi}a + e^{i\phi}a^\dagger\rangle + 2\kappa_{oc}\Delta t\langle e^{-i\phi}a + e^{i\phi}a^\dagger\rangle^2$$
$$+ 2\kappa_{oc}\left(\Delta W^2 - \Delta t\right)\langle a^\dagger a\rangle$$

so that

$$\rho_{\Delta Q} = \bar{\rho}_{\Delta Q}/\mathrm{Tr}\bar{\rho}_{\Delta Q}$$
$$= \rho + \sqrt{2\kappa_{oc}}\Delta W \mathcal{H}[e^{-i\phi}a]\rho + \kappa_{oc}\Delta t\left(2a\rho a^\dagger - a^\dagger a\rho - \rho a^\dagger a\right)$$
$$+ (\Delta W^2 - \kappa_{oc}\Delta t)(a\rho a^\dagger - \langle a^\dagger a\rangle\rho + \langle e^{-i\phi}a + e^{i\phi}a^\dagger\rangle\mathcal{H}[e^{-i\phi}a]\rho).$$

$$(42)$$

This recursion relation is the analog of Eqs. (36) and (38) we derived for photon counting. In this case also the nonlinearity in the equations is solely due to normalizing the density matrix after the measurement. All that remains is to consider the limit $\Delta t \to 0$. If we consider iterating this relation for a finite time τ, taking the limit requires taking more and more steps Δt. It turns out that in this limit not only the mean value but also the variance of the terms involving $\Delta W^2 - \Delta t$ goes to zero. This results from the relations between the variance and the higher order moments that are obeyed by the Gaussian distribution of ΔW.[112] So the final line of Eq. (42) has zero mean square in this limit and can be disregarded.

As in the case of photon counting it is usual to adopt the language and notation of stochastic differential equations when considering homodyne detection quantum trajectories. So we replace ΔW by the Wiener increment dW and of course dW has mean zero so $E[dW] = 0$. From the mean square limit of the previous paragraph we get the property $dW^2 = dt$ appropriate to the Itô stochastic calculus.[112,113] Just as we did in the recursion relations we must always consider terms up to dW^2 in any expression, so these stochastic differential equations do not obey some of the standard rules of calculus such as the chain rule. This results from the fact that the solutions to stochastic differential equations are continuous but not in fact differentiable. We then obtain the following stochastic master equation

$$d\rho_{dQ} = \kappa_{oc} \left(2a\rho a^\dagger - a^\dagger a\rho - \rho a^\dagger a \right) dt + \sqrt{2\kappa_{oc}} \mathcal{H}[e^{-i\phi}a]\rho dW. \qquad (43)$$

As before this equation preserves trace and positivity of the density matrix and averages to give the master equation. We can formally define the measurement current

$$I(t) = \frac{dQ}{dt} = 2\kappa_{oc}\langle e^{-i\phi}a + e^{i\phi}a^\dagger\rangle + \sqrt{2\kappa_c}\frac{dW}{dt} \qquad (44)$$

so the quadrature field amplitude is recovered in a white noise background. We see that the sensitivity for detecting the quadrature amplitude is $1/\sqrt{2\kappa_{oc}}$. Recall that in each timestep it is the current increment dQ which is directly available in the experiment and the innovation dW is calculated by Eq. (44) using $\rho(t)$. In order to really describe the state conditioned on measurement it is necessary to deal with issues such as detector efficiency and bandwidth, see Refs. 99–101.

Again the case of a coherent state in the cavity is an important special case. Since $a|\alpha\rangle = \alpha|\alpha\rangle$ the conditioned state is in fact independent of unravelling and measurement outcomes, remaining a coherent state at all times, so we get $|\psi(t)\rangle = |\alpha e^{-\kappa_{oc}t}\rangle$ and the measurement current is $dQ(t) = 4\kappa\text{Re}[e^{-i\phi}\alpha e^{-\kappa_{oc}t}]dt + \sqrt{2\kappa}dW(t)$.

Once again, if the initial state is pure it remains so at all times. It is possible to show that an initial pure state evolves according to the (unnormalized) stochastic Schrödinger equation

$$d|\psi\rangle_{dQ} = \left(-\kappa_{oc} a^\dagger a \, dt + dQa\right)|\psi\rangle.$$ (45)

The normalized version can be calculated in a straightforward but tedious calculation applying the Itô calculus. Again this equation can be used to simulate the dynamics according to a master equation by drawing dW at random for a series of pure state trajectories and then averaging over trajectories. As before the pure state unravelling is possible only if there are no additional couplings to the bath, in the cavity QED model we are considering the spontaneous emission term will lead to mixed states and require the solution of the stochastic master equation. Although we may write down pure state trajectories that unravel the coupling to both the transmitted field and the light modes outside the cavity that cause spontaneous emission, such unravellings have the status of simulations rather than representing information available in any given experimental run.

Finally in heterodyne detection, as described above, the phase $\phi = \Delta_{\text{LO}} t$ is a linearly increasing function of time. Nothing in our derivation is changed so Eq. (43) still holds. It is most useful though to consider the limit in which the detuning of the local oscillator is larger than the dynamical rates appearing in the master equation.[114] In this limit we may coarse grain over times $\Delta t'$ that are very short compared to the dynamical timescales but very long compared to $1/\Delta_{\text{LO}}$ the period of the beat note between the local oscillator and the cavity field. This leads to the replacement

$$dW \;\rightarrow\; \int_0^{\Delta t'} \cos(\Delta_{\text{LO}} t') dW(t') + i \int_0^{\Delta t'} \sin(\Delta_{\text{LO}} t') dW(t')$$
$$\equiv (\Delta W_x(t) + i\Delta W_y(t))/\sqrt{2}.$$

In the limit $\Delta t' \gg 1/\Delta_{\text{LO}}$ we find that the new noise variables are again Wiener increments satisfying $\mathrm{E}[\Delta W_x] = 0 = \mathrm{E}[\Delta W_y]$ and $\mathrm{E}[(\Delta W_x)^2] = \Delta t' = \mathrm{E}[(\Delta W_y)^2]$. Then taking the limit $\Delta t' \ll 1/\kappa_{oc}$ gives the heterodyne detection stochastic master equation

$$d\rho_{dQ} = \kappa_{oc}\left(2a\rho a^\dagger - a^\dagger a\rho - \rho a^\dagger a\right) dt + \sqrt{\frac{1}{2}}\left(\mathcal{H}[a]\rho dW_x + \mathcal{H}[-ia]\rho dW_y\right).$$ (46)

Once again, as long as no additional environmental couplings are added to the master equation, an initial pure state will remain pure under the heterodyne stochastic master equation (46).

The relevant parts of the difference photocurrent in the heterodyne experiment are at frequencies close to Δ_{LO} and the demodulated signal has both sine and cosine components given by

$$I_x(t) \equiv \frac{1}{\Delta t'} \int_0^{\Delta t'} \cos(\Delta_{\mathrm{LO}} t') I(t') dt'$$

$$I_y(t) \equiv \frac{1}{\Delta t'} \int_0^{\Delta t'} \sin(\Delta_{\mathrm{LO}} t') I(t') dt'.$$

In the limit $\Delta t' \gg 1/\Delta_{\mathrm{LO}}$ we get

$$I_x(t) = \kappa_{\mathrm{oc}} \langle a + a^\dagger \rangle + \sqrt{\kappa_{\mathrm{oc}}} \frac{dW_x}{dt}$$

$$I_y(t) = \kappa_{\mathrm{oc}} \langle -ia + ia^\dagger \rangle + \sqrt{\kappa_{\mathrm{oc}}} \frac{dW_y}{dt}.$$

By cycling through the different local oscillator phases, it is possible to extract information about both quadratures of the cavity field. However, the sensitivity with which the quadratures are measured is now $1/\sqrt{\kappa_{\mathrm{oc}}}$, reflecting the need for longer averaging time as compared to homodyne detection to get the same signal to noise for any given quadrature. The extra noise results from measuring two noncommuting observables of the cavity mode simultaneously.

Simultaneous recovery of orthogonal quadrature amplitudes has been demonstrated in an experiment by Mabuchi et al.[72] Using laser-cooled atoms, they were able to track changes in the optical amplitude and phase response of the atom–cavity system as g varied during the transit of a single atom through the cavity. "Phasor" plots of the complex-valued response versus g exhibit distinctly quantum-mechanical features of the cavity QED master equation. The ability to track both phase and amplitude of the transmitted optical field during atom transits also made it possible to fully explore the crossover from absorptive to dispersive coupling in cavity QED. For a driving field always resonant with the cavity mode, the influence of an intracavity atom on the cavity transmission is mainly in the amplitude quadrature when the cavity resonance coincides (in frequency) with the atomic resonance, but moves to the phase quadrature with increasing atom-cavity detuning.

Atom detection and position monitoring via cavity QED are most sensitive under resonant conditions, but then the atomic motion is strongly heated by absorption-emission recoils and by dipole force fluctuations. With sufficiently large ($\gg \gamma_\perp$) detuning of the cavity mode from atomic resonance, however, motional heating is suppressed and the atomic position measurement can have minimal excess back-action.

Homodyne or heterodyne measurement of the phase of the transmitted field is also of interest for proposed experiments on phase "bistability" in cavity QED with strong coupling.[115,26] Here one expects to observe a stochastic switching between two metastable phases for the transmitted field, where the switching is associated with spontaneous emission events of the intracavity atom. As noted in Ref. 26, these switching events appear to have a curious "retroactive" nature when viewed in the context of quantum trajectory theory; this analysis of single atom phase bistability highlights intriguing possibilities for utilizing cavity QED to explore fundamental issues in quantum measurement theory.

3.5. *Quantum Feedback Control*

The quantum trajectory theory that we discussed in the previous section has the great advantage that it is straightforward to describe experiments with feedback. Such experiments have fundamental interest because any description of feedback or control of quantum mechanical systems must properly take into account the backaction noise in designing and assessing the performance of the servo. These backaction effects are expected to be significant when the sensitivity of the measurement is such that the conditioned state given by the appropriate stochastic master equation has very low entropy. That is we are interested in situations where the noise on the measurement signal is quantum-limited, hence the term "quantum feedback control." For the class of experiments we envisage it is also necessary to feed back onto the dynamics of the single quantum system being measured in real time. This is in distinction to experiments that are sometimes termed quantum feedback in the literature where ensembles of systems are measured and the system is reset after each measurement meaning that backaction effects are not relevant. Experiments in cavity QED are uniquely placed to realize this new and interesting regime of control.

The performance of classical technology such as cars and aeroplanes is guaranteed through the use of feedback control and there is a very advanced general theory of control of classical systems.[116,117] An analog theory based on stochastic master equations has to some extent been developed. In particular, Belavkin was the first to describe quantum feedback in this way see Refs. 86 and 118 and references therein. There were independent developments in the quantum optics literature. By taking the continuous limit of a process in which a position measurement is made and then a displacement operator that depends on the measurement result is applied to the system, Caves and Milburn derived a master equation describing the continuous feedback of a position measurement record.[89] Wiseman and Milburn developed a more general theory of quantum limited feedback for Markovian open quantum systems.[97,119,120]

Wiseman and Milburn wished to be able to derive a Markovian master equation for the evolution of the quantum system alone including the effects of the feedback loop. This drove them to consider the limit of instantaneous, infinite bandwidth feedback of the photocurrent. They also specialized to the case of linear feedback, since in this Markov limit it is hard to regularize nonlinear functions of the photocurrent I. In effect an extra term is added to the system Hamiltonian which depends on the measured photocurrent I

$$H_{\text{fb}}(t) = I(t - \tau)F, \tag{47}$$

for some operator F, in the limit $\tau \to 0$. The effects of finite delay or of a finite feedback bandwidth can be taken into account, particularly for linear systems, see for example Refs. 120–122.

In the work of Wiseman and Milburn the feedback equations take slightly different forms depending on whether it is photon-counting or homodyne detection that is being performed. In measurements based on photon-counting the experimenter turns on the Hamiltonian

$$H_{\text{fb}}(t) = \hbar \lambda F, \tag{48}$$

for time $1/\lambda$ and the limit that λ is much larger than any system frequency is taken. (Wiseman in fact considers the more general case of evolution under a master equation during the time $1/\lambda$.) So if there is a photodetection the resulting instantaneous time evolution may be written $U_{\text{fb}} = \exp[-iFdN]$. Recalling that the stochastic increment is $dN = 1$ when there is a photodetection and zero otherwise, we get from (40)

$$\rho_{\text{c}}(t + dt) = dN(t)U_{\text{fb}}\left(\rho_{\text{c}}(t) + \mathcal{G}[a]\rho_{\text{c}}(t)\right)U_{\text{fb}}^{\dagger} - \kappa_{\text{oc}}dt\mathcal{H}[a^{\dagger}a]\rho_{\text{c}}(t). \tag{49}$$

This equation just expresses the idea that if there is a photon detection, the feedback evolution U_{fb} takes place immediately afterwards. In this Markovian example it is straightforward to average over the measurement results using $\mathrm{E}[dN(t)] = 2\kappa_{\text{oc}}\langle a^{\dagger}a(t)\rangle dt$ and the definitions of $\mathcal{G}$ and $\mathcal{H}$ to find

$$\frac{d}{dt}\rho(t) = 2\kappa_{\text{oc}}U_{\text{fb}}a\rho a^{\dagger}U_{\text{fb}}^{\dagger} - \kappa_{\text{oc}}\left(a^{\dagger}a\rho + \rho a^{\dagger}a\right). \tag{50}$$

The case of homodyne detection is similar. Now the appropriate feedback in each measurement step is $U_{\text{fb}} = \exp[-iFdQ]$ and we have

$$\rho_c(t + dt) = U_{\text{fb}} \left(\rho_c(t) + d\rho_{dQ} \right) U_{\text{fb}}^\dagger.$$

It is important that although we are in the Markov limit the feedback always acts after the evolution due to measurement. Substituting from the equation for homodyne detection (43) an explicit formula for $d\rho_c$ is found by expanding the exponentials and using the Itô rule $dQ^2 = dW^2 = dt$. When averaged over the measurement results we obtain the homodyne detection feedback master equation

$$\frac{d}{dt}\rho(t) = \kappa_{\text{oc}} \left(2a\rho a^\dagger - a^\dagger a\rho - \rho a^\dagger a \right) - i\sqrt{2\kappa_{\text{oc}}}[F, a\rho + \rho a^\dagger] + F\rho F - \rho. \quad (51)$$

This theory has been applied to noise reduction in a number of systems in quantum optics. Feedback had already been used to reduce the noise of a homodyne photocurrent below the shot noise level[123,124] by negative feedback of the photocurrent onto the light beam. The theory of Wiseman and Milburn is able to describe these experiments and the observed fact that any linear element, such as a beam splitter, couples out light with noise above the shot-noise level.[125] A free-running beam with noise below the shot-noise level in one quadrature can be generated by feedback if the measurement is made on one of two beams from a parametric oscillator with correlated noise and the feedback is onto the other beam. This system has been realized experimentally.[126,127] The recent teleportation experiment of Furusawa *et al.*[128] also uses instantaneous feedback of homodyne photocurrents on entangled beams from a parametric oscillator. A nonlinear element, such as a two-level atom, can couple nonclassical light out of the feedback loop. For example, the spontaneous emission spectrum of an atom inside the "squashed" in-loop light field will be altered in very similar ways to the spontaneous emission spectrum inside a squeezed vacuum.[122]

There are several proposals to use quantum limited feedback of this kind to control the motion of atoms conditioned on optical measurements of their position in cavity QED. Dunningham *et al.*[129] used the information about the atomic position in the cavity output field of a cavity QED system to modify the cavity driving and thus alter the effective potential seen by the atom with the aim of creating a potential which is nearly harmonic. Mancini *et al.* used a similar feedback scheme taking an analytic approach in the Raman–Nath regime with the aim of improving the atomic localization where cavity decay, while the atom is passing through the cavity, cannot be ignored.[130]

Direct feedback of the measured photocurrent of the kind we have been discussing is rather limiting compared to classical controller synthesis and analysis techniques in which a linear controller may have a nontrivial frequency response. One approach to controller design is to divide the controller into an estimation and a control phase. In the case of quantum feedback control this approach corresponds to a controller that integrates (an approximation of) the stochastic master equation in real time and bases the control Hamiltonian on the current conditioned state. Recall that the conditioned state may be calculated from the initial state of the system and the measurement results so the limitation here is the speed of the classical computation of the appropriate solution to the stochastic master equation. This possibility was discussed in the work of Belavkin[86,118] and in Refs. 131 and 132 and corresponds to classical optimal control techniques. In fact for linear quantum systems (such as position measurement of an oscillator) there is a direct mapping onto the so-called Linear, Quadratic, Gaussian problem of classical control theory.[133,134] There is a comparison of these instantaneous and estimation based control strategies for a particular system in Ref. 135. A similar idea is present in adaptive phase measurement[136] where an estimate of the phase of an optical beam is fed back to the local oscillator phase in a homodyne detection resulting in reduced noise in measurement of optical phase. This has recently been demonstrated experimentally.[137]

3.6. Semi-Classical Limits and the Quantum–Classical Transition

While one of the main attractions of cavity QED is the possibility of creating highly nonclassical states for the atom and intracavity field, there has long been comparable interest in understanding its "semi-classical" limit. In this limit, it is presumed permissible to treat the atomic internal degrees of freedom quantumly (the atomic center-of-mass is typically ignored), while the state of the cavity field is essentially replaced by a c-number corresponding to the (time-dependent) complex amplitude of a coherent state. Many early theoretical treatments of the single mode atom–cavity system were motivated by investigations of optical bistability,[2] in which the intracavity atoms are thought of as an optically-active medium whose dynamical role is mainly to induce nonlinear input–output maps for the complex amplitude of a laser field.

While optical nonlinearity necessarily implies atom–field entanglement in systems with strong coupling, it can be shown that systems with many weakly coupled atoms can achieve a high degree of input–output nonlinearity with only minor perturbations to the coherent-state character of the transmitted field.[138] In the early days of experimental cavity QED large numbers of intracavity atoms were required to achieve strong coupling, but even seminal works in the field recognized the interest of exploring the transition from semi-classical to fully quantum behavior with decreasing atom number. The first theoretical explorations of this modeling transition were based on the cavity QED Master Equation, and mechanisms for the preservation of coherent-state field character were analyzed using a type of mean field limit that assumes large atom number.[139] Such results established rigorous connections between the open quantum systems perspective and existing semi-classical models, based on self-consistent solution of the optical Bloch equations and Maxwell's equations with periodic boundary conditions.[2]

Since the advent of quantum trajectory theory, however, it has become possible to ask whether qualitative features of the semi-classical model are preserved in single-atom scenarios with strong coupling, but where a coherent-state character for the field is induced by continuous observation via heterodyne or homodyne photodetection. (Studies[115] have so far also assumed conditions of strong driving, such that the average intracavity photon number is kept large compared to the critical photon number.) As mentioned above, such scenarios have recently become experimentally realizable, and this measurement-based perspective has motivated studies such as Refs. 26 and 27. Current work focuses on the bifurcation set of the optical input–output map as an aspect of cavity QED phenomenology that could plausibly be robust to the differences between a mean field limit and measurement-induced conditioning as mechanisms for limiting the intracavity field to coherent states.

Tight correspondence between the semi-classical Maxwell–Bloch equations and the heterodyne/homodyne stochastic Master Equation would provide an interesting case study of the emergence of classical phenomena in quantum systems. In the former case, "sharp" trajectories for the intracavity field appear to result from a mean-field limit in which fluctuations become negligible as the system size is increased; the derivation is statistical in nature. In the latter case, localization of the field state is a result of conditioning due to continuous observation, and the reduction from a quantum to a classical description of the field relies essentially on quantum measurement theory. It seems natural to ask whether the emergence of classical behavior might depend generically on the existence of a mechanism for localizing trajectories in phase space,[140] and whether the primary features of the emergent dynamics could be independent of the details of this mechanism. At a more formal level, it is interesting to note that the passage from a linear model (the usual cavity Master Equation, which always has a unique and stationary steady-state) to a nonlinear model (the Maxwell Bloch equations, which contain multistability and nontrivial attractors) is accomplished in the traditional setting via a quasi-classical representation of the field degrees of freedom plus a Gaussian *ansatz*. In the quantum trajectory approach, however, nonlinearity enters the evolution equations through terms associated with conditioning and the model necessarily takes on a stochastic character. It will be interesting to elucidate the relation between these two perspectives on the emergence of nonlinear dynamics (such as bifurcation and limit cycles) in cavity QED, and to explore their experimental predictions.

4. Conclusions

The open quantum systems theory described in this chapter provides a firm basis for the formulation of proposals to utilize strong-coupling cavity QED in novel scientific and technological applications. Many recent proposals have focused on nonequilibrium statistical mechanics (including decoherence and quantum measurement theory), or on quantum information science. During the next few years it will be interesting to see which of these proposals can be brought to fruition in the laboratory, and also to see which aspects of the existing theoretical and experimental infrastructure of cavity QED can be adapted to new systems in the solid state.

Acknowledgments

The authors wish to thank Jeff Kimble and the late Dan Walls for introducing them to the subject of cavity QED, and for guiding them through their early research in the field. They also want to acknowledge Howard Wiseman for numerous enlightening discussions.

References

1. K. H. Drexhage, *Prog. Opt.* **12**, 163, 1974.
2. L. A. Lugiato, *Prog. Opt.* **21**, 69, 1984.
3. E. A. Hinds, *Adv. Atom. Mol. Opt. Phys.* **28**, 237, 1990.
4. C. Santori *et al.*, *Nature* **419**, 594, 2002.
5. G. Khitrova *et al.*, *Rev. Mod. Phys.* **71**, 1591, 1999.

6. R. Blatt, J. I. Cirac, A. S. Parkins and P. Zoller, *Phys. Scripta* **T59**, 294, 1995.
7. S. Girvin and R. Schoelkopf, A. Blais *et al.*, *Phys. Rev.* **A69** 062320, 2004.
8. R. G. Hulet, E. S. Hilfer and D. Kleppner, *Phys. Rev. Lett.* **55**, 2137, 1985.
9. H. J. Kimble, *Phys. Scripta* **T76**, 127, 1998.
10. S. Haroche, *Philos. T. Roy. Soc.* **A361**, 1339, 2003.
11. H. Walther, *Adv. Chem. Phys.* **122**, 167, 2002.
12. C. Greiner, T. Wang, T. Loftus and T. W. Mossberg, *Phys. Rev. Lett.* **87**, 253602, 2001.
13. M. Keller *et al.*, *Appl. Phys.* **B76**, 125, 2003.
14. J. Eschner *et al.*, *Fortschr. Phys.* **51**, 359, 2003.
15. H. Mabuchi and A. C. Doherty, *Science* **298**, 1372, 2002.
16. P. R. Berman (ed.), *Cavity Quantum Electrodynamics*, Academic Press, San Diego, 1994.
17. R. J. Thompson, G. Rempe and H. J. Kimble, *Phys. Rev. Lett.* **68**, 1132, 1992.
18. F. Bernardot *et al.*, *Europhys. Lett.* **17**, 33, 1992.
19. Q. A. Turchette *et al.*, *Phys. Rev. Lett.* **75**, 4710, 1995.
20. M. Brune *et al.*, *Phys. Rev. Lett.* **76**, 1800, 1996.
21. C. J. Hood, M. S. Chapman, T. W. Lynn and H. J. Kimble, *Phys. Rev. Lett.* **80**, 4157, 1998.
22. H. Mabuchi, Q. A. Turchette, M. S. Chapman and H. J. Kimble, *Opt. Lett.* **21**, 1393, 1996.
23. C. J. Hood *et al.*, *Science* **287**, 1447, 2000.
24. J. McKeever *et al.*, *Phys. Rev. Lett.* **90**, 133602, 2003.
25. A. C. Doherty, A. S. Parkins, S. M. Tan and D. F. Walls, *Phys. Rev.* **A57**, 4804, 1998.
26. H. Mabuchi and H. M. Wiseman, *Phys. Rev. Lett.* **81**, 4620, 1998.
27. M. A. Armen and H. Mabuchi, in preparation, 2004.
28. C. D. Fidio and W. Vogel, *J. Opt. B-Quantum S. O.* **5**, 105, 2003.
29. L. M. Duan and H. J. Kimble, *Phys. Rev. Lett.* **90**, 253601, 2003.
30. L. K. Thomsen, S. Mancini and H. M. Wiseman, *J. Phys.* **B35**, 4937, 2002.
31. J. E. Reiner, H. M. Wiseman and H. Mabuchi, *Phys. Rev.* **A67**, 042106, 2003.
32. S. Zippilli, D. Vitali, P. Tombesi and J. M. Raimond, *Phys. Rev.* **A67**, 052101, 2003.
33. D. J. Berkeland *et al.*, *Phys. Rev. Lett.* **80**, 2089, 1998.
34. A. S. Parkins *et al.*, *Phys. Rev.* **A51**, 1578, 1995.
35. J. I. Cirac, P. Zoller, H. J. Kimble and H. Mabuchi, *Phys. Rev. Lett.* **78**, 3221, 1997.
36. S. J. van Enk, J. I. Cirac and P. Zoller, *Phys. Rev. Lett.* **78**, 4293, 1997.
37. W. Dur, H. J. Briegel, J. I. Cirac and P. Zoller, *Phys. Rev.* **A59**, 169, 1999.
38. T. Pellizzari, S. A. Gardiner, J. I. Cirac and P. Zoller, *Phys. Rev. Lett.* **75**, 3788, 1995.
39. H. Mabuchi *et al.*, *Quant. Inform. Comput.* **1**, 7, 2001.
40. C. Ahn, A. C. Doherty and A. J. Landahl, *Phys. Rev.* **A65**, 042301, 2002.

41. C. Cohen-Tannoudji, J. Dupont-Roc and G. Grynberg, *Photons and Atoms: Introduction to Quantum Electrodynamics*, Wiley, New York, 1989.
42. C. W. Gardiner and P. Zoller, *Quantum Noise*, 2nd ed., Springer-Verlag, Berlin, 2000.
43. D. F. Walls and G. J. Milburn, *Quantum Optics*, Springer-Verlag, Berlin, 1994.
44. H. J. Carmichael, *Statistical Methods in Quantum Optics 1: Master Equations and Fokker-Plank Equations*, Springer-Verlag, Berlin, 1999.
45. A. O. Caldeira and A. J. Leggett, *Physica* **A121**, 587, 1983.
46. A. S. Holevo, *Statistical Structure of Quantum Theory*, Springer-Verlag, Berlin, 2001.
47. C. W. Gardiner and M. J. Collett, *Phys. Rev.* **A31**, 3761, 1985.
48. B. J. Dalton, S. M. Barnett and P. L. Knight, *J. Mod. Opt.* **46**, 1107, 1999.
49. B. J. Dalton, S. M. Barnett and P. L. Knight, *J. Mod. Opt.* **46**, 1315, 1999.
50. B. J. Dalton and P. L. Knight, *J. Mod. Opt.* **46**, 1817, 1999.
51. G. Barton, *Proc. R. Soc. Lond.* **A453**, 2461, 1997.
52. R. Graham and H. Haken, *Z. Phys.* **213**, 420, 1968.
53. I. H. Deutsch and J. C. Garrison, *Phys. Rev.* **A43**, 2498, 1991.
54. J. Anglin and S. Habib, *Mod. Phys. Lett.* **A11**, 2655, 1996.
55. M. R. da Costa, A. O. Caldeira, S. M. Dutra and H. W. Jr., *Phys. Rev.* **A61**, 022107, 2000.
56. G. W. Ford and R. F. O'Connell, *Phys. Rev.* **D64**, 105020, 2001.
57. W. H. Louisell, *Quantum Statistical Properties of Radiation*, Wiley, New York, 1973.
58. F. Haake, *Quantum Statistics in Optics and Solid-State Physics*, Springer Tracts in Modern Physics, Vol. 66, Springer-Verlag, Berlin, 1973, p. 117.
59. I. R. Senitzky, *Phys. Rev.* **119**, 670, 1960.
60. I. R. Senitzky, *Phys Rev.* **124**, 642, 1961.
61. G. Lindblad, *Comm. Math. Phys.* **48**, 119, 1976.
62. J. Dalibard and C. Cohen-Tannoudji, *J. Phys.* **B18**, 1661, 1985.
63. J. Javanainen and S. Stenholm, *Appl. Phys.* **21**, 35, 1980.
64. D. Kleppner, *Phys. Rev. Lett.* **47**, 233, 1981.
65. H. J. Carmichael and D. F. Walls, *J. Phys.* **A6**, 1551, 1973.
66. R. Alicki, M. Horodecki, P. Horodecki and R. Horodecki, *Phys. Rev.* **A65**, 062101, 2002.
67. G. Rempe, R. J. Thompson and H. J. Kimble, *Phys. Rev. Lett.* **67**, 1727, 1991.
68. Q. A. Turchette, R. J. Thompson and H. J. Kimble, *Appl. Phys.* **B60**, S1, 1995.
69. R. J. Thompson, Q. A. Turchette, O. Carnal and H. J. Kimble, *Phys. Rev.* **A57**, 3084, 1998.
70. P. W. H. Pinkse, T. Fischer, P. Maunz and G. Rempe, *Nature* **404**, 365, 2000.
71. J. Ye, D. W. Vernooy and H. J. Kimble, *Phys. Rev. Lett.* **83**, 4987, 1999.
72. H. Mabuchi, J. Ye and H. J. Kimble, *Appl. Phys.* **B68**, 1095, 1999.

73. A. C. Doherty, A. S. Parkins, S. M. Tan and D. F. Walls, *J. Opt. B-Quant. S. O.* **1**, 475, 1999.

74. H. J. Kimble *et al.*, in preparation, 2003.

75. D. W. Vernooy and H. J. Kimble, *Phys. Rev.* **A55**, 4287, 1997.

76. R. Quadt, M. Collett and D. F. Walls, *Phys. Rev. Lett.* **74**, 351, 1995.

77. P. Munstermann *et al.*, *Phys. Rev. Lett.* **82**, 3791, 1999.

78. A. C. Doherty and K. Jacobs, *Phys. Rev.* **A60**, 2700, 1999.

79. C. W. Gardiner, A. S. Parkins and P. Zoller, *Phys. Rev.* **A46**, 4363, 1992.

80. M. J. Collett, R. Loudon and C. W. Gardiner, *J. Mod. Opt.* **34**, 881, 1987.

81. H. J. Carmichael, *J. Opt. Soc. Am.* **B4**, 1588, 1987.

82. E. B. Davies, *Quantum Theory of Open Systems*, Academic, London, 1976.

83. A. Barchielli, L. Lanz and G. M. Prosperi, *Nuovo Cimento* **72B**, 79, 1982.

84. A. Barchielli, *Phys. Rev.* **A34**, 1642, 1986.

85. A. Barchielli and V. P. Belavkin, *J. Phys.* **A24**, 1495, 1991.

86. V. P. Belavkin, in *Modelling and Control of Systems*, ed. A. Blaquière, Springer, Berlin, 1988, pp. 245–265.

87. V. P. Belavkin, *Phys. Lett.* **A140**, 355, 1989.

88. V. P. Belavkin, *J. Multivariate Anal.* **42**, 171, 1992.

89. C. M. Caves and G. J. Milburn, *Phys. Rev.* **A36**, 5543, 1987.

90. L. Diósi, *Phys. Lett.* **A129**, 419, 1988.

91. H. Carmichael, *An Open Systems Approach to Quantum Optics*, Springer-Verlag, Berlin, 1993.

92. K. Mølmer, Y. Castin and J. Dalibard, *J. Opt. Soc. Am.* **B10**, 524, 1993.

93. R. J. Cook and H. J. Kimble, *Phys. Rev. Lett.* **54**, 1023, 1985.

94. C. Cohen-Tannoudji and J. Dalibard, *Europhys. Lett.* **1**, 441, 1986.

95. P. Zoller, M. Marte and D. F. Walls, *Phys. Rev.* **A35**, 198, 1987.

96. M. B. Plenio and P. L. Knight, *Rev. Mod. Phys.* **70**, 101, 1998.

97. H. M. Wiseman, Ph.D. thesis, University of Queensland, 1994.

98. T. Brun, *Am J. Phys.* **70**, 719, 2002.

99. P. Warszawski, H. M. Wiseman and H. Mabuchi, *Phys. Rev.* **A65**, 023802, 2002.

100. P. Warszawski and H. M. Wiseman, *J. Opt.* **B5**, 1, 2003.

101. P. Warszawski and H. M. Wiseman, *J. Opt.* **B5**, 15, 2003.

102. S. M. Tan, *J. Opt.* **B1**, 424, 1999.

103. P. Marte *et al.*, *Phys. Rev. Lett.* **71**, 1335, 1993.

104. Y. Castin and K. Mølmer, *Phys. Rev. Lett.* **74**, 3772, 1995.

105. H. J. Carmichael, S. Singh, R. Vyas and P. R. Rice, *Phys. Rev.* **A39**, 1200, 1989.

106. W. P. Smith *et al.*, *Phys. Rev. Lett.* **89**, 133601, 2002.

107. H. Mabuchi, *Equant. Semicl. Opt.* **8**, 1103, 1996.

108. H. Mabuchi and P. Zoller, *Phys. Rev. Lett.* **76**, 3108, 1996.

109. P. Kochan, H. J. Carmichael, P. R. Morrow and M. G. Raizen, *Phys. Rev. Lett.* **75**, 45, 1995.

110. H. P. Yuen and J. H. Shapiro, *IEEE Trans. Inform. Th.* **IT-26**, 78, 1980.

111. H. M. Wiseman and G. J. Milburn, *Phys. Rev.* **A47**, 642, 1993.

112. C. W. Gardiner, *Handbook of Stochastic Methods for Physics, Chemistry and the Natural Sciences*, 2nd ed., Springer-Verlag, Berlin, 1985.
113. B. Øksendal, *Stochastic Differential Equations: An Introduction with Applications*, 5th ed., Springer-Verlag, Berlin, 2002.
114. H. M. Wiseman and G. J. Milburn, *Phys. Rev.* **A47**, 1652, 1993.
115. P. Alsing and H. J. Carmichael, *Quant. Opt.* **3**, 13, 1991.
116. J. Doyle, B. Francis and A. Tannenbaum, *Feedback Control Theory*, Macmillan, London, 1990.
117. K. Zhou, J. C. Doyle and K. Glover, *Robust and Optimal Control*, Prentice Hall, Upper Saddle River, NJ, 1996.
118. V. P. Belavkin, *Rep. Math. Phys.* **43**, 405, 1999.
119. H. M. Wiseman and G. J. Milburn, *Phys. Rev. Lett.* **70**, 548, 1993.
120. H. M. Wiseman, *Phys. Rev.* **A49**, 2133, 1994; *Phys. Rev.* **A49** 5159, **B50**, 4428 (Errata), 1994.
121. V. Giovannetti, P. Tombesi and D. Vitali, *Phys. Rev.* **A60**, 1549, 1999.
122. H. M. Wiseman, *Phys. Rev. Lett.* **81**, 3840, 1998.
123. J. G. Walker and E. Jakeman, *Proc. Soc. Photo-Opt. Instrum. Eng.* **492**, 274, 1985.
124. S. Machida and Y. Yamamoto, *Opt. Commun.* **57**, 290, 1986.
125. M. S. Taubman, H. Wiseman, D. E. McClelland and H.-A. Bachor, *J. Opt. Soc. Am.* **B12**, 1792, 1995.
126. P. R. Tapster, J. G. Rarity and J. S. Satchell, *Phys. Rev. Lett.* **37**, 2963, 1988.
127. J. Mertz *et al.*, *Phys. Rev. Lett.* **64**, 2897, 1990.
128. A. Furusawa *et al.*, *Science* **282**, 706, 1998.
129. J. A. Dunningham, H. M. Wiseman and D. F. Walls, *Phys. Rev.* **A55**, 1398, 1997.
130. S. Mancini and P. Tombesi, *Phys. Rev.* **A56**, 2466, 1997.
131. A. C. Doherty and K. Jacobs, *Phys. Rev.* **A60**, 2700, 1999.
132. A. C. Doherty *et al.*, *Phys. Rev.* **A62**, 012105, 2000.
133. A. C. Doherty, S. M. Tan, A. S. Parkins and D. F. Walls, *Phys. Rev.* **A60**, 2380, 1999.
134. A. C. Doherty and H. M. Wiseman, in preparation, unpublished.
135. H. M. Wiseman, S. Mancini and J. Wang, *Phys. Rev.* **A66**, 013807, 2002.
136. H. M. Wiseman, *Phys. Rev. Lett.* **75**, 4587, 1995.
137. M. A. Armen *et al.*, *Phys. Rev. Lett.* **89**, 133602, 2002.
138. P. R. Rice and H. J. Carmichael, *IEEE J. Quant. Elect.* **24**, 1351, 1988.
139. H. J. Carmichael, *Phys. Rev.* **A33**, 3262, 1986.
140. T. Bhattacharya, S. Habib and K. Jacobs, *Phys. Rev.* **A67**, 042103, 2003.

CHAPTER 9

PROGRESS IN ASYMMETRIC RESONANT CAVITIES: USING SHAPE AS A DESIGN PARAMETER IN DIELECTRIC MICROCAVITY LASERS

H. G. L. Schwefel, H. E. Tureci, A. Douglas Stone and R. K. Chang

Department of Applied Physics, Yale University,
P.O. Box 208284, New Haven, CT 06520, USA

We report on progress in developing optical microresonators and microlasers based on deformations of dielectric spheres and cylinders. We review the different semiconductor and polymer dye microlasers which have been developed and demonstrated using this approach. All the lasers exhibit highly directional emission despite the presence of ray chaos in the system. Lasing has been demonstrated using both optical pumping and electrical pumping in the case of InGaP quantum cascade lasers and very recently in GaN MQW lasers. Lasing modes based on stable and unstable periodic orbits have been found as well as modes based on chaotic whispering gallery orbits; the lasing mode depends on the material, shape and index of refraction. The lasing from modes based on unstable orbits dominated for certain shapes in the GaN cylinder lasers, and is related to the "scarred" states known from quantum chaos theory. Extreme sensitivity of the emission pattern to small shape differences has been demonstrated in the polymer microlasers. Large increases in output power due to optimization of the resonator shape have been demonstrated, most notably in the quantum cascade "bowtie" lasers. Efficient numerical approaches have been developed to allow rapid calculation of the resonant modes and their directional emission patterns for general resonator shapes. These are necessary because the lasing modes are not usually amenable to standard analytic techniques such as Gaussian optical or eikonal theory. Theoretical analysis of the directional emission from polymer lasers has shown that highly directional emission is compatible with strongly chaotic ray dynamics due to the nonrandom character of the short-term dynamics. Very recently unidirectional emission and electrical pumping have been demonstrated in the GaN MQW system using a spiral-shaped resonator design, bringing this general approach in which shape is used as a design parameter closer to useful applications.

1. Introduction

1.1. *Overview*

Microresonators based on spherical, cylindrical and disk-shaped dielectrics have been studied for applications in lasers and integrated optical devices for at least two decades.[1-3] These devices exploit the (nearly) total internal reflection which confines whispering gallery modes of such structures and leads to high Q resonances. However obtaining useful output from such devices, both in terms of lasing power and in terms of controllable directional out-coupling has always been challenging due to their intrinsic isotropy and the fact that out-coupling is typically dominated by random features such as surface roughness. Nine years ago Nöckel, Stone and Chang[4] proposed that smooth deformations of such resonators, termed as *asymmetric resonant cavities* (ARCs), could achieve both usefully high Q modes and controlled out-coupling which might be optimized by varying the shape as a design parameter. In the previous volume of this series the basic physical concepts and theory behind ARCs were presented along with very preliminary experimental results obtained from deformed spherical microdroplet lasers.[5] The ARC concept is of theoretical interest because such resonators are examples of wave-chaotic systems, similar to systems studied in the field of quantum chaos.[6] The motion of a light ray confined in such a resonator is in many cases chaotic in the technical sense that this motion exhibits exponential sensitivity to small differences in initial conditions; as a result, the analysis of such resonators can be related to a well-known class of problems in nonlinear physics, that of classical and quantum billiards, as discussed in detail in the initial and subsequent work.[5,6] Since that initial work at least four different realizations of semiconductor ARC lasers have been developed and studied,[7-11] as have polymer ARC dye lasers.[12] We will review much of that experimental work in the current chapter. Overall this work has deepened our understanding of wave-chaotic resonators and of using shape as a design parameter in optimizing the performance of microresonators. It has also shown that a number of assumptions of the initial theoretical work have a limited range of validity and that the properties of these resonators are more diverse and complex than initially anticipated.

To summarize the major new results prior to a detailed exposition:

- One can get highly directional emission from smoothly-deformed (ARC) resonators and also from dielectric resonators with abrupt deformations from circular symmetry (such as the "spiral" laser discussed below).[13]

- The lasing modes can have a wide range of geometries and properties: these include chaotic whispering gallery modes, modes based on stable periodic orbits, modes based on unstable periodic orbits ("scarred modes"), and chiral whispering gallery modes (modes strongly favoring one sense of rotation).

- Fully chaotic laser resonators (i.e. those with no stable or marginally stable periodic orbits) can still have highly directional emission due to nonrandom short-term dynamics.

- The high emission directions are extremely sensitive to the shape of the resonator and its index of refraction in a manner which can be understood by analysis of the phase space for ray motion.

- The lasing mode selected also depends on the shape of the resonator and its index of refraction and gain, however in a manner which is not yet fully understood.

- Theoretical analysis of the passive cavity based on efficient new computational algorithms allows one to identify the lasing mode based on comparison with experiment.

- Deformation of the resonator from circular symmetry can lead to a substantial improvement in the peak power output (several orders of magnitude) for lasing media with the same gain.

- An efficient electrically-pumped microlaser in the GaN materials system with unidirectional emission has been demonstrated using the shape design approach.

We will review the experimental and theoretical work leading to these conclusions below.

2. Review of Theoretical Techniques

2.1. *Background*

The use of mirror-based "open" resonators was a key step in the development which led from the maser to the laser. The theory of mirror-based resonators is well developed for standard Fabry–Perot and ring resonator configurations, and in itself fills several hundred pages in standard textbooks.[14] In such a case the location of the mirrors defines an optical path which leads to high Q resonances and feedback (in the case of an active cavity); in most cases this path is a simple linear motion between parallel mirrors. The shape and spacing of the mirrors defines the stability of the ray motion between them and other properties of the output beam. Another important type of resonator for semiconductor lasers is based on distributed Bragg reflectors, dielectric layers which are spaced to cause destructive interference, hence acting as efficient mirrors for light at normal incidence. Again for this case the light path is simply a linear back and forth motion with reflection at normal incidence. The ray motion in such resonators can be fitted into the general framework of paraxial optics and "ABCD" matrices which describe propagation through a series of optical elements such as lenses and mirrors. This ray description is easily translated into solutions of the wave equation using the methods of Gaussian optics if the ray path is stable and periodic.[14,15] In contrast, dielectric resonators allow trapping of many different light trajectories for long times and mode geometries which are much more complex. The paradigm of a simple correspondence between a periodic ray orbit and a set of resonant modes of the cavity fails. Even for the case of simple whispering gallery orbits of a perfectly-reflecting cylinder the resonant modes are determined by zeros of the Bessel function which are not in general equally spaced in wavevector as are the modes of resonances based on stable periodic ray orbits. However in this case of a perfectly circular cylinder it is possible to write down approximate analytic solutions of the wave equation based on ray trajectories using eikonal theory.[16] A much more fundamental problem arises in generically deformed cylinders. In this case both familiar analytic methods for treating resonators, Gaussian optics and eikonal theory, are simply not applicable to a large fraction of the spectrum due to the possibility of chaotic ray motion. This is a crucial point which does not seem to have been appreciated anywhere in the optics literature: *all conventional methods of geometric or Gaussian optics fail in a resonator which has chaotic ray motion.* We shall explain the origin of this failure shortly. Several analytic methods for making short-wavelength approximations to such chaotic wave problems have been developed in the recent past for the Schrödinger equation in the study of "quantum chaos", but these methods do not allow one to construct *individual* solutions as one can for regular ray motion using Gaussian or eikonal methods. Therefore it is particularly important to have efficient numerical approaches to these problems; we present such an approach and some representative results from it below.

2.2. *Failure of Conventional Geometric Optics*

We begin by explaining how the presence of ray chaos leads to the failure of conventional methods of eikonal theory and Gaussian optics for treating the wave equation in a dielectric resonator analytically in the short-wavelength limit. Both eikonal theory and Gaussian optics apply only in the limit in which the wavelength of the modes is much less than typical geometric features of the resonator, such as relevant chord lengths and radii of curvature of the boundary. Gaussian optics, based on the parabolic equation method,[15,17] only allows one to determine the modes of the resonator that are localized in the vicinity of isolated *stable* periodic orbits, and may be regarded as an improved version of the eikonal method; the latter method works both for such stable orbit modes and for a more general class of marginally stable modes to be discussed below. Hence we will focus mainly on the failure of eikonal methods in our initial discussion and at the end explain the relationship to Gaussian optics. As noted, both methods are based on approximations to the exact wave equation which are only valid when $kR \gg 1$, where we shall use R to refer to a typical linear dimension of the resonator and assume all radii of curvature are of order R. It should also be noted that all of the problems we will be interested in correspond to nonseparable boundary conditions on the wave equation and hence cannot be solved exactly by separation of variables or by any other known analytic method. Therefore short-wavelength approximations are the natural method to use to attempt a solution. For almost all microresonators that have been studied experimentally the resonator is indeed in the limit $kR \gg 1$ which would appear to be sufficient to validate such approaches. Nonetheless the methods fail; a more detailed version of the argument explaining this failure has been given recently in Ref. 18. Interestingly, the basic argument goes back to a little-known paper by Einstein in 1917.[19]

For simplicity, throughout this article we will be dealing with the resonances of an infinite uniform dielectric rod of index of refraction n and arbitrary cross-section ∂D, and will focus on planar solutions for which the z-component of conserved momentum, $k_z = 0$ (see Fig. 1). For this system the solutions have either the electric field (TM solution) or magnetic field (TE solution) solely in the z-direction and the amplitude of this field $E_z(x, y), B_z(x, y)$ satisfies the scalar wave (Helmholtz) equation:

$$(\nabla^2 + n^2 k^2)\, \psi(x, y) = 0. \tag{1}$$

Here $\psi(x, y)$ refers to the electric or magnetic field for the TM, TE cases respectively and we assume a harmonic time-dependence with frequency $\omega = ck$. This is a reasonable model for a micropillar resonator with a large aspect ratio of height to radius. The highest Q modes will be the planar

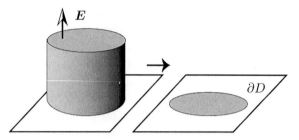

Fig. 1. Illustration of the reduction of the Maxwell equation for an infinite dielectric rod of general cross-section to the 2D Helmholtz equation for the TM case (E field parallel to axis) and $k_\parallel = k_z = 0$.

modes which do not escape through the top and bottom. The correct boundary conditions for the problem are the continuity of ψ and $\partial_{\hat{n}} \psi$ across the boundary, where the wavevector changes from nk (inside) to k (outside). These conditions will describe the physics of near total internal reflection and Fresnel refraction and reflection at the interface. Below we will present briefly a new numerical method to solve Eq. (1) with these boundary conditions very efficiently, and we will show many numerical results obtained with this method. However to illustrate why such a problem is not amenable to analytic description using ray optical or eikonal methods, it is sufficient and simpler to consider perfectly reflecting boundary conditions corresponding to $\psi = 0$ everywhere on the boundary. It can be straightforwardly shown that dielectric matching boundary conditions do not remove the fundamental limitation which we now describe.[20] For the perfectly reflecting case we need only consider $\psi(x, y)$ inside the domain D of uniform index n and hence we can set $n = 1$ for convenience.

The eikonal method consists of attempting an asymptotic solution of Eq. (1) (now with $n = 1$) of the form

$$\psi(x, y) = A(x, y) e^{ikS(x,y)} \qquad (2)$$

where $kR \gg 1$ and S, A are real functions independent of k and $A \equiv A_0$ is the first term in a power series in k^{-1}. This ansatz is used in Eq. (1) and terms of lower order in k^{-1} are initially neglected to yield the eikonal and transport equations:

$$(\nabla S)^2 = n^2(x, y) \qquad (3)$$
$$2\nabla S \cdot \nabla A + A\nabla^2 S = 0 \qquad (4)$$

for $S(x, y)$ and $A(x, y)$. $S(x, y)$ is a scalar field whose level curves describe the "wavefronts" of the solution which are assumed to be slowly varying in space, as the factor k in the exponent takes care of the rapid variation on the scale of the wavelength; the unit vector field ∇S describes the direction of ray motion at a given wavefront. For a uniform medium a ray originating on one wavefront must "move" in a straight line, even though the wavefronts themselves cannot be straight lines if the confining boundary of the medium is curved. For the current discussion the properties of the amplitude A and the transport equation which determines its properties once S is known are not crucial except for one property. The Dirichlet boundary conditions we are assuming require that ψ vanish on the boundary; one can easily see that if an attempt to satisfy this by setting $A = 0$ on the boundary with the transport equation will give $A = 0$ everywhere in the domain D. Thus the boundary conditions must be satisfied by the cancellation of two or more terms of the form Eq. (2) and for eikonal solutions within a bounded region the solution must have the form:

$$\psi(x, y) = \sum_n^N A_n(x, y) e^{ikS_n(x,y)} \tag{5}$$

where $N \geq 2$. Thus any eikonal solution must involve at least $N \geq 2$ sets of wavefronts defined by $S_1(x, y), S_2(x, y), \ldots, S_N(x, y)$ and N sets of rays determined by $\nabla S_1, \nabla S_2, \ldots, \nabla S_n$. A further implication of the boundary conditions is that the functions S_n must be pairwise equal on the boundary and their gradients on the boundary must satisfy "specular reflection" pairwise, i.e. $\hat{n} \cdot \nabla S_1 = -\hat{n} \cdot \nabla S_2$ and so on, for each pair. The key question raised by these constraints is whether such a set of wavefronts and associated ray vector fields can be consistently constructed for a given boundary ∂D?

We can reduce the previous question to a very specific question about ray dynamics in a perfectly reflecting "billiard" (the term for the problem of a point mass specularly reflecting from hard boundaries in two dimensions). Modern research in non-linear dynamics then allows us to answer this question generically. Consider a point $r_0 = (x_0, y_0)$ in the domain which is arbitrary except that the solution we are seeking $\psi(x_0, y_0)$ and $\nabla \psi(x_0, y_0)$ are nonzero at that point and in a small neighborhood around it. Then at this point there are N different ray directions defined by $\nabla S_1, \nabla S_2 \ldots$. Choose one of these directions e.g. ∇S_1 and follow it in a straight line to the boundary (for a uniform medium this line will run exactly along ∇S_1 as noted). The specular reflection boundary condition just mentioned implies that upon reaching the boundary and specularly refecting this ray the new direction corresponds to a ray of another of the vector fields, e.g. ∇S_2. We can thus be assured that each ray we follow from (x_0, y_0) will remain on one of the allowed ray directions determined by the N sets of wave-fronts defined by the S_n. By a well-known property of classical mechanics in a bounded system this ray must eventually return to the neighborhood of (x_0, y_0) over and over. If such eikonal solutions exist, such a ray by assumption must pass through this neighborhood each time in one of the N allowed ray directions defined at r_0. However this is a special dynamical behavior which need not hold. It turns out that three situations are possible.

- If the ray dynamics of this system allows this special behavior to occur for *all* initial ray choices at r_0, then we can construct a full spectrum of consistent eikonal solutions with a finite number of terms N. The quantization condition on k arises from requiring the single valuedness of ψ at each point r_0; an elegant means to implement this condition is described in the classic paper by Keller and Rubinow[16] following the suggestion by Einstein. Such cases are referred to as *integrable*. A solvable example for which the method works is an elliptical boundary. The relevant ray fields and wavefronts are illustrated in Fig. 2.

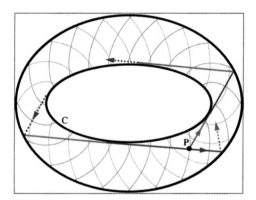

Fig. 2.   The wavefronts and the corresponding set of rays generated by an elliptic bound-ary. There are two sets of wavefronts S_1 = const. and S_2 = const. (and their corre-sponding rays), drawn in red and blue respectively, which together satisfy the boundary conditions on the elliptic boundary, provided the caustic curve and the wavevector k is chosen according to the EBK quantization conditions. In this figure, a (red) ray is started at point P, towards the caustic C. Subsequent iterations according to the specular reflec-tion rule generate rays which are always tangent to the elliptic caustic C. Irrespective of the starting position P, there can be one and only one other return direction (blue) at P. This is the unique hallmark of integrable ray motion. Note that we consider only one sense of rotation for rays; the other sense of rotation is disjoint from this set (ray dynamics conserves the "chirality") and the corresponding wavefronts generate a second linearly independent eikonal solution.

- If the number of ray return directions tends to infinity as $t \to \infty$ for all choices of initial ray directions at r_0 then no consistent eikonal solutions exist. This will be the case for systems which are completely chaotic.
- If the number of ray return directions at r_0 is sometimes finite and sometimes tends to infinity depending upon the initial ray direction then the system is referred to as *mixed* and in principle it will be possible to find eikonal solutions for only a subset of the spectrum. In practice, for the mixed case, eikonal solutions are only easily found near stable periodic orbits and quasi-periodic KAM tori.

Modern research in nonlinear dynamics tells us that the third, mixed case is the generic case. For example, any smooth deformation of a circular boundary which is not exactly elliptical will lead to the mixed case. We shall see below that a simple smooth quadrupolar deformation of the circle generates a very high degree of chaos and makes it impossible in practice to use the eikonal method except near the few remaining short stable periodic orbits.

To summarize the basic point of the previous argument: in order to generate a resonant mode within a given boundary one has to be able to launch a finite set of waves from each point which bounce around in the cavity and return so as to constructively interfere and form a standing wave. Only certain very special symmetric boundaries allow one to do this from all points in the cavity using waves propagating in an arbitrary direction. When it is not possible to do this the eikonal method does not apply.

It is important to realize that even when the eikonal approach fails there exists the same average density of modes as for the more symmetric shapes; this is guaranteed by various theorems, such as the Weyl theorem.[21] However, these solutions do not have wavefronts that are smooth on the scale of the wavelength and hence cannot be obtained by the asymptotic (eikonal) method.

Having explained this fundamental limitation on the use of the eikonal method for chaotic or mixed systems, we now briefly discuss how similar considerations apply to the method of Gaussian optics. In this method we search for solutions localized near periodic ray orbits of the problem. It is helpful to note that such localized solutions can always be found using the eikonal method as well. The eikonal solutions near stable periodic orbits are of the type found near the stable two-bounce orbit of the ellipse billiard illustrated in Fig. 3. A ray emanating from a point in the neighborhood of the periodic orbit moving in approximately the same direction will bounce

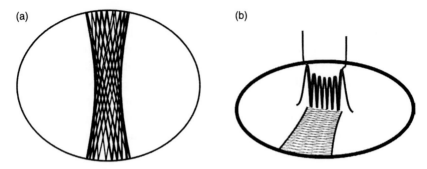

Fig. 3. (a) Real space plot of the simulation of a ray initially started in the stable bouncing ball region of an elliptical billiard of deformation $\epsilon = 0.12$. (b) The EBK wavefronts corresponding to a bouncing ball mode generated by an elliptic boundary. For simplicity we have only plotted two of the four sets of the wavefronts above the major axis. The other two sets of wavefronts represent the time-reversed motion. We also plot the transverse variation of the resulting eikonal solution in gray. Note that this solution has a singularity on its hyperbolic caustic. In black is plotted the Gaussian-optical approximation to the bouncing ball mode, which is uniformly valid over the whole transverse cross-section of the wavefield.

back and forth in the vicinity of the orbit indefinitely; each segment of its trajectory will be tangent to a hyperbolic caustic curve and hence will satisfy the property of only returning to the initial point in four possible directions. Therefore we can build up consistent wavefronts leading to standing waves in the vicinity of the periodic orbit and quantize the wavevector by imposing periodicity. One finds the following quantization rule:

$$kL = 2\pi m + (q + 1/2)\phi + \pi \qquad (6)$$

where L is the total length of the two-bounce orbit, m, q are integers, ϕ is the phase velocity of the orbit as it rotates around the fixed point, and is directly obtainable in terms of the radius of curvature at the bounce points and length of the orbit. The additional phase π is the specific value for the two-bounce orbit with Dirichlet boundary conditions of the Maslov phase which appears for any such periodic orbit; the general value of this phase for an arbitrary periodic orbit depends on the boundary conditions, number of bounces and topological properties of the orbit.[15] Thus we have two characteristic constant modes spacings: the longitudinal mode spacing or free spectral range (FSR), $\Delta k_L = 2\pi/L$ and the transverse mode spacing $\Delta k_T = \phi/L$. This quantization ruleGaussian optics!quantization rule, obtained from the eikonal method,[16] is identical to that obtained by the Gaussian optics method[15] when specialized to this two-bounce case. The actual solution $\psi(x, y)$ one constructs via the eikonal method will however have a diverging amplitude at the caustic of the ray motion, which is a standard limitation of the eikonal method, analogous to the well-known divergence of WKB solutions at a classical turning point.

The Gaussian optical solution is somewhat different as it begins from the reduction of the wave equation to a parabolic differential equation in the large k limit, which it then solves by the Gaussian ansatz. The solutions, while having the same quantized k-values as the eikonal solution, provide a more accurate description of the mode in space, which does not diverge at the caustic. Instead the Gaussian solution has a finite peak at the caustic and is welldefined everywhere in space (see Fig. 3(b)). In fact the Gaussian optics method can be regarded as an improved eikonal ansatz in which the phase function $S(x, y)$ is complex (something we excluded earlier) leading to a uniform approximation which allows continuation across the caustic (or classical turning point).

However the Gaussian approach does *not* provide a solution to the fundamental problem of quantizing chaotic motion. Chaotic motion occurs in the vicinity of *unstable* periodic orbits. A ray emanating from a point near an unstable periodic orbit will not remain near that orbit, confined by a caustic. Instead it will propagate far away from the original orbit and generically will return to the original neighborhood in a random direction (see Fig. 4). Therefore both eikonal and Gaussian methods will fail here. Technically, in the derivation of the Gaussian solutions, stability of the associated periodic orbit is required in order for the transverse behavior to have a Gaussian decay; if the orbit is unstable the transverse solution oscillates without decay and this violates the assumptions one makes in

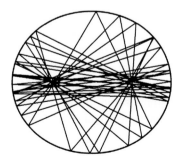

Fig. 4. Real-space plot of the simulation of a ray initially started close to the long diameter of the quadrupole of deformation $\epsilon = 0.07$.

defining the approximation.[15] Thus, like eikonal theory, Gaussian optics fails to provide an analytic description of modes associated with regions of chaotic ray motion.

However in generic (mixed) systems there always exist stable periodic orbits and the Gaussian method is a convenient analytic method to extract the *subset* of the resonant modes which *are* associated with stable periodic orbits. The application to dielectric resonators has been worked out in detail in Ref. 15 and some results using the method applied to stable "bowtie" resonances are reviewed in Sec. 5.1.

2.3. *The Phase Space Method for Ray Dynamics*

In the previous section we argued for the failure of analytic short-wavelength approximations to describe a finite fraction of the spectrum in generic resonators. We based the argument on modern results describing the motion of a point mass moving freely within a perfectly reflecting two-dimensional boundary or billiard, which is mathematically identical to ray motion within a closed resonator. In introducing the phase space methods for treating such systems we will initially treat only this closed case; afterwards we will note the change in the picture necessitated by the possibility of ray escape. The crucial result we quoted was that for a billiard which is a smooth deformation of a circle the ray dynamics is mixed, meaning that some initial conditions lead to regular motion tangent to a caustic curve and other initial conditions lead to chaotic motion which is pseudo-random at long times. This statement has an important meaning in phase space. The phase space for a point mass in two dimensions is four dimensional, but as the energy is assumed conserved, any given trajectory must lie on a three-dimensional subspace of phase space. If there is a second constant of the motion, such as the angular momentum for the circle (or its generalization for the ellipse), then each trajectory lies on a two-dimensional subspace of the constant energy surface with the topology of a torus. However if there is no second global constant of motion, as for generic deformations of a circular billiard, then the results of Kolmogorov–Arnold–Moser (KAM) theory[22,23] imply that for such mixed systems some initial conditions result in trajectories which explore a finite fraction of the three-dimensional constant energy surface and other initial conditions result in trajectories which remain on a two-dimensional subspace of this surface with the topology of a torus (a "KAM" torus). To get an overall view of the phase space dynamics for a given shape it is very convenient to use the tool of the Poincaré surface of section, introduced in this context sometime ago,[5] which we now briefly review.

As the behavior we are describing is known from KAM theory to be generic for smooth deformations of the circle we will restrict ourselves in the subsequent discussion to the simple example of the quadrupole billiard described by the boundary shape:

$$R(\phi) = 1 + \epsilon \cos 2\phi \qquad (7)$$

which in the zero deformation limit $\epsilon = 0$ reduces to a circular billiard and is integrable. Variation of the parameter ϵ starting from zero induces a transition to chaos, meaning a fraction of finite measure of the initial conditions lead to chaotic motion. This measure increases with increasing deformation but does not reach unity for any known smooth deformation (there are nonsmooth deformations of a circle, such as the stadium billiard for which the fraction is known to be unity). Real-space ray-tracing is not helpful to analyze this transition since chaotic trajectories tend to fill the entire real-space even if they do not fill the constant energy surface uniformly; thus much of the structure is not visible. Instead, to visualize the increase in the chaotic fraction of phase space we imagine a set of trajectories each time hitting the boundary and plot the result in a two-dimensional graph known as the surface of section[21,24] (see Fig. 5).

In this two-dimensional phase space representation, the internal ray motion is conveniently parametrized by recording the pair of numbers $(\phi_i, \sin \chi_i)$ at each reflection i, where ϕ_i is the polar angle denoting the position of the ith reflection on the boundary and $\sin \chi_i$ is the corresponding angle of incidence of the ray at that position (see Fig. 5). Each initial point is then evolved in time through the iteration of the SOS map $i \to i+1$, resulting in two general classes of distributions. If the iteration results in a one-dimensional distribution (an *invariant curve*), the motion represented is *regular*. On the other hand, exploration of a two-dimensional region is the signature of *chaotic* motion which covers a finite fraction of the constant energy surface in phase space.

The transition to ray chaos in the quadrupole billiard is illustrated in Fig. 6. At zero deformation the conservation of $\sin \chi$ results in straight line trajectories throughout the SOS and we have globally regular motion. These are the well-known whispering gallery (WG) orbits. As the deformation is increased (see Fig. 6) chaotic motion appears (the areas of scattered

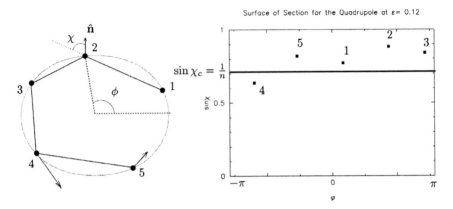

Fig. 5.   The construction of the surface of section plot. Each reflection from the boundary is represented by a point in the SOS recording the angular position of the bounce on the boundary (ϕ) and the angle of incidence with respect to the local outward pointing normal ($\sin \chi$). For a standard dynamical billiard there is perfect specular reflection and no escape. For "dielectric billiards" if $\sin \chi > \sin \chi_c > 1/n$, total internal reflection takes place, but both refraction and reflection according to Fresnel's law results when a bounce point (bounce #4 in the figure) falls below the "critical line" (shown in gray) $\sin \chi > \sin \chi_c$. Note that $\sin \chi < 0$ correspond to clockwise sense of circulation. We do not plot the $\sin \chi < 0$ region as the SOS has reflection symmetry. Below we will plot the SOS for ideal billiards without escape unless we otherwise specify.

points in Fig. 6) and a given initial condition explores a larger range of values of $\sin \chi$. Simultaneously, islands of stable motion emerge (closed curves in Fig. 6), but there also exist extended "KAM curves" (the SOS projection of KAM tori)[23] (open curves in Fig. 6), which describe a deformed WG-like motion close to the boundary. These islands and KAM curves cannot be crossed by chaotic trajectories in the SOS. As the transition to chaos occurs, a crucial role is played by the *periodic orbits* (POs), which appear as fixed points of the SOS map. The local structure of the islands and chaotic layers can be understood through the periodic orbits which they contain. Thus, the center of each island contains a stable fixed point, and close to each stable fixed point the invariant curves form a family of rotated ellipses. The Birkhoff fixed point theorem[24] guarantees that each stable fixed point has an unstable partner, which resides on the intersection of separatrix curves surrounding the elliptic manifolds. Chaotic motion sets in at separatrix regions first, and with increasing deformation pervades larger and larger regions of the SOS. Already at $\epsilon = 0.1$, much of the phase space is chaotic and a typical initial condition in the chaotic sea explores a large range of $\sin \chi$, eventually reversing its sense of rotation. For $\epsilon = 0.18$ the

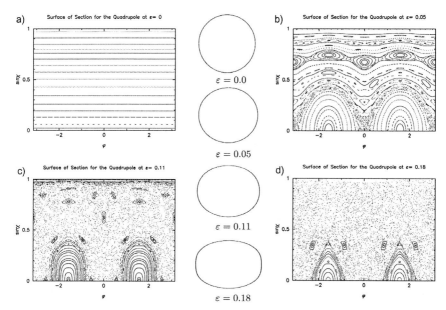

Fig. 6. The SOS of a quadrupole at fractional deformations $\epsilon = 0, 0.05, 0.11, 0.18$. The closed curves and the curves crossing the SOS represent two types of regular motion, motion near a stable periodic orbit and quasi-periodic motion, respectively. The regions of scattered points represent chaotic portions of phase space. A single trajectory in this "chaotic component" will explore the entire chaotic region. With increasing deformation the chaotic component of the SOS (scattered points) grows with respect to regular components and is already dominant at 11% deformation. Note in (b) the separatrix region associated with the two-bounce unstable orbit along the major axis where the transition to chaotic motion sets in first.

entire SOS above $\sin \chi \approx 0.4$ appears chaotic. There is an important practical implication of these results. For rather smooth shapes and relatively small deformations chaotic motion dominates and the failure of analytic methods for such shapes is a barrier to understanding the spectrum of the corresponding wave problem.

We now comment briefly on the relevance of these results for the closed behavior the dielectric billiards. As we are always assuming $kR \gg 1$ we are in the limit for which ray optics describes a light ray interacting with the dielectric boundary. Therefore internal reflection for light rays hitting at $\sin \chi > 1/n$ is almost total and the closed billiard description for the portion of trajectories which remain for some time in this region of phase space is quite accurate. However any portion of the trajectory which involves reflection with $\sin \chi < 1/n$ will be subject to refractive escape, typically within a few bounces. Therefore in the context of the analysis of dielectric resonators we introduce a *critical line* $\sin\chi = 1/n$; portions of the SOS below this line are not to be regarded as supporting long-lived resonances even though they might do so in the closed resonator. Typical resonances interest emit most of their radiation from the vicinity of the critical line as we shall see below. Certain aspects of these resonances can be modeled by following ray bundles in the SOS and allowing rays to escape with the probability given by the Fresnel law when a ray passes below the critical line. The phase space "flow pattern" then determines the directional emission from these resonances as shown in detail in Sec. 3.

The reason that ray models have some relevance to the wave solutions even in the chaotic case is that it is possible to associate wave solutions with different regions of phase space and (neglecting interference effects) hence with bundles of rays. In the next sections we will illustrate this fact by formulating the resonance problem and a numerical method for its solution, and then show how such solutions can be projected onto the SOS and interpreted in terms of ray dynamics.

2.4. *The Resonance Problem*

We now briefly review the formulation of the exact resonance problem, specializing to an infinite uniform dielectric rod of arbitrary cross-section D. For this geometry, the Maxwell equations for the problem reduce to the Helmholtz equation (1) for the E-field (TM), and B-field (TE) polarizations, which we have denoted $\psi(x,y)$, assuming a uniform solution in the z-direction. As also noted, the electromagnetic boundary conditions reduce to continuity of ψ and the normal derivative on the boundary ∂D. Assuming the boundary is a smooth deformation of the circle it is convenient to expand the solutions inside and outside the rod in terms of the solutions of the Helmholtz equation at a given wavevector in polar coordinates:

$$\psi_1(r,\phi) = \sum_{m=-\infty}^{\infty} \left(\alpha_m H_m^+(nkr) + \beta_m H_m^-(nkr)\right) e^{im\phi} \quad r < R(\phi) \quad (8)$$

$$\psi_2(r,\phi) = \sum_{m=-\infty}^{\infty} \left(\gamma_m H_m^+(kr) + \delta_m H_m^-(kr)\right) e^{im\phi} \qquad r > R(\phi) \quad (9)$$

where H_m^-, H_m^+ are the incoming and outgoing Hankel functions respectively. If we assume a single incoming wave with unit amplitude for angular momentum m, the matching conditions are sufficient to find a solution for all k values and the coefficients of the outgoing waves define a unitary S-matrix $S_{m,m'}(k)$. However we are only interested in the special values of k at which a long-lived resonance of the system exists, as these will correspond (approximately) to the emitting modes of the active cavity. We could look for rapid variations in the S-matrix as we vary k, which indicate resonant scattering, but this is inconvenient for various reasons. Instead we define the quasi-bound states as the solutions of this matching problem with no incoming wave from infinity ($\delta_m = 0$ in Eq. (9)). Due to the violation of flux conservation, no such solution exists for real wavevectors k, but a discrete set of solutions exist at complex wavevectors $k = q - i\gamma$, known as the quasi-bound states or quasi-normal modes of the system. Long-lived resonances have $q \gg \gamma$ and the Q value can be defined as $Q = 2q/\gamma$. After the resonance wavevectors are found, the corresponding mode can also be determined and plotted both within the resonator and in the farfield. The farfield solutions have the unphysical feature that they increase in intensity as $\exp[2\gamma r]$, reflecting the decay from the cavity, but this unphysical dependence does not affect the angular distribution of radiation, which is the farfield quantity we are interested in. Introducing an imaginary part of the index into the problem (representing amplification in the cavity) would lead to solutions for real k with the same angular dependence and no exponential growth at infinity.

An approach to solving this problem termed the "S-matrix method" has been developed over a number of years;[18,20,25,26] the most recent version of this approach[18] is highly efficient for the specific problem of ARC resonators. The numerical results plotted in the remainder of this paper were all obtained by this method. The approach begins by integrating the matching conditions over the azimuthal angle ϕ with $r = R(\phi)$; this eliminates the spatial dependence and transforms the matching conditions into an infinite set of linear relations for the coefficients $\{\alpha_m\}, \{\beta_m\}, \{\gamma_m\}$. This infinite set of relations can be truncated because for $m \gg nkR$ the corresponding Hankel functions have negligible weight in the cavity. We thus end up with order $2nkR$ linear relations which must be satisfied by the coefficients. These relations and the regularity condition on the solution at the origin yield a determinantal equation the form:

$$\zeta(k) = \det[1 - \mathcal{S}(k)]. \tag{10}$$

The matrix $\mathcal{S}(k)$ is not the unitary scattering matrix of this problem, but it is nearly unitary for real k; the complex values of k which make the determinant zero (eigenvalue of $\mathcal{S} = 1$) are the quasi-bound wavevectors we seek. Once they are known the coefficients $\{\alpha_m\}, \{\gamma_m\}$ can be determined and the quasi-bound state can be constructed from Eqs. (8) and (9).

At this level of description a complex root search of this determinantal equation is needed in order to actually find the solutions of interest, which does not appear to make the method more efficient than various other brute force methods one might employ, such as point matching on the boundary. However there are two reasons the current method is much more powerful. First, as has been known for some time, the eigenvectors of $\mathcal{S}$ do not change much over a range of k corresponding to the mode spacing; hence the "unquantized" solutions have the same physical content as the true quasi-bound solutions. Therefore basic physical properties such as directional emission patterns and distributions of Q values can be obtained without the root search. Second, quite recently it was shown[18,20] that there exists an efficient extrapolation method to find the roots once the eigenvalues of $\mathcal{S}$ are found at two values of k, so that no true root search is necessary. The technical details supporting and expanding on these statements can be found in Ref. 18.

Once we have obtained numerical solutions to the resonance problem we would like to interpret their "classical" (ray dynamical) meaning and use our knowledge of the phase space structure and flow to explain the properties of the resonance spectrum, such as the directional emission patterns and distribution of Q values. Despite the fact that there exists no simple classical construction of individual solutions, the correspondence between solutions and properties of the ray phase space is quite helpful in extracting the physical properties interest as demonstrated below. The technique for extracting ray dynamical information from a real-space solution is known as *Husimi projection*;[18,27] this technique allows us to represent a solution within the ray phase space of the problem and ultimately on the surface of section. Such a representation attempts to extract both momentum and position information simultaneously and just as for quantum mechanics, we cannot have full information about real-space and momentum space at the same time due to the analog of the uncertainty principle for the electromagnetic wave equation (often written as $\Delta x \Delta k \geq 1$, this is a basic property of Fourier transforms). Phase space coordinates involve both position and momentum and our resolution in phase space will be limited by this uncertainty relation.

The specific procedure which is widely followed to project a real-space solution into phase space is a version of a "windowed" Fourier transform known as Husimi projection which involves integrating the real-space solution $\psi(\boldsymbol{x})$ against windowing functions peaked around position $\bar{\boldsymbol{x}}$

$$Z_{\bar{\boldsymbol{x}}\bar{\boldsymbol{p}}}(\boldsymbol{x}) = \left(\frac{1}{\pi\eta^2}\right)^{1/4} \exp(ik\bar{\boldsymbol{p}} \cdot \boldsymbol{x}) \exp\left(-\frac{1}{2\eta^2}|\boldsymbol{x} - \bar{\boldsymbol{x}}|^2\right) \qquad (11)$$

where the width parameter $\eta = \sigma_0/\sqrt{k}$, σ_0 is a parameter with dimensions of square-root of a length which can be chosen for convenience, and we note that the momentum vector has been factorized as $k\bar{\boldsymbol{p}}$ so that $\bar{\boldsymbol{p}}$ is a unit vector denoting the direction of the wavevector. The Fourier transform of this windowing function will also be a Gaussian in the unit vector $\boldsymbol{p}$ peaked around $\bar{\boldsymbol{p}}$. In these scaled variables, which are the appropriate choice for projecting onto the billiard SOS, the uncertainty relation takes the form:

$$\Delta x \cdot \Delta p \geq \frac{1}{2k}. \qquad (12)$$

The function $Z_{\bar{\boldsymbol{x}}\bar{\boldsymbol{p}}}(\boldsymbol{x})$ and its Fourier transform $\tilde{Z}_{\bar{\boldsymbol{x}}\bar{\boldsymbol{p}}}(\boldsymbol{p})$ have standard deviations which satisfy,

$$\Delta x = \frac{\sigma_0}{\sqrt{2k}} = \frac{\eta}{\sqrt{2}} \qquad \Delta p = \frac{1}{\sqrt{2k}\sigma_0}, \qquad (13)$$

hence they saturate this inequality and represent a "minimum uncertainty" basis for projecting the solutions onto phase space (these functions are the "coherent states" often used in quantum mechanics). The Husimi density in phase space is then defined as:

$$\rho_\psi(\bar{\boldsymbol{x}}, \bar{\boldsymbol{p}}) = \left|\int d^2\boldsymbol{x} Z^*_{\bar{\boldsymbol{x}}\bar{\boldsymbol{p}}}(\boldsymbol{x})\psi(\boldsymbol{x})\right|^2, \qquad (14)$$

which is positive semi-definite on the phase space of the problem. Since we have scaled $\bar{\boldsymbol{p}}$ to be a unit vector this phase space is already confined to the three-dimensional constant energy surface of the four-dimensional phase space, but we now wish to project it down one dimension further onto the surface of section. For this purpose it is useful to introduce windowing functions in cylindrical coordinates[18] and calculate the Husimi distribution at a fixed radius $r = R_c$. Careful limiting procedures must be observed to get a meaningful result as described in Ref. 18. The resulting Husimi–SOS distribution at $r = R_c$ is given as

$$H_\psi(\bar\phi, \sin\bar\chi) = \left| \sum_{-\infty}^{\infty} \alpha_m H_m^+(nkR_c) e^{-inkR_c(\sin\chi - \sin\bar\chi)\bar\phi} \right.$$

$$\left. \times e^{-\sigma_0^2(nkR_c)^2(\sin\chi_m - \sin\bar\chi)^2/2} \right|^2 \qquad (15)$$

where $\sin\chi_m = \frac{m}{nkR_c}$ and equal spatial and momentum resolution in the SOS is achieved by choosing $\sigma_0 \sim 1/\sqrt{nkR_c}$. Note that the numerical real-space wavefunction enters this expression through the coefficients α_m which are assumed known.

Equation (15) is a perfectly good Husimi–SOS distribution, but it does not correspond to our conventional choice of the SOS at the boundary. However, for each value of $(\bar\phi, \sin\bar\chi)$ we can simply calculate the values of $(\phi, \sin\chi)$ that would result from following this ray to the boundary and assign the corresponding point on the boundary the values of the circle Husimi–SOS at $(\bar\phi, \sin\bar\chi)$ corrected by a Jacobian factor for the Gaussian propagation between the two sections. This is the quantity we use to compare and interpret wave solutions in the classical SOS of the problem. Again, a detailed recipe for constructing these Husimi distributions is given in Ref. 18. Note that for the Husimi–SOS the uncertainty relation (12)

$$\Delta\phi \cdot \Delta\sin\chi \geq \frac{1}{2nkR} \qquad (16)$$

is saturated at its lower bound, i.e. $\Delta\phi \sim \Delta\sin\chi \sim 1/\sqrt{2nkR}$, where R is the average radius of the billiard and we have used the approximation that the arc length along the boundary is $R\Delta\phi$. Hence EM wave solutions only resolve the classical structures in the SOS on a scale of area $(2nkR)^{-1}$; this is the EM analog of the statement in quantum chaos theory that only wavefunctions are sensitive to classical structures of order $1/\hbar$.

To illustrate how the ray-wave correspondence works for billiards in the mixed regime we present three examples of numerical solutions for resonance wavefunctions of the quadrupole billiard calculated by the S-matrix approach and their corresponding Husimi–SOS transforms. In Figs. 7(a) and 7(b) we show a whispering gallery mode of a slightly deformed quadrupole billiard ($\epsilon = 0.03$); this is a typical wavefunction corresponding to quasi-periodic ray motion which could be calculated analytically (in principle) using the eikonal method. Projection of the state onto the SOS shows it follows closely an invariant curve of the problem, but smeared out to agree with the uncertainty relation just noted. In Figs. 7(c) and 7(d) we

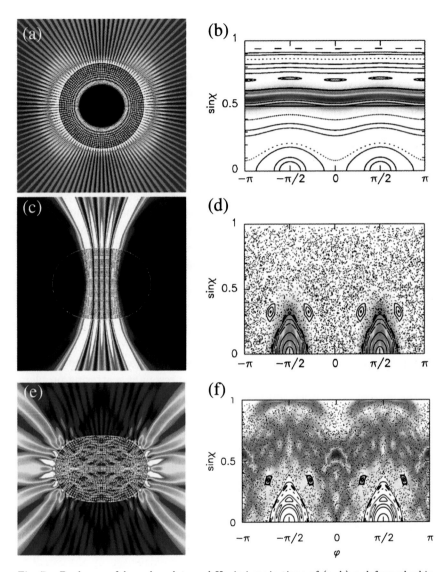

Fig. 7.   Real-space false color plots and Husimi projections of (a, b) a deformed whispering gallery mode at $\epsilon = 0.03$ and $n = 2$, (c, d) transverse excited bouncing ball mode at $\epsilon = 0.16$ and $n = 2$, (e, f) mode localized on the chaotic portion of the phase space at $\epsilon = 0.18$ and $n = 2.65$.

show a two-bounce stable orbit mode of the type one could calculate using the Gaussian optical method. In Husimi it is well-localized on the stable island and is relatively insensitive to the existence of chaos elsewhere in the system. Thus these quite conventional modes can coexist with chaotic modes at the same deformation, as will be discussed further in Sec. 5.1. In Figs. 7(e) and 7(f) we show a highly chaotic mode of the strongly deformed quadrupole ($\epsilon = 0.18$). Note the "tangled" wavefronts in much of the resonator which vary in direction on the scale of the wavelength. The Husimi projection shows this mode lives completely in the chaotic portion of the SOS, although it is not completely spread out on the chaotic component. From experience we find that "chaotic" solutions are still not fully randomized on the chaotic component at the values $nkR \sim 100$–300 which can be treated by our numerical method. Nonetheless the ray-wave correspondence is clearly present in these chaotic resonators and we will use it as the primary tool for interpreting the exciting experiments which have been done on ARC and spiral microlasers.

3. Ray Dynamics and Shape-Dependent Directional Emission from ARCs

In the previous sections we have reviewed the phase space formulation of ray dynamics in ARCs and the formulation of the resonant scattering problem. In the current section we will begin to present the most recent experimental and theoretical developments relating to ARC resonators and lasers. First, we will review some of the experimental techniques used in the studies we report. Then we will briefly review the ray model for the directional emission of ARCs and present experimental data from two different sets of experiments on low index ARC resonators. The first of these studied lasing emission from differently-shaped polymer ARC cylindrical microlasers[12] and the second of these studied resonant scattering from passive ARC silica microspherical cavities.[28] The first part of this section focuses on the lasing experiments and how they can be understood in terms of the phase space ray-dynamical method for ARCs.

3.1. *The Imaging Technique for the Study of Microcavity Resonators*

The detection part of the experiment was designed in accordance with the information contained in the SOS diagram. The detector must be able to extract two pieces of information: (1) where along the sidewalls the light is emerging from the microcavity, that is, the angle ϕ, and (2) what is the angle of the emitted ray which is related to the internal incident angle χ by Snell's law of refraction. A detector that can only measure the farfield radiation pattern is insufficient because it misses where the light emerges along the sidewall. The farfield pattern alone is not unique in that the same pattern can occur for different sidewall distributions of emission. Any detection system ought to be able to distinguish between the two different emission types shown in Fig. 8, where the farfield patterns are similar, but the image patterns along the sidewalls are different.

Figure 9 shows the detection system that was settled upon as the best compromise between obtaining the farfield pattern while maintaining spatial resolution along the sidewalls. In this setup, microlaser devices are optically pumped normal to the plane of the pillar and light emitted from the side-walls is collected through an aperture in the farfield, passed through a lens and collected on a CCD camera. The key element in this detection system is the aperture placed before the collection lens. The aperture accomplishes two purposes: (1) it limits the solid angle of the collected light; and (2) it extends the depth of field that the light is collected from. The aperture sets a solid-angle limitation and restricts the farfield profile to an angular resolution of 5°. The small aperture extends the depth of field to be larger than the longest diameter of the microstructure. The depth of field associated with the numerical aperture (NA = 0.047) is 200 μm. The largest microcavity being imaged has the longest dimension of 120 μm. Thus the entire microcavity is in focus at the same time, regardless of the rotational alignment of the microcavity with respect to the camera.

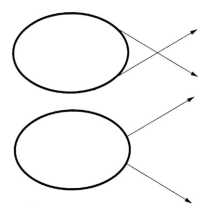

Fig. 8. Two possible emission patterns with different emitting points on the boundary, which can result in identical farfield distributions.

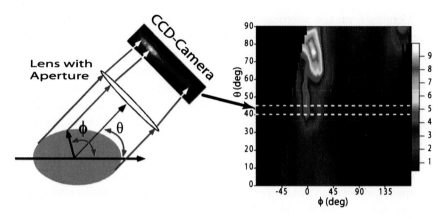

Fig. 9.   Experimental setup for measuring simultaneously farfield intensity patterns and images of the sidewall emission.

The relative angle between the CCD camera and the major axis of the quadrupolar shaped microcavity is designated as θ. The relative angle is accurate to plus or minus 5° and is determined by making laser-emission measurements from a square-shaped microcavity, specifically designed on the photographic mask to serve the purpose of alignment. All the other microstructures, during the mask designing time, are aligned relative to the square. The square, acting as a calibration marker, emits laser radiation (8 beams) only at its four corners and propagates parallel to its edges. Therefore, when the CCD camera is normal to one edge of the square, two equally bright spots should appear from the two edges. The relative angle is varied by either rotating the sample while keeping the CCD camera stationary or vice versa.

At any given camera angle, the horizontal axis corresponds to different locations along the sidewalls. That horizontal strip gives false color coded intensity information as a function of pixels on the CCD camera, which can be converted to a position ϕ on the resonator boundary. The next angle forms another strip which is placed directly under the former strip. Measurement of the intensity is made every 5° from 0 to 360. This yields a two-dimensional plot, called the *imagefield*, where a given data point $I(\phi, \theta)$ denotes the intensity emitted from sidewall position ϕ towards the farfield angle θ. The latter can easily be converted to an incidence angle $\sin \chi$, using Snell's law and basic trigonometry. Hence, what is recorded is actually a phase space plot of the emitted radiation. This correspondence is put into a rigorous basis in Ref. 18. Such 2D imagefield plots will be presented throughout the text for many of the experiments.

The farfield intensity at any angle is obtained by summing up all the pixels within the horizontal strip. This sum at a given angle is called the farfield intensity at that camera angle; we show many such plots below. This way of obtaining the farfield intensity is subtly different from placing a photomultiplier (with a pin hole) to define the angular resolution. Similarly, the *boundary image field* is obtained by integrating over all farfield angles for a fixed point ϕ on the boundary. This allows us to identify the brightest emission points on the sidewall (we rarely show these plots below, but they are used in our interpretation of the data).

A few comments are in order here. The aperture has an important role of defining a window in the direction space $(\Delta \sin \chi)$, so that a given pixel on the camera can be identified upto a diffraction limited resolution with a pair $(\phi, \sin \chi)$. Mathematically, the effect of the lens-aperture combination is equivalent to a windowed Fourier transform of the incident field on the lens.[29] Note that infinite aperture limit is simply a Fourier transform of the incident field and we lose all the information about direction $\sin \chi$, consistent with our intuition with conjugate variables. It has to be emphasized that we are only probing the farfield, and hence the image-data does not contain the "nearfield" details we would see in a typical numerical solution, nor does it contain information about the internally reflected components of the internal cavity field (see Ref. 20 for further details). On the other hand, it provides us with valuable information as to the $\sin \chi - \phi$ correlations in the emitted field, allowing us to put forward a ray interpretation of the emission and hence the internal resonance.

3.2. *Phase Space Ray Escape Model for Emission from ARCs*

In Sec. 2.3 we discussed the ray dynamics of ARCs using the surface of section to illustrate the generic properties of mixed phase space and contrast them with integrable dynamics. In the section, ARC was treated as a closed two-dimensional billiard with specular reflection and zero loss. We saw that the ray dynamics is qualitatively different for an integrable billiard shape, such as the circle or ellipse, as compared to a generic, partially-chaotic billiard shape such as the quadrupole. An implication of that difference (illustrated in Fig. 14 below) is that for the ellipse, which is integrable for any eccentricity, phase space flow occurs on a one-dimensional curve in the SOS and the variations in angle of incidence $\sin \chi$ are bounded for any initial condition. For generic shapes there are regions of phase space corresponding to chaotic motion for which motion in the SOS fills a two-dimensional region in a diffusive manner, and for deformations above 10% these chaotic regions typically make up a large fraction of the phase space. A dielectric cavity differs from an ideal metallic cavity in that rays at angles of incidence below the critical angle $\sin \chi_c = 1/n$ are partially refracted out of the cavity providing a new mechanism for emission into the farfield which differs from the evanescent coupling of whispering gallery modes. In a series of papers beginning in 1994, Nöckel, Stone and Chang[4,5,30,31] proposed to model the resonant emission from ARCs by a ray escape model in which an initial bundle of rays was propagated in phase space and allowed to escape the ARC according to a physically-motivated "escape rule"; the mean rate of escape and the distribution in angle of the outgoing rays were used to predict the Q values and emission patterns from ARCs. The escape rule reduces to the Fresnel law of refraction from a flat interface for angles of incidence below the critical angle but takes into account the tunneling (evanescent) leakage which occurs for a curved interface when the angle of incidence is above the critical angle for total internal reflection. It should be pointed out that these tunneling corrections are unimportant when the ray dynamics is highly chaotic and the critical angle is rapidly crossed, but become crucial for small perturbations where initial rays remain above the critical angle (this situation will be relevant to the silica ARC experiments reported below).

A challenging point for the general definition of such a model is that in the case of chaotic dynamics there is no simple correspondence between a set of rays and a set of modes of the wave equation (as there is in the integrable case — see Sec. 2.2). Nöckel and Stone proposed[5,31] that an appropriate set of initial conditions for ARCs would be to start a uniform distribution of rays on an adiabatic curve of the boundary,[23,32] which can be thought of as the curve in the SOS that a ray *would* follow in the absence of chaos (this approximation describes the exact flow in the ellipse, see Fig. 10 for an example). Using this model they were able to predict a striking difference in the emission patterns from quadrupole resonators with index $n = 1.5$ as opposed to index $n = 2.0$. They also noted that this difference was not highly sensitive to the choice of initial conditions. The theoretical analysis we present here indicates that the adiabatic model does not apply over most of the experimental range but that the ray escape model still gives good results, because its predictions for high deformations are almost completely independent of initial conditions for ARCs with index of refraction $n \approx 1.5$.

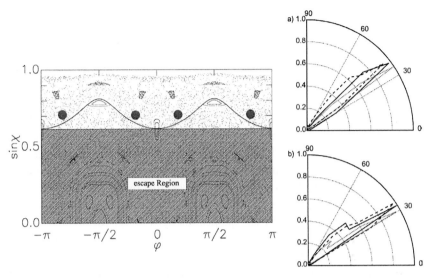

Fig. 10. (Left) Surface of section of the quadrupole ARC with $\epsilon = 0.12$ and index of refraction $n = 1.5$. The portion of the SOS below $\sin \chi_c = 1/n$ is shaded to indicate that in this region rays escape rapidly by refraction. In color we have indicated two possible types of initial conditions used in the ray escape model; the blue curve represents one of the possible adiabatic curves which were used as initial conditions in the Nöckel–Stone model and the gray circles initial conditions localized on the unstable four-bounce orbit. A third initial condition used extensively below is simply to start randomly on all possible points in the SOS originating in the trapped region above the critical line. (Right) Farfield emission plots calculated using the ray escape model for this system with the three possible choices of initial conditions just described. The qualitative and semi-quantitative features of the emission patterns are seen to be independent of the choice of initial conditions for this system. (Right) Ray simulations of the farfield emission patterns for the quadrupole with $\varepsilon = 0.12$ (a), $\varepsilon = 0.18$ (b) with different types of initial conditions. The solid curve is the result of choosing random initial conditions about the critical line $\sin \chi = 1/n$, the dashed curve is for initial conditions on the adiabatic curve with minimum value at the critical line. The dotted curve is for initial conditions localized around the unstable fixed point of the rectangle periodic orbit. In each of the ray simulations 6000 rays were started with unit amplitude and the amplitude was reduced according to Fresnel's law upon each reflection, with the refracted amplitude "collected" in the farfield. The emission pattern found by the ray model agrees well with microlaser experiments.

The ray model and its independence of initial conditions for a strongly deformed quadrupole ARC is illustrated in Fig. 10.

3.3. *Tests of the Ray Model in Polymer ARC Lasers*

Note that the emission pattern for the quadrupole at $\epsilon = 0.12$ and index $n = 1.5$ is predicted by the ray model to be highly directional with a peak in roughly the 35° direction in the farfield. Below we will see that this emission directionality for this shape is observed experimentally and is also found in numerical solutions of the wave equation. As pointed out in the initial work,[30,5,31] this emission pattern contradicts the intuitive expectation that the resonator should emit from the points of highest curvature ($\phi = 0, \pi$) in the tangent direction (critical emission) which would lead to peaks at $\theta = \pm\pi/2$ in the farfield. Moreover in later work[33] it was shown that an ellipse with the same index of refraction and the same major to minor axis ratio emits in the $\pm\pi/2$ direction as intuitively expected. This suggested a very dramatic shape sensitivity of the emission patterns, as the ellipse and the quadrupole are identical shapes to leading order in ϵ. In that same work[33] experiments on deformed spherical lasing droplets were interpreted in terms of the ray model for the quadrupole ARC (Fig. 10). While suggestive, those experiments did not have the ability to study a specific defined shape and were complicated by the three-dimensional nature of the modes of the droplet. Here we focus on experiments on polymer microcylinder lasers which do not suffer from these drawbacks.

In the polymer lasing experiments deformed cylindrical lasers were fabricated with shapes defined by a mask to approximate closely shapes which would exhibit these different behaviors. In the specific experiments we now discuss the shapes studied were cylinders with elliptical, quadrupolar and hexadecapolar deformations of between 10 and 20%[12] (see precise definitions, caption of Fig. 11). As noted, the ellipse for any eccentricity gives integrable ray dynamics and the quadrupole and hexadecapole are two simple examples of generic shapes with mixed dynamics. We will see that the

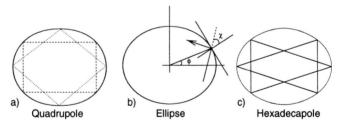

Fig. 11. Cross-sectional shapes of micropillar resonators studied: (a) The quadrupole, defined in polar coordinates by $R = R_0(1 + \varepsilon \cos 2\phi)$, (b) The ellipse, defined by $R = R_0(1 + ((1+\varepsilon)^4 - 1)\sin^2\phi)^{-1/2}$ and (c) The Quadrupole–Hexadecapole, defined by $R = R_0(1 + \varepsilon(\cos^2\phi + \frac{3}{2}\cos^4\phi))$ all at a deformation of $\varepsilon = 0.12$. Note that all shapes have horizontal and vertical reflection symmetry and have been defined so that the same value of ε corresponds to approximately the same major to minor axis ratio. In (a) we show short periodic orbits ("diamond, rectangle") relevant to the discussion below.

boundary shape of the microlaser does indeed have a dramatic influence on the emission patterns. Here we will only discuss in detail the comparison of the ellipse and the quadrupole; a more detailed study including the quadrupole–hexadecapole shaped ARCs is reported by Schwefel *et al.*[12]

As already noted, earlier work[13] had predicted that quadruple ARCs with index $n = 1.5$ and deformation in the range of 10–12% would emit primarily in the $\theta = 35°$–$45°$ direction in the farfield while an ellipse with the same major–minor axis ratio emits primarily in the $\theta = 90°$ direction (as might have been expected). It was argued that the origin of this effect is the presence in the quadrupole of a stable four-bounce periodic ray orbit which prevents emission from the highest curvature points in the tangent direction, an effect termed "dynamical eclipsing".[30,31] This finding was supported by numerical solutions of the linear wave equation for the quasi-bound states and their farfield emission patterns. Mode selection and non-linear lasing processes were not treated in the theory. This earlier work on ARCs did not look extensively at deformations above $\varepsilon = 0.12$ for the case of low index materials such as polymers or glass. The belief was that the adiabatic model would become questionable at higher deformations as the phase space became more chaotic and the ray motion departed from the adiabatic curves very rapidly. A natural expectation was that due to increased chaos the emission patterns in the farfield would become less directional and more fluctuating. The experimental data we now review[12] strongly contradicts this expectation.

3.4. *Experimental Results*

The experiments we report were performed by Rex *et al.*[10,12] on differently shaped dye (DCM)-doped polymer (PMMA) samples that are fabricated on top of a spin-on-glass buffer layer coated over a silicon substrate via a sequence of micro-lithography and O_2 reactive ionic etching steps. The effective index of refraction of these microcavities is 1.49, much lower than for other experiments (discussed below) which were performed using a similar set-up on GaN, where the index of refraction is $n = 2.65$.[11,34] The cavities are optically pumped by a Q-switched Nd:YAG laser at $\lambda = 532$ nm incident normal to the plane of the micropillar. Light emitted from the laser is imaged through an aperture subtending a $5°$ angle and lens onto a ICCD camera which is rotated by an angle θ in the farfield from the major axis. A bandpass filter restricts the imaged light to the stimulated emission region of the spectrum. The ICCD camera records an image of the intensity profile on the sidewall of the pillar as viewed from the angle ϕ which

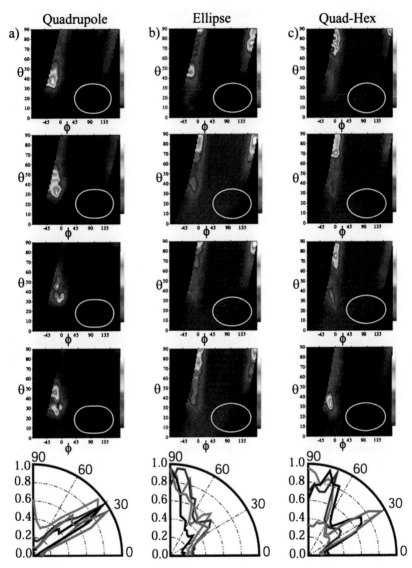

Fig. 12.   Two-dimensional display of the experimental data shown in false color scale the emission intensity as a function of sidewall angle ϕ (converted from ICCD images) and of the farfield angle θ (camera angle). Columns from left to right represent the quadrupole, ellipse and quadrupole-hexadecapole, respectively. Insets show the cross-sectional shapes of the pillars in each case (for definitions see Fig. 11). The graphs at the bottom show the farfield patterns obtained by integration over ϕ for each θ, normalized to unity in the direction of maximal intensity. The deformations are $\varepsilon = 0.12, 0.16, 0.18, 0.20$ (red, blue, black and green, respectively).

is converted from pixels to angular position ϕ. Here we show microcavities with elliptic and quadrupolar shapes of an average radius $R_0 = 100\,\mu\text{m}$ (see formulas in Fig. 11 caption). Each shape was analyzed at eccentricities of $\varepsilon = 0.12, 0.14, 0.16, 0.18$ and 0.20.

In Fig. 12 we show the experimental results in the form of a color scale 2D imagefield (ϕ, θ) plot as discussed previously. We omit the data for $\varepsilon = 0.14$ deformation as it does not indicate any effects captured by the data at the other deformations. As insets we show the exact shape of each of the microcavities. Although the shapes appear very similar to the eye, we find dramatic differences in the farfield emission patterns, which in the case of the ellipse versus the quadrupole, persist over a wide range of deformations. Specifically, the farfield emission intensity for the quadrupole exhibits a strong peak at $\theta = 34°–40°$ which remains rather narrow over the observed range of deformations. Over the same range of deformation the boundary image field (not shown) for the quadrupole changes substantially and does not exhibit one localized point of emission. In contrast, the ellipse emits into the $\theta \sim 90°$ direction in the farfield, but with a much broader angular intensity distribution, while the boundary image field remains well-localized around $\phi \sim 0°$ (the point of highest curvature in the imaged field). Thus we see qualitatively different behaviors for the two shapes studied over the same range of variation of the major to minor axis ratios. The hexadecapole shape shows yet a third behavior with a cross-over between ellipse-like and quadrupole-like patterns with increasing deformation; the origin of this is discussed in Ref. 12; this shape will not be analyzed further here.

Several different samples with the same boundary shape were measured in each case and confirmed that the basic features of this data set are reproduced within each class (with small fluctuations).[10] This shows that the effects measured are a property of the boundary shape and not of uncontrollable aspects of the fabrication process. Moreover the theoretical calculations, which will be presented next, are based on uniform dielectric rods with the ideal cross-sectional shape specified by the mask; therefore the agreement of these calculations with the measurements also confirms that the differences are due to controllable shape differences.

3.5. *Ray and Wave Simulations of Polymer Experiments*

The experiment is performed well into the short-wavelength limit, and we employ the ray escape model of Nöckel and Stone described above (caption,

Fig. 10) to calculate the emission patterns. To compare with experiment we collected the emitted rays in 5° bins. As shown above (Fig. 10), for the quadrupole the basic results are independent of initial conditions over a wide range and we employ a uniform random set of ray initial conditions above the critical line for escape. In the case of the nonchaotic ellipse the results are not independent of initial conditions and we chose initial conditions localized on invariant curves near the critical angle (this improves agreement with experiment and is more physically reasonable).

In addition to these ray simulations we also performed exact numerical calculations of the resonances of the passive cavity using the S-matrix method described in Ref. 18 and reviewed in Sec. 2.4. This method generates the entire range of high Q and low Q resonances for such a cavity and thus there is some arbitrariness in choosing which resonance to compare with the experiment. Previous experiments have indicated that mode selection is complicated in these dielectric resonators and that there is no simple rule relating the observed lasing mode to the Q value of the mode in the passive cavity. Due to their low output coupling and the multimode nature of these lasers, high Q modes are not necessarily the observed lasing modes in the farfield. This set of experiments did not allow the collection of spectral data and the Q values of the lasing modes are not directly measured. Thus from the set of calculated resonances we chose the resonance which coincides well with the observed farfield pattern and has a relatively high $Q = -2\text{Re}[k]/\text{Im}[k]$. We also confirmed that theoretical 2D imagefield data coincides well with the experimental results. Moreover in all cases discussed here, there were many resonances which gave good agreement with the data, indicating the existence of a robust class of modes any of which could be the lasing mode.

In Fig. 13 we compare the experimental results for the farfield emission patterns for the two shapes measured at $\varepsilon = 0.12, 0.18$ for both the ray model and the wave calculations. The agreement in both cases is quite good. In Fig. 13 we show in red the numerical farfield by calculating the asymptotic expansion of our wavefunction in the farfield. Numerical limitations prevent us from performing the calculations at the experimental values of $kR_0 \sim 1000$ but the major features of the emission pattern are not sensitive to kR_0 over the range we can study numerically. The finding (discussed next) that we can reproduce these patterns from ray escape simulations also suggests that the wavelength is not a relevant parameter for the features we are studying. In green we show the experimental results.

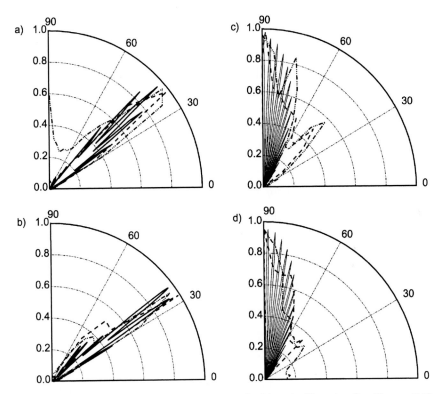

Fig. 13. Farfield intensity for the quadrupole (a, b) and ellipse (c, d) with $\varepsilon = 0.12$ (a, c) and 0.18 (b, d). The dash-dotted curve is the experimental result, (dashed) the ray simulation and (solid) a numerical solution of the wave equation. The ray simulation for the quadrupole was performed starting with 6000 random initial conditions above the critical line and then propagated into the farfield in the manner described in the text. The ray simulation for the ellipse was performed starting with 6000 initial conditions spread over seven caustics separated by $\Delta \sin \chi = 0.02$ below the critical caustic (the caustic that just touches the critical line). The numerical solutions selected for the quadrupole have $kR_0 = 49.0847 - 0.0379i$ with a $Q = -2\mathrm{Re}[kR_0]/\mathrm{Im}[kR_0] = 2593.05$ and $kR_0 = 49.5927 - 0.0679i$ with $Q = -2\mathrm{Re}[kR_0]/\mathrm{Im}[kR_0] = 1460.72$ for $\varepsilon = 0.12$ and 0.18 respectively. The numerical wave solutions for the ellipse shown correspond to $kR_0 = 49.1787 - 0.0028i$ with $Q = -2\mathrm{Re}[kR_0]/\mathrm{Im}[kR_0] = 17481.38$ and $kR_0 = 49.2491 - 0.0110i$ with $Q = -2\mathrm{Re}[kR_0]/\mathrm{Im}[kR_0] = 4488.20$ for $\varepsilon = 0.12$ and 0.18 respectively.

4. Surprising Features of the Data

The strong sensitivity of the emission patterns to small differences in boundary shape is quite striking. This sensitivity was predicted in the earlier work of Refs. 30, 31 and 33 and was therefore not unexpected. However there are major aspects of the experimental data which *are* quite surprising even in the light of the earlier work on ARCs. In particular, the persistence of highly directional emission in the quadrupolar shapes at quite high deformations was not predicted theoretically and was unexpected for reasons we will now discuss. In order to understand the unexpected features of the data and to develop principles to predict the emission patterns for untested boundary shapes we now present recent theoretical arguments about phase space flow in these systems which can account for the persistence of directional emission to high deformations and high degree of chaos.

4.1. *Dynamical Eclipsing Effect*

We begin by briefly reviewing the adiabatic picture used previously to discuss the directional emission from the quadrupole. In Sec. 2.3 we reviewed the concept of phase space flow and the Poincaré surface of section. In Fig. 14 we exhibit the difference in phase space structure between the ellipse and the quadrupole. While the behavior of the quadrupole shown in Fig. 14(A) is generic there do exist special billiards that exhibit the two extremes of dynamical behavior. One limiting case already noted is the integrable billiard exemplified by the ellipse billiard, the SOS of which is shown in Fig. 14(B). Due to its integrability, phase space flow in the ellipse is particularly simple: every initial condition lies on one of the invariant curves given by Eq. (17) below, and the trajectory retraces this curve indefinitely (see Fig. 14(B)). Curves which cross the entire SOS correspond to real-space motion tangent to a confocal elliptical caustic Fig. 14(B)(a); curves which do not cross the entire SOS represent motion tangent to a hyperbolic caustic in real-space Fig. 14(B)(b). The ellipse was conjectured to be the only convex deformation of a circular billiard which is integrable,[35] and a recent proof of this was given by Amiran.[36] At the opposite extreme is the Bunimovich stadium billiard (see inset in Fig. 19) for which it is proven that there exist no stable periodic orbits and the entire phase space (except sets of measure zero) is chaotic. We will study theoretically the emission from stadium-shaped resonators in Sec. 4.4.

Phase space flow in mixed systems is much more complex and is ergodic on each chaotic region. However a key property of mixed dynamical systems is that the different dynamical structures in phase space are disjoint; this implies that in two dimensions KAM curves and islands divide phase space into regions which cannot be connected by the chaotic orbits. This puts constraints on phase space flow despite the existence of chaos in a significant fraction of the phase space. For small deformations (~5%) most of phase

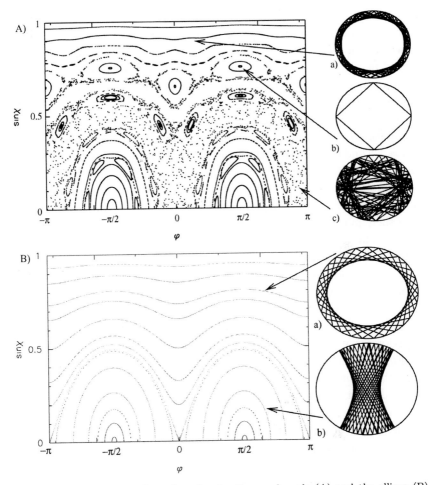

Fig. 14. The Poincaré surface of section for the quadrupole (A) and the ellipse (B) with $\varepsilon = 0.072$. The schematics (A)(a–c) on right show three classes of orbits for the quadrupole, (A)(a) a quasi-periodic orbit on a KAM curve, (A)(b) a stable period-four orbit, (the "diamond"), and (A)(c) a chaotic orbit. Schematic (B)(a, b) show the two types of orbits which exist in the ellipse, the whispering gallery type, with an elliptical caustic (B)(a) and (B)(b), the bouncing ball type, with a hyperbolic caustic.

space is covered by KAM curves the form of which can be estimated using an adiabatic approximation.[31] This approximation gives the exact result for all deformations in the case of the ellipse; it can be written in the following form:

$$\sin \chi(\phi) = \sqrt{1 + (S^2 - 1)\kappa^{2/3}(\phi, \varepsilon)} \qquad (17)$$

where κ is the radius of curvature along the boundary and S is a constant. Plotting this equation for different values of S, ε gives an SOS of the type shown in Fig. 14(B). For the mixed case, exemplified by the quadrupole billiard in Fig. 14(A), Eq. (17) describes quite accurately the behavior for values of $\sin \chi$ near unity, but does not work well at lower $\sin \chi$ where chaos is more prevalent.

Nöckel and Stone used the adiabatic curve picture to give a qualitative explanation for the difference in emission patterns between the quadrupole at $n = 1.5$ and the ellipse for the same index (or the quadrupole for $n = 2.0$). The idea was that for some range of deformations the phase space flow in the quadrupole could be seen as rapid motion along adiabatic curves and slow diffusion between them. The adiabatic invariant curves for the quadrupole have their minimum values of $\sin \chi$ at the points of highest curvature on the boundary $\phi = 0, \pm \pi$, just as they do in the ellipse. If the diffusion in phase space is sufficiently slow, emission would be near these points of highest curvature and at the critical angle, i.e. in the tangent direction, as in the ellipse. This reasoning held as long as the escape points $\sin \chi = 1/n$, $\phi = 0, \pm \pi$ occurred in the chaotic region and were reachable from the totally-internally-reflected region of $\sin \chi > 1/n$. This is the calculated behavior for $n = 2$ quadrupole.[31] However for $n = 1.5$ quadrupole and deformations around 10%, these expected emission points are enclosed by the stable island corresponding to the four-bounce "diamond" orbit and due to the disjoint nature of the dynamics, "chaotic" rays cannot escape there. Instead they will escape at higher or lower values of ϕ leading to a large change in the emission pattern from that of the ellipse with similar minor-major axis ratio. This phenomenon was termed "dynamical eclipsing".

Figure 15 contrasts the phase space for the ellipse and the quadrupole for $\varepsilon = 0.12$. The island associated with the stable diamond orbit is smaller than at $\varepsilon = 0.072$, but is still present for the quadrupole; there is no such island at any deformation for the ellipse. Note that in the experimental data for the quadrupole at $\varepsilon = 0.12$ we do not see a bright spot at the boundary at $\phi = 0$, consistent with the dynamical eclipsing model in which the island structure forces the chaotic WG modes to emit away from the point of highest curvature. In contrast, the bright spot in the ellipse which emits $\theta = 90°$ clearly is at $\phi = 0$ for $\varepsilon = 0.12$. Thus the adiabatic model of Refs. 31 and 30 does seem consistent with the data for $\varepsilon = 0.12$ in the quadrupole.

The surprising feature of the data shown in Fig. 13 is that the farfield peak remains just as narrow and displaced from the 90° direction for

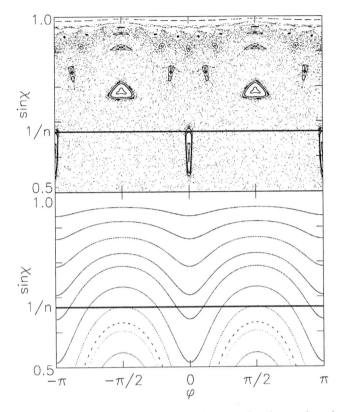

Fig. 15.   Comparison of the Poincaré surface of section for the quadrupole and the ellipse with $\varepsilon = 0.12$ showing mostly chaotic behavior in the former case and completely regular motion in the latter. The red line denotes $\sin \chi_c = 1/n$, the critical value for total internal reflection; rays above that line are trapped and those below escape rapidly by refraction. The quadrupole still exhibits stable islands at $\phi = 0, \pi$ and $\sin \chi = \sin \chi_c$ which prevent escape at the points of highest curvature in the tangent direction.

much larger deformations of $\varepsilon = 0.18, 0.20$. For these deformations the diamond orbit of Fig. 15 is unstable and generates no island in the surface of section (see Fig. 16 and inset). A simple view of chaotic motion as diffusion in phase space would suggest that escape at the critical angle in the 90° direction is now quite possible, and should lead to a broadened and less directional emission pattern in the farfield, contrary to the experimental observation and the results of ray simulations. Thus there must be a more robust mechanism to get highly directional emission than that identified in the adiabatic and dynamical eclipsing

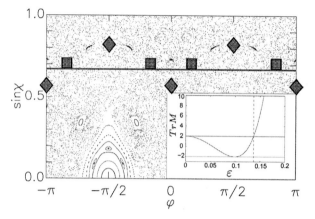

Fig. 16. Poincaré surface of section for the quadrupole with $\varepsilon = 0.18$. The gray line indicates the critical angle of incidence. The diamonds indicate the location of the fixed points of the (now) unstable "diamond" orbit and the squares the fixed points of the unstable rectangular orbit. In the inset we show the trace of the monodromy (stability) matrix (see Eq. (20)) for the diamond orbit versus deformation. When the magnitude of the trace of the monodromy matrix is larger than two its eigenvalues become real, the periodic motion becomes unstable and the associated islands vanish. For the diamond this happens at $\varepsilon = 0.1369$ (see dashed vertical line in the inset) and the simple dynamical eclipsing picture of Fig. 15 does not apply at larger deformations.

models. We discuss in the next section and improved model for the phase space flow that explains the experimental observations at higher deformations and predicts nontrivial directional emission from completely chaotic resonators.

4.2. Short-Time Dynamics and Unstable Manifolds

At higher deformations chaotic diffusion is fast and rays tend to escape rapidly even if they are initially well confined (i.e. far away from the critical angle). It is not at all clear that the motion in phase space is equivalent to slow diffusion between adiabatic curves. An iteration of random initial conditions above the critical angle for 50 steps in the closed billiard is shown in Fig. 17. It reveals a structure in the short-term dynamics which is not similar to the adiabatic curve model and actually breaks the reflection symmetry of these curves (and of the infinite time SOS) around $\phi = 0$. The colored curves which actually determine this flow pattern are the unstable manifolds of short periodic orbits in the system. We now review briefly this fundamental concept in nonlinear dynamics.

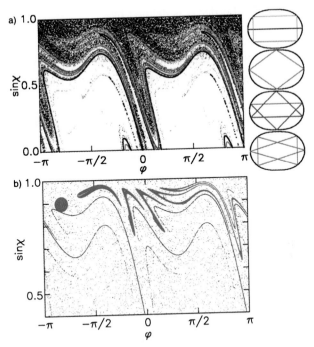

Fig. 17. (a) Ray simulations of short-term dynamics for random initial conditions above the critical line, propagated for 10 iterations, plotted on the surface of section for the quadrupole with $\varepsilon = 0.18$. The areas of the SOS covered are delineated very accurately by the unstable manifolds of the short periodic orbits which are indicated in the schematics on the right. These manifolds are overlaid in the figure with appropriate color coding. (b) Flow of phase space volume in the surface of section of the quadrupole with $\varepsilon = 0.18$. A localized but arbitrary cloud of initial conditions (red) is iterated six times to illustrate the flow. The initial volume is the circle at the far left, successive iterations are increasingly stretched by the chaotic map. The stretching clearly follows closely the unstable manifold of the rectangle orbit which we have plotted in blue.

4.3. Unstable Manifolds

The SOS is defined by a discrete map of the billiard dynamics. One can get a good idea of the short-term dynamics of a chaotic region of such a map by linearizing it in the neighborhood of unstable fixed points (corresponding to unstable periodic orbits in real space). If we take the initial position and direction/momentum of one ray at the boundary to be $(s, u) = (\phi, \sin \chi)$ we define the map which projects the ray to the next position and direction to be

$$T : (\phi, \sin \chi) \to (\phi', \sin \chi'). \tag{18}$$

A set of fixed points of order N is defined by

$$T^N: (\phi, \sin\chi) = (\phi, \sin\chi). \tag{19}$$

We can propagate an initial ray corresponding to a small deviation from the fixed point values by linearizing the map around the fixed point.

$$T(\phi, \sin\chi) \sim M(\phi, \sin\chi) = \begin{pmatrix} \frac{\partial s'(s,u)}{\partial s} & \frac{\partial u'(s,u)}{\partial s} \\ \frac{\partial s'(s,u)}{\partial u} & \frac{\partial u'(s,u)}{\partial u} \end{pmatrix} (\phi, \sin\chi)^T. \tag{20}$$

The nature of the nearby motion can then be characterized by calculating the eigenvalues and eigenvectors of M. For Hamiltonian flows M is always an area-preserving map, i.e. $\det M = 1$. The matrix M is also known as the *monodromy*, or *stability* matrix. The eigenvalues can be either complex on the unit circle or purely real and reciprocal to each other. If the eigenvalues are complex, the fixed points are stable (elliptic) and nearby points oscillate around the fixed points tracing an ellipse in the SOS. In this case the long-time dynamics is determined by the linearized map to a good approximation. In the case of real eigenvalues there will be one eigenvalue with modulus larger than unity (unstable) and one with modulus less than unity (stable) and there will be two corresponding eigendirections (not usually orthogonal). In the stable direction, deviations relax exponentially towards the fixed points; in the unstable direction deviations grow exponentially away from the fixed points. Generic deviations will have at least some component along the unstable directions and will also flow out along the unstable direction. Therefore, in a short time generic deviations move out of the regime of validity of the linearized map and begin to move erratically in the chaotic "sea". Hence the linearized map is not a good tool for predicting long-term dynamics in a chaotic region of phase space. However, in open billiards, rays will escape if they wander away from the fixed points into the part of the chaotic sea which is below the critical angle for total internal reflection. Therefore we find the unstable eigenvectors of the short periodic orbits useful in predicting ray escape. For the short periodic orbits in our shapes it is possible to calculate the matrix M giving the linearized map around all of the short periodic orbits. Thus we can calculate the eigendirections and determine the unstable directions analytically. For deviations away from the fixed points which are outside the range of validity of the linear approximation to the map one can still define generalized curves known as the stable and unstable *manifolds* of the periodic orbit which describe the set of points which would approach the fixed points asymptotically closely as $t \to \infty, t \to -\infty$ respectively. Each unstable fixed point has associated with it stable and unstable manifolds which coincide with the eigendirections as one passes through the fixed point. Note that for integrable systems there is only one asymptotic manifold for both past and future and it coincides exactly with the invariant curves, which can be calculated analytically in some cases (e.g. the ellipse). For the nonintegrable case, e.g. the quadrupole, we can only calculate the eigendirections near the fixed point analytically and must trace out the full manifolds numerically. As the unstable manifolds deviate further from the fixed points, generically they begin to have larger and larger oscillations. This is necessary to preserve phase space area while at the same time have exponential growth of deviations. This tangling of the unstable manifolds has been used to devise a mathematical proof of chaotic motion.[37] Strikingly, we see in Fig. 17 that the phase space flow at large deformations is perfectly predicted by the shape of the unstable manifolds of the short periodic orbits in that region of phase space.

One can argue qualitatively that the unstable manifolds of the short periodic orbits ought to control the ray escape dynamics at large deformations. The manifolds of short periodic orbits are the least convoluted as they are typically the least unstable; hence the unstable direction is fairly linear over a large region in the SOS. A typical ray will only make small excursions in phase space until it approaches one of these manifolds and then it will rather rapidly flow along it. If the direction leads across the critical line for escape, that crossing point and the portion just below will be highly favored as escape points in phase space. Note further that the different unstable manifolds must fit together in a consistent manner and cannot cross one another; if they did such a crossing point would define a ray which asymptotically in the past approaches two different sets of fixed points, which is not possible. Because of this noncrossing property the unstable manifolds define just a few major flow directions in the SOS. To see this more explicitly, in Fig. 17(b) we propagate an arbitrary but localized set of initial conditions and see that they are stretched along and parallel to nearby unstable manifolds. Thus it appears that for the highly deformed case the phase space flow of a generic ray is much better predicted by simply plotting these manifolds.

As a confirmation that these manifolds do control escape we perform a further ray simulation for the "open" billiard. We propagate, as before, an ensemble of rays with a uniform random distribution above the critical angle. As we have done in calculating the ray emission pattern, we associate to every starting ray in the surface of section an amplitude which decreases as the ray propagates forward in time according to Fresnel's law (if the point falls below the critical line). Instead of following the refracted amplitude into the farfield, in this case we plot the *emitted* amplitude onto the surface of section, as shown in Fig. 18(a). The emission amplitude is almost completely confined within the two downwards "fingers" created by the unstable manifold of the four-bounce rectangular orbit. As noted earlier, the availability of the two-dimensional data obtained from the imaging technique (see Fig. 12), gives us a unique ability to reconstruct the emitting part of the lasing mode both in real-space and momentum space directly from experimental data. It is therefore possible to check directly this ray simulation in phase space against experimental data. The intensity data is sorted into intensity pixels according to both its sidewall location (the angle

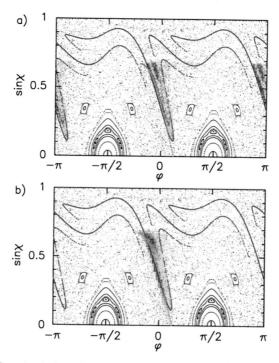

Fig. 18.   (a) Ray simulation of emission: emitted ray amplitude (color scale) overlaid on the surface of section for the quadrupole with $\varepsilon = 0.18$. (b) Farfield intensity from experimental image data Fig. 12 projected in false color scale onto the surface of section for the quadrupole with $\varepsilon = 0.18$. The blue line is the unstable manifold of the periodic rectangle orbit. In green we have the line of constant 34° farfield (see the discussion in Sec. 4.3). Absence of projected intensity near $\phi = \pm\pi$ in (b) is due to collection of experimental data only in the first quadrant.

ϕ from which emitted intensity originated) and its farfield angle, which by geometric considerations and Snell's law can be converted to the internal angle of incidence $\sin \chi$. Therefore we can project this data "back" onto the SOS for emission. In Fig. 18(b) we show this projection for the same deformation as in Fig. 18(a); we find remarkable agreement between the projected data and the ray simulation. As noted above, this is a much more demanding test of agreement between theory and experiment than simply reproducing the experimental farfield patterns.

Although the phase space flow along the unstable manifolds leads to a highly nonuniform emission pattern in phase space, this alone does not fully explain the very narrow farfield emission peak observed in the data. We see in Fig. 18(a) that there is still a significant spread of angles of incidence for escape. In fact the spread of escape angle we see in Fig. 18 would lead to an angular spread of nearly 80° in the farfield if all the escape occurred from the same point on the boundary. However as we see from Fig. 18, the point of escape and the angle of incidence are correlated and vary together according to the shape of the unstable manifold. Because the boundary is curved, different angles of incidence can lead to the same angle of emission in the farfield. It is straightforward to calculate the curves of constant farfield for a given shape; for the quadruple at $\varepsilon = 0.18$ this curve for the peak observed emission angle of 34° is plotted in Fig. 18. The curve tends to lie remarkably close to the unstable manifold. Therefore we find that the curvature of the boundary tends to compensate almost completely for the dispersion in the angle of incidence at escape.

4.4. *Directional Emission from Completely Chaotic Resonators*

The existence of highly directional emission for the highly deformed quadrupole ($\varepsilon = 0.20$) suggests that the slow diffusion in phase space, characteristic of mixed systems, is not essential to get this effect. Therefore we decided to study theoretically resonators for which the corresponding billiard is completely chaotic and for which there exist no stable periodic orbits at all. The Bunimovich Stadium (see inset in Fig. 19), mentioned above, was a natural choice due to its similarity to the quadrupole. As before we did both ray escape simulations and numerical solutions of the wave equation. In Fig. 19 we show our predictions. We find again highly directional emission with a peak direction (55°) slightly shifted from the quadrupole; the narrowness of the farfield peak in the stadium is comparable

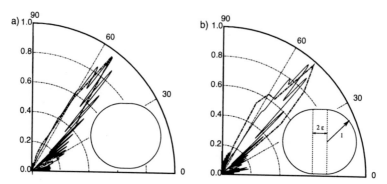

Fig. 19. Farfield emission patterns for the stadium with $\varepsilon = 0.12, 0.18$. The dash-dotted curve is the ray simulation and the solid a numerical solution of the wave equation; no experimental data was taken for this shape. The ray simulation was performed with random initial conditions exactly as in Fig. 10. The numerical solutions were for resonances with $kR = 50.5401 - 0.0431i$ with $Q = -2\text{Re}[kR]/\text{Im}[kR] = 2342.71$ and $kR = 48.7988 - 0.1192i$ with $Q = -2\text{Re}[kR]/\text{Im}[kR] = 818.83$ for $\varepsilon = 0.12$ and 0.18, respectively. The inset shows the shape of the stadium; it is defined by two half circles with radius one and a straight line segment of length 2ε.

to that of the farfield peak in the quadrupole. We can associate this peak with the slope and position of the manifold of the unstable rectangular orbit in the stadium, Fig. 20(a). The noticeable shift between the $\varepsilon = 0.12$ and $\varepsilon = 0.18$ deformation (see inset in Fig. 19) originates from the change in the slope of the unstable manifold of the rectangular orbit, Fig. 20(b). The discontinuities of slope in the unstable manifolds of the periodic orbits in the stadium result from its nonsmooth boundary. These results indicate clearly that a fully chaotic dielectric resonator can nonetheless sustain highly directional lasing modes. It would be interesting to test this in future experiments.

4.5. Tunneling versus Refractive Directional Emission

We have seen in the previous sections that low index ARC polymer lasers exhibit highly directional emission at high deformations with emission patterns that are extraordinarily sensitive to the specific shape of the boundary. The observations and simulations of directional emission at such high deformations in the quadrupole ARC indicate clearly that such modes are supported by the chaotic component of the ray phase space and emit in a manner determined by the chaotic phase space flow (specifically determined by the unstable manifolds of short periodic orbits as just discussed). Very recent experiments by Lacey et al.[28] on fused silica microspherical ARCs

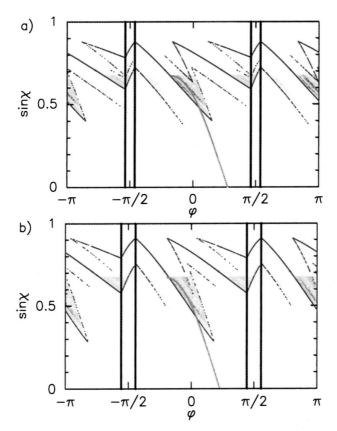

Fig. 20. Ray emission amplitude (color scale) overlaid on the surface of section for the stadium with (a) $\varepsilon = 0.12$ and (b) $\varepsilon = 0.18$. Solid blue curve is the unstable manifold of the periodic rectangle orbit. The green curve is the line of constant (a) 55° and (b) 48° emission direction into the farfield. The thick black lines mark the end of the circle segments of the boundary and coincide with discontinuities in the manifolds.

complement the lasing experiments nicely by looking at low deformations in which the phase space flow is nonuniversal and one can have either refractive emission from chaotic modes or tunneling emission from regular modes. The two types of modes have very different Q values and farfield emission patterns, with the chaotic modes showing a kind of symmetry-breaking which would be quite surprising for standard resonators. The index of refraction of these systems is $n = 1.45$ and the phase space structures determining the behavior are the same as in the polymer experiments, i.e. the stable and unstable diamond and rectangle orbits in the quadrupole and motion in their vicinity. These experiments are also important as they directly probe resonant elastic scattering from the passive cavity, for which the wave calculations are essentially exact, as opposed to lasing emission from ARCs for which issues of nonlinearity and mode selection may contribute to the observed behavior.

In this experimental work deformed spherical ARCs are fabricated by fusing two silica spheres with a CO_2 lasers at different durations of exposure, leading to nearly spherical silica "beads" on a stem with a range of deformations with major to minor axis ratios corresponding to $\varepsilon = 0.01$–0.07. Light is scattered from the spheres using frustrated total internal reflection coupling via an adjacent prism. This allows strong coupling into relatively high Q modes not accessible via a focused beam input. These shapes lack axial symmetry, which implies that the overall phase space motion is in five dimensions, not three as in the cylindrical or axial-symmetric spherical case. However for small deformations the authors argue that the escape is dominated by the same phase space structures as in the 2D quadrupole, with the crucial difference that phase space barriers such as KAM tori are not impassable classically, but can be "crossed" by Arnold diffusion in the higher dimensional phase space. The authors admit that surface scattering and other effects may play a major role as well, but the upshot is that while the prism injection excites modes with $\sin \chi_0 \approx 1$ the emission takes place from lower values of $\sin \chi$ determined by the fastest escape channel. The experiments find qualitatively different emission patterns for different deformations which can be explained by the assumption that different 2D resonances dominate the emission in the different cases.

Experimental results are shown in Fig. 21 for prism excitation; note that in such a scattering experiment only one sense of rotation of the waves is excited (see inset Fig. 21(a)). For relatively large deformation ($\varepsilon = 0.067$ is still much smaller than in the polymer experiments reported above) there is a single emission peak in $\theta = 45°$ (the other symmetric one at $\theta = 225°$ is not visible due to the presence of the prism). This pattern is essentially the same behavior as the polymer ARC lasers at higher deformation when one takes into account the presence of only clockwise circulating waves (note the authors' opposite convention on the farfield angle such that θ is positive in the fourth quadrant). There is no similar bright emission in the $\theta = 135°$ directions (dashed versus bold arrows in Fig. 21(a)); hence the pattern breaks the symmetry one might have expected for emission from modes based on the rectangular four-bounce orbit. Note that we have already seen this symmetry-breaking in the ARC polymer lasers and understand that it arises from the asymmetry of the stable and unstable directions

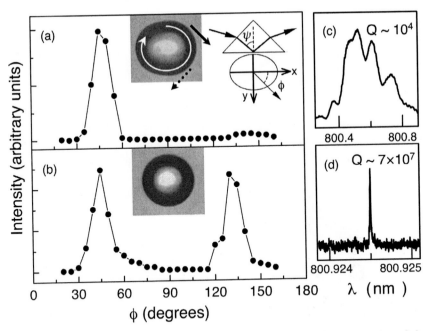

Fig. 21.   (a, b) Farfield emission patterns of WG modes. (Insets) Bottom view of the resonators showing the progression of shapes in the $x - y$ cross section $\varepsilon = 6.7\%$ and $\varepsilon = 1.2\%$ respectively. WG modes were launched at $\sin \chi_o \approx 1$. (c, d) the spectra corresponding to the modes in (a, b), from which we deduce the Q factors.

near this unstable orbit. A very different pattern is seen for samples with lower deformations ($\varepsilon = 0.012$); here two equal peaks are seen symmetrically situated around $\theta = 90°$, coincident with much higher Q factors [Fig. 22(d)]. This is consistent with emission from the points $\phi = \pm 45°$ on the boundary as one might expect from a mode based on the rectangle orbit.

The authors[28] provide a simple explanation of this difference based on the phase space structure of the quadrupole near the critical line. First, as they emphasize, any nonelliptical deformation of the circle is nonperturbative in that it destroys an infinity of symmetry-related periodic orbits (e.g. squares and rhombi) and replaces them by two periodic orbits (the unstable rectangle and the stable diamond). The size of the stable island of the diamond scales as $\sqrt{\epsilon}$ and not as ϵ and hence is much larger than naive expectations from perturbation theory. Similarly, the separatrix region of chaos near the unstable rectangle orbit will extend over a large range of $\sin \chi \sim \sqrt{\epsilon}$. Hence even very small deformations can give highly directional emission patterns. The authors assume that Arnold diffusion or some other

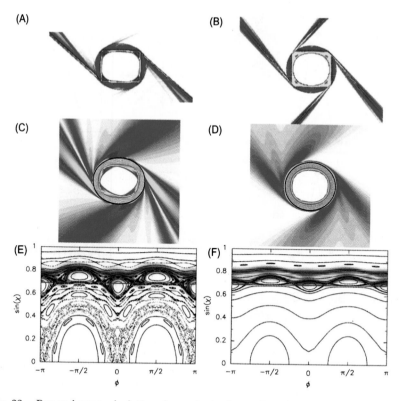

Fig. 22.   Ray and wave calculations for modes in the quadrupole with index of refraction
$n = 1.45$. Left column has a deformation of $\varepsilon = 0.065$, right column of $\varepsilon = 0.034$.
Emission patterns with qualitatively different symmetry are found in agreement with the
silica ARC experiments just reported. (Top) Intensity pattern of escaping rays with a ray
simulation based on a Gaussian bundle of rays around the unstable period four fixed point
(A) and a Gaussian bundle above the separatrix (B). In both cases we used the modified
Fresnel formula[20,38] with $kR = 1000$ which takes into account tunneling corrections
due to curvature. We propagate a Gaussian bundle of 6000 rays for 600 reflections. (C)
and (D) show intensity pattern of two associated WG modes with $kR = 112.63$ and
$kR = 112.452$ respectively. (E) and (F) show the associated Husimi distribution and the
SOS. The purple line indicates the critical angle of incidence $\sin \chi_c = 1/n$.

mechanism allows injected rays to emit lower down in the surface of section
than their injection angle. With this assumption they offer the following
explanation of the data. The nonsymmetric emission patterns observed at
$\epsilon = 0.067$ are based on refractive emission from separatrix states near the
rectangular orbit; the symmetric patterns at lower deformation are based
on states slightly higher in the surface of section which does not reach the
critical angle and emit by tunneling (evanescent leakage) from points on

the boundary symmetrically placed around $\phi = 0$ (see Fig. 22). Such states should have much higher Q (as observed) because they involve leakage from modes which would be totally-internally reflected classically. In Fig. 22 we see examples of such symmetric (tunneling) modes and nonsymmetric (refractive) modes calculated numerically, with their respective Q factors differing as in the experiment. These experiments provide another dramatic indication of the influence of nonperturbative phase space structure on the emission patterns from deformed spheres and cylinders. They are complementary to the polymer ARC laser experiments as the passive cavity mode emission patterns and Q values are directly measured, whereas the cavity shapes are not precisely controlled as in the polymer ARCs and the phase-space diffusion mechanism in the 3D cavity is not fully understood.

4.6. *Overview of Low Index ARCs*

To summarize the results of this section: low index ARC lasers and resonators show dramatic differences between the emission patterns of similarly-shaped devices which can be understood by analysis of the phase-space ray dynamics. These differences are particularly dramatic when comparing integrable shapes such as the ellipse with nonintegrable billiard shapes such as the quadrupole. The persistence of highly directional emission in strongly deformed quadrupole ARC lasers was not consistent with the earlier adiabatic model,[31] and a more recent model[12] in which emission directionality at large deformations is controlled by the geometry of the unstable manifolds of short periodic orbits gave a much better account of the data. Calculations indicate that fully chaotic ARC laser resonators should also give highly directional emission. Recent experiments using prism coupling to passive ARC cavities also indicate the coexistence of modes of different Q value and very different directional coupling. The nature of this difference arises from the different out-coupling mechanisms (tunneling versus refractive emission). In all cases, study of the phase space structures in the surface of section gives qualitative explanations of the observed emission patterns and ray models can reproduce semi-quantitatively the experimental and wave-optical results.

5. Semiconductor ARC Lasers

In the previous section we reported in some detail experiments on polymer and silica ARC resonators and lasers with index of refraction $n \approx 1.5$.

There have been, in addition, several experiments on deformed liquid droplet lasers.[39-41] In these experiments a universal phase space flow (i.e. a property insensitive to initial conditions in phase space) determined the directional output of the laser. Moreover, all of these have been optically pumped systems and hence not suitable as prototypes for microcavity lasers of technological interest. During roughly the same time period a number of ARC semiconductor lasers have been fabricated and measured, and in two cases these have been electrically-pumped lasers of some potential technological interest. In these systems the index of refraction ranged between $n = 2.65 - 3.3$ leading to stronger confinement of light rays by (near) total internal reflection, and hence increasing the fraction of phase space accessible to long-lived modes for lasing. In these lasers the nature of the lasing modes were more diverse and the issue of how mode selection occurs was strongly raised (but not yet fully answered). We will review the relevant experiments in roughly chronological order: electrically-pumped quantum cascade ARC lasers first, then optically-pumped GaN ARC lasers, and finally optically and electrically-pumped "spiral" GaN multiple-quantum-well lasers, which are not really ARCs by our definition, but which were a natural outgrowth of the shape design program which began with ARCs.

5.1. Quantum Cascade ARC Lasers

The development of efficient semiconductor microlasers is of primary importance for current as well as future photonic or optoelectronic applications. A major milestone in this direction was the development of quantum cascade (QC) lasers by the Bell Labs team of Faist, Cappasso et al.[42] Typical semiconductor lasers are bipolar in character (i.e. are diodes), meaning that the laser action is fed by the *interband* electron-hole transitions of a semiconducting heterostructure involving doped and undoped III–IV semiconductors such as GaAs, InAs, GaN. Quantum Cascade lasers on the other hand are *unipolar* and employ the electronic inter-subband transitions between quantized conduction band states in a multiple quantum-well structure. The unipolar character of QC lasers excludes nonradiative combination of electrons and holes which is a major problem with diode lasers when one wants to access shorter or longer wavelengths other than the typical communication window of about 1.3–1.5 μm. Another attractive feature of the QC devices is their versatility in emission wavelength. Different transition energies can be realized by adjusting the individual layer thicknesses without changing the composition of the constituent materials, covering virtually the complete infrared spectral region ($\lambda \sim 3$–$25\,\mu$m).

While the ever-improving semiconductor quantum-engineering technology is an important factor in increasing the efficiency of these miniature lasers, the optimization of the geometric shape of the resonator also plays a major role. The study of ARC microcylinder QC lasers was undertaken along these lines in a 1997 collaboration between Yale and Bell labs leading to the exciting results reported by Gmachl *et al.*[7]

5.1.1. *Directional Emission from Stable Bowtie Modes*

To test the effect of deformation on the lasing properties, a set of microcylinder lasers of increasing quadrupolar deformation were fabricated and tested on the same chip.[7] It was found that while for lower deformations the emission is more or less isotropic, above a deformation of about $\epsilon = 0.14$ a large anisotropy rapidly developed. At $\epsilon = 0.16$, the emission pattern peaked at about 45° with a maximum to minimum ratio of about 30 : 1 (see Fig. 23). While the low deformation data were consistent with emission from whispering gallery modes, it was clear that a mode of a different character was dominating the emission at higher deformations. There were two experimental clues which were available for the determination of the

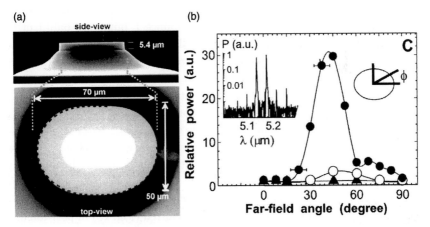

Fig. 23. (a) Scanning electron micrographs of the top and side views of one of the deformed cylindrical quantum cascade microlasers. (b) Angular dependence of the emission intensity for deformations $\epsilon = 0$ (triangles), $\epsilon = 0.14$ (open circles), $\epsilon = 0.16$ (filled circles). The right inset shows the coordinate system used and the left inset shows the the logarithmic plot of the measured power spectrum. The FSR of the peaks is found to agree with the calculated bowtie FSR (after Ref. 7).

lasing mode:

- Sudden onset of directionality above a deformation of $\epsilon = 0.14$.
- The existence of six equally spaced peaks in the measured spectrum at maximum power.

From the standpoint of ray dynamics, the main difference between semi-conductor ARCs and lower index polymer ARCs analyzed in the previous section is their typically higher index of refraction. For the material system used in the above experiment (InGaAs/InAlAs), the effective index of refraction was $n \sim 3.3$. Such a large index of refraction makes available a larger portion of the SOS available for the support of long-lived resonances. Particularly, the lower portion of the phase space close to the critical line ($\sin \chi \sim 0.3$) is a regime which is subject to more (classical) nonlinearity and hence chaos than the higher lying WG region (as a simple measure, the effective "kick strength" of the quadrupole billiard map is $(1-\sin^2 \chi)^{3/2}$).[38] Therefore the difference between stable and chaotic motion is enhanced for intermediate deformations in this part of the SOS, where large islands of stable motion and strong chaos coexist.

The observation of rather sudden onset of directional emission suggested a mode which (unlike the whispering gallery modes) is not continuously connected to zero deformation. In fact, there is one clear candidate, the stable bowtie orbit, that does not exist below a deformation of $\epsilon = 0.11$. For all ARCs there is a large stable two-bounce ("bouncing ball") orbit which appears nonperturbatively for small deformations and which typically destabilizes as the deformation increases. The bowtie orbit arises as a by-product of the destabilization of the bouncing ball orbit; it is "born" through a period-doubling bifurcation of this orbit at $\phi = \pm\pi/2$ (see Fig. 24). Below a deformation of about $\epsilon = 0.14$, the corresponding bowtie resonances are too leaky to provide efficient laser feedback, because as seen from Fig. 24, the stability island of the bowtie orbit is far below the critical line. However, further deformation slowly moves the island up in $\sin \chi$ until at about $\epsilon = 0.16$, when the critical line is reached and the data indicate that its associated mode becomes the favored lasing mode. Since the stable bowtie motion is strongly localized (Fig. 25) in position and space, it leads to highly directional emission. Properties of modes based on stable ray motion close to a periodic orbit are amenable to an analytic study through a generalized Gaussian optical description.[15] Quite similar to the sequence of Gauss–Hermite modes found in a stable Fabry–Perot cavity, it is possible to associate a sequence of modes with an island of stable motion.

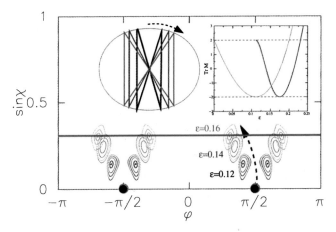

Fig. 24. Illustration of the motion of the bowtie island with changing deformation. Drawn in red is the critical line. Inset shows the stability diagram of the bouncing ball and bowtie orbit as given by the variation of the trace of the monodromy matrix M. For $|TrM| > 2$ an orbit is unstable, hence one sees that motion in the vicinity of the bowtie orbit is stable for $0.11 < \epsilon < 0.23$. The bouncing ball restabilizes after the bifurcation and only becomes unstable at $\epsilon \approx 0.2$; a nongeneric behavior which is possible due to the discrete symmetry of the billiard.

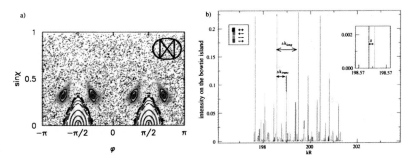

Fig. 25. (a) Surface of section for quadrupole billiard at deformation of $\epsilon = 0.17$ which supports a stable bowtie orbit (see inset) leading to four islands in the SOS. A numerical solution of the resonance problem is Husimi-projected onto the SOS and is plotted in color scale; it localizes on the island and is a stable bowtie mode. (b) Spectrum weighted by overlap of the Husimi projection of the solutions in a spectral range with the bowtie island. Note the emergence of regularly spaced levels with two main spacings Δk_{long} and Δk_{trans}. These spacings, indicated by the arrows, are calculated from the length of the bowtie orbit and the associated stability angle (see Eq. (21)). The color coding corresponds to the four possible symmetry types of the solutions, as explained in the text. In the inset is a magnified view showing the splitting of quasi-degenerate doublets. Note the pairing of the $(++)$ and $(+-)$ symmetry types. The different symmetry pairs alternate every FSR (Δk_{long}).

To lowest order, the resonance frequencies and lifetimes of these modes are functions of the index of refraction n, the length of the orbit L and the radius of curvature at each bounce b of the periodic ray orbit, $\{\rho_b\}$, and the corresponding impact angles $\{\sin \chi_b\}$:

$$\mathrm{Re}[nkL] = 2\pi m + \mathrm{mod}_{2\pi} \left[\left(\frac{1}{2}N + N_\mu\right)\pi\right] + \left(n + \frac{1}{2}\right)\varphi + \varphi_f \quad (21)$$

$$\mathrm{Im}[nkL] = -\gamma_f \quad (22)$$

where $\varphi_f = \mathrm{Re}\left[-i\sum_b^N \log\left[\frac{n\mu_b-1}{n\mu_b+1}\right]\right]$ and $\gamma_f = \mathrm{Im}\left[-i\sum_b^N \log\left[\frac{n\mu_b-1}{n\mu_b+1}\right]\right]$. μ_b is the ratio of incidence angle to transmitted angle $\cos\chi_i / \cos\chi_t$ calculated from Snell's law at each bounce b, φ is the stability angle for the particular periodic orbit, N is the number of bounces and N_μ is the Maslov index which depends on the topology of the phase space motion.[15] The resulting spectrum contains two distinct constant spacings. Here, the longitudinal mode index m gives rise to a FSR $\Delta k_{\mathrm{long}} = 2\pi/L$ and the transverse index n results in a shorter FSR of $\Delta k_{\mathrm{trans}} = \varphi/L$. Because of the symmetry of the resonator shape under reflections with respect to its minor and the major axes, the solutions can be classified into four different classes. Each class is distinguished by the parity of the corresponding solutions under the reflection operations, denoted by $++, +-, -+$ and $--$. For instance, the solution $+-$ has even parity under reflection with respect to the long axis, and odd with respect to the short axis. The Gaussian theory predicts[15] that for the bowtie mode, solutions with identical parity with respect to the *short axis* form degenerate doublets, and that these two parity types ($[++, +-]$ and $[--, -+]$) alternate in the spectrum every FSR. A group theoretical analysis of the symmetry properties of the *exact* solutions however shows that the degeneracy is not exact. In fact, the exact quasi-degenerate solutions display an exponentially small (in k) splitting. As seen in Fig. 25, these arguments reproduce the exact (numerical) spectrum quite well.

The observed lasing spectrum in the experiments shows six equally spaced peaks (see Fig. 23), with a FSR of $\Delta\lambda = 40.4\,\mathrm{nm}$. This spacing is approximately equal to the longitudinal FSR ($\Delta\lambda_{\mathrm{long}} = 39.5\,\mathrm{nm}$) calculated from Eq. (21), indicating that it was a particular transverse mode which was lasing for different values of m. For comparison, the transverse mode spacing expected from Eq. (21) is about $\Delta\lambda_{\mathrm{trans}} = 2.2\,\mathrm{nm}$ and the splitting of quasi-degenerate resonances is about $\Delta\lambda_{\mathrm{split}} \approx 0.1\,\mathrm{nm}$. Just by looking at the spectrum it is not possible to discern which transverse mode or which symmetry class is lasing. At this point, the farfield emission pattern can be used to pinpoint the lasing mode. A crucial point here is that each spectral peak is expected to be formed through locking of the quasi-degenerate modes corresponding to two different symmetry classes. It has been shown both theoretically[43,44] and experimentally[45] that if eigen-frequencies of the modes participating in the nonlinear lasing process are close enough, locked behavior can occur leading to stationary output intensity pattern. This is called *cooperative frequency locking*. The resulting non-linear modes are linear combinations of the participating modes and may display asymmetric emission patterns despite the fact that the resonator is symmetric (a good example of spontaneous symmetry breaking). To compare with the experiments, we plot in Fig. 26, linear combinations of the quasi-degenerate transverse modes close to the central lasing frequency. The farfield emission pattern observed in the experiments is found to be in conformity with that of the transverse doublet $m = 2$. Note that the resulting emission pattern peaks at a point away from what is expected from Snell's law applied to the bowtie orbit. These findings were reinforced by experiments by the TU-Wien group[8] on QC GaAs/AlGaAs microlasers emitting around $\lambda = 10\,\mu$m. The boundary deformations investigated in this work were again quadrupolar and the effective index of refraction around

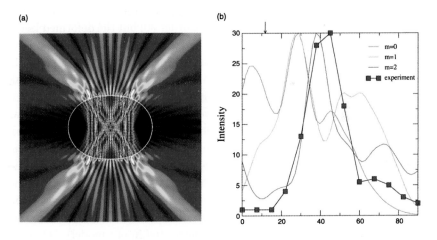

Fig. 26. (a) The numerically calculated resonance corresponding to the $m = 2$ bowtie-mode at $\epsilon = 0.16$ that is consistent with the observed spectrum and emission pattern. (b) Comparison of experimental data for $\epsilon = 0.16$ to numerically determined farfield patterns for the $m = 0$, $m = 1$ and $m = 2$ transverse modes of the bowtie resonance. We have plotted linear combinations of quasi-degenerate doublets close to $nkR = 120$. The peak tranmission is in good agreement with the $m = 2$ tranverse mode. The arrow points to the peak angle expected from ray optics, i.e. refracting out of the bowtie-orbit using Snell's law.

$n_{\text{eff}} \approx 3.2$. Emission patterns observed were of two distinct types, one of which was found to correspond to that of the bowtie mode. The robustness of the emission pattern to a change in lasing wavelength is a testament to the effectiveness of inferences based on short-wavelength approximations and ray-optical phase space. As long as the resulting phase space (and the index of refraction) is identical, similar farfield patterns are expected regardless of the wavelength and the size of the resonator (provided $nkR \gg 1$; in the Bell Labs–Yale experiment $nkR \sim 120$ and in the TU Wien experiment $nkR \sim 100$).

Although numerical work and physical arguments allowed us to identify this mode after the experiment, we have no predictive theory for the mode selection in this case. The second excited bowtie mode is not the highest Q mode of the cold cavity, nor is it particularly selected by the peak of the gain curve, which is broad enough to allow other modes to lase.

5.1.2. *Power Increase and Mode Selection in ARCs*

The issue of mode selection in ARCs is made more salient because of the second major finding of the Bell Labs experiment. Not only did the bowtie laser provide highly directional emission, improving the brightness (power into a given solid angle) of the laser by a factor of order 30; but the deformed lasers produced more than a thousand times the output power of the identically-fabricated circular lasers.[7] The demonstrated high output power and directional emission solve major problems with earlier semiconductor microdisk lasers and make their use technologically promising.

The difference in peak output power between the ARC and circular cylinder lasers is interesting theoretically and not yet explained. The peak output power certainly depends on the nonlinear properties of the lasing and is not a property of a given mode of the linear wave equation. The standard and well-verified model of the power output of a single-mode Fabry–Perot laser[14] finds that the power output is optimized for a given pump power when the external cavity loss (which is the width we are calculating) equals the internal cavity loss (neglected in our model). One may conjecture then that the bowtie optimizes the power output, even though it is not the highest Q mode. However this observation is not sufficient to explain the experiment. The circular lasers measured have a range of Q values corresponding to different radial quantum numbers for a given angular momentum. In particular, there should exist modes with Q very near that of the bowtie. The main difference between circular and ARC lasers is that in the circular case this mode will strongly overlap in space with other higher Q modes; whereas in the ARC strong chaos has wiped out the competing higher Q modes in the vicinity of the bowtie. Therefore we conjecture that it is the lack of mode competition in bowtie lasers which allows it to optimize its output power. The experiment provides some modest support for this conjecture as the circular lasers are consistently found to lase on several modes simultaneously, whereas the ARC lasers are single mode unless they are pumped very hard.[7] Work is underway to analyze mode competition in chaotic lasers and verify these speculations.

Both experiments and the accompanying theoretical analysis indicate that the geometrical shaping of the resonator is an important parameter in the design of efficient miniature laser devices. It was demonstrated that by merely optimizing the shape of the resonator, it is possible to increase the optical output power by three order of magnitude and simultaneously obtain a directionality emission asymmetry of 30:1.

5.1.3. *Anomalous Q Values of Stable ARC Modes: A New Signature of Chaos*

We close this section by reporting a dramatic prediction of the wave chaos theory relevant to stable orbit modes, but not yet tested experimentally. This prediction is not relevant to the QC ARC lasers just mentioned for which the stable orbit is right at the critical angle, but will be very relevant for modes which are totally internally reflected (TIR). In Sec. 5.1.1 above we have noted that the discrete symmetry of quadrupole ARCs (as opposed to the continuous symmetry of circular resonators) precludes the existence of exactly degenerate stable (or unstable) modes, despite the fact that such degeneracies are predicted to occur within the Gaussian optics approximation for the stable modes. This is a well-known shortcoming of semiclassical methods which at leading order do not resolve exponentially small effects due to tunneling, and this short-coming applies to both integrable ARCs, such as the ellipse, as well as the (generic) nonintegrable ARCs. However it has been predicted[15,47] that the absolute size of these tunneling effects are dramatically sensitive to the presence or absence of chaos in parts of the phase space away from the stable islands which give rise to Gaussian modes. We will not attempt to go into any details of the theory here; they can be found in Refs. 15 and 47. The qualitative physics is the following. Consider any stable periodic orbit which is not self-retracing; there will exist an independent Gaussian mode series with a distinct field pattern corresponding to the two senses of traversing the orbit, and within the Gaussian theory these two series will be exactly two-fold degenerate. However there is some small tunneling probability for a ray circulating in one sense to eventually reverse its sense of circulation. This tunneling between different stable motions at the same energy has been termed *dynamical tunneling*.[48] More formally, taking symmetric and anti-symmetric combinations of the two wavefunctions just mentioned will lead to two states of slightly different energy/frequency (differing by twice the tunneling rate as in the familiar double-well problem). For objects without continuous symmetries this rate will always be nonzero, but its value will depend strongly on the nature of the surrounding phase space. Specifically for stable island modes surrounded by a chaotic "sea" the tunneling rate is parametrically larger than in a comparable integrable system, leading to a much large splitting of the Gaussian modes in the closed system.[15] However these splittings may still be difficult to resolve experimentally.

Very recently, Narimanov[49] has pointed out that this enhanced tunneling rate, known as "chaos-assisted tunneling" (CAT), will have dramatic observable consequences for ARCs with totally internally reflected stable orbit modes (TIRSO modes). Here the signature of the effect is in the Q values of the modes and not in the splittings, and the effect can be seen in comparison to circular resonators with zero splitting of the comparable modes. As reported above,[28] it is possible to couple to totally-internally reflected modes of passive dielectric cavities using prism coupling. According to the new prediction of CAT theory for ARCs the measured Q values of TIRSO modes will be orders of magnitude smaller than for the same angle of insertion in the circular cavities. Moreover, the Q values of TIRSO modes will fluctuate rapidly between different resonances in the same longitudinal sequence while no such fluctuations will be observed for the circular case. This is the signature that escapes from the stable orbit modes of ARCs mediated by tunneling into the chaotic states of the resonator prior to escape into the farfield, whereas no such escape mechanism exists for circular resonators. As the phenomenon of chaos-assisted tunneling in wave-chaotic systems has had few if any experimental demonstrations, it is hoped that experiments of this type will be attempted in the future.

5.2. *Diode ARC Lasers*

Our arguments and examples up to this point indicates that a key requirement for a resonant mode to be the dominant observed mode of a laser is that its weight in the SOS (as measured e.g. by Husimi projection) be concentrated near the critical line for refractive escape. Hence, the investigation of the classical phase space structures in the vicinity of critical reflection gave us a quick and crude estimate of the emission directionality. Particularly, islands of stable motion in the vicinity of critical incidence can lead to localized modes with highly directional emission, as was the case for the bowtie modes of the previous section. One of the intriguing and well-studied results of quantum chaos theory is the existence of localized modes based on *unstable* periodic orbits, known as *scarred states*.[50-52] If such an orbit is trapped by near total internal reflection near the critical line it may lead to an "unstable" lasing mode. Exactly such a situation was realized in semiconductor diode microlasers studied by Rex *et al.*[11] We will now review this experiment and discuss its theoretical interpretation.

In these experiments, microlaser devices were produced by growing GaN on a sapphire substrate, and etching the resulting wafer using a mask and standard photolithography into a $2\,\mu\text{m}$ high pillar with a quadrupolar deformation of the cross-section, $r(\phi) = R_0(1+\epsilon\cos 2\phi)$ where $R_0 = 100\,\mu\text{m}$ (see Fig. 27). The resulting structure, which has an index of refraction $n = 2.65$, is optically pumped at 355nm normal to the plane of the pillar and emits at 375nm. Light emitted from the side-walls is collected through an aperture subtending a $5°$ angle, passed through a lens and detected by a CCD camera in the farfield to yield two-dimensional image data plots of the type described above in Sec. 3.1).

We will now focus on the data collected for $\epsilon = 0.12$ quadrupoles. Note that this is the deformation at which the polymer cylinder lasers discussed

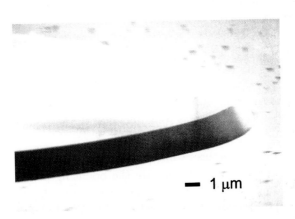

Fig. 27. A scanning electron micrograph of the GaN resonators used in the experiments. The device in the figure has a diameter of $200\,\mu\text{m}$.

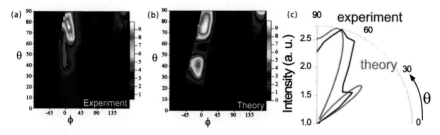

Fig. 28.   (a) Experimental data shown in color scale for the CCD images (converted to sidewall angle ϕ) as a function of camera angle θ. Three bright spots are observed on the boundary for camera angles in the first quadrant, at $\phi \approx 17°, 162°, -5°$. (b) Calculated image field corresponding to the scarred mode shown in Fig. 31. (c) Calculated and experimental farfield patterns obtained by integrating for each θ over ϕ.

earlier emitted at roughly 35° direction to the major axis, with a boundary image showing a single bright spot on the boundary at negative values of sidewall angle ϕ which correlated well with the farfield peak if one assumed tangent emission. The imagefield for the GaN quadrupole lasers is displayed in Fig. 28(a) and the farfield is shown in Fig. 28(b).

The data show a very different emission pattern than the polymer lasers of the same shape, demonstrating the crucial role of the refractive index in determining the lasing mode selected. For the GaN lasers the maximum intensity in the first quadrant is observed at angle $\theta \approx 74°$ and correlates with emission from the region of the sidewall around $\phi \approx +17°$. The data also show a secondary bright spot at slightly negative $\phi \approx -5°$ and another one at $\phi \approx 162°$ which do not lead to strong maxima in the first quadrant in the farfield. The observation of a small number of well-localized bright spots on the sidewall suggests a lasing mode based on a short periodic ray trajectory. In Fig. 29, we have indicated the approximate positions of the four bright spots on the boundary (the imagefield can be unfolded to the range $\theta = 0 \ldots 2\pi$ using the symmetry of the quadrupole). The imagefields for the polymer lasers showed brighter spots and a variable number of them, inconsistent with as single short periodic orbit. In the same figure is shown a view of the SOS at this deformation.

The only stable structures which would result in localized modes in the framework of the previous section, are the bouncing ball and the bowtie islands. For comparison, the stable bouncing ball mode would emit from $\phi = 90°$ in the direction $\theta = 90°$. The stable four-bounce bowtie mode, dominant in the devices of Ref. 7, is also ruled out by our data. It is very low Q at this deformation due to its small angle of incidence and would

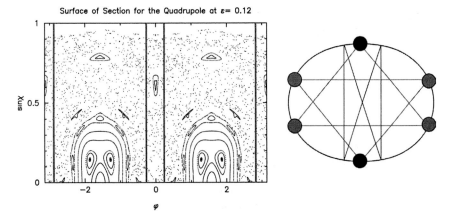

Fig. 29.   The SOS of the quadrupolar billiard at a deformation of $\epsilon = 0.12$. The red vertical lines indicate the values of ϕ at which the bright spots in the imagefield are observed. On the right is a schematic indicating in red the experimental bright spots in the real-space. The location of these spots is strongly inconsistent with the bowtie orbit at this deformation but is consistent with modes based on the two triangle orbits shown. These orbits would have the two "dark" bounce points (indicated in black) that are well above total internal reflection for the index $n = 2.65$.

give bright spots at $\phi = 90° \pm 17°$, far away from the brightest spot at $\phi = 17°$ (see Fig. 29). There is however a pair of symmetry-related isosceles triangular periodic orbits with bounce points very close to the observed bright spots (see Fig. 29). The two equivalent bounce points of each triangle at $\phi = \pm 17°$ and $180° \pm 17°$ have $\sin \chi \approx 0.42$, very near to the critical value, $\sin \chi_c = 1/n = 0.38$, whereas the bounce points at $\phi = \pm 90°$ have $\sin \chi = 0.64$ and should emit negligibly (Fig. 29). This accounts for the three bright spots observed experimentally in Fig. 28(a) (the fourth spot at $\phi \approx 197°$ is completely blocked from emission into the first quadrant). Note furthermore the proximity of the four emitting bounce points to critical incidence; a simple application of Snell's law to these rays would lead to farfield maxima in reasonable agreement with the observed peaks in the farfield distribution Fig. 28(c) (however not with the imagefield data, see below).

These basic observations could be explained with generalized Gaussian modes of the previous section, were it not for the fact that the triangular periodic orbit is unstable at this deformation. In Fig. 30, we plot the trace of the monodromy matrix as a function of deformation, which shows that at deformation $\epsilon = 0.12$, $\mathrm{Tr}(M) = -5.27$. The triangular periodic orbit is unstable with a Lyapunov exponent of $\lambda \approx 1.62$ (see Fig. 30). We have remarked in Sec. 5.1 that our method of constructing a localized mode on a

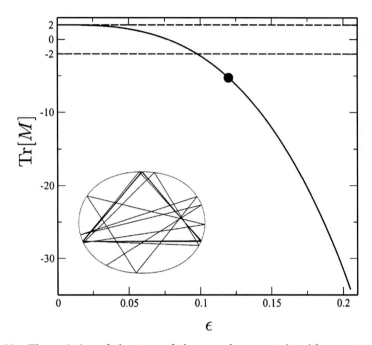

Fig. 30. The variation of the trace of the monodromy matrix with respect to the quadrupolar deformation ϵ. The black circle indicates the experimental value $\epsilon = 0.12$, at which $\text{Tr}(M) = -5.27$. The two dashed lines delimit the regime $-2 < \text{Tr}(M) < 2$ at which the triangular orbit is stable. In the inset is shown real-space simulation of a ray orbit started with initial conditions which are away from the triangle fixed point at least by $\delta\phi = 10^{-3}$, $\delta \sin \chi = 10^{-4}$, followed for 20 bounces.

periodic orbit fails, if the orbit is unstable. Failure of the method however does not mean that localized modes do not exist. In fact, numerical solution of the quasi-bound states at this deformation, using the method of Sec. 2.4, finds modes localized on the triangular orbit, as seen in the configuration space plot in Fig. 31(a). A much clearer picture, free of interference fringes, is provided by the Husimi plot of this mode in Fig. 31 projected onto the SOS. The brightest spots clearly coincide with the triangular fixed points, and the whole density is localized in the midst of the chaotic sea. This mode is an instance of a *scarred state* and is one of the most surprising and esoteric objects of quantum chaos theory.

Note however that the numerical calculations are performed at $nkR \approx 129$, whereas the experimental lasing frequency corresponds to $nkR \approx 4400$. Despite this difference of more than one order of magnitude, the agreement between experimental results (farfield and imagefield) and

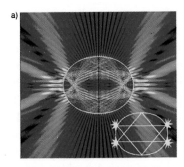 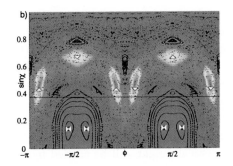

Fig. 31.   (a) Real-space false color plot of the modulus of the electric field for a calculated quasi-bound state of $nkR_0 \approx 129$ (n is the index of refraction, k is the real part of the resonant wavevector) and $\epsilon = 0.12$ which is scarred by the triangular periodic orbits shown in the inset (M. V. Berry has termed this the "Scar of David"). The four points of low incidence angle which should emit strongly are indicated. (b) Husimi (phase-space distribution) for the same mode projected onto the surface of section of the resonator. The x-axis is ϕ_W and the y-axis is $\sin\chi$, the angle of incidence at the boundary. The surface of section for the corresponding ray dynamics is shown in black, indicating that there are no stable islands (orbits) near the high intensity points for this mode. Instead the high intensity points coincide well with the bounce points of the unstable triangular orbits (triangles). The black line denotes $\sin\chi_c = 1/n$ for GaN; the triangle orbits are just above this line and would be strongly confined whereas the stable bowtie orbits (bowtie symbols) are well below and would not be favored under uniform pumping conditions.

simulation is quite good. To understand this we need to discuss some further aspects of scarred modes.

"Scarring" refers strictly to the imprint left by *unstable* periodic orbits in a *group* of states.[50] There is still a lot of discussion about the quantification of this imprint,[52] but for our purposes here, the simplest and most intuitive of such measures is the enhancement of eigenstate intensity along (a tube surrounding) a given unstable periodic orbit and its invariant manifolds (the latter is best measured in the SOS). Understanding the phenomenon of scarring requires a major departure from the approaches of EBK and Gaussian optics which allow construction of individual modal solutions of the wave equation. Scarring on the other hand refers to a statistical phenomenon. It is a statistically significant correction to Berry's conjecture[53] that for ergodic systems (in the short-wavelength limit), individual eigenstates will cover uniformly all the available energy hypersurface (translated to optical resonators, this means that the local angular spectrum calculated at any point in the resonator will contain all the possible directions uniformly), up to uncorrelated Gaussian fluctuations. In the extreme short-wavelength limit no individual mode of the wave equation will localize on an unstable periodic orbit. Instead, the effect of a short periodic orbit and its associated hyperbolic manifold will be seen in a group of eigenstates in an energy range Δk, where $\Delta k \sim u/L$ (here u is the Lyapunov exponent and L is the length of the orbit). The broadened spectral peaks corresponding to these states repeat periodically with a period of $2\pi/L$. Since the average density of states of a 2D or 3D optical resonator increases with k once $kR \gg 1$ there will be many states under this broadened peak and no single state will be strongly localized on the periodic orbit; the additional statistical weight will be carried by many of them. In systems of the type we are considering there are an infinite number of periodic orbits, however the effect of longer orbits on the spectrum and modes is less significant, because they are more unstable: $\Delta k \sim u/L \sim$ constant whereas the spacing of the peaks $\sim 1/L$ so the broadening becomes much larger than their spacing and no local density of state oscillations are observable. For the short and not very unstable periodic orbits one finds that the averaged wavefunction magnitude over this range Δk (and space) is found to display a strong enhancement in space in the vicinity of the particular periodic orbit with a form depending only on the parameters of the ray orbit and a simple scaling with k.[54] Along similar lines, for optical resonator modes, the farfield averaged over a given wavelength (spectral resolution of the spectrum analyzer) and spatial range, would display clearly the effect of a single short periodic orbit and its linear manifolds. Our numerical solution in Fig. 31 at low nkR is a good representation of how an *averaged* wavefunction and emission at higher nkR would manifest, because at lower nkR the range $\Delta k \sim u/L$ will contain only a few modes and a single mode can be a "strong scar". One may also conjecture that the effect of nonlinear mode locking may work to create a single nonlinear scar out of many nearby modes in the actual lasing system. Therefore a scarred mode, or multimode lasing emission consistent with scarring is a reasonable explanation of the data.

However closer inspection of the imagefield in Fig. 28(a) presents an intriguing puzzle from the point of view of ray optics. A mode localized on these triangular orbits might be expected to emit from the four-bounce points approximately in the tangent direction according to Snell's law; this means that the bright spot at $\phi = 17°$ should emit into the direction $\theta \approx 115°$ (Fig. 32), whereas the data clearly indicate that the $\phi = 17°$ bright spot emits in the direction $\theta = 72°$. (Note that the Snell's law argument worked well for the polymer lasers studied above.) Thus the emission pattern here violates the intuitive expectations of ray optics by $43°$, a huge

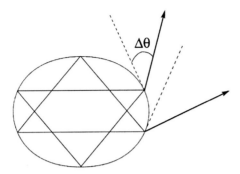

Fig. 32. Schematics showing the three emitted "beams" detected in the experiment (solid lines) and illustrate their strong deviation from Snell's law (dashed tangent lines).

discrepancy (see Fig. 32). Moreover, $\lambda/nR = 2.8 \times 10^{-3}$, so we are far into the regime in which the wavelength is small compared to the geometric features of the resonator and ray optics would be expected to be a good approximation. To ensure that this discrepancy did not arise from some error in the experimental imagefield we simulated the full experimental set-up, starting with our numerically-determined scarred solution [Fig. 31(a)] inside the resonator and propagating it through an aperture and lens into the farfield, reproducing the expected imagefield. The way to do this is described in Ref. 20. The imagefield corresponding to the numerical resonance calculated in this manner is reproduced in Fig. 28(b). The good agreement with the experimental data in Fig. 28(a) indicates that the effect is real and is robust over a range of wavelengths.

The physical mechanism for this surprisingly large violation of ray optics was given in Refs. 11 and 55. From the numerical data of Fig. 31(b), it is clear that the scarred mode, while localized around the triangle orbit, has a significant spread in angle of incidence, $\Delta \sin \chi \approx 0.2$. This means that we must regard the scarred mode as a bounded beam with a large angular spread, with some components almost totally reflected and other components transmitted according to the Fresnel transmission law. It was shown[11,55] that such a beam incident on a dielectric interface is strongly deflected in the farfield away from the tangent direction expected from a naive application of Snell's law resulting in Fresnel Filtering (FF). The farfield peak-shift, $\Delta\theta_{FF}$, depends on the beam width Δ and on n; analysis of the stationary phase solution gives the result that, for incidence at critical angle χ_c,

$$\Delta\theta_{FF}^c \approx (2/\tan\chi_c)^{1/2}\Delta^{-1/2} \tag{23}$$

It is clear this effect will be crucial for analysis of the emission patterns of dielectric microlasers, because the dominant lasing mode will be often based on long-lived and localized quasi-normal modes which always involve ray components close to the critical incidence. Furthermore, even if the experiments are performed deep in the short-wavelength limit $nkR \to \infty$, localization (in SOS, per coordinate) will be of order $1/\sqrt{nkR_0}$ (as is for stable island modes for example) leading to a spectral width $\Delta \propto \sqrt{nkR_0}$, so from Eq. (23), the deviation angle at critical incidence $\theta_{FF}^c \propto (nkR)^{-1/4}$ and hence may be large for $nkR \sim 10^2 - 10^3$, as in recent experiments on semiconductor ARC lasers.[7,8,11]

Evidence for laser action on a triangular scarred state was also obtained in the Bell Labs experiments on a series of quadrupolarly deformed semiconductor diode (GaInAs quantum wells embedded in GaAs/GaInP waveguide) lasers with an effective index of refraction $n_{\text{eff}} = 3.4$. There were three interesting features of this experiment. First, with the given index of refraction and a quadrupolar boundary deformation $\epsilon \approx 0.16$, one would expect to observe emission from the stable bowtie modes, because the situation is almost identical to the QC ARC experiment of the previous section. However, this is not the case because of the preference of the device towards TE polarized modes. In the QC laser intersubband optical transitions lead to a selection rule which allows light emission only in the 2D plane with TM polarization normal to the quantum well layers.[7] Because of the existence of the Brewster angle for TE polarization, the reflectivity of the boundary drops practically to zero for the TE version of the bowtie mode before again rising sharply to TIR close to critical angle. Therefore, the bowtie modes in diode lasers are considerably leakier than their TM counterparts in QC lasers and are not apparently selected as the lasing modes. The second notable observation was the nature of transition from lasing via regular modes to chaotic modes with increasing deformation. Because imaging or farfield data was not available, they looked for a signature of this transition in the spectral data. Above a deformation of about $\epsilon = 0.03$–0.06 (which coincides approximately with the KAM transition close to the critical line), the spectrum displayed evenly spaced mode doublets with large splitting, which are absent in low deformation data. The FSR of the doublets are found to be consistent with a triangular periodic orbit of the type seen in the Yale GaN experiments and the authors attribute the relatively large splittings to boundary roughness. Considered as a perturbation, boundary roughness affects localized modes to a greater extent than extended whispering gallery modes of smaller deformation. The final significant point is that this experiment found a persistence of the lasing characteristics through the stability–instability transition of the triangular orbit, which happens at about $\epsilon \approx 0.1$. It is worth reiterating that while the nature of the mode on the stable side is of Gaussian type, the individual modes based on unstable motion does not yield to a simple analytic description and the resulting scarred states can only be obtained numerically. The numerical solutions are found to yield localized modes on both sides of the transition.

A more direct demonstration of the last observation was provided by Rex *at al.*[10] in an extension of the GaN laser experiments to a range of quadrupole–hexadecapolar deformations, defined by the mathematical equation:

$$r(\phi) = 1 + \epsilon \left(\cos^2 \phi + \frac{3}{2} \cos^4 \phi \right). \tag{24}$$

Because the laser operated in the visible ($\lambda \sim 400 \, \text{nm}$), it was possible to image the emission from the cavity with conventional optics. The farfield emission data are reproduced in Fig. 33. All the deformations above $\epsilon = 0.12$ show a well-localized emission which has approximately the same character over a wide range of deformations. Looking at the SOS of the quadrupole–hexadecapole at $\epsilon = 0.12$, we note that there is a stable orbit — again a triangular one — this time however rotated 90° from the one observed in the quadrupole, with bounce points at $\phi = 0°$ and $\phi = 180°$, at which points the trajectory is incident just above the critical angle for TIR. These are also the points where the maximum emission is emanating from the boundary, according to the imaging data. Unlike the quadrupole lasers then, for this

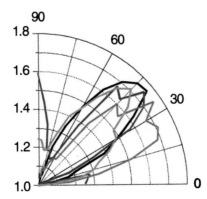

Fig. 33. Farfield emission data for quadrupole–hexadecapole for deformations of $\epsilon = 0.12$ (blue), $\epsilon = 0.16$ (red), $\epsilon = 0.18$ (green), and $\epsilon = 0.2$ (black).

shape there is a stable orbit with bounce points near the critical angle
for this deformation and index of refraction and this will support Gaussian
modes similar to the bowtie modes seen in the QC laser. An example of such
a mode localized on the stable triangular orbit and resulting in emission
consistent with the experimental results is shown in Fig. 34.

For this shape, these triangular orbits become unstable above a deformation of $\epsilon = 0.13$, and despite this change, the farfield data do not change
in any appreciable way (Fig. 34). Figure 35 shows results of calculations
for the deformation $\epsilon = 0.16$. The Husimi projection of this mode reveals
that it is localized on the triangular periodic orbit, in the vicinity of which
complete chaos reigns. Thus, the laser operates on a mode which is based
on one and the same classical periodic orbit, insensitive to whether it is
stable or unstable.

It is clear from the review of these experiments that lasing modes of
cylindrical microresonators need not be based on "regular" modes such as

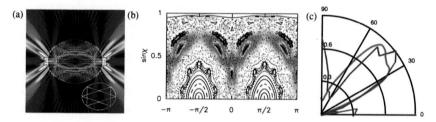

Fig. 34.   (a) A numerically calculated mode for a quadrupole–hexadecapolar deformation of $\epsilon = 0.12$ and $n = 2.65$. (b) Husimi projection of the mode in (a). Clearly, the
projection is localized on a reflection symmetric pair of stable triangular periodic orbits.
(c) The calculated farfield emission pattern.

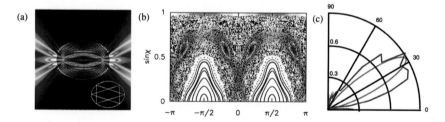

Fig. 35.   (a) A numerically calculated mode for a quadrupole–hexadecapolar deformation of $\epsilon = 0.16$ and $n = 2.65$. (b) Husimi projection of the mode in (a). The projection
is localized on a triangular orbit of the same geometry as the one in Fig. 35, but at
this deformation the motion in its vicinity is unstable, leading to chaotic motion. The
resulting mode is hence a scarred state. (c) The calculated farfield emission pattern.

stable orbits or whispering gallery modes, but also can get feedback from unstable ray trajectories. It is worth remarking that unstable Fabry–Perot laser resonators have been known since almost the initial conception of the laser[56,57] and for many purposes are the best design for high-gain laser devices[14] because of their large modal volumes. In ARC microlasers such unstable lasing action arises naturally as one increases the deformation, with the SOS being dominated by more and more chaotic motion. Whether there is any advantage of microlasers based on unstable modes remains to be seen. One crucial point that needs emphasis is that there do exist many complicated chaotic modes which are not related to any single periodic orbit. Indeed in the passive cavity these modes dominate the spectrum as $kR \to \infty$. However, it may be that the nonlinear effects in lasing cavities, either by averaging over fluctuating modes or by mode-locking, enhance the role of modes based on short periodic orbits, whether stable or unstable. It is striking that all of the experiments on semiconductor ARC lasers can be interpreted as demonstrating lasing from such modes.

6. Unidirectional GaN Spiral Microlasers

Since the ARC concept was introduced to obtain directional emission from planar dielectric microlasers one important challenge has been to obtain a single, directed beam out of these devices. As reviewed in previous sections, by using the shape of the resonator as a design parameter it is possible to achieve highly directional emission and other desirable characteristics such as much improved output power. ARC studies, based on smoothly deformed cylindrical resonators, have been supplemented by studies of hexagonal,[58] triangular,[59] and square[60] microcavity lasers. However all of these designs were found to emit in multiple directions and exhibit farfield patterns with multiple lobes. In fact it would seem that any lasing mode based on non-normal incidence rays (required for near total internal reflection and hence high Q) would generate at least two output beams due to the possibility of time-reversed motion on the same trajectory.

Particularly for the development of compact, high power UV emitters, where GaN-based semiconductor compounds are the materials of choice, a planar emitter with unidirectional output coupling is of great interest. Current GaN based lasers use distributed Bragg reflectors (DBRs) in a VCSEL arrangement to achieve high Q. However DBRs which are also good conductors for the injection current are extremely difficult to fabricate because of the well-known material challenges surrounding the growth of GaN-based layers. Dielectric microcavities can greatly simplify these material problems associated with GaN-based lasers by using the sidewalls for high Q feedback while current is fed from the top of the structure. Such structures would of course be planar emitters as opposed to the DBR-based lasers which emit vertically, and would be preferable for most integrated optics applications. However, none of the planar emitters has so far been able to provide a single, directed beam which is a desirable characteristic peculiar to the Fabry–Perot configuration.

Very recently, this difficulty has been overcome with the introduction of spiral-shaped micropillar structures that provide unidirectional emission.[13] These devices were based on an InGaN multiple quantum-well (MQW) active region, sandwiched between waveguide layers for transverse modal confinement and etched into spiral cross-sections (see Fig. 36), defined by the following equation

$$r(\phi) = R_o \left(1 + \frac{\epsilon}{2\pi}\phi \right) \tag{25}$$

Here ϵ is the deformation parameter, R_0 is the radius of the spiral at $\phi = 0$, and ϕ is measured in radians. This equation implies that the spiral has a discontinuity in radius at $\phi = 0$ where the radius changes abruptly from $r = R_0(1+\epsilon)$ back to R_0. Imaging of the emission from the pillar sidewalls shows that the unidirectional emission is from this "notch" (see e.g. inset, Fig. 37(b)). Due to its boundary discontinuity, with a sharp corner on the order of the wavelength at the notch, this structure is not an ARC and is not expected to be well-described by any pure ray-optical description. The discontinuity in the boundary of the spiral at the notch clearly scatters whispering gallery ray trajectories (at least for counter-clockwise rotating

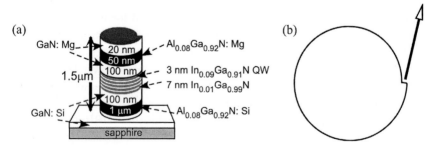

Fig. 36.   (a) Structure of the InGaN MQW sample. (b) Schematic showing notch emission seen in lasing data.

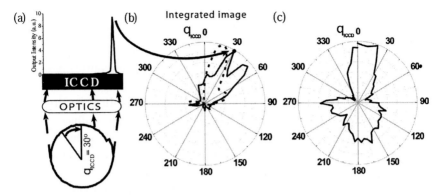

Fig. 37. (a) Image profile of the emission from the spiral microcavity ($\epsilon = 0.10$ and $d = 500\,\mu\text{m}$) sidewall at camera angle $\theta = 30°$. Integrating the image profile at each camera angle θ gives (b) unidirectional farfield pattern of the spiral. Experimental data (solid line) are in good agreement with numerical calculation (dotted line). (c) Farfield pattern obtained when the spiral cavity is pumped uniformly.

waves), and it was not clear in advance that such a structure would support any high Q modes for lasing. Sharp corners are known to give rise to diffraction, and diffractive effects are in practice thought to be destructive for device performance. As will be explained shortly, the mechanism of this novel design which enables uni-directional out-coupling is primarily based on diffractive effects which couple out a non-emitting (counter-clockwise rotating) WGM.

Initial studies providing the proof of principle were performed under optical pumping conditions. Spirals of various deformations ($\epsilon = 0.05, 0.10,$ and 0.15) and sizes ($d = 100\,\mu\text{m},\ 300\,\mu\text{m}$ and $500\,\mu\text{m}$) were examined. Results showed lasing of the structures at a wavelength of 404 nm, and that larger sized spirals possess lower thresholds. The spiral with $\epsilon = 0.10$ had the most unidirectional and narrow emission lobe, thus we will focus on the emission characteristics of $\text{In}_{0.09}\text{Ga}_{0.91}\text{N}$ MQWs with $\epsilon = 0.10$ deformation, optically pumped with 355 nm radiation.

Unidirectionality of the observed lasing emission is shown in the farfield image of Fig. 37(b). This polar plot is obtained by integrating over image profiles taken at 5° intervals of the camera angle θ defined such that at $\theta = 0°$ detected emission is normal to the notch (note that this differs from our standard definitions above for which this direction would be $\theta = 90°$ if the notch is along the x-axis). The image field in Fig. 37(a) shows that at $\theta = 30°$, the majority of the emission comes from the spiral notch.

A crucial feature in the success of spiral lasers is the *selective pumping method* employed. Based on previous experiments[61] on circular micropillars, which have shown that lower thresholds can be attained by a spatially selective pumping of the microcavity, the spiral cavities were optically excited using an axicon lens to form a ring shaped beam. The aim was to achieve an optimal overlap with the high Q mode, following theoretical results which show that the spiral resonator supports long-lived modes confined close to the perimeter. This also serves to *suppress* other unwanted modes and scattering from the discontinuity, which are then dissipated mainly due to the absorption of the material. The effectiveness of selective pumping method is demonstrated in Fig. 37(c), which shows a broad emission pattern under uniform pumping conditions. The spatial disposition of the numerically calculated high Q modes together with the results of the selective pumping method can be exploited in the design of the electrodes under current injection conditions to optimize the emission.

Numerical calculations show that there are indeed such dominantly notch-emitting quasi-bound states of the spiral microcavity. Figure 38(a) plots the modulus of the electric field of such a resonance in real-space, showing that it is concentrated close to the boundary and exhibiting properties associated with a WG-like mode. However, these modes display a crucial difference from the regular WG-modes of a circular resonator: the high Q resonances of the spiral exhibit a pronounced *chirality* and are predominantly composed of clockwise rotating components (corresponding to

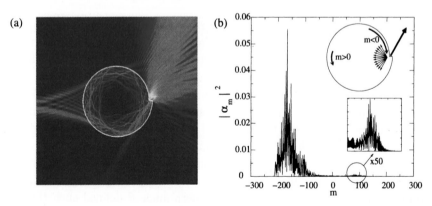

Fig. 38.   (a) Real-space false color plot of the modulus of the electric field for a calculated quasi-bound state at $nkRo \approx 200$ at $\epsilon = 0.10$ deformation. (b) Distribution of angular momenta for the resonance plotted in (a). Note the peak at negative m corresponding to clockwise rotation (see inset) and the small weight at positive m which constitute the diffracted waves emitting from the notch.

ray motion which could not escape at the notch). This can be easily deduced from the decomposition of the mode into its circular harmonics

$$E(\mathbf{r}) = \sum_{m=-\infty}^{+\infty} \alpha_m J_m(r) e^{im\varphi} \qquad (26)$$

given in Fig. 38(b).

Interpreting each component m in the sum as representing ray motion incident on the boundary at an angle of incidence $\sin \chi = m/nkR$ in the short-wavelength limit, we get a distribution of incidence angles. The strong weighting of the distribution to negative components $(-m)$ peaked around $|m| \approx 160$ corresponds to a mode with mostly clockwise-rotating waves having angle of incidence $\sin \chi \approx 0.8$ and which are hence totally internally reflected (the critical angle χ_c is defined by $\sin \chi_c = \pm\frac{1}{n} = \pm 0.38$ for $n = 2.6$). Such a stationary distribution is counter-intuitive from the point of view of ray dynamics inside this spiral-shaped dielectric billiard. Ray simulations show that if we start a ray bundle which is predominantly composed of clockwise rotating rays close to the boundary ($\sin \chi \approx -1$), the average impact angle diffuses monotonically towards 0, where the rays strike the boundary at normal incidence, and further towards positive $\sin \chi$ (see Fig. 39).

For a dielectric billiard, the intensity would then refract out at some point away from the notch long before it can reverse its sense of rotation. Furthermore, there are no periodic orbits whose bounce points are entirely on the smooth portions of the boundary, simply because the curvature is

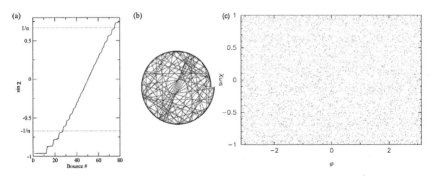

Fig. 39. (a) The diffusion of a ray started at $\varphi = 2\pi - 0.01, \sin \chi = -0.96$. At the uppermost point of the curve, the ray leaves the resonator from the notch, (b) the trajectory plotted in real-space, (c) surface of section of the spiral with $\varepsilon = 0.1$.

monotonic. This is clear e.g. for the bouncing ball orbit as due to the monotonic curvature there exist no two points on the boundary with colinear and opposite normal vectors. More generally any periodic orbit would have to have pairs of bounces for which $\Delta \sin \chi$ has both signs, whereas the monotonic variation in $\sin \chi$ away from the notch does not permit this. Therefore the only periodic orbits of this billiard have at least one bounce point on the notch area. Their associated islands of stability are quite small as shown by the surface of section in Fig. 39 which appears completely chaotic. Modes based on such long periodic orbits are rarely important. If they are unstable they are unlikely to have strong scars for the reasons discussed in the previous section; if they are stable the islands of stability are typically very small and the resulting modes have small mode volumes. More importantly, our numerical calculations on the passive cavity indicate no weight in the emitting mode near zero angular momentum, which cannot happen for periodic orbits in this structure.

In systems where periodic orbits are rare, diffractive contributions can be significant. It is known for instance from the study of open microwave resonators, that in the regime where there are no stable periodic orbits, orbits that diffract off the sharp edges of the resonator can have a strong influence on the spectrum and wavefunctions.[62] Similarly, the analysis of the spiral billiard suggests that for the formation of a notch-emitting high Q mode, a mechanism beyond geometric optics (such as diffraction) is necessary. If a wavepacket of cylindrical waves with dominantly negative components were injected into the system, the notch, which is discontinuous on the scale of a wavelength, would diffract a small part of the amplitude into positively rotating components above the critical angle, which would eventually be emitted from the notch. Hence we interpret the small amount of counter-clockwise (ccw) rotating waves found numerically as responsible for emission at the notch; and these components are due to diffraction of the clockwise-waves as they pass the inner corner of the notch (see inset Fig. 38(b)). Attempts to reproduce the experimental emission patterns with any reasonable initial ray bundle that only reflects specularly (and refracts out according to Snell's Law) fail to reproduce the observed emission behavior for the reasons alluded to above: clockwise ray bundles escape through the smooth part of the boundary by refraction before they reverse direction, solely counter-clockwise ray bundles do escape from the notch but would correspond to an unphysically low Q value (and contradict wave solutions). Therefore we believe diffraction effects are crucial to the unidirectional lasing from this device.

Another counter-intuitive feature of the experimental results is that the farfield emission lobe is not maximum at $\theta_{ICCD} = 0°$ corresponding to normal emission from the notch, but has two "lobes" peaked at angles: $\theta_{ICCD} \approx 30°$ and $\theta_{ICCD} \approx 50°$ (see Fig. 37(b)). This tilt in the vicinity of the notch arises because the ccw-component of the resonance cannot be viewed as a Gaussian beam incident on the notch interface; instead there is a distribution of wavevectors determined by the specific resonance. By an angular decomposition of the incident field on the notch, we can numerically propagate it to the farfield.

We assume first that the emitting (part of the) mode is composed of a single angular momentum component, which is a reasonable starting point given that the distribution of the positive components in the numerically calculated quasi-bound states is narrow, and becomes narrower for higher wavenumbers. Let the notch interface be at $z = 0$. At the interface, infinitesimally below it, the angular spectrum of the field can be expressed as

$$\tilde{\Psi}(z = 0^-, s) = \int_0^\infty dx J_m(nkx) e^{-inksx}. \tag{27}$$

Then we can write

$$\Psi(z = 0^+, x) = \int_{-\infty}^\infty t(s)\tilde{\Psi}(z = 0^-, s)e^{iksx}. \tag{28}$$

So that if we propagate this field to points $z > 0$

$$\Psi(z > 0, x) = \int_{-\infty}^\infty t(s)\tilde{\Psi}(z = 0^-, s)e^{ik(sx+\sqrt{1-s^2}z)}. \tag{29}$$

Consider the transform in Eq. (27). Only part of the incident radiation will be incident on the notch interface, so we change the limits of the integral accordingly. Scaling the variables inside the integral as well, we get:

$$\tilde{\Psi}(s) = \frac{1}{nk} \int_{nkR_0}^{nkR_0(1+\varepsilon)} d\tilde{x} J_m(\tilde{x}) e^{-is\tilde{x}}. \tag{30}$$

In this way, the size scale of the notch (with respect to the wavelength) enters the calculation. Calculating the field asymptotically at infinity, we obtain after a saddle point calculation[55]:

$$I(\theta) \propto \left| \frac{\sqrt{n^2 - \sin^2\theta}\cos\theta}{\sqrt{n^2 - \sin^2\theta} + n\sqrt{1 - \sin^2\theta}} \tilde{\Psi}\left(\frac{\sin\theta}{n}\right) \right|^2. \tag{31}$$

The results in Fig. 40(a) show a fluctuating emission pattern with no strong directional peak when $R_c = m/nk \ll R_0$, i.e. the caustic of the emitting

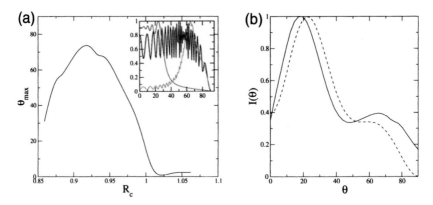

Fig. 40.   (a) The variation of the angular shift of the peak emission from the notch with the position of the caustic of the counterclockwise rotating cylindrical wave (smoothed). (Inset) Samples of the behavior of the field in the farfield for three values of R_c; $R_c = 0.86$ (black), $R_c = 0.95$ (red) and $R_c = 1.05$ (blue). All intensities are scaled to be 1 at maximum. The calculations are performed for the experimental size parameter $nkR \approx 9300$. (b) Comparison of the numerically calculated farfield for the resonance in Fig. 38 (solid curve) and the propagated farfield (cf. Eq. (31)) (dashed curve).

circular wave is between the edge of the notch at $\phi = 0, r = R_0$ and the origin. Directional emission sets in as the caustic approaches the notch and a single (though rather broad) peak results, which is displaced from the normal direction. The angular displacement from the notch normal vanishes as the mode intensity moves fully into the notch area and becomes confined closer to the outer rim $r = R_0(1+\epsilon)$, as expected from rays incident increasingly at normal angles to the notch interface.

Finally, it is possible to check whether this asymptotic propagation method gives a good representation of the actual farfield by calculating the farfield emission profile of the numerically calculated internal mode in Fig. 38 using Eq. (31). The result is shown in Fig. 40(b), and is in good agreement with the numerically determined farfield. The reliability of Eq. (31), of course, is better for higher size parameter nkR.

If a diffractive mechanism is responsible for the unidirectional emission from the notch of the spiral resonator then the output characteristics are expected to be sensitive to the shape and sharpness of the notch. Further experiments and modeling are required to see if this is the case. It remains possible that a rather different physical picture of the lasing mode will emerge under further study. For example a "ray model" of diffraction, based on isotropic diffraction from the corner of the notch does not reproduce the experimental behavior. This suggests that there are additional physical effects which allow this device to perform as well as it does; we hope to elucidate these mechanisms through further research.

Very recently new experiments have been performed in a PARC–Yale collaboration on spiral-shaped InGaN MQW lasers with electrical injection pumping demonstrating low current thresholds and output power of more than 25 mw at 400 nm.[63] The heterostructure used was identical to the one reported in the above work. The design (placement and shape) of the electrode on the top-face of the p-GaN layer, which we pointed out is an important issue, followed the results of the optical injection investigation. It is to be noted that while selective optical pumping is limited to having simple shapes, e.g. in the form of a ring or a line at various tilt angles, selective injection current pumping can have rather complex designs simply by incorporating the shape complexities in the mask design, e.g. designing the opaque electrode patterned after the perimeter of the notch. Future work is expected to make optimal use of this technique in more complex scenarios.

In conclusion, the work summarized in this section demonstrates that the spiral geometry is a viable design of GaN-based laser devices for integrated optical applications.

7. Summary and Outlook

A wide range of experiments combined with ray and wave modeling has shown that nonsymmetric dielectric microresonators and lasers have a rich set of properties. Low index ARC lasers and resonators are remarkably well described by ray escape models based on flow patterns in partially chaotic phase space. Semiconductor ARCs show interesting lasing effects based on short periodic orbits. Basic phenomena of interest in the field of quantum/wave chaos are observed, such as chaos-assisted tunneling and scarring. Classical concepts such as the unstable and stable nonlinear manifolds of periodic orbits lie at the heart of the observed directional emission in low index ARCs, which occurs despite the presence of strong chaos. The semiconductor ARC and spiral lasers are typically not well described by ray models but can still be analyzed fruitfully by looking at the high Q modes in the ray phase space.

From the point of view of applications these nonsymmetric dielectric resonators have shown two striking advantages. First, at least in the case of the quantum cascade ARC lasers, deformation from a symmetric boundary shape led to an enormous increase in output efficiency. Second, for the spiral laser, chiral whispering gallery modes appear to provide both relatively high Q and unidirectional emission. This is of particular importance for the GaN-based blue and UV lasers for which conventional approaches are difficult or impossible to implement at the same size scale. It is hoped that during the next period of research and development of asymmetric resonant cavities these promising device characteristics will allow such cavities to become part of useful technologies.

Acknowledgments

We would like to acknowledge the contributions to this research by our collaborators. On the theory side these include Jens Nöckel, Evgenii Narimanov, Gregor Hackenbroich and Philippe Jacquod. On the experimental side these include Nathan Rex, Grace Chern, Andrew Poon and Seongsik Chang from Yale, Tahar Ben Massoud and Joseph Zyss from ENS Cachan, Noble Johnson and Michael Kneissl from PARC, Lou Guido from Virginia Polytechnic and Claire Gmachl and Federico Capasso from Bell Labs. Various aspects of the research here were supported by NSF grants DMR-008541, PHY-9612200, AFOSR (F49620-00-1-0182), DARPA SUVOS program under SPAWAR systems grant No. N666001-02-C-8017. We thank Scott Lacey and Jens Nöckel for providing us with figures and calculations related to the research in Ref. 28.

References

1. Y. Yamamoto and R. E. Slusher, "Optical processes in microcavities," *Phys. Today* **46**, 66–73, 1993.
2. R. K. Chang and A. K. Campillo (eds.) *Optical Processes in Microcavities*, World Scientific, Singapore, 1996.
3. K. J. Vahala, "Optical microcavities," *Nature* **424**, 839–846, 2003.
4. J. U. Nöckel, A. D. Stone and R. K. Chang, "Q-spoiling and directionality in deformed ring cavities," *Opt. Lett.* **19**, 1693–1695, 1994.
5. J. U. Nöckel and A. D. Stone, *Chaotic Light: A Theory of Asymmetric Cavity Resonators*, World Scientific, Singapore, Chap. 11, 389–426, 1996.
6. A. D. Stone, "Wave-chaotic optical resonators and lasers," *Phys. Scr.* **T90**, 248–262, 2001.
7. C. Gmachl, F. Capasso, E. E. Narimanov, J. U. Nöckel, A. D. Stone, J. Faist, D. L. Sivco and A. Y. Cho, "High-power directional emission from microlasers with chaotic resonators," *Science* **280**, 1556–1564, 1998.
8. S. Gianordoli, L. Hvozdara, G. Strasser, W. Schrenk, J. Faist and E. Gornik, "Long-wavelength $\lambda = 10\,\mu m$ quadrupolar-shaped GaAs-AlGaAs microlasers," *IEEE J. Quant. Electron.* **36**, 458–464, 2000.
9. C. Gmachl, E. E. Narimanov, F. Capasso, J. N. Baillargeon and A. Y. Cho, "Kolmogorov–Arnold–Moser transition and laser action on scar modes

in semiconductor diode lasers with deformed resonators," *Opt. Lett.* **27**, 824–826, 2002.

10. N. B. Rex, *Regular and Chaotic Orbit Gallium Nitride Microcavity Lasers*, PhD thesis, Yale University, 2001.

11. N. B. Rex, H. E. Tureci, H. G. L. Schwefel, R. K. Chang and A. D. Stone, "Fresnel filtering in lasing emission from scarred modes of wave-chaotic optical resonators," *Phys. Rev. Lett.* **88**, 094102, 2002.

12. Harald G. L. Schwefel, Nathan B. Rex, Hakan E. Tureci, Richard K. Chang, A. Douglas Stone, Tahar ben Massoud and J. Zyss, "Dramatic shape sensitivity of directional emission patterns from similarly deformed cylindrical polymer lasers," *J. Opt. Soc. Am.* **B21**, 923–934, 2004.

13. G. D. Chern, H. E. Tureci, A. D. Stone, R. K. Chang, M. Kneissl and N. M. Johnson, "Unidirectional lasing from InGaN multiple-quantum-well spiral-shaped micropillars," *Appl. Phys. Lett.* **83**, 1710–1712, 2003.

14. A. E. Siegman, *Lasers*, University Science Books, Mill Valley, California, 1986.

15. H. E. Tureci, H. G. L. Schwefel, A. D. Stone and E. E. Narimanov, "Gaussian-optical approach to stable periodic orbit resonances of partially chaotic dielectric micro-cavities," *Opt. Express* **10**, 752–776, 2002.

16. J. B. Keller and S. I. Rubinow, "Asymptotic solution of eigenvalue problems," *Ann. Phys.* **9**, 24–75, 1960.

17. V. M. Babič and V. S. Buldyrev, *Asymptotic Methods in Shortwave Diffraction Problems*, Springer, New York, USA, 1991.

18. H. E. Tureci, H. G. L. Schwefel, Ph. Jacquod and A. D. Stone, "Modes of wave-chaotic dielectric resonators," *Prog. Opt.*, Vol. 47.

19. A. Einstein, "Zum Quantensatz von Sommerfeld und Epstein," *Verhandl. Deut. Physik. Ges.* **19**, 82–92, 1917.

20. Hakan E. Türeci, *Wave Chaos in Dielectric Resonators: Asymptotic and Numerical Approaches*, PhD thesis, Yale University, 2003.

21. L. E. Reichl, *The Transition to Chaos in Conservative Classical Systems: Quantum Manifestations*, Springer, NY, USA, 1992.

22. V. I. Arnold, *Mathematical Methods of Classical Mechanics*, Springer, NY, USA, 1989.

23. V. F. Lazutkin, *KAM Theory and Semiclassical Approximations to Eigenfunctions*, Springer, NY, USA, 1993.

24. A. J. Lichtenberg and M. A. Lieberman, *Regular and Chaotic Dynamics*, Springer, New York, USA, 1992.

25. E. Doron and U. Smilansky, "Semiclassical quantization of chaotic billiards — a scattering theory approach," *Nonlinearity* **5**, 1055–1084, 1992.

26. B. Dietz, J. P. Eckmann, C. A. Pillet, U. Smilansky and I. Ussishkin, "Inside-outside duality for planar billiards — a numerical study," *Phys. Rev. E* **51**, 4222–4231, 1995.

27. K. Husimi, "Some formal properties of the density matrix," *Proc. Phys. Math. Soc. Jpn.* **22**, 264–314, 1940.

28. S. Lacey, H. Wang, D. H. Foster and J. U. Nöckel, "Directional tunnel escape from nearly spherical optical resonators," *Phys. Rev. Lett.* **91**, 033902, 2003.

29. J. W. Goodman, *Introduction to Fourier Optics*, McGraw-Hill, New York, USA, 1996.
30. J. U. Nöckel, A. D. Stone, G. Chen, H. L. Grossman and R. K. Chang, "Directional emission from asymmetric resonant cavities," *Opt. Lett.* **21**, 1609–1611, 1996.
31. J. U. Nöckel and A. D. Stone, "Ray and wave chaos in asymmetric resonant optical cavities," *Nature* **385**, 45–47, 1997.
32. M. Robnik and M. V. Berry, "Classical billiards in magnetic-fields," *J. Phys. A-Math. Gen.* **18**, 1361–1378, 1985.
33. S. Chang, R. K. Chang, A. D. Stone and J. U. Nöckel, "Observation of emission from chaotic lasing modes in deformed microspheres: displacement by the stable-orbit modes," *J. Opt. Soc. Am. B-Opt. Phys.* **17**, 1828–1834, 2000.
34. S. Chang, N. B. Rex, R. K. Chang, G. B. Chong and L. J. Guido, "Stimulated emission and lasing in whispering gallery modes of GaN microdisk cavities," *Appl. Phys. Lett.* **75**, 3719–3719, 1999.
35. H. Poritsky, "The billiard ball problem on a table with convex boundary — an illustrative dynamical problem," *Ann. Math.* **51**, 446–470, 1950.
36. E. Y. Amiran, "Integrable smooth planar billiards and evolutes," *New York J. Math.* **3**, 32–47, 1997.
37. S. Smale, "Differentiable dynamical systems," *Bull. Amer. Math. Soc.* **73**, 747–817, 1967.
38. J. U. Nöckel, *Resonances in Nonintegrable Open Systems*, PhD thesis, Yale University, 1997.
39. A. Mekis, J. U. Nöckel, G. Chen, A. D. Stone and R. K. Chang, "Ray chaos and Q-spoiling in lasing droplets," *Phys. Rev. Lett.* **75**, 2682–2685, 1995.
40. S. Chang, R. K. Chang, A. D. Stone and J. U. Nöckel, "Observation of emission from chaotic lasing modes in deformed microspheres: displacement by the stable-orbit modes," *J. Opt. Soc. Am. B-Opt. Phys.* **17**, 1828–1834, 2000.
41. S. B. Lee, J. H. Lee, J. S. Chang, H. J. Moon, S. W. Kim and K. An, "Observation of scarred modes in asymmetrically deformed microcylinder lasers," *Phys. Rev. Lett.* **8803**, 033903, 2002.
42. F. Capasso, C. Gmachl, D. L. Sivco and A. Y. Chou, "Quantum cascade lasers," *Phys. Today* **55**, 34–40, 2002.
43. L. A. Lugiato, C. Oldano and L. M. Narducci, "Cooperative frequency locking and stationary spatial structures in lasers," *J. Opt. Soc. Am. B-Opt. Physics* **5**, 879–888, 1988.
44. V. Zehnle, "Theoretical analysis of a bimode laser," *Phys. Rev.* **A57**, 629–643, 1998.
45. C. Tamm, "Frequency locking of 2 transverse optical modes of a laser," *Phys. Rev.* **A38**, 5960–5963, 1988.
46. V. A. Podolskiy and E. E. Narimanov, "Universal level-spacing distribution in quantum systems," preprint server nlin.CD/0310034.

47. M. J. Davis and E. J. Heller, "Quantum dynamical tunneling in bound-states," *J. Chem. Phys.* **75**, 246–254, 1981.

48. E. E. Narimanov, private communication, October 2003.

49. E. J. Heller, "Bound-state eigenfunctions of classically chaotic hamiltonian-systems — scars of periodic-orbits," *Phys. Rev. Lett.* **53**, 1515–1518, 1984.

50. L. Kaplan and E. J. Heller, "Linear and nonlinear theory of eigenfunction scars," *Ann. Phys.* **264**, 171–206, 1998.

51. L. Kaplan, "Scars in quantum chaotic wavefunctions," *Nonlinearity* **12**, R1–R40, 1999.

52. M. V. Berry, "Regular and irregular semiclassical wavefunctions," *J. Phys. A-Math. Gen.* **10**, 2083–2091, 1977.

53. E. B. Bogomolny, "Smoothed wave-functions of chaotic quantum-systems," *Physica* **D31**, 169–189, 1988.

54. H. E. Tureci and A. D. Stone, "Deviation from snell's law for beams transmitted near the critical angle: application to microcavity lasers," *Opt. Lett.* **27**, 7–9, 2002.

55. A. E. Siegman, "Laser beams and resonators: The 1960s," *IEEE J. Sel. Topics Quant. Electron.* **6**, 1380–1388, 2000.

56. A. E. Siegman, "Laser beams and resonators: Beyond the 1960s," *IEEE J. Sel. Top. Quant. Electron.* **6**, 1389–1399, 2000.

57. I. Braun, G. Ihlein, F. Laeri, J. U. Nöckel, G. Schulz-Ekloff, F. Schuth, U. Vietze, O. Weiss and D. Wohrle, "Hexagonal microlasers based on organic dyes in nanoporous crystals," *Appl. Phys. B-Lasers Opt.* **70**, 335–343, 2000.

58. Y. Z. Huang, W. H. Guo and Q. M. Wang, "Influence of output waveguide on mode quality factor in semiconductor microlasers with an equilateral triangle resonator," *Appl. Phys. Lett.* **77**, 3511–3513, 2000.

59. A. W. Poon, F. Courvoisier and R. K. Chang. "Multimode resonances in square-shaped optical microcavities," *Opt. Lett.* **26**, 632–634, 2001.

60. N. B. Rex, R. K. Chang and L. J. Guido. "Threshold lowering in GaN micropillar lasers by means of spatially selective optical pumping," *IEEE Photon. Technol. Lett.* **13**, 1–3, 2001.

61. J. S. Hersch, M. R. Haggerty and E. J. Heller, "Diffractive orbits in an open microwave billiard," *Phys. Rev. Lett.* **83**, 5342–5345, 1999.

62. M. Kneissl, M. Teepe, N. Miyashita, N. M. Johnson, G. D. Chern and R. K. Chang. "Current-injection spiral-shaped microcavity disk laser diodes with uni-directional emission," *Appl. Phys. Lett.* **84**, 2485–2487, 2004.

INDEX

Index